Practical Handbook of

GENETIC ALGORITHMS

Complex Coding Systems
Volume III

Edited by Lance D. Chambers

CRC Press
Taylor & Francis Group
Boca Raton London New York

CRC Press is an imprint of the
Taylor & Francis Group, an **informa** business

CRC Press
Taylor & Francis Group
6000 Broken Sound Parkway NW, Suite 300
Boca Raton, FL 33487-2742

First issued in paperback 2020

© 1999 by Taylor & Francis Group, LLC
CRC Press is an imprint of Taylor & Francis Group, an Informa business

No claim to original U.S. Government works

ISBN-13: 978-0-367-45572-9 (pbk)
ISBN-13: 978-0-8493-2539-7 (hbk)

Visit the Taylor & Francis Web site at
http://www.taylorandfrancis.com

and the CRC Press Web site at
http://www.crcpress.com

Library of Congress Cataloging-in-Publication Data

Catalog record is available from the Library of Congress.

Preface

This is the third and probably last book in the *Practical Handbook of Genetic Algorithm* series. The first volume dealt with applications, the second with new work in the field, and the present volume with computer code. I believe this set of three volumes completes main sections of the field that should be addressed, but I would be happy to be corrected on that score by anyone patient enough to let me know what other sectors should be covered.

Given the feedback I have received from readers of the two previous volumes, I am happy to present this third volume in the series to those who have an interest in the field. It presents computer code segments that can be downloaded from the CRC Press website and used by those undertaking research and/or are building real-world applications in the GA arena. It also contains detailed chapters covering the code and offering in-depth explanations and examples on the use of that code. A few chapters have been included because of their new/quirky uses of GAs; I believe readers will benefit from their content and perspectives. The remaining chapters contain material that is of more traditional value. They cover "problems" that are real and often encountered in the field.

I have not tested the code. You will need to do that either on your own or in collaboration with the author/s or others, if necessary.

The main reason for development of this volume was the apparent paucity of material containing GA code in a single reference volume. There are a few books, papers, and articles that have code segments but not a single work that is dedicated to it. There was perceived to be a need for a reference that offered an array of code covering a number of applications. This book responds to that need and is designed to be of value to programmers who do not want to start from the beginning in developing solutions to particular problems but would rather have a work to which they can refer for code to start the process in a serious manner or where a particular problem solution or computation approach is sought.

The code presented by each author is available at the CRC Press website at: www.crcpress.com and is available for downloading at no cost.

There is much to absorb and understand in this volume. If you have any particular queries on any of the material, contact details are given for each author on the first page of each contribution. Don't hesitate to get further help from the "horses mouth." Each of the authors would be happy to help; they are as proud of their contribution as I am to have edited them.

Hopefully, the chapters will also lead readers to new understandings and new ventures. There is still much work to be done in the application of GAs to the solution of problems that have, until now, been considered relatively intractable. The value that the GA fraternity has brought to academic, business, scientific, and social systems in the few years we have been in practice is something we should all view with pride. I do!

We are a young field with far to go and much to do and contribute. In less than three decades we have made a significant impact in many areas of human struggle and achievement and will continue to do so into the foreseeable future. It is my hope that this book will contribute to that progress.

I am very proad to have been able, with the help of CRC Press and their excellent and very supportive staff (I must particularly thank Nora and Bob – many thanks for all your help and support), to have had the opportunity to edit these three volumes and to have developed associations with authors of the chapters and some readers.

Lance Chambers
Perth, Western Australia
26th Sept, 1998

lchambers@transport.wa.gov.au

NOTE: Against the advice of my editor (even editors have editors) I have not Amercianised (sic) the spelling of English spelling contributors. So as you read you will find a number of words with s's where many may have expected z's and you may also find a large number of u's where you might least expect them as in the word 'colour' and 'behaviour'. Please don't be perturbed. I believe each author has the right to see their work in a form they recognise as theirs. I have also not altered the referencing form employed by the authors.

Ultimately, however, I am responsible for all alterations, errors and omissions.

For Jenny, Lynsey, and Adrian
Thanks for the time and support.

All my love, Lance.

LIST OF TABLES

LIST OF FIGURES

Chapter 1 A Lamarckian Evolution Strategy for Genetic Algorithms

Brian J. Ross
Brock University
Department of Computer Science
St Catharines, Ontario, Canada
email: bross@cosc.brocku.ca

1.1. Introduction

Prior to Charles Darwin's theory of evolution by natural selection, Jean Baptiste Lamarck (1744-1829) proposed a multifaceted theory of evolution [Dawkins 1996, Gould 1980, Cochrane 1997]. One aspect of his theory is the notion that characteristics acquired by an organism during its lifetime are inheritable by its offspring. Lamarck proposed this as the means by which organisms passed on specialized traits for surviving in the environment, and this has since become known as Lamarckian evolution or Lamarckism. For example, if a horse developed especially adept leg muscles for negotiating mountainous terrain, Lamarckian evolution suggests that its offspring would inherit similarly muscular legs. Lamarckian inheritance of acquired characteristics maintains that the acquired development of strong legs, through years of exercise in a mountainous environment, will influence the actual genetic makeup of the horse. This altered genetic information is inheritable by the horse's offspring. This contrasts with the Darwinian tenet that a horse that is genetically predisposed to having muscular legs will probably have offspring with a similar genetic tendency.

Lamarckism has been universally rejected as a viable theory of genetic evolution in nature. There are hundreds of millions of genetic variations in a typical DNA of an organism. Physical characteristics of a phenotype are the result of combined interactions between many separate components of the DNA. In addition, the acquired characteristics of a phenotype are manipulations of the organism's tissues as permitted within their range of possible forms as dictated by the genotype. Darwinian evolution proposes that the complex structure of DNA results in sensible phenotypes due to the accumulation of millions of years of minute genetic mutations, and their subsequent success through natural selection. On the other hand, the mechanism required by Lamarckian evolution is that physical characteristics of the phenotype must in some way be communicated backwards to the DNA of the organism. The biological mechanism by which such communication would arise is unknown. Furthermore, the inversion of gross physical characteristics of the phenotype into the many genetic codes needed to reproduce them in subsequent generations seems to be an impossibly complex task. It is worth mentioning that Lamarckian effects have been conjectured in the context of some experiments with cellular life forms [Cochrane 1997]. These recent results are somewhat of a reprieve for the disrepute that Lamarck's evolutionary theory has suffered over the centuries.

Although discredited in biology, Lamarckian evolution has proven to be a powerful concept within artificial evolution applications on the computer. Unlike life in the natural world, computer programs use very simple transformations between genotypes and phenotypes, and the inversion of phenotypes to their corresponding genotypes is often tractable. In the case that genotypes are their own phenotypes, there is no transformation whatsoever between them. The overall implication of this with respect to Lamarckian evolution is that it is possible to optimize a phenotype in the context of a particular problem environment, and have this optimization reflected in the corresponding genotype for subsequent inheritance by offspring. Therefore, the problems encountered by Lamarckism in nature is solved on the computer: inverted communication from the phenotype to the genotype is typically a simple computation.

Lamarckian evolution and genetic search have been combined together in genetic algorithms (GA) [Hart and Belew 1996, Ackley and Littman 1996, Li et al. 1996, Grefenstette 1991]. Lamarckism typically takes the localized-search form of a phenotype's structure space within the context of the problem being solved. [Hart and Belew 1996] found that the traditional genetic algorithm is most proficient for searching and reconciling widely separated portions of the search space caused by scattered populations, while Lamarckian localized search is more adept at exploring localized areas of the population that would be missed by the wide global swath of the genetic algorithm. Lamarckism may be especially practical when the population has converged into pockets of local minima that would not be thoroughly explored by a standard genetic algorithm. The contribution of Lamarckism is a noticeable acceleration in overall performance of the genetic algorithm.

1.2. Implementation

1.2.1 Basic Implementation

The appendix contains Prolog code that introduces Lamarckian evolution to a genetic algorithm and this code is discussed in detail in the remainder of this section. The language is Quintus Prolog 3.2 and the code was implemented in the Silicon Graphics Irix 5.3 environment.

The code presumes the following is done by the main genetic algorithm. First, the individual genotypes are saved as separate Prolog clauses in the following form: *individual(ID, Fitness, Expression)*.

ID is an identification label such as an integer that uniquely identifies each individual in the population. *Fitness* is the evaluated fitness score for the individual, where the lower scores denote higher fitness. *Expression* is the individual's chromosome as used by the genetic algorithm for the particular problem at hand.

Second, two clauses are used to communicate control and population information from the genetic algorithm and user to the Lamarckian module. A clause *population_size(PopSize)* should contain the current population size as its single numeric argument. A clause *lamarckian(Percent, K)* is used to control the Lamarckian evolution itself. *Percent* is a fractional value between 0.0 and 1.0 denoting the percentage of the population upon which Lamarckian evolution should be applied. *K* is the number of iterations used by the search strategy, which is discussed below.

Finally, the user should have the following three predicates defined somewhere in the main genetic algorithm (change as appropriate): (i) *select(ID)*. This returns a single ID from the population to the calling code. This predicate should use a selection technique of choice, such as tournament selection or fitness-proportional selection. (ii) *mutation(Expr, Mutant)*. This applies an appropriate mutation on an *Expression*, resulting in its *Mutant*. (iii) *eval_fitness(Individual, Fitness)*. This applies the problem-specific fitness evaluation function on an *Individual*, resulting in its Fitness value, such that lower values of Fitness denote fitter individuals.

The top-level predicate of the module is *lamarckian_evolution/0*. This predicate should be called within the main generational loop of the genetic algorithm, at a point after the main genetic reproduction has occurred for a generation. For example,

> *loop :- ... {perform reproduction and mutation for entire population}*
> ...
> *% (a new generation has now been formed...)*
> *lamarckian_evolution,   loop.*
> *% iterate to next generation*

Although the code is written to be used in a generational style of genetic algorithm, as opposed to a steady-state genetic algorithm [Mitchell 1996], it is easily adapted to a steady-state approach if desired (see Section 1.2.2).

The first clause of *lamarckian_evolution* takes care of the case when the user wants to perform Lamarckian evolution on the entire population (*Percent* is 1.0). The clause collects the entire set of individual identifiers from the population, and calls the Lamarckian algorithm on them. Otherwise, if a fraction of the population greater than 0.0 is to be processed, the second clause is used. Here, the rough number of individuals to process is computed from the percentage parameter and the population size, and an appropriately sized list of unique IDs is obtained and processed.

The second clause of *lamarckian_evolution* calls the predicate *get_unique_IDs/3* to select a set of unique identifiers of a desired size between 1 and the size of the population as a whole. To do this, *get_unique_IDs* loops repeatedly until a list of the desired size has been obtained. The selection routine from the main genetic algorithm is used to select individuals. Note that if a high percentage of the population is to be selected for Lamarckian processing, this routine can become slow, since the selection of individuals unique to an already large list of selections can take time. It is recommended that the entire population be processed in such cases, as the first clause of *lamarckian_evolution* will efficiently process such selections. Finally, if the user does not wish to use Lamarckism, the final clause will activate.

The predicate *lamarck_loop*/2 controls the Lamarckian evolution of a list of selected individuals. For each individual as referenced by an ID in the list, its current expression and fitness are obtained from the database. Then, a localized search is performed on it, for the number of iterations specified by the parameter *K*. The result of the localized search is possibly a new individual *NewExpr*, with fitness *NewFit*. If the new fitness is better than that of the original, then the original individual's clause in the database is updated with its newly optimized expression and fitness. Otherwise, the original individual is not altered. The calls to the database manipulation routines *retract* and *assert* within *lamarck_loop*/2 perform the actual Lamarckian evolution. The phenotype is explored using the search algorithm and, should a better result be obtained, the corresponding genotype (which is identical to the phenotype in this problem) is revised.

The localized search routine is *hill_climb*/3. It takes the iteration value, the individual and its fitness, and performs a hill-climbing search for a fitter individual. The search presumes that the genetic encoding is complex enough that cycling visits to the same individuals are unlikely. This also presumes that the mutation operator is adequately robust to permit exploration of the entire search space. The first clause of *hill_climb* activates when the iteration count has been depleted, and returns the best individual discovered during the search. The second clause takes the current fittest expression, and then mutates it with the genetic algorithm's mutation predicate. Its fitness is measured with the fitness evaluator, and the fitter of the old and new individuals is used for subsequent hill-climbing.

An alternate hill-climbing predicate, *hill_climb_nocycle*/3, is also given. This version prevents cycling during the search, which can arise in some problem domains in which the number of mutations of a given individual can be less than the desired iteration value. The *hill_climb2* predicate keeps a list of the visited individuals and does not permit repeated visits to members of this list. In addition, the list is constructed so that the best individual(s) seen are stored at its head. This permits a shunting procedure to be applied when a cycle has been detected, in which the front item, which must be one of the fittest, is moved to the right of its fellow least-valued siblings, if any. *insert_list*/2 places the mutant into the correct position of the list of saved individuals.

1.2.2 Options and Enhancements

The code described in Section 2.2 is simple and compact and is easily specialized and enhanced if necessary. As is discussed later, a determining criterion in applying Lamarckian evolution is its relative computational cost in terms of the genetic algorithm as a whole. The need for alteration will critically depend upon the computational costs associated with the application being investigated.

The predicate *get_unique_IDs* uses the selection routine from the main genetic algorithm. This can be enhanced in a number of ways, for example, by permitting a subset of the most fit individuals to be selected, or even a random set of individuals. Empirical experimentation should be used to determine the best selection strategy for a given problem.

The best-first search procedure is not the only local search algorithm possible [Winston 1992], and others may be applicable depending on the shape of the search space and nature of the mutation operator. In addition, the hill-climber uses mutation to generate new individuals to explore. A possible unorthodox enhancement is to include crossover and other reproduction operators as methods for generating individuals. The frequency of use of these different operators might be specified by the user. Lamarckian evolution can be circumvented if an optimal individual replaces another weaker individual in the population, using an appropriate selection criterion.

The code can be easily adapted for use within a steady-state genetic algorithm. To do this, Lamarckian localized search should be treated as a reproduction operator along with crossover and the others. The simplest means to do this is to call the following on a selected individual ID: *lamarck_loop([ID], K)*.

This applies best-first search on the selected ID for K iterations, and replaces that individual in the database if a better variation is discovered.

1.3. Example Application

The application of the Lamarckian evolution system on a representative problem is now given. The Traveling Salesman Problem (TSP) is a NP-complete problem that has attracted much attention in the theoretical computer science and genetic algorithm communities. The approach to the TSP taken in this section is not particularly advanced, as it uses an inefficient denotation that has been proven to hinder genetic algorithm performance [Tamaki et al. 1994]. Rather, the experiment described here is designed to clearly illustrate the overall effect of Lamarckian evolution on a straight-forward problem.

The TSP is defined as follows. A map of cities and paths between them is given. There is a distance associated with each inter-city path. The TSP is to determine a tour order for the cities such that all the cities are visited, the tour finishes where it began, and the overall tour distance is minimal. In the experiment attempted, we consider 64 cities laying on a square 8 by 8 grid (see Figure 1.1). The vertical and horizontal distance between neighbour cities is taken to be 1. The cities are maximally connected with one another, making it possible to travel between any two cities on the grid. For convenience, we label each city with the row and column label from the grid, for example, *aa* and *cf*. Table lookup is used to determine the coordinates of a city, and path distances are directly computed from the city's grid coordinates. Routes must explicitly refer to the connected cities: a path between *aa* and *ac* that happens to pass through *ab* is not considered to connect to *ab*.

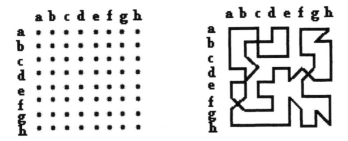

Figure 1.1 City grid, and one route with distance of 70.14

The TSP denotation chosen for the genetic algorithm is a simple one in which a list of 64 city names representing a particular order of cities in a route: *aa bf fh gh ah ... cc* where *cc* at the end is presumed to connect to *aa*. Each city should reside on the list once and only once. This denotation has ramifications on the crossover operation. During crossover, two random crossover points are determined for two selected paths. The intervening strings between the crossover points are then swapped between the routes. This will often result in illegal routes, as cities may be duplicated and missing after the transfer. Corrective post-processing is therefore performed on the offspring to remove these anomalies, for example, with the OX strategy in [Goldberg 89]. Mutation does not require correction. It simply takes a city and moves it to a random alternate location in the tour, which is correctness-preserving.

Other parameter settings are in Figure 1.2. The genetic algorithm uses a hybrid steady-state approach, in which pseudo-generations are defined for the purposes of applying Lamarckism and collecting statistics. Population changes occur to a single population. After a set number of population changes (taken to be the population size), a generation is said to have passed. At this point, the *lamarckian_evolution/0* routine is invoked. Tournament selection is performed for both reproduction and replacement of individuals. Since the use of tournament selection tends to result in populations saturated around the fittest individual, the algorithm disallows duplicates in the population.

Population size:	200
# generations:	150
# runs:	35
Probability of crossover:	1.0
Tournament size, reproduction:	2
Tournament size, replacement:	3
Lamarckian evolution	
% population processed:	0.50
# iterations in search (K):	(i) 5, (ii) 20
Selection criteria:	tournament reproduction
Unique population:	yes

Figure 1.2 Lamarckian experiment parameters

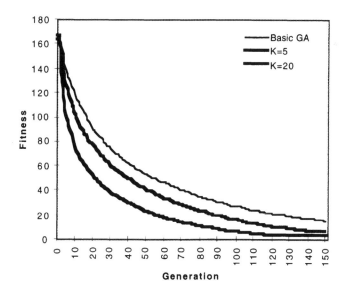

Figure 1.3 Performance graph (avg. 35 runs)

Figure 1.3 is a performance graph of the average best fitness per generation, for 35 runs with Lamarckian evolution (iteration values of 5 and 20), and 35 runs without any Lamarckism. The use of Lamarckism clearly accelerates performance with respect to generational time. The asymptotic nature of the curves shows that the gains become most significant during later generations, since the Lamarckian-influenced populations are concentrated on fitter areas of the search space. As is expected, the run using 20 iterations yields the best results. Even with the seemingly negligible iteration value of 5, however, by generation 100 the Lamarckian runs have a lead of approximately 50 generations over the basic genetic algorithm.

Figure 1.4 shows the average number of optimized individuals found with localized search over the run. The 20-iteration search does a better job of finding individuals to optimize for the first 100 or so generations. After 100 generations, it performs only marginally better than the 5-iteration search. This implies that the number of iterations must be increased dramatically if mutation is to make significant progress during later stages of a run.

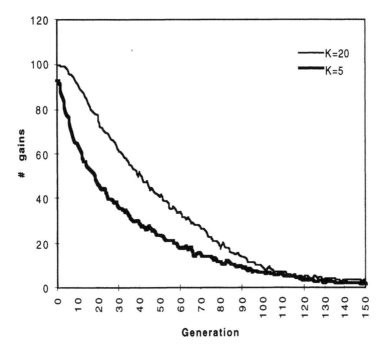

Figure 1.4 Optimized individuals

The above performance results must be examined in light of the overall computational overhead used by Lamarckian evolution. In order to further compare the performance of the basic genetic algorithm with the Lamarckian ones, extended runs of the non-Lamarckian case were done to 300 generations (see Figure 1.5). In general, the non-Lamarckian runs would need to run to generation 150 before obtaining the K = 5 performance at generation 100, and to almost generation 300 to reach the K = 20 results at generation 100.

Consider the table in Figure 1.6. The Lamarckian runs are much more expensive than the non-Lamarckian in terms of the total number of individuals processed. This measurement would initially indicate that Lamarckism is not an economical enhancement to the overall processing. On the other hand, the total processing time (Silicon Graphics O2 workstation, R5000 180 MHz CPU) is less than proportional to the number of individuals processed. One reason for this is that the basic route encoding used for the TSP is such that crossover is more computationally expensive to perform than mutation. Crossover requires the detection and correction of illegally repeated cities, whereas mutation's swapping of two cities in the list can be done quickly without correction. (See [Tamaki et al. 1994] for more advanced encodings of the TSP that circumvent the need for correction after crossover.) In addition, the cost of fitness evaluation in this experiment is negligible. Therefore, the time overhead of using Lamarckism was not proportional to the number of individuals processed, and the processing benefits from it, especially after later generations have lapsed.

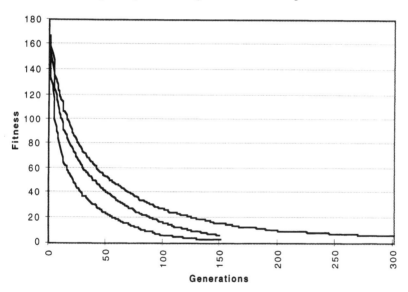

Figure 1.5 Extended performance graph (avg. 35 runs)

1.4. Discussion

The practicality of Lamarckian evolution within a genetic algorithm depends upon the problem being investigated. Lamarckian localized optimization will typically boost the relative fitness of the population, and hence accelerate search performance of the parent genetic algorithm. This acceleration, however, can be costly if the fitness evaluation of phenotypes during localized searches is computationally expensive. Thus the benefits of Lamarckism must be weighed with the costs inherent in the problem being studied.

Experiment	Generations	Individuals processed	Avg. CPU time
K=5	150	105,000	00:07:52
K=20	150	330,000	00:12:38
Basic GA	150	30,000	00:06:01
Basic GA	300	60,000	00:12:18

Figure 1.6 Individuals processed and CPU processing time

The performance of Lamarckian evolution can vary during a run. Further research is needed to determine strategies for dynamically adapting Lamarckism during execution. For example, the iteration values and percentage of the population processed by Lamarckism could automatically adapt to the performance of the run. Perhaps the most promising approach is to adaptively evolve various control parameters of Lamarckian learning, in the style of [Grefenstette 1986, Back 1992]. The code described in this chapter may be a stepping stone for more sophisticated enhancements in this direction.

One other issue is the nature of mutation operator used by a Lamarckian strategy. The minimal requirement is that the mutation used is robust enough to permit exploration of the entire search space. The experience with the TSP experiment is that the simple mutation operator had difficulty making optimizations during later generations. The population was already relatively fit, and random mutation of very fit phenotypes resulted in a blind search. Processing may benefit, therefore, with the use of a library of varied mutation operators, as well as more intelligent next-state operators for generating new individuals.

Bibliography

Ackley, D.H. and Littman, M.L. 1994. "A Case for Lamarckian Evolution." *Artificial Life III*, Ed. C.G. Langton, Addison-Wesley.

Back, T. 1992. "Self-Adaptation in Genetic Algorithms." *Proc. 1st ECAI*, Dec. 1991, MIT Press.

Cochrane, E. 1997. "Viva Lamarck: A Brief History of the Inheritance of Acquired Characteristics". [http://www.ames.net/aeon/].

Dawkins, R. 1996. *The Blind Watchmaker*. Norton.

Goldberg, D.E. 1989. *Genetic Algorithms in Search, Optimization and Machine Learning*. Addison-Wesley.

Gould, S.J. 1980. *The Panda's Thumb*. New York: Norton.

Grefenstette, J.J. 1986. "Optimization of control parameters for genetic algorithms." *IEEE Transactions on Systems, Man and Cybernetics*, 16(1):122-128.

Grefenstette, J.J. 1991. "Lamarckian Learning in Multi-agent Environments." *Proc. 4th Intl. Conference on Genetic Algorithms*, Morgan Kaufman.

Hart, W.E. and Belew, R.K. 1996. "Optimization with Genetic Algorithm Hybrids that Use Local Search." In *Adaptive Individuals in Evolving Populations*. Ed. R.K. Belew and M. Mitchell, Addison-Wesley.

Li, Y., Tan, K.C., and Gong, M. 1996. "Model Reduction in Control Systems by Means of Global Structure Evolution and Local Parameter Learning." *Evolutionary Algorithms in Engineering Applications*, Eds. D. Dasgupta and Z. Michalewicz, Springer Verlag.

Mitchell, M. 1996. *An Introduction to Genetic Algorithms*. MIT Press.

Tamaki, H., Kita, H., Shimizu, N., Maekawa, K., and Nishikawa, Y. 1994. "A Comparison Study of Genetic Codings for the Traveling Salesman Problem." *1st IEEE Conference on Evolutionary Computation*, June 1994.

Winston, P.H. 1992. *Artificial Intelligence (3e)*. Addison-Wesley.

Appendix: Code listing

% Author: Brian J Ross

% Dept. of Computer Science, Brock University, St Catharines, ON, Canada

% July 1997

% lamarckian_evolution

%

% Performs Lamarckian evolution on P% of population, iterating

% each K times using hill-climbing.

% lamarckian(P, K) is set by user. Population exists as facts in

% database of form: individual(ID, Fitness, Expression).

% Fitness is standardized (0 is best, higher values are less fit).

% An improved individual, if found, is replaced in the program database.

% First clause efficiently processes entire population.

% Second case is if less than entire population to be used, in which case

% selection must be performed.


```
lamarckian_evolution :-  lamarckian(Percent, K),Percent >= 1.0,setof(ID,
F^E^individual(ID, F, E), IDs),write('Lamarckian evolution...'),
nl,lamarck_loop(IDs, K),!.

lamarckian_evolution :-lamarckian(Percent, K),Percent > 0.0,Percent <
1.0,population_size(PopSize),N is integer(Percent *
PopSize),write('Lamarckian evolution...'), nl,get_unique_IDs(N, [],
IDs),lamarck_loop(IDs, K),!.

lamarckian_evolution.
```


% get_unique_IDs(N, SoFar, IDs)

% N - Number of unique individual ID's to collect

% SoFar - ID's collected so far

% IDs - final set of unique IDs

```
%

% Retrieves a list of N unique individual ID's,

% selecting each one via a tournament selection routine defined elsewhere.


get_unique_IDs(0, IDs, IDs) :- !.

get_unique_IDs(N, SoFar, IDs) :-repeat, select(ID),\+ member(ID, SoFar),
M is N - 1,get_unique_IDs(M, [ID|SoFar], IDs),!.


% lamark_loop(IDs, K)

% IDs - individuals to perform Lamarckian evolution upon

% K - number of iterations for best-first search

%

% The individuals in IDs have hill-climbing performed on them.

% If the result of this search is a fitter individual, then that new

% individual replaces the original in the database. A side effect is that

% individual/3 clauses are replaced with improved individuals when found.

% The + and - characters written to screen to give overview of

% Lamarckian performance.


lamarck_loop([ ], _) :- !.

lamarck_loop([ID|Rest], K) :-individual(ID, Fit, Expr),hill_climb(K, (Fit,
Expr), (NewFit, NewExpr)),% replace previous line with following if
cycling is a problem% hill_climb_nocycle(K, (Fit, Expr), (NewFit,
NewExpr)),(NewFit >= Fit ->    write('-')    ;    write('+'),
retract(individual(ID, _, _)),        assert(individual(ID, NewFit,
NewExpr))),lamarck_loop(Rest, K),!.


% hill_climb(K, Item, Soln)

%  K - iterations to perform on individual for search

%  Item - individual to perform localized search upon

%  Soln - The most improved solution found. Worst case is an expression

%  having same fitness as original.

%
```

% Does hill-climbing search for K iterations on an individual.

% Expression is mutated, and if mutant is fitter than individual, it is the new

% expression to mutate. Otherwise, the original will be used again.

```
hill_climb(0, Item, Item) :- !.
hill_climb(K, (TopFit, TopExpr), Soln) :- mutation(TopExpr,
NewExpr),eval_fitness(NewExpr, NewFit),select_best((NewFit, NewExpr),
(TopFit, TopExpr),  BestSoFar),K2 is K - 1,hill_climb(K2, BestSoFar,
Soln),!.
```

% select_best(Item1, Item2, BestItem)

%

% BestItem is the one of the Items with the better (lower) fitness value.

```
select_best((F1, E1), (F2, _), (F1, E1)) :- F1 =< F2, !.
select_best(_, X, X).
```

% hill_climb_nocycle(K, Item, Soln)

% hill_climb2(K, Items, Soln)

% K - iterations to perform on individual for search

% Item - individual to perform localized search upon

% Items -list of individuals searched thus far; first items are the fittest

% Soln - The most improved solution found. Worst case is an expression

% having same fitness as original.

%

% An alternate hill-climbing predicate. This version is useful if the problem

% domain is such that cycling (repeated visits) during the search is likely.

% It disallows cycling, and permits fairer search from the best individuals.

```
hill_climb_nocycle(K, Item, Soln) :-hill_climb2(K, [Item], Soln), !.
```

```
hill_climb2(0,[Item|_], Item) :- !.
```

```
hill_climb2(K, [(TopFit, TopExpr)|Rest], Soln) :- mutation(TopExpr,
NewExpr),\+ member((_, NewExpr), Rest),eval_fitness(NewExpr,
NewFit),insert_list((NewFit, NewExpr), [(TopFit, TopExpr)|Rest],
List2),K2 is K - 1,hill_climb2(K2, List2, Soln),!.
```

```
hill_climb2(K, List, Soln) :-K2 is K - 1, shunt(List,
List2),hill_climb2(K2, List2, Soln).
```

```
% insert_list(Item, List, NewList)
```

```
% Item - Expression/Fitness to add to the list
```

```
% List - list of Expr/Fit pairs found so far in search
```

```
% NewList - List with Item appropriately inserted
```

```
%
```

```
% Item is placed into List, such that the List retains all the fittest members
      at
```

```
% the front. If Item is a fittest individual, it will go at the front. Otherwise
      it
```

```
% is placed behind them, and before an item in the list that is less-fit than
```

```
% itself (convenient to code - could be optimized if desired).
```

```
insert_list(Item, [ ], [Item]).
```

```
insert_list((Fit, Expr), [(Fit2, Expr2)|Rest], [(Fit, Expr), (Fit2,
Expr2)|Rest]) :-Fit =< Fit2.
```

```
insert_list((Fit, Expr), [(Fit2, Expr2)|Rest], [(Fit2, Expr2)|Rest2]) :-Fit >
Fit2,insert_list((Fit, Expr), Rest, Rest2).
```

```
% shunt(Items, NewItems)
```

```
% Items - ordered list
```

```
% NewItems - Items, but with first item shunted to the right-most
```

```
% end of its sibling expressions of same fitness
```

```
%
```

```
% Optional utility. To decrease chance of thrashing on a poor expression
```

```
% in list, shunt the best individual at the start of the list to the far right of
```

% equally best-valued siblings (if any).

shunt([(Fit, Expr1), (Fit, Expr2)|Rest], [(Fit, Expr2)|Rest2]) :-!,shunt([(Fit, Expr1)|Rest], Rest2).

shunt([A, B|Rest], [A, B|Rest]):- !.

shunt(List, List).

% etc...

member(A, [A|_]).

member(A, [_|B]) :- member(A, B).

Chapter 2 The Generalisation and Solving of Timetable Scheduling Problems

Colin D. Green

colin@zuyuva.demon.co.uk

http://www.zuyuva.demon.co.uk

2.1. Introduction

Timetabling at present is still very much a manual process, usually involving a draft timetable which is modified over a period of time as problems and limitations are uncovered. The final timetable from this method is often acceptable to the people who have to follow the timetable; however, much time and effort is spent and some improvements may still be possible [10].

2.1.1 Aims

With this chapter I hope to create an application on a PC which will allow the user to build a model of their timetabling problem; the application will then use the model to produce a usable timetable.

Timetable schedules are used to coordinate use of resources in many areas, some of these are

Transport timetables: Buses, trains, planes, etc.

Manufacturing: Planning the use of resources in a manufacturing environment, be it machines, raw material, and/or personnel.

University, School timetables, etc.

In this project I will attempt to look into as many types of timetable problems as possible and build a general model capable of representing each of them.

I will then be looking at known techniques used to solve or, at least, create suitable solutions to the problems and assess the relevance of each technique to the general model I have created.

Finally, I will choose a technique or a number of techniques most suited to the problem and implement them in an application on a PC. The application will request all relevant data from the user to build up a complete timetabling problem. The user will then be able to output the finished timetable solution in a number of differing formats relevant to different entities in the timetable, e.g., in the case of a bus timetable, there are timetables for passengers but the data could be represented from the point of view of a driver or a bus, stating where a driver or bus should be at key times.

The possibility of modifying the timetable at a later date to accommodate changes should be included.

2.2. Introduction to Timetabling

2.2.1 Definition of a Timetable

A timetable can be loosely defined as being something which describes where and when people and resources should be at a given time. Most people are familiar with the school timetable which can be presented as a table of days of the week and time slots.

	9 am – 10 am		10 am – 11 am	11 am –12 pm
Monday	MATHS	ENGLISH	SCIENCE	
	Mr. Green	Mrs. White	Mrs. Teacher	
	Room 1	Room 2	Room 2	
Tuesday				

Figure 2.1 Example of a Timetable

You can see that each day is split into timeslots; each timeslot has a list of what subjects are being taught, by whom and where. The timetable can be represented in a number of different ways. Each student will have their own timetable depending upon which subjects they study, as will each teacher and each room; these are all different perspectives on the same timetable.

Other situations where timetables are required are :

1. Manufacturing: production lines, project planning.

2. Travel: trains, buses, etc.

3. University/school examinations.

4. University lecturing.

5. School timetable.

6. T.V./radio/media schedules.

7. Conferences/meetings.

These situations require timetables of varying complexity depending upon the number of resources we are trying to timetable, the number of timeslots and locations.

2.2.2 Problems Related with Producing Timetables

At first sight, timetable problems can seem simple. Using the school as an example, we can use a simple algorithm to group subjects with no common students and assign classes to rooms in whatever order is convenient. We can then assign teachers to classes depending upon what qualifications they have.

Some problems become apparent. If a student is given a choice of subjects to study, we increase the chance that clashes will occur between the subjects [7]. A clash is defined as an instance when a student or another person is scheduled to be in more than one place at the same time.

A clear explanation of this problem would be a situation where a student is taught a group of subjects and each subject is taught only once a week; if you miss the lesson, there is no repeated lesson. If two or more of these subjects occur at the same time, the student will be forced to miss a lesson, and there is a clash in the schedule. This problem worsens when we increase the number of students and subjects or the number of subjects taken by each student, and also in modularised courses where subjects are taught to students in different courses.

Other problems that could occur are :

- Two or more subjects which have no students common between them require a single teacher so that, again, we cannot place them in the same time slot.

- Some classes may need specific limited resources, such as lab equipment, or a class may be so large that it will be limited to a few large rooms.

These problems are described for a teaching environment. The same problems occur in other timetabling problems and there may well be additional domain specific constraints. When combined, the constraints can make the production of a timetable far from trivial.

2.2.3 NP-Hardness of a Problem

For every problem that can be solved with a well-defined algorithm, there is a relationship between the size of the problem and the amount of time required to solve it [1, 19]. Obviously, if we implement an algorithm within a computer, the speed at which the computer runs the algorithm directly affects how fast we find a solution. New computer hardware may well run an algorithm faster than before and find solutions in less time, but the problem has remained the same.

We could calculate the number of computer processor cycles needed to find a solution, assuming that a faster computer can run more cycles in a given amount of time but will need to complete the same number of cycles as a slower machine. Unfortunately, this type of measurement is also prone to variations dependent upon computer hardware.

The solution is to use the concept of time complexity of an algorithm. A time complexity function exists for any algorithm. The function describes how the algorithm's time requirements change with respect to the size of the problem the algorithm is solving, but it does not explicitly state how long it will take to find a solution [1,19].

An example time complexity function could be $t = v$, where v is problem size and t is time complexity; this simply means that if we double the problem size then the time required to solve it will double.

Simple algorithms such as sorting algorithms generally have time complexity functions such as $t = v, t = 2v, t = v^2$.

As we increase v (the problem size), t does not increase too dramatically. These algorithms are said to have 'polynomial time complexity' because they have the same order as a given polynomial p(v), e.g., v+v2+v3 [1,19].

In problems such as the travelling salesman problem, graph colouring (see Section 2.4.2), scheduling, and timetabling problems, as we increase the size of the problem, the time required to solve it increases exponentially. These problems are said to have non-polynomial time complexity and are described as being NP-hard [1,19].

NP-hard problems become effectively unsolvable as we increase the problem size. This is because the amount of time required to solve them becomes astronomical. This is demonstrated in Figure 2.2.

Time Complexity Function	Problem size					
	10	20	30	40	50	60
v	0.00001 sec.	0.00002 sec.	0.00003 sec.	0.0004 sec.	0.0005 sec.	0.0006 sec.
v^2	0.0001 sec.	0.0004 sec.	0.0009 sec.	0.0016 sec.	0.0025 sec.	0.0036 sec.
v^5	0.1 sec.	3.2 sec.	24.3 sec.	1.7 min.	5.2 min.	13 min.
v^{10}	2.7 hrs.	118.5 days	18.7 yrs.	3.3 centuries	30.1 centuries	192 centuries
2^v	0.001 sec.	1 sec	17.9 min.	12.7 days	35.7 yrs.	366 centuries
3^v	0.59 sec.	58 min	6.5 yrs.	3855 yrs	$2.28*10^8$ centuries	$1.3*10^{13}$ centuries
$v!$	3.6 sec.	771 centuries	$8.4*10^{16}$ centuries	$2.6*10^{32}$ centuries	$9.6*10^{48}$ centuries	$2.6*10^{66}$ centuries

Figure 2.2 Time Complexity Function

Although time complexity functions don't actually tell us how long a problem will take to solve, for any given problem size they will have only one value which can be scaled to real time depending on how fast we know we can implement the algorithm. In the following table we will assume that the time complexity function returns the number of computer processor cycles required to find a solution. If we also assume that one cycle takes 1 microsecond (0.000001 seconds), we can now calculate how long an algorithm of a given time complexity will take to solve a problem of a given size.

For each of the time complexity functions in the left-hand column, the table shows the time taken to solve various sizes of problems. The problem size is just a Figure and could represent the number of parameters to an algorithm or the range of an input value. What these values represent is not important. If we look at the way in which the amount of time required to solve each problem increases, we can see the difference between NP-hard problems and those which are not.

For the first four time complexity functions, the increase in time needed is much lower than that for the later three functions. Although the fourth function does seem to be borderline, if we were to increase the problem size further, we would see the time required to solve the NP-hard problems increase at a much higher rate [1].

2.3. A General Model for Timetabling Problems

In this section I will look into different types of timetables, identify similarities, and from this produce a general model for the representation of timetable problems. This model will define what data we need to give an algorithm before we can even attempt to find a solution.

2.3.1 Generalising Timetable Problems

Timetabling problems I have identified are :

1. Manufacturing: production lines, project planning.

2. Travel: trains, buses, etc.

3. University/school examinations.

4. University lecturing.

5. School timetable.

6. T.V./radio/media schedules.

7. Conferences/meetings.

There are some key commonalties which can be immediately identified. Each of the above has a resource or number of resources which need to be assigned to a time and/or location.

Although I am attempting to build a general model, I would like to describe the scope of the model more accurately by omitting some of the above timetabling problem types. First, there are many specific techniques for building manufacturing/project schedules which go into far more detail than is possible in a general model [19, 22, 23]. Travel timetables are also hard problems to fit into a general model because of the large amount of data needed before we can build a solution, e.g., varying number of passengers during the day and data pertaining to locations such as travel times, positions and route information.

T.V./radio scheduling is not really a problem for automation. When assigning a program, there is usually no choice to be made regarding on which channel it will be broadcast. The time of the broadcast is chosen to fit social patterns, not something easily reduced to a set of data or mathematical functions!

The remaining set of timetable problems that I will be building a model to represent are as follows:

1. University/school examinations.

2. University lecturing.

3. School timetabling.

4. Conference/large meeting timetables.

More generally, any timetable problem has different groups of entities that need to come together one or more times with varying resources available each time. Also, the number of times and locations available for timetabling is limited.

For now, I will refer to the above list of timetable problems. For each of these problems, I have identified the main entities.

Problem   Type Entities

The exam timetable	- Supervisor, Student, Subject, Location, Time slot.
University lecturing	- Lecturer, Student, Subject, Location, Time slot.
School timetable	- Teacher, Pupil, Subject, Location, Time slot.
Conferences/meetings	-Speaker, Attendees, Topic, Location, Time slot.

These are the entities that will need to be directly assigned in a completed timetable. Other entities, their attributes and relationships between them, will influence the assigning of the above entities but are not actually visible on the completed timetable. I will identify these other entities, but first I will examine these main entities.

It is easy to see the similarities between the four types of timetabling problems I have listed; the entities can be grouped and given a general name as follows :

	Problem Type			
General Entity Name	**Exam**	**Lecturing**	**School**	**Conference**
Speaker	Supervisor	Lecturer	Teacher	Speaker
Attendee	Student	Student	Pupil	Attendee
Context	Subject	Subject	Subject	Topic
Location	Location	Location	Location	Location
Time slot	Time slot	Time slot	Time slot	Time slot

We can apply some simple relationships to the entities.

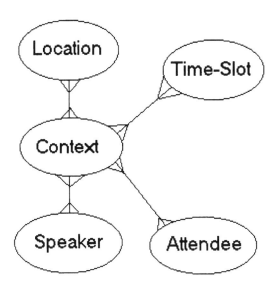

Many people could be qualified to be the speaker for a given subject or context; also, a speaker may speak on many different topics or contexts. Similarly, many attendees will want to be present for a meeting of a particular context and may also want to be present at meeting of many different contexts, etc.

This is an overall relationship model where all the relationships are many to many. Here is the model for one timetable assignment.

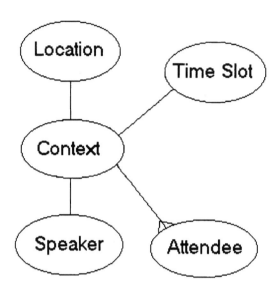

This is a more useful model. It shows limitations which should be enforced for each timetable assignment. An assignment in the timetable must somehow reference all of the above entities either directly or indirectly. The speaker and the attendees are meeting under a specific context at a single location and single time slot. These limitations must be met and are therefore called hard constraints; these are discussed in more detail in Section 3.2.

Earlier I stated that other entities not referenced in the end timetable will affect how we can assign the main entities described above. Actually, I will introduce just one more entity, a **Resource.**

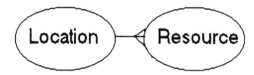

Each location has an associated list of resources. These are the resources available at each location, such as seats or machinery.

It is envisaged that these simple relationships are sufficient for the general model. More complex relationships can be built around this, e.g., where a subject is taught by a team made up of lecturers in a university, data pertaining to teams is not needed by the scheduler, only that a lecturer has been marked as a teacher of a subject.

2.3.2 Hard Constraints on Resource Allocation

Each of the timetable problem types on which I am basing the general model have common constraints limiting how we allocate the speakers and attendees to a location and time slot. The most obvious of these is that no one can be at more than one location at any given time. A physical constraint such as this is know as a 'Hard' constraint; a timetable that violates hard constraints is not feasible in the real world [7].

The hard constraints are :

A speaker can be assigned to only one location at any given time.

An attendee can be assigned to only one location at any given time.

A location can have only one context assigned to it at any given time.

4. Attendees cannot be allocated so as to exceed a location's maximum capacity.

Note that the third constraint is not necessarily applicable to the exam timetable, where exams for many subjects may be held in the same location.

2.3.3 Soft Constraints on Resource Allocation

Soft constraints are defined as being limits on allocation of resources which, if violated, still result in a feasible timetable, but we would like to avoid them if possible [7].

For example, we do not want to set consecutive exams for students, and we want to spread exams out over the whole examination period as much as possible.

Some of the things we might want to achieve in a timetable with respect to any one person are:

1. Minimise the occurrence of consecutive timetable assignments.

2. If consecutive assignments do occur, minimise the number of assignments in a row.

3. Maximise gaps between timetable assignments (especially applicable to examinations).

4. Minimise gaps between timetable assignments.

5. Minimise travel times/distances.

The desirable characteristics in a timetable depend upon the type of timetable we are dealing with (points 3 and 4 conflict). Point 3 is desirable for examinations, point 4 for day-to-day timetables.

2.3.4 Timetable Templates

So far we have assumed that before any allocation was performed all speakers, attendees and contexts were available for any time slot and could be assigned to any location. In reality, people have other commitments and resources may be unavailable for any number of reasons. We need a method of describing when people and resources are available.

To service this need I will define a template which describes when and where people can be allocated.

The template will directly describe unavailability and also allow daily/weekly slots to be made unavailable. A more implicit description, de-allocating slots dependent on which slots have already been (de)allocated will not be used on the grounds that it would over complicate the final application. Also, it is not feasible to describe a template for every resource and attendee, so I will be providing templates only for speakers.

2.4. *An Investigation Into Techniques for Producing Timetables*

2.4.1 Job-Shop Scheduling

A lot of work on scheduling is concerned with manufacturing processes and much of the terminology, in this field, reflects this. The job-shop scheduling problem refers to the assignment of jobs to machines in a workshop; however, the model also applies to many scheduling problems. For instance, a job could be a patient in a hospital and the machine could be a doctor. The basic template for a job-shop problem is that we have a number of entities which need processing (the jobs) and some other entities which do the processing (the machines) [1].

The general case where there are n jobs to be processed on m machines with each job having it's own sequence of processing is called the general job-shop. This model can be used to represent all job-shop problems, but more specifically defined problems can be represented using four parameters (from [1]).

n : The number of jobs

m: The number of machines

A: Describes the flow of jobs through machines. When m = 1 (only one machine), this parameter is omitted. A can be any one of the following;

F: The flow shop. Each job goes through the same sequence of machines.

P: The permutation flow-shop. As the flow-shop, but here the job order for each machine is also the same.

G: The general job-shop

B: Describes the performance measure by which the schedule is evaluated.

The last parameter is used to specify the performance measures we wish to minimise or maximise when generating a schedule, e.g., total time the schedule takes, elimination of quiet periods in the workshop, or keeping an expensive machine working as much as possible, etc. The job-shop, as it stands, cannot always completely represent a timetable. For instance, timetabling in a school needs to address not only which room a group of pupils should be in but also who will teach them and what subject they should be taking.

If we assume that all class sizes are the same and that all subjects can be taught in all rooms, then we could represent the problem as classes being jobs and timetable slots being machines. Allocating the classes to a teacher can then be treated as a separate job-shop problem in which classes (now allocated a timetable slot) are jobs and teachers are machines.

So, in the last case, we have broken the timetable problem into two job-shop problems; however, the case still assumes that classes are made up of a fixed group of pupils and that all subjects can be taught in all rooms. These extra variables limit the way we can allocate jobs to machines. They are constraints that vary with each instance of a timetable problem, otherwise known as domain-dependent constraints.

These domain-dependent constraints are not all allowed for in standard job-shop scheduling techniques; some modification would be needed. Also, by splitting the timetabling problem into two separate job-shop problems, we may not find feasible timetables.

For example, if we again take the example of the school timetable, the first problem was to allocate classes to timetable slots. We assume we have successfully solved this problem and now go on to the second problem, allocating the class a teacher. We may find that this problem cannot be solved because some classes may need specific teachers, and these teachers may have already been allocated to a class. If this occurs we need to go back to the first job-shop problem and alter the allocation of timetable slots to classes.

So, the two problems are not completely independent and, therefore, we really need a single technique which can deal with all the variables.

2.4.2 Graph Colouring

If we take a university timetable as an example, the problem of grouping subjects so that student clashes are eliminated is similar to the mathematical problem of colouring the vertices of a graph [2].

Graph colouring algorithms will group together subjects so that no student has to be in more than one place at any time.

The graph itself represents subjects as vertices and the clashes between subjects as connections between the vertices. Here is a simple graph.

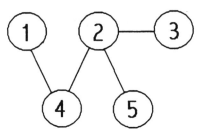

A connection between two vertices indicates there are students who study both of these subjects; it could be just one student or a hundred. The process of colouring the graph refers to assigning colours or symbols to vertices indicating that they can be assigned the same timetable slot. This can be done by merging vertices that do not clash. For our simple graph, we can merge vertices 1, 3 and 5. This results in the following graph:

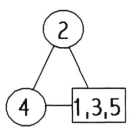

We have now coloured vertices 1, 3 and 5 to the same colour. No more vertices will merge with 1, 3 or 5 so we can remove the new vertex and continue to colour the remaining vertices.

Clearly only vertices 2 and 4 are left and they are connected, so they are given different colours.

Applying the colours or groupings to the original graph, we get:

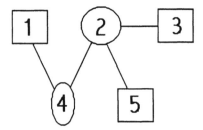

Vertices of the same colour (here I have used different shapes) represent subjects which can be time tabled at the same time. In the above graph there are three different symbols; therefore, we need a minimum of three time slots.

In colouring the graph, we have calculated the minimum number of time slots needed to prevent any student clashes. The number of time slots needed may be more than are available. Also, if many subjects are grouped together there may not be sufficient locations or other resources to actually allocate all of these subjects in one time slot.

The graph colouring problem is NP-complete [2]. Various algorithms exist for colouring graphs of different structures or with different colouring requirements which can be utilised [2]. This is a useful technique but it does not address the remaining problem of allocating the subjects to timetable slots which is also an NP-complete problem.

2.4.3 Genetic Algorithms

The genetic algorithm works on the same principal as natural evolution. We have a population of candidate solutions to a problem, we select solutions based on how good they are, and then we create new solutions from those which were selected [15]. A genetic algorithm acts as a directed search algorithm or a heuristic algorithm, with the search being performed on the set of possible solutions to a problem.

Genetic algorithms have proven effective at tackling many problems where no other algorithms exist, usually where the problem is NP-Hard [26, 28]. Among these are timetabling problems [5, 6, 7, 8, 9, 15].

2.4.4 Choice of Algorithm for the General Timetable Problem

I have decided to use a genetic algorithm to solve the general timetabling problem. The general timetable problem has many different parameters and therefore has a particularly complex set of possible solutions. A genetic algorithm is capable of searching through a complex search space [16]. Also, if extra constraints on the timetable are required at a later date, all that is required is a new evaluation function which describes the new search space.

2.5. Genetic Algorithms

2.5.1 Introduction to Evolution and Genetics.

Genetic algorithms or evolutionary computing is based upon biological genetics and natural evolution. In nature every living organism has a unique set of chromosomes which describes how to build the organism. A chromosome is just one long molecule of DNA; however, different parts of a chromosome describe how to make different things, i.e., one part may describe how to build an eye or a limb or just a simple protein. These different sections of the chromosome are called genes [29].

When an organism reproduces, it passes on copies of its chromosomes. The process of copying the chromosomes can result in errors (mutations) which occasionally can give rise to useful features and better offspring [29].

In sexual reproduction, chromosomes from two parents are combined producing offspring with a mixture of the two parents chromosomes. This may lead to combinations of good genes coming together, again producing better offspring. This key process is known as crossover, which refers to the crossing over of genes between chromosomes [15, 29].

In the previous paragraphs I have used the phrase 'better offspring,' so what is it that makes some organisms 'better' than others? In nature, a species which generates offspring with good genetic information will survive, but organisms of the same species will not all have the same genetic information and there will be variations due to mutation and crossover described above. These variations will make some individuals better able to survive and reproduce; they are better or fitter.

The result is that fitter individuals reproduce more and increase in population. Similarly, individuals that are not as fit will not reproduce as much, if at all, and will have a smaller population and may die off altogether. Over time the build up of good genes, maybe coming together via sexual reproduction, will result in individuals which have adapted very well to their environment.

Evolution can be broken down into two basic steps:

1. Selection of individuals based on their fitness.

2. Reproduction of the selected individuals.

These two steps are continuously repeated, one cycle being a generation.

2.5.2 Evolutionary Computing

We can use the concept of natural evolution to solve problems in the computer. Possible solutions make up the population and better solutions, equivalent to fitter organisms in nature, are more likely to reproduce and pass on their genetic information to the next generation. Over time we could expect that good solutions evolve just as organisms have evolved in nature.

To solve a problem using genetic algorithms, otherwise know as evolutionary computing, we need the following:

1. A system of encoding the possible solutions or chromosome structure,

2. An initial population of solutions.

3. A function to evaluate a solution's fitness.

4. A method of selecting solutions to be used to produce new solutions.

5. Recombination and mutation operators to create new solutions from those existing.

2.5.2.1 Solution Representation: The Chromosome Structure

Typically a bit-string or a string of some other kind is used to represent a solution in evolutionary computing. The string data structure is most similar to the natural chromosome and therefore we can manipulate the strings in ways similar to natural chromosomes [15].

2.5.2.2 Initial Population

Before we can begin evolving solutions we must have an initial population of solutions which we can evolve.

Initial solutions are sometime randomly created or by using an algorithm to produce solutions. Previously evolved solutions can also be used. It is possible to use any combination of the these three techniques together, since increasing the diversity of the solutions in a population will give a better chance of finding good solutions.

2.5.2.3 Evaluation Functions

This function takes a single solution as a parameter and returns a number indicating how good the solution is. The number can be integer or real and have any range. By itself the number returned means nothing; only when we compare the values returned by all of the possible solutions can we select the better solutions.

2.5.2.4 Selection

Each generation must produce new solutions from the current population of solutions. How many of the current population are used to generate the new solutions, which ones we select, and which of the current population we should erase to make room for the new solutions all need to defined.

Some types of selection are:

Roullette wheel: Imagine that each solution in the population has a slice of a roulette wheel. Better solutions have proportionately larger sections of the wheel and therefore there is a greater chance they will be selected. If we want to select two solutions, we spin the wheel twice. We then have the choice of removing selected solutions from the wheel before we run again or we can leave them on the wheel and, therefore, good solutions may be selected more than once.

Tournament Selection: Picks 2 or more solutions at random and uses a tournament strategy, where a number of rounds are played. In each round a number of solutions come together to compete, and only the fittest solution will win and be selected.

Elitism: Not really a selection strategy, but other strategies can employ elitism. This is where the best solution in the population is guaranteed to survive to the next generation.

2.5.2.5 Recombination and Mutation Operators

After selecting solutions we need a technique or techniques of combining these solutions in a manner that will produce good, new solutions. These are known as recombination operators.

Mutation operators alter the new solutions in a totally random manner. This helps maintain diversity in the population and results in solutions that otherwise could not have been produced by recombination alone.

2.5.2.5.1 Crossover Operators

Crossover operators will normally take two parents and create offspring with a mixture of both parents' genetic information. The common forms of crossover are 1-point, 2-point, n-point and uniform crossover.

In 1-point crossover a random point is chosen along each parent's chromosome, each half of the chromosomes is then swapped over, e.g., if we have two parents:

parent1   1234|5678

parent2   8765|4321

If we perform crossover at the mid-points (show by the bar), we get two new offspring:

offspring1   12344321

offspring2   87655678

1-point crossover is inspired by biological processes [15].

2-point crossover is similar to 1-point, but there are two crossover points. This results in one parent retaining the head and tail of its chromosome, gaining a new mid-section. The other parent will retain its mid-section and gain a new head and tail.

N-point crossover uses the same technique but we select N crossover points, swapping parts of the chromosomes between every other crossover point, starting at the head and the first crossover point for parent one and parent two, respectively.

Uniform crossover selects x number of points in the chromosomes, where x is a random number less than the chromosome length. Each selected point is then swapped over.

2.5.2.5.2 Mutation

When we create offspring using whatever technique, we must copy genetic information from the parents. In nature, duplicating DNA can sometimes result in errors. DNA is also prone to damage in day-to-day existence which also results in errors. These errors or mutations can sometimes result in good features and these features can then come together via sexual reproduction and eventually lead to new species [29]. Therefore, errors in DNA perform a vital role in natural evolution.

In Evolutionary computing we can imitate mutation by generating errors in the offspring. A point in the offspring's chromosome can be set to a random value; however, this may be an invalid value and so it can be beneficial to replace the value with one which is valid.

2.6. A Genetic Algorithm For Our General Timetabling Problem

2.6.1 Chromosome Structure

Each chromosome represents a complete timetable. The chromosome is made up from a string of genes and a gene consists of four numbers. These four numbers represent a subject, location, lecturer and a time slot respectively. One gene therefore assigns a subject and a lecturer to a time and location.

So the gene 2, 5, 7, 23 assigns lecturer 7 to teach subject 2 at location 5 in time slot 23.

2.6.2 The Initial Population

Each chromosome in the initial population is generated by a simple algorithm which allocates as many timetable slots as possible without creating subject, lecturer or room clashes in a time slot. If a lecturer defined as being able to teach a subject is available, he will be selected; otherwise, any other available lecturer is chosen.

Subject clashes are not considered.

2.6.3 Evaluation Function

The evaluation function calculates a punishment value for each chromosome. Good chromosomes have low punishment values. A value of zero indicates the chromosome meets all the criteria described in the evaluation function.

The evaluation function is made up from many different evaluations which all produce a punishment value which accumulates to give the final punishment.

The following sections describe how a chromosome is evaluated.

2.6.3.1 Timetable Clashes

The following clashes between entities are evaluated and will increase the punishment value if they are violated.

Subject clashes

Where two subjects are allocated at the same time the number of students common to both subjects is added to the punishment value.

Location clashes

Where two or more subjects are allocated to a location at the same time punishment is increased by (subjects allocated-1).

Lecturer clashes

Where two or more lecturers are allocated to a location at the same time punishment is increased by (allocations-1).

Lecturer template violations

If a lecturer is assigned during a time slot for which they are not available, punishment is increased by 1000.

2.6.3.2 Subject Slot Satisfaction

Here we evaluate how well a chromosome satisfies a subjects' requirements. These requirements are the number of time slots each subjects requires and also the resources that are required for each slot.

Surplus and Insufficient Timetable Allocations

If a chromosome allocates too many or too few timetable slots to a subject punishment is increased by the quantity of surplus resources or the quantity of resources not allocated, multiplied by the number of time slots the surplus/shortfalls effect.

Surplus and Insufficient Resources

For all slots that a chromosome assigns to a subject, we punish for any surplus of resources that are available at the assigned location and also for any resources that are required but are not available at the assigned location.

2.6.3.3 Weighting of Evaluations

Running the genetic algorithm with just the clash evaluations resulted in a population of zero length chromosomes, because if nothing is assigned, then no clashes can occur. Similarly, using just the slot satisfaction evaluations resulted in long chromosome with many clashes, but most of the subject slots where assigned a timetable slot.

By adding together the punishment values from all of the evaluations, the genetic algorithm tended towards short chromosomes, avoiding clashes at the expense of not assigning timetable slots.

To alleviate this problem some of the evaluation punishment values are multiplied by an independent value. This increases or weights the importance of some evaluations with respect to the other evaluations. This prevents important factors from being 'drowned' by less important factors.

2.6.4 Selection

We use roulette wheel selection since chromosomes with small punishment values have a greater chance of being selected. This is done by finding the highest of the punishment values and a new value is then calculated for each chromosome which is its punishment value subtracted from the highest punishment value. This value divided by the highest punishment value now gives the probability a chromosome will be selected.

The selection is weighted so that a very good chromosome cannot have a selection probability of more than 90%. This is called selection pressure – the pressure to select good chromosomes. By reducing the selection pressure we prevent a few good chromosomes from dominating the population of solutions and reducing the effectiveness of the genetic algorithm.

The offspring produced from the selected chromosomes replace the worst chromosomes in the population. Therefore, this strategy guarantees the best solution survives in each generation; it is elitist.

2.6.5 Recombination Operator Incorporating Mutation

When two chromosomes have been selected as parents, they are used to produce offspring using the crossover operator. Here we will use one-point crossover.

The length of a chromosome is variable so we must choose the crossover point in each of the parents separately since one parent may be longer than the other. We then produce two offspring. The first offspring has the first half of parent one (up to the crossover point) and the second half of parent two. The second offspring has the first half of parent two and the second half of parent one.

As we copy each gene from a parent to the offspring, each part of the gene has a 1% chance of being mutated. If a value is mutated, it is replaced with a valid value for that part of the gene, so if we decide to mutate the subject value it will be replaced with a random value between 0 and the number of subjects minus 1. Any clashes that may occur as a result are not considered; a clash will result in a higher punishment value from the evaluation function.

2.7. Conclusions

2.7.1 Assessment of the Applications Usefulness

To test the genetic algorithm, I initially created a very simple problem. The problem consists of nine subjects, each requiring one timetable slot. There are 21 students each studying two subjects. The subjects have been selected so that they form three main groups (1, 2, 3), (4, 5, 6) and (7, 8, 9). Subjects within a group have common students and therefore should not be assigned in the same time slot.

Three hours, three locations and three lecturers are available for timetabling, effectively giving 9 hours of teaching time. This is the exact amount required and so there should be no free timetable slots. Finally, each lecturer is available for teaching all of the subjects.

A satisfactory solution to this problem is indeed found within a few minutes. A population of 100 was satisfactory.

I then created a more complex problem in test file 'bc4.ttd'. This file describes the 23 subjects currently being taught in the final year computer science/information technology degree courses at Staffordshire University computing school. All 270+ students and their options are listed in the STUDENT_DATA block.

From then on I constructed requirements for the timetable. Each subject requires 2 hours a week and only enough seats for all students on a subject must be present; therefore, we need 46 timetable slots. I defined 50 timetable slots made up from 2 days, 5 slots a day, 5 lecturers and 5 locations. Again, each lecturer is available for all subjects and are always available.

The performance of the genetic algorithm is shown in the following graphs:

First 6000 Generations for Test Problem BC4

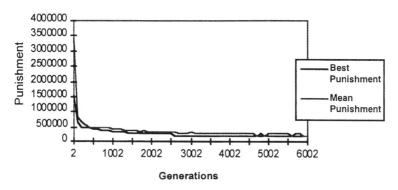

6000+ Generations for Test Problem BC4

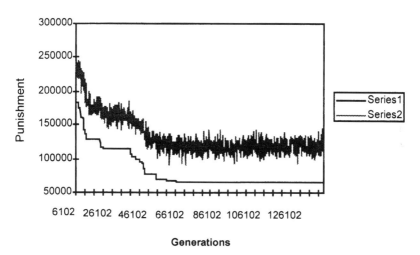

We can see that both the current best weight and the mean weight fall rapidly early on. This indicates that the genetic algorithm is finding fitter solutions resulting in a fitter population as a whole.

The actual timetable produced described by the best solution from this run is show in file 'BC4.MST' in Appendix D. The timetable contains no location clashes and no lecturer clashes (although these are not visible on a finished timetable). The two problems with the time table are that only 43 slots from the 46 that are required are time-tabled and there are 41 subject clashes from a possible 73. On the positive side, the subject clashes do tend to have low numbers of common students, less than ten in most cases. To see if this problem can be alleviated, I increased the punishment on student clashes by a factor of ten.

The resulting timetable is shown in file BC4_2.MST in Appendix D. The time-table does have less subject clashes (31 from a possible 58), but also only time-tables 39 hours from the 46 that are required. Of course, we do not know if it is possible to timetable all of the subjects 5 at a time so that no clashes occur. A graph colouring algorithm could be employed at the initialisation to estimate the possibility that a good solution will be found and also to seed the initial population with better solutions.

2.7.2 Assessment of the Bibliography

[1] "Sequencing and Scheduling - An introduction to the Mathematics of the Job shop" by S. French (1982) provided a good introduction to the job-shop problem and traditional techniques used to produce solutions and systems of evaluating problems. Also a good introduction to the mathematics behind scheduling problems, the NP-Hardness, etc. There is very little reference to timetabling problems, however it has given an insight into scheduling which has shown that the use of job-shop scheduling techniques alone are not appropriate to the timetabling problem.

There are a number of papers on timetabling by the Timetabling Research group at the Department of Computer Science, University of Nottingham. These papers almost exclusively deal with examination timetable problems, with a view to expanding the examination techniques to day-to-day university timetabling.

There are also a number of papers from the Evolutionary Computing group at Edinburgh university. These deal with the use of evolutionary computing for a range of applications, some of which are timetabling problems.

These two sets of papers have been the main source of information regarding the implementation and abilities of genetic algorithms.

Another paper from Edinburgh worth mention is [15] "Investigating Genetic Algorithms for Scheduling" by Hsiao-Lan Fang. This paper also deals with examination timetabling. The early sections on genetic algorithms provide a concise description of the various components of a genetic algorithm.

2.7.3 Conclusions

Although the solutions found above are not particularly good, they do show promise. There is a lot of scope for investigation of the genetic algorithm. Some areas which could help the search for solutions are:

Chromosome structure – The structure can affect how sections of good solution come together.

Recombination operators – In this project I used only one-point crossover, the other types of crossover discussed in Section 5.2.5.1 may be beneficial. Also, more intelligent crossover/mutation operators which avoid producing non-viable timetables effectively reduce the search space by eliminating the non-viable timetables. Reducing the search space should increase the chance of finding a good solution.

Selection methods – Tournament selection and also using different selection pressures could be effective.

There are many variables within the genetic algorithm which affect the path of the search through the set of possible solutions, i.e., those specified above and also simple variables such as mutation and crossover rates and different values or even variable rates could be used.

Bibliography

[1] S French (1982) "Sequencing and Scheduling. An Introduction to the Mathematics of the Job Shop."

[2] EK Burke, DG Elliman and RF Weare, "A University Timetabling System based on Graph Colouring and Constraint Manipulation," Journal of Research on Computing in Education, vol 27, no 1, pp 1-18, 1994.

[3] EK Burke, DG Elliman and RF Weare, "Automated Scheduling of University Exams," proceedings of the IEE Colloquium on Resource Scheduling for Large Scale Planning Systems (10th June 1993), digest No. 1993/144, Section 3, 1993.

[4] EK Burke, DG Elliman and RF Weare, "Extensions to a University Exam Timetabling System," proceedings of the IJCAI-93 Workshop on Knowledge-Based Production, Planning, Scheduling and Control (Chambery, France, 29th Aug 1993), pp 42-48.

[5] EK Burke, DG Elliman and RF Weare, "A Genetic Algorithm for University Timetabling," proceedings of the AISB Workshop on Evolutionary Computing (University of Leeds, U.K., 11th-13th April 1994), Society for the Study of Artificial Intelligence and Simulation of Behaviour (SSAISB).

[6] EK Burke, DG Elliman and RF Weare, "A Genetic Algorithm based University Timetabling System," Proceedings of the 2nd East-West International Conference on Computer Technologies in Education (Crimea, Ukraine, 19th-23rd Sept. 1994), vol 1, pp 35-40, 1994.

[7] EK Burke, DG Elliman and RF Weare, "The Automation of the Timetabling Process in Higher Education," Journal of Educational Technology Systems, vol 23, no 4, pp 257-266, Baywood Publishing Company, 1995.

[8] EK Burke, DG Elliman and RF Weare, "A Hybrid Genetic Algorithm for Highly Constrained Timetabling Problems," Proceedings of the 6th International Conference on Genetic Algorithms (ICGA'95, Pittsburgh, U.S.A., 15th-19th July 1995), pp 605-610, Morgan Kaufmann, San Francisco, CA, U.S.A.

[9] EK Burke, DG Elliman and RF Weare, "The Automated Timetabling of University Exams using a Hybrid Genetic Agorithm," proceedings of the AISB (Artificial Intelligence and Simulation of Behaviour) workshop on Evolutionary Computing (University of Sheffield, U.K., 3rd-7th April 1995).

[10] EK Burke, DG Elliman, PH Ford and RF Weare, "Examination Timetabling in British Universities – A Survey," proceedings of the 1st

International Conference on the Practice and Theory of Automated Timetabling (ICPTAT'95, Napier University, Edinburgh, U.K., 30th Aug - 1st Sept 1995), pp 423-434.

[11] EK Burke, JP Newall and RF Weare, "A Memetic Algorithm for University Exam Timetabling," Proceedings of the 1st International Conference on the Practice and Theory of Automated Timetabling (ICPTAT'95, Napier University, Edinburgh, U.K., 30th Aug-1st Sept 1995), pp 496-503.

[12] EK Burke, DG Elliman, PH Ford and RF Weare, "Specialised Recombinative Operators for the Timetabling Problem," Proceedings of the AISB (Artificial Intelligence and Simulation of Behaviour) Workshop on Evolutionary Computing (University of Sheffield, U.K., 3rd-7th April 1995), Lecture Notes in Computer Science, Springer-Verlag.

[13] RF Weare, "Automated Examination Timetabling," PhD Thesis, Department of Computer Science, University of Nottingham, U.K., June 1995.

[14] Paul Field (1996) "A Multary Theory for Genetic Algorithms: Unifying Binary and Non-binary Problem Representations," Submitted to the university of London for the degree of Doctor of Philosophy in Computer Science.

[15] Hsiao-Lan Fang (1992) "Investigating Genetic Algorithms for Scheduling," M.Sc. Dissertation, Department of Artificial Intelligence, University of Edinburgh.

[16] Dave Corne, Peter Ross, Hsiao-Lan Fang "Evolutionary Timetabling: Practice, Prospects and Work in Progress," Department of Artificial Intelligence, University of Edinburgh. Presented at the U.K. Planning and Scheduling SIG Workshop, Strathclyde, September 1994.

[17] "Dynamic Training Subset Selection for Supervised Learning in Genetic Programming," Chris Gathercole, Peter Ross. PPSN-III, Springer-Verlag, 1994.

[18] "Applications of Genetic Algorithms," Peter Ross and Dave Corne, AISB Quarterly No. 89, Autumn 1994, Special Theme on Evolutionary Computing, Terry Fogarty (Ed.), ISSN 0268-4179, pp 23-30.

[19] "Resource Allocation Problems. Algorithmic Approaches," Toshihide Ibaraki and Naoki Katoh. (1988).

[20] "Computers and Intractability: A guide to the theory of NP-Completeness," Garey M.R and Johnson D.S. (1979)

[21] "Solving the Module Exam Scheduling Problem with Genetic Algorithms," Dave Corne, Hsiao-Lan Fang, Chris Mellish, Proceedings of the Sixth International Conference on Industrial and Engineering

Applications of Artificial Intelligence and Expert Systems, Gordon and Breach Science Publishers, Chung, Lovegrove, Ali (eds), pp 370-373, 1993. This is also available as DAI Research Report No. 622.

[22] "A Promising Genetic Algorithm Approach to Job-Shop Scheduling, Rescheduling, and Open-Shop Scheduling Problems," Hsiao-Lan Fang, Peter Ross, Dave Corne, Proceedings of the Fifth International Conference on Genetic Algorithms, S. Forrest (Ed), San Mateo: Morgan Kaufmann, pp 375-382, 1993.

[23] "Successful Lecture Timetabling with Evolutionary Algorithms," Peter Ross, Dave Corne, Hsiao-Lan Fang, Appears in Workshop Notes, ECAI'94 Workshop W17: Applied Genetic and other Evolutionary Algorithms.

[24] "Improving Evolutionary Timetabling with Delta Evaluation and Directed Mutation," Peter Ross, Dave Corne, Hsiao-Lan Fang, PPSN III, Springer Verlag, 1994. Also, DAI Research Paper No. 707.

[25] "Fast Practical Evolutionary Timetabling," Dave Corne, Peter Ross, Hsiao-Lan Fang, in Evolutionary Computing: AISB Workshop 1994, Selected Papers, Springer Verlag Lecture Notes in Computer Science 865, T. Fogarty (Ed), Springer Verlag, 1994. Also DAI Research Paper No. 708.

[26] Spears, William M. (1990). Using Neural Networks and Genetic Algorithms as Heuristics for NP-Complete Problems. Masters Thesis, George Mason University, Fairfax, Virginia.

[27] Spears, William M. (1995). Simulated Annealing for Hard Satisfiability Problems. Proceedings of the 2nd DIMACS Implementation Challenge, David S. Johnson and Michael A. Trick (Eds.), DIMACS Series in Discrete Mathematics and Theoretical Computer Science.

[28] De Jong, Kenneth A. and William M. Spears (1989). Using Genetic Algorithms to Solve NP-Complete Problems. In Proceedings of the Int'l Conference on Genetic Algorithms, 124-132.

[29] "The Evolution of Life," Macdonald Educational Ltd. (1980).

Appendix A: User Documentation

1. Overview of Programs

There are two programs which make up the complete application, they are called "TT_GA" and "TT_OUT." Both are text based and run under Microsoft Windows.

The TT_GA program reads in a description file (with extension .ttd) and actually runs the genetic algorithm to find a solution to the timetable problem. When the genetic algorithm is terminated, the user has the option of saving the best solution to a timetable save file (with extension .tts). This file contains the raw data describing the timetable.

The TT_OUT program reads in a timetable save file and allows the user to save a text file containing a readable form of the timetable solution. The user is then free to load this text file into any other application for output to a printer or to perform further manipulation of the timetable.

The following sections will assume that all files and programs pertaining to the application are installed under MS-Windows.

2. The Timetable Problem Definition File (.ttd file)

Before you can run TT_GA to find a solution to your timetable problem, it is necessary to create a file describing the timetable problem. We will call this file the 'timetable definition file' and it will always have the filename extension '.ttd.' Example .ttd files can be found in Appendix C.

The timetable problem files contain key words which denote what is being described at that point in the file. The primary key words are :

DAYS, SLOTS, RESOURCE, LOCATION, SUBJECT, LECTURER, STUDENT_DATA.

DAYS, SLOTS

These two keywords must be present, and before any other text. Both should be followed by an integer value like:

DAYS 5

SLOTS 8

This defines how many days the timetable covers and how many distinct time slots there are in each day. Time slots are therefore all the same length and are not allowed to overlap.

RESOURCE

We use this keyword to describe different types of resources. These resources are later used in the LOCATION definitions to define what resources are present at each location. Therefore, resources must be declared before locations.

Typical resource definitions would be:

```
RESOURCE SEAT          {    "    "    }
RESOURCE    TERMINAL    {    "A    standard    dumb
terminal"   }
```

After the resource keyword, you must specify a mnemonic which will be used to identify the resource throughout the rest of the file. After the mnemonic, a text string between speech marks and parenthesis must be specified. This string is a description of the resource which the user can refer to in later editing of the file.

LOCATION

A typical LOCATION definition would be:

```
LOCATION   KC15
{
    TERMINAL  20
    SEAT  20
}
```

The LOCATION keyword must be followed by a mnemonic which is used to refer to the location in the final output timetable produced by the second program TT_OUT (see Section 2.4). Within the brackets you can now list the resources that are present at the location using the resource mnemonics defined earlier. Each resource mnemonic must be followed by an integer which defines the quantity of each resource at the location.

SUBJECT

A typical subject definition would be:

```
SUBJECT   AF
{
    SLOT
```

```
    [    RESOURCE  SEAT  0

         HOURS  2     ]

    SLOT

    [    RESOURCE  TERMINAL  0

         RESOURCE  SEAT  0

         RESOURCE  DEC5500  0

         HOURS  1

    ]

}
```

The SUBJECT keyword must be followed by a mnemonic which is used to refer to the subject in the STUDENT_DATA section of the .ttd file and also in the final timetable output produced by the second program TT_OUT (see Section 2.4).

Within a set of curly brackets we now define how many timetable slots the subject requires and the resources that are required on each occasion.

A slot or set of slots which require the same set of resources can be defined with one SLOT statement. After the SLOT keyword you can define the resources that are required using the RESOURCE keyword followed by a resource mnemonic and finally the quantity of the resource that is required. If a resource quantity of zero is given then it is assumed that the quantity of the resource that is required is equal to the number of students for that subject.

Also in the SLOT statement you must use the HOURS keyword to specify for how many time slots in the timetable this slot definition should be assigned. So, e.g., if a subject requires two hours of lectures a week, we would write something like:

SLOT

[

RESOURCE SEAT 0

HOURS 2

]

Finally each SLOT definition must be with a set of square brackets as shown above.

LECTURER

Defines a lecturer, a list of subjects he is available for teaching, and when he is unavailable for teaching.

A typical lecturer definition would be:

```
LECTURER    JFB
{
    NAME   "Joe  F.  Bloggs"
    SUBJECT  CNDS  MTOS  ;
    TEMPLATE  0  0  EVERY  1  0
    TEMPLATE  1  0  THROUGH  2  8
}
```

After the LECTURER keyword, you must specify a mnemonic that is used to refer to the lecturer in the final output timetable produced by TT_OUT (see Section 2.4).

Within a set of curly brackets, you must specify the name of the lecturer using the NAME keyword followed by the lecturers name within speech marks. This should be followed by the SUBJECT keyword which is followed by a list of subject mnemonics (defined earlier) and terminated with a semicolon.

Finally, the optional keyword TEMPLATE can be used to describe when a lecturer is unavailable. A single TEMPLATE statement can take one of two forms:

TEMPLATE <int> <int> EVERY <int> <int>

or

TEMPLATE <int> <int> THROUGH <int> <int>

where <int> refers to an integer value.

In both cases the first integer specifies a day and the second integer a time slot within that day. In the first case the lecturer is made unavailable for this first time slot and then repeatedly after the number of days and time slots defined after the EVERY keyword.

So, if we wanted to make a lecturer unavailable for the first slot of every day, we would use the template:

TEMPLATE 0 0 EVERY 1 0

If we were to use the template:

TEMPLATE 0 0 EVERY 1 1

then the lecturer is unavailable for the 1st slot on day one, the 2nd slot on day two and so on.

In the second form of the TEMPLATE statement, a second set of integers after the THROUGH statement also describes a day and time slot; the first and second day/time-slot are made unavailable and also all of the time slots between.

Multiple TEMPLATE statements can be used to build a detailed template of a lecturer's availability.

STUDENT DATA

This section defines which subjects each student is learning. From this a great deal of useful information is calculated, such as the total number of students, the number of students in each subject and the number of students common to any given set of subjects.

An example STUDENT_DATA block:

STUDENT_DATA

{

   SUB1 SUB2 SUB3 ;

   SUB2 SUB4 SUB5 ;

}

Each line here contains a list of subjects terminated by a semicolon. These are the subjects taken by one student. Each student's subjects must be given in order to build an accurate model of the timetable problem. Each line must be terminated by both a space and a semicolon.

3. The TT_GA Program

To run the program simply click on the TT_GA program icon in the relevant window. A new window will appear in which text will be displayed and text can be typed.

The new window will immediately show the following:

```
Enter name of timetable problem definition file,

.ttd extension will be added if not specified

:
```

You must enter the name of a prewritten timetable problem definition file before the genetic algorithm can be run. Upon pressing <return> the file will be parsed, as the various sections of the input file are parsed messages indicating the parsing of each section will be displayed. If the input file has parsed successfully you will see the following:

```
Input file parsed successfully

Do you wish to load a previously saved solution? (Y/N)
```

If you ran the genetic algorithm previously and saved a solution in a timetable save file (with .tts filename extension), then you can read in the solution. It will be used in the search for a better solution by the genetic algorithm.

Also, if a good solution has been found but the problem definition file has since been altered, the previous solution may provide a good starting point in the search for a new solution.

So, if you now press 'Y' for yes to load a solution, you will be asked to give the name of a timetable save file, enter an appropriate filename and press return, or just press return to use the same name as the problem definition file but with a .tts extension rather than a .ttd extension.

If you have successfully read in a previous solution or selected 'no' at the previous prompt, the genetic algorithm will begin to run.

The window should now show a number of values which are being continually updated:

Now Searching for timetable				
I Generation	I Best weight	I Mean weight	I Mean Length	I Length of best I
I 10	I 200000	I 400000	I 21	I 34

These values are:

Generation – How many generations have been performed so far.

Best Weight – The weight of the best solution that has been found so far. As calculated by the evaluation function.

Mean weight – The mean weight of the entire population of solutions.

Mean length – The mean length of the entire population of solutions.

Length of best – The chromosome length of the best solutions.

As the algorithm progresses through the generations, the best and mean weights should be seen to fall. When they no longer fall, then the genetic algorithm is struggling to find better solutions. At this point, press 'Q' to stop searching.

You will now be asked if you wish to save the best solution. If you select 'yes,' you will be asked for a save timetable filename. At this point you can press <return> to use the same file name as the timetable description file, but with a .tts file extension rather than a .ttd.

4. The TT_OUT Program

To run the program simply click on the TT_OUT program icon in the relevant window. A new window will appear in which text will be displayed and into which text can be typed.

The new window will immediately show the following:

```
Enter name of timetable problem definition file,

.ttd extension will be added if not specified

:
```

As with the TT_GA program, you must enter the name of a prewritten timetable problem definition file. The file will be parsed as before and, if successful, you will be prompted for the name of a timetable save file. Again, by pressing <return> the timetable definition file name will be used but with the relevant .tts extension. Otherwise, you can type in any other file name.

```
Select an output format for the timetable
:-

1) Complete master timetable

2) Quit
```

If everything has been successful, the window will show the following text:

Currently there is only one output format for the timetable. Press 1 and you will be asked to enter a filename to save the timetable. The timetable will be saved as a text file with days and slots across the top and lecturers listed on the left-hand column. An example master timetable can be found in Appendix D.

Appendix C: Example Timetable Problem

Definition   Files

QUICK1.TTD - A small test problem

DAYS 1

SLOTS 3

RESOURCE SEAT   { " Most resources are obvious "   }

LOCATION RED

{

  SEAT 3

}

LOCATION BLUE

{

  SEAT 3

}

LOCATION GREEN

{

  SEAT 3

}

SUBJECT SUB1

{

  SLOT

  [RESOURCE SEAT 0

    HOURS 1   ]

}

SUBJECT SUB2
{
  SLOT
  [RESOURCE SEAT 0
   HOURS 1]
}

SUBJECT SUB3
{
  SLOT
  [RESOURCE SEAT 0
   HOURS 1]
}

SUBJECT SUB4
{
  SLOT
  [RESOURCE SEAT 0
   HOURS 1]
}

SUBJECT SUB5
{
  SLOT
  [RESOURCE SEAT 0
   HOURS 1]
}

SUBJECT SUB6
{

    SLOT
    [RESOURCE SEAT 0
      HOURS 1]
}

SUBJECT SUB7
{
  SLOT
  [RESOURCE SEAT 0
    HOURS 1]
}

SUBJECT SUB8
{
  SLOT
  [RESOURCE SEAT 0
    HOURS 1]
}

SUBJECT SUB9
{
  SLOT
  [RESOURCE SEAT 0
    HOURS 1]
}

LECTURER LEC1
{
  NAME "Lec1"
  SUBJECT SUB1 SUB2 SUB3 SUB4 SUB5 SUB6 SUB7 SUB8 SUB9 ;

}

LECTURER LEC2

{

  NAME "Lec2"

  SUBJECT SUB1 SUB2 SUB3 SUB4 SUB5 SUB6 SUB7 SUB8 SUB9 ;

}

LECTURER LEC3

{

  NAME "Lec3"

  SUBJECT SUB1 SUB2 SUB3 SUB4 SUB5 SUB6 SUB7 SUB8 SUB9 ;

}

STUDENT_DATA

{

SUB1 SUB2 ;

SUB1 SUB2 ;

SUB1 SUB2 ;

SUB2 SUB3 ;

SUB2 SUB3 ;

SUB1 SUB3 ;

SUB1 SUB3 ;

SUB4 SUB5 ;

SUB4 SUB5 ;

SUB4 SUB5 ;

SUB5 SUB6 ;

SUB5 SUB6 ;

SUB4 SUB6 ;

SUB4 SUB6 ;

SUB7 SUB8 ;

SUB7 SUB8 ;

SUB7 SUB8 ;

SUB8 SUB9 ;

SUB8 SUB9 ;

SUB7 SUB9 ;

SUB7 SUB9 ;

}

Test file BC4.TTD - Student data from the 4th year computing degree course at Staffordshire University.

DAYS 2

SLOTS 5

MINFLAG 0

RESOURCE SEAT { " Most resources are obvious "   }

RESOURCE TERMINAL { " see what I mean" }

RESOURCE DEC5500 { "DEC Vaxstation 5500's" }

RESOURCE PC-BCC { "PC with BorlandC++" }

RESOURCE PC-LINUX { "PC with linux" }

LOCATION RED

{

    SEAT 220

}

LOCATION BLUE

{

    SEAT 180

}

LOCATION GREEN

{

    SEAT 180

}

LOCATION D116

{

    SEAT 200

}

LOCATION D118

{

  SEAT 200

}

SUBJECT AF

{

  SLOT

  [RESOURCE
SEAT 0

    HOURS 2]

}

SUBJECT AI

{

  SLOT

  [RESOURCE
SEAT 0

    HOURS 2]

}

SUBJECT ASAD

{

  SLOT

  [RESOURCE
SEAT 0

    HOURS 2]

}

SUBJECT CBO

{

  SLOT

  [RESOURCE
SEAT 0

    HOURS 2]

}

SUBJECT CLTI

{

  SLOT

  [RESOURCE
SEAT 0

    HOURS 2]

}

SUBJECT CNDS

{

  SLOT

  [RESOURCE
SEAT 0

    HOURS 2]

}

SUBJECT DB

{

  SLOT

  [RESOURCE
SEAT 0

    HOURS 2]

}

SUBJECT DP

{

  SLOT

  [RESOURCE
SEAT 0

    HOURS 2]

}

SUBJECT FR

{

  SLOT

  [RESOURCE
SEAT 0

    HOURS 2]

}

SUBJECT GE

{

  SLOT

  [RESOURCE
SEAT 0

    HOURS 2]

}

SUBJECT GR

{

  SLOT

  [RESOURCE
SEAT 0

    HOURS 2]

}

SUBJECT HCI
{
  SLOT
  [RESOURCE
SEAT 0
    HOURS 2]
}

SUBJECT IDE
{
  SLOT
  [RESOURCE
SEAT 0
    HOURS 2]
}

SUBJECT IP
{
  SLOT
  [RESOURCE
SEAT 0
    HOURS 2]
}

SUBJECT IRM
{
  SLOT
  [RESOURCE
SEAT 0
    HOURS 2]
}

SUBJECT MI
{
  SLOT
  [RESOURCE
SEAT 0
    HOURS 2]
}

SUBJECT MTOS
{
  SLOT
  [RESOURCE
SEAT 0
    HOURS 2]
}

SUBJECT NN
{
  SLOT
  [RESOURCE
SEAT 0
    HOURS 2
    ]
}

SUBJECT OR
{
  SLOT
  [RESOURCE
SEAT 0
    HOURS 2]
}

SUBJECT PIO
{
  SLOT
  [RESOURCE
SEAT 0
    HOURS 2]
}

SUBJECT RTES
{
  SLOT
  [RESOURCE
SEAT 0
    HOURS 2]
}

SUBJECT SE
{
  SLOT
  [RESOURCE
SEAT 0
    HOURS 2]
}

SUBJECT SIM
{
  SLOT
  [RESOURCE
SEAT 0
    HOURS 2]
}

LECTURER PVB

{

NAME "Pete Best"

SUBJECT AF
AI ASAD CBO
CLTI CNDS DB
DP FR GE GR
HCI IDE IP IRM
MI MTOS NN
OR PIO RTES
SE SIM ;

}

LECTURER DKB

{

NAME "Di Bishton"

SUBJECT AF
AI ASAD CBO
CLTI CNDS DB
DP FR GE GR
HCI IDE IP IRM
MI MTOS NN
OR PIO RTES
SE SIM ;

}

LECTURER CM

{

NAME "Chris Mills"

SUBJECT AF
AI ASAD CBO
CLTI CNDS DB
DP FR GE GR
HCI IDE IP IRM
MI MTOS NN
OR PIO RTES
SE SIM ;

}

LECTURER PG

{

NAME "Pete Gittins"

SUBJECT AF
AI ASAD CBO
CLTI CNDS DB
DP FR GE GR
HCI IDE IP IRM
MI MTOS NN
OR PIO RTES
SE SIM ;

}

LECTURER PC

{

NAME "Phil Cornes"

SUBJECT AF
AI ASAD CBO
CLTI CNDS DB
DP FR GE GR
HCI IDE IP IRM
MI MTOS NN
OR PIO RTES
SE SIM ;

}

STUDENT_DATA

{

IRM HCI AF DB ;

AI NN HCI MTOS ;

HCI IRM SE DB ;

AI MTOS CNDS HCI ;

ASAD PIO HCI IRM ;

ASAD PIO DB AI ;

HCI DB IRM CBO ;

DB AF IRM HCI ;

GR IDE CNDS SIM ;

CNDS MTOS DB HCI ;

IRM HCI ASAD CBO ;

ASAD PIO HCI IRM ;

MTOS IP CNDS RTES ;

MTOS IDE SE CNDS ;

DP AI IP SIM ;

ASAD PIO DB IRM ;

CNDS ASAD DB HCI ;

ASAD PIO IRM DB ;

ASAD HCI IRM CBO ;

DB CNDS AF IRM ;

IDE MTOS RTES CNDS ;

MI MTOS CNDS RTES ;

RTES DP DB SIM ;

IRM HCI ASAD CBO ;

ASAD PIO IRM HCI ;

ASAD DB IDE DP ;

DB AF HCI IRM ;

IRM DB ;

ASAD PIO HCI IRM ;

DP DB RTES SIM ;

FR CNDS IP MTOS ;

SE CNDS RTES AI ;

DB MTOS IDE CNDS ;

MTOS IP DB IDE ;

CNDS AF DB IRM ;

HCI CNDS MI IP ;

IRM CNDS DB OR ;

HCI CNDS GRAI ;

HCI CNDS DB OR ;

ASAD CBO OR IRM ;

DB CNDS SE IRM ;

HCI CNDS DB IRM ;

HCI OR DB ASAD ;

ASAD CBO DB IRM ;

AF CNDS DB IRM ;

ASAD GE CBO IDE ;

CNDS IP RTES GR ;

IP AF HCI IRM ;

MTOS CNDS DB HCI ;

DB IRM HCI CBO ;

ASAD PIO HCI IRM ;

GR MI MTOS IP ;

MTOS GR CNDS IP ;

MTOS CNDS HCI RTES ;

HCI IRM DB AF ;

DB IRM HCI CNDS ;

MTOS GRIP AI ;

DB GR HCI CNDS ;

DB HCI IRM CBO ;

AI MI MTOS CNDS ;

ASAD PIO HCI IRM ;

RTES IDE MTOS DP ;

DB HCI CNDS AF ;

HCI CNDS DB OR ;

CNDS MI MTOS CBO ;

GR AI MTOS HCI ;

IDE AF SIM CLTI ;

AF GR DB CLTI ;

ASAD FR CBO DB ;

CNDS MTOS IRM SE ;

ASAD PIO DB HCI ;

CNDS IRM   HCI
  DB ;

NN   AI DP SIM
  ;

NN   CNDS
  MTOS GR ;

DB ASAD SIM
  CBO ;

CNDS DB RTES
  MTOS ;

IRM   AF HCI
  DB ;

ASAD CBO IRM
  DB ;

DB CNDS RTES
  MTOS ;

RTES CNDS CLTI
  SIM ;

ASAD DB IP OR ;

DB IRM   HCI
  CBO ;

DB HCI   ASAD
  IRM ;

CNDS HCI   OR
  IRM ;

IRM HCI DB
  PIO ;

CNDS IP AF SIM
  ;

MTOS GR IP IDE
  ;

AI MTOS CNDS
  NN ;

HCI   IRM MI
  CBO ;

MTOS RTES IP
  MI ;

ASAD GE CBO
  IDE ;

SIM MTOS GR
  CNDS ;

CNDS DB IRM
  ASAD ;

GR IP CNDS
  HCI ;

ASAD CBO DB
  IRM ;

DB RTES MI GR ;

CNDS MTOS IRM
  HCI ;

ASAD PIO   DB
  IRM ;

HCI   GR IP MI ;

IP OR DB HCI ;

AI CNDS DB OR ;

AF DB OR IP ;

HCI   IRM DB
  CBO ;

SIM IP RTES
  GR ;

ASAD PIO HCI
  DB ;

DB CNDS AF IP ;

AF CNDS DB HCI
  ;

GR   CNDS IP
  DB ;

CNDS IRM OR
  DB ;

IP CLTI DB AI ;

AI DP GR IP ;

DB AF HCI   OR ;

CNDS OR ASAD
  IRM ;

ASAD CBO DB
  HCI ;

HCI   CNDS DB
  AF ;

OR   CNDS DB
  ASAD ;

ASAD CBO DB
  IRM ;

ASAD HCI   DB
  OR ;

AF CNDS DB IRM
  ;

ASAD CBO HCI
  MI ;

GR   CNDS AF
  IRM ;

DB HCI   AF IRM
  ;

OR   IRM   DB
  HCI ;

GR IP MI
  MTOS ;

HCI   IRM DB
  CBO ;

GR IDE   DB
  RTES ;

DB GR RTES
  MTOS ;

OR   AF DB HCI
  ;

PIO   ASAD IRM
  HCI ;

HCI   DB SE GR ;

GR   IDE   RTES
  IP ;

FR AI NNIP ;

IP GR MI NN;

ASAD CBO HCI
   IRM ;

GR   IP HCI
   MI ;

ASAD PIO HCI
   IRM ;

GR   NNCNDS
   DP ;

CNDS MTOS CLTI
   MI ;

AI DP MTOS IP ;

CNDS ASAD IDE
   DP ;

ASAD PIO NN
   SE ;

ASAD PIO DB
   HCI ;

MI GR MTOS IP ;

MTOS DB RTES
   CNDS ;

IRM DB;

DB HCI   IP SIM
   ;

IP GR MTOS
   CNDS ;

HCI AF DB IRM
   ;

DB ASAD IP OR;

ASAD DB CNDS
   IRM ;

CNDS DB HCI
   ASAD ;

HCI CNDS DB
   IRM ;

ASAD CBO IRM
   HCI ;

CNDS AF HCI
   ASAD ;

HCI CNDS DB
   OR;

DB OR HCI IRM
   ;

OR   CNDS DB
   IRM ;

MTOS CNDS MI
   DB ;

IRM HCI DB
   AF ;

ASAD IRM HCI
   DB ;

IRM CNDS DB
   OR;

MI CNDS MTOS
   RTES ;

AI NNIP GR;

GR   IDE DB
   RTES ;

MTOS GRIP MI ;

IP MTOS CNDS
   GR;

AF CNDS DB IRM
   ;

IRM DB ASAD
   CBO ;

DB HCI MTOS
   CNDS ;

IP GR HCI RTES
   ;

HCI MI IRM
   AF;

ASAD PIO DB
   HCI ;

CNDS MTOS DB IP
   ;

ASAD PIO DB
   HCI ;

IP CNDS GR DB;

IRM DB HCI
   CBO ;

DP GR MI MTOS ;

IP CLTI MTOS AI
   ;

ASAD CBO IRM
   HCI ;

HCI DB IRM
   CNDS ;

CNDS DB HCI
   AF ;

CNDS MI DP IP ;

GR   CNDS MI IP
   ;

MTOS CNDS DB
   CLTI ;

DB SE HCI   GR;

OR   IP DP AI ;

GR   HCI   IP
   DB ;

MTOS CNDS AF
   DB ;

FR CNDS AI GR;

IP MTOS CNDS
   GR;

GR   IP MI
   MTOS ;

CNDS SIM AI
   DP ;

GR   CNDS IP
    NN ;

ASAD PIO   DB
    IDE   ;

CNDS IDE   CLTI
    DB ;

GR   CNDS IP
    IRM   ;

ASAD CBO   IRM
    HCI   ;

AF HCI   DB IRM
    ;

ASAD PIO   HCI
    IRM   ;

HCI   CNDS DB
    OR ;

OR   HCI   DB IP
    ;

CNDS DB HCI
    AF ;

DB CNDS IRM
    HCI   ;

ASAD CBO   IRM
    HCI   ;

GR   HCI
    CNDS IP   ;

ASAD CBO   IRM
    HCI   ;

ASAD CBO   IRM
    HCI   ;

GR   MTOS RTES
    SIM   ;

GR   SIM   IP
    RTES ;

HCI   CBO   IP
    IRM   ;

GR   CNDS
    MTOS IRM   ;

NN   CNDS
    MTOS SIM   ;

GR   SIM   IP AI
    ;

GR   SIM
    MTOS IDE   ;

OR   DB CNDS IP
    ;

MTOS GR NN RTES
    ;

ASAD CBO   HCI
    MI ;

CNDS DB AI HCI
    ;

HCI   CNDS IRM
    CBO   ;

CNDS DB IRM
    OR ;

IP IRM   MI
    CNDS ;

MTOS CNDS MI
    RTES ;

OR   CNDS DB
    IRM   ;

CNDS DB OR IRM
    ;

ASAD CBO   IRM
    OR ;

CNDS HCI   CBO
    DB ;

MTOS DB IP
    CNDS ;

CNDS AF DB IRM
    ;

ASAD HCI   AF
    IRM   ;

HCI   CNDS DB
    OR ;

ASAD PIO   HCI
    IRM   ;

SE AF DB IRM   ;

ASAD DB IRM
    OR ;

MTOS CNDS IP
    DB ;

HCI   IP GR DB ;

CNDS DB CBO   IP
    ;

MTOS DB HCI
    MI ;

MTOS HCI   IRM
    MI ;

ASAD PIO   IRM
    DB ;

IP SIM   NN GR ;

HCI   AI CNDS
    SIM   ;

HCI   IRM   DB
    PIO   ;

ASAD CBO   HCI
    DB ;

HCI   IP MTOS
    DB ;

MTOS CNDS IDE
    DB ;

IRM   HCI   DB
    CBO   ;

HCI   SIM   AF
    DB ;

ASAD PIO   HCI
  IRM ;

ASAD PIO   IRM
  HCI ;

IP IDE   CNDS
  RTES ;

FR DB AF HCI ;

SE CNDS IDE
  MTOS;

ASAD PIO   HCI
  IRM ;

ASAD SE HCI
  DB ;

CNDS MTOS AI
  DP ;

ASAD CBO   DB
  IRM ;

HCI   CNDS DB
  OR;

DB IRM   IP OR;

DB CNDS IRM
  HCI   ;

CNDS DB SIM   IP
  ;

DB CNDS OR AF;

DB IRM   IP OR;

ASAD DB OR HCI
  ;

HCI   IRM   CBO
  SE ;

RTES DB IP
  CNDS ;

CNDS ASAD DB
  OR;

OR   CNDS DB
  IRM   ;

ASAD CBO   IRM
  HCI   ;

HCI   DB IRM
  AF ;

CNDS AF DB HCI
  ;

ASAD CBO   DB
  IRM   ;

CNDS DB IRM
  AF ;

}

Appendix D: Master Timetable Files

BC4.MST

	DAY 0 0	1	2	3	4	DAY 1 0	1	2	3	4
PVB	OR BLUE	IP RED	CBO RED		CLTI D116	DB D118				CLTI D116
DKB	AF D116	SIM GREEN	OR D118	ASAD D118	MTOS BLUE	RTES BLUE	IRM D116	MI GREEN	AI GREEN	SIM BLUE
CM	MTOS D118	MI D118	RTES D116	CNDS BLUE	DP D118		GR BLUE	DB BLUE	AF D118	IDE RED
PG	ASAD GREEN	IRM BLUE	IP BLUE	FR GREEN	HCI GREEN		AI RED	PIO D116	CNDS RED	HCI D118
PC	NN RED	SE D116	PIO GREEN	SE D116		NN GREEN	IDE GREEN	GR D118	CBO BLUE	DP GREEN

BC4_2.MST

	DAY 0 0	1	2	3	4	DAY 1 0	1	2	3	4
PVB		CBO D118	CNDS D116	SIM GREEN	GR RED	MI BLUE			AF GREEN	AF D118
DKB	MI RED			MTOS D118		IRM RED	NN D116	HCI RED		IP BLUE
CM	DB GREEN	CNDS BLUE	CBO BLUE			IDE D118	IP D118	SIM D116	ASAD D118	
PG	GE BLUE		PIO GREEN	HCI BLUE	DB BLUE	RTES D116	IDE GREEN	OR GREEN	OR D116	ASAD GREEN
PC	AI D118	PIO GREEN	DP RED	GE D116	DP GREEN	AI GREEN	IRM BLUE	MTOS BLUE	GR BLUE	RTES D116

Chapter 3 Implementing Fast and Flexible Parallel Genetic Algorithms

Erick Cantu-Paz
Illinois Genetic Algorithms Laboratory
University of Illinois at Urbana-Champaign
117 Transportation Building
104 S. Mathews Avenue
Urbana, IL 61801

Office: (217) 333-0897

Fax: (217) 244-5705

Abstract

Genetic algorithms are used to solve harder problems and it is becoming necessary to use more efficient implementations to find good solutions fast. This chapter describes the implementation of a fast and flexible parallel genetic algorithm. Since our goal is to help others to implement their own parallel codes, we describe some of the design decisions that we faced and discuss how the code can be improved even further.

3.1. Introduction

Genetic algorithms (GAs) are moving forward from universities and research centers into commercial and industrial settings. In both academia and industry genetic algorithms are being used to find solutions to hard problems, and it is becoming necessary to use improved algorithms and faster implementations to obtain good solutions in reasonable amounts of time. Fortunately, parallel computers are making a similar move into industry and GAs are very suitable to be implemented on parallel platforms. This chapter describes a software tool that is being used to systematically study parallel GAs. Our goal is to help developers implement their own parallel GAs, so we explain some of the design decisions we made and a few ideas to optimize the code.

The code that we describe here was designed to be used in a research environment, where both flexibility and efficiency are indispensable to the experiment. The code is designed to allow easy addition of new features and, at the same time, it is very efficient so the same experiment can be repeated multiple times to obtain statistically significant results. This quest for flexibility and efficiency should also be appealing to practitioners outside academia who want to run parallel GAs on their applications.

In the design of software there is usually a tradeoff between making programs more flexible or making them efficient and easy to use. In our design we decided to make this tradeoff as small as possible by incorporating enough flexibility to be able to experiment with many configurations, but

0-8493-2539-0/98/$0.00+$.50

at the same time trying to keep the code as simple as possible to maximize its efficiency and its usability. In our view, it was pointless to design a system with as much flexibility as possible because it would have made the programs too complicated to understand and execute.

Another important concern in our design was portability because access to parallel computers changes over time. For this reason, we chose PVM to manage processes and all the communications between them. PVM is a message-passing parallel programming environment that is available in many commercial (and experimental) parallel computers. The parallel GA code is implemented in C++ and the design follows an object-oriented methodology. The choice of language allows a fast execution and the object-oriented design allows easy maintenance and extensions.

In the next section we present some background information on GAs, parallel computers, and different kinds of parallel GAs. Section 3.3 contains a detailed description of the implementation of some critical portions of the software. In Section 3.4 we present the results of some tests that show the efficiency of the parallel GA working on different problems. Finally, this contribution ends with a summary and a discussion of possible extensions to the system.

3.2. GAs and Parallel Computers

How do GAs relate with parallel computers? A very short, but true, answer would be that they get along very well. A very long answer would look into the history of GAs and parallel computers and explore their relationship across several decades, but it is not the intention of this chapter to review the rich literature on parallel GAs (an interested reader can refer to the survey by Cantú-Paz (1995)). Instead, in this section we focus on the present and the future of parallel computers and on how GAs relate to them.

It is difficult to discuss parallel computers in general when one considers the rich history of this field and the variety of architectures that have appeared over time. However, there is a clear trend in parallel computing to move toward systems built of components that resemble complete computers interconnected with a fast network. More and more often, the nodes in parallel computers consist of off-the-shelf microprocessors, memory, and a network interface. Parallel and supercomputer manufacturers are facing a reducing market for their computers and commercial success is more difficult. This makes it almost impossible to compete successfully using custom components, because the cost of designing and producing them cannot be amortized with the sale of only a few machines.

Today many parallel computer vendors spend most of their engineering efforts in designing software that exploits the capabilities of their machines and that makes them easy to use. Similarly, the design goals of the parallel GA code that is presented here are efficiency, flexibility, and ease of use.

The trend to build parallel machines as networks of essentially complete computers immediately brings to mind *coarse-grained* parallelism, where there is infrequent communication between nodes with high computational capabilities. There are two kinds of parallel GAs that can exploit modern coarse-grained architectures very efficiently: multiple-population GAs (also called coarse-grained or island model GAs) and master-slave (or global) parallel GAs. We shall review these algorithms in detail in later sections.

There is a third kind of parallel GA that is suitable for fine-grained massively parallel computers, but because these machines are not very popular and it is highly unlikely that they will be in the near future, we shall not discuss fine-grained parallel GAs any further. Instead, our attention shifts now to the (apparent) problem of fine-tuning GAs to find good solutions to a particular problem.

GAs are very complex algorithms that are controlled by many parameters and their success largely depends upon adequately setting these parameters. The problem is that no single set of parameter values will result in an effective and efficient search in all cases. For this reason, the fine tuning of a GA to a particular application is still viewed as a black art by many, but in reality we know a great deal about adequate values for most of the parameters of a simple GA. For example, a critical factor for the success of a simple GA is the size of the population, and there is a theory that relates the problem length and difficulty of some class of functions to the population size (Harik, Cantú-Paz, Goldberg, & Miller, 1997).

Another critical factor for the success of GAs is the exchange of valuable genetic material between strings, and there are some studies that explore the balance that must exist between crossover and selection (Thierens & Goldberg, 1993; Goldberg & Deb, 1991). If selection is very intense, the population will converge very fast and there might not be enough time for good mixing to occur between members of the population. When this premature convergence occurs the GA may converge to a suboptimal population. On the other hand, if the selection intensity is very low, crossover might disrupt any good strings that may have already been found, but that have not had time to reproduce. If this is the case, then the GA will not likely find a good solution.

The problem of choosing adequate parameter values is worse in multiple population parallel GAs as they have even more parameters than simple GAs to control their operation. Besides deciding on all the GA parameters, one has to decide on migration rates, subpopulation sizes, interconnection topologies, and a migration schedule.

The research on parallel GAs spans several decades, but we are still a long way from understanding the effects of the multiple parameters that control them. For example, should we use high or low migration rates? Worse yet, what are considered to be low and high migration rates? What is an adequate way to connect the subpopulations on a parallel GA? How many subpopulations and of what size should we use? Studies are underway to answer at least some of these questions and to develop simple models that may help practitioners through this labyrinth of options.

The next two sections review some important aspects of two kinds of parallel GAs that are suitable for the architecture of current parallel computers. We review first the simpler master-slave GA and then proceed to examine multiple-population parallel GAs.

3.2.1 Master-Slave Parallel GAs

Master-slave GAs are probably the simplest type of parallel GAs and their implementation is very straightforward. Essentially, they are a simple GA that distributes the evaluation of the population among several processors. The process that stores the population and executes the GA is the master, and the processes that evaluate the population are the slaves (see Figure 3.1).

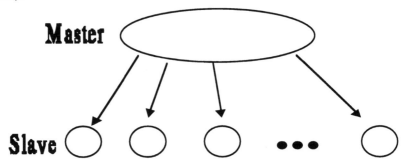

Figure 3.1 A schematic of a master-slave GA. The master process stores the population and each slave evaluates a fraction of the individuals.

Just like in the serial GA, each individual competes with all the other individuals in the population and also has a chance of mating with any other. In other words, selection and mating are still *global,* and thus master-slave GAs are also sometimes known as global parallel GAs.

The evaluation of the individuals is parallelized assigning a fraction of the population to each of the slaves. The number of individuals assigned to any processor can be constant, but in some cases (like in a multiuser environment where the utilization of processors is variable) it may be necessary to balance the computational load among the processors using a dynamic scheduling algorithm (e.g., guided self-scheduling).

Regardless of the distribution strategy (constant or variable), if the algorithm stops and waits to receive the fitness values for all the population before proceeding into the next generation, then the algorithm is synchronous. A synchronous master-slave GA searches the space in exactly the same manner as a simple GA. However, it is also possible to implement an asynchronous master-slave GA where the algorithm does not stop to wait for any slow processors. Slaves receive individuals and send their fitness values at any time, so there is no clear division between generations. Obviously, asynchronous master-slave GAs do not work exactly like a simple GA. Most global parallel GA implementations are synchronous, because they are easier to implement, but asynchronous GAs might better exploit any computing resources that might be available.

Master-slave GAs were first proposed by Grefenstette (1981), but they have not been used too extensively. Probably the major reason is that there is very frequent interprocessor communication, and it is likely that the parallel GA will be more efficient than a serial GA only in problems that require considerable amounts of computation. However, there have been very successful applications of master-slave GAs like the work of Fogarty and Huang (1991), where the GA evolves a set of rules for a pole-balancing application. The fitness evaluation uses a considerable amount of computation as it requires a complete simulation of a cart moving back and forth on a straight line and a pole attached to the top of the cart with a hinge. The goal is to move the cart so that the pole stands straight.

Another success story is the search of efficient timetables for schools and trains by Abramson (Abramson & Abela, 1992; Abramson, Mills, & Perkins, 1993). Hauser and Manner (1994) also show a successful implementation of a master-slave GA on three different computers.

As more slaves are used, each of them has to evaluate a smaller fraction of the population and therefore we can expect a reduction of the computation time. However, it is important to note that the performance gains do not grow indefinitely as more slaves are used. Indeed, there is a point after which adding more slaves can make the algorithm slower than a serial GA. The reason is that when more slaves are used, the time that the system spends communicating information between processes increases, and it may become large enough to offset any gains that come from dividing the task. There is a recent theoretical model that predicts the number of slaves that maximizes the parallel speedup of master-slave GAs depending upon the particular system used to implement them (Cantú-Paz, 1997a).

It is also possible to parallelize other aspects of GAs besides the evaluation of individuals. For example, crossover and mutation could be parallelized using the same idea of partitioning the population and distributing the work among multiple processors. However, these operators are so simple that it is very likely that the time required to send individuals back and forth will offset any possible performance gains. The communication overhead is also a problem when selection is parallelized because several forms of selection need information about the *entire* population and thus require some communication.

One straightforward application of master-slave GAs is to aid in the search for suitable parameter values to solve a particular problem. Even though there is enough theory to guide users to choose parameter values, it is still necessary to refine the parameters by hand. A common way to fine tune a GA empirically is to run it several times with a scaled-down version of the problem to experiment and find appropriate values for all the parameters. The parameters that give the best results are then scaled to the full-size problem and production runs are executed. With a master-slave GA, both the experimental and the production runs should be faster and there can be considerable savings of time when solving a problem.

In conclusion, global parallel GAs are easy to implement and it can be a very efficient method of parallelization when the fitness evaluation needs considerable computations. Besides, the method has the advantage of not changing the search strategy of the simple GA, so we can directly apply all the theory that is available for simple GAs.

3.2.2 Multiple-Population Parallel GAs

The second type of parallel GAs that are suitable to exploit coarse-grained computer architectures are multiple-population GAs. Multiple-population GAs are also called Island Model or coarse-grained parallel GAs and consist of a few subpopulations that infrequently exchange individuals (see Figure 3.2). This is probably the most popular type of parallel GAs, but it is controlled by many parameters and a complete understanding of the effect of these parameters on the quality and speed of the search still escapes us. However, there have been recent advances to determine the size and the number of subpopulations that are needed to find solutions of a certain quality in some extreme cases of parallel GAs. It is well known that the size of the population is critical to find solutions of high quality (Harik, Cantú-Paz, Goldberg, & Miller, 1997; Goldberg, Deb, & Clark, 1992) and it is also a major factor in the time that the GA takes to converge (Goldberg & Deb, 1991), so a theory of population sizing is very useful to practitioners.

It has been theoretically determined that as more subpopulations (or demes, as they are usually called) are used, their size can be reduced without sacrificing quality of the search (Cantú-Paz & Goldberg, 1997a). If we assume that each deme executes on a node of a parallel computer, then a reduction on the size of the deme results directly on a reduction of the wall-clock time dedicated to computations. However, using more demes also increases the communications in the system. This tradeoff between savings in computation time and increasing communication time causes the existence of an optimal number of demes (and an associated deme size) that minimizes the total execution time (Cantú-Paz & Goldberg, 1997b). The deme-sizing theory and the predictions of the parallel speedups have been validated using the program described in this contribution.

One of the major unresolved problems in multiple-deme parallel GAs is to understand the role of the exchange of individuals. This exchange is called migration and is controlled by: (1) a migration rate that determines how many individuals migrate each time, (2) a migration schedule that determined when migrations occur, and (3) the topology of the connections between the demes.

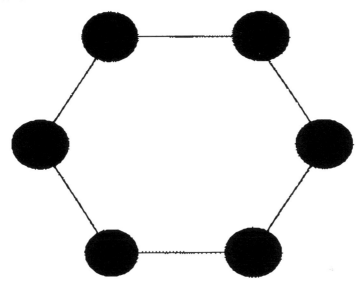

Figure 3.2 A schematic of a multiple-deme parallel GA. The subpopulations exchange individuals with their logical neighbours on the connectivity graph.

Migration affects the quality of the search and the efficiency of the algorithm in several ways. For instance, frequent migration results in a massive exchange of potentially useful genetic material, but it also negatively affects the performance because communications are expensive. Something similar occurs with densely connected topologies where each deme communicates with many others. The ultimate objective of parallel GAs is to find good solutions fast and, therefore, it is necessary to find a balance between the cost of using migration and increasing the chances of finding good solutions.

Our program was designed to experiment with multiple-deme GAs, so most of our discussion in later sections will center on the features of the program that permit experimentation with the parameters of these types of algorithms. We begin describing some critical parts of the implementation of parallel GAs in the next section.

3.3. Implementation

As we mentioned in the introduction, our parallel GA is written in C++ and it has an object-oriented design. This section first discusses the overall design of the program and then gives a detailed description of the most important objects in the system.

Object-oriented design is well suited for genetic algorithms. Many elements of GAs have the three main elements that characterize an object: a defined state, a well-determined behavior, and an identity (Booch, 1994). For example, a population consists, among a few other things, upon a collection of certain individuals (state), it has some methods to gather statistics about them (behavior), and it can be differentiated from all the other objects in the program (identity). An individual is also a well-defined object. Its state is determined by the contents of a string and a fitness value, it can be precisely identified, and it has a very well-defined behavior when it interacts with other objects.

There are many ways to design an object-oriented GA and there are many programmiug styles. For example, one could implement all the genetic operators as methods of the individuals aud use a population object to control the global operations like selection. Another option is to design individuals and populations as objects with a very simple behavior and to implement the GA operations at a higher level. The latter was the approach used in our system.

The general idea in our design is to view the simple GA as a basic building block that can be extended and used as a component in a parallel GA. After all, the behavior of the GA is very similar in the serial and the parallel cases. In particular, to make a master-slave parallel GA it is necessary to only modify the evaluation procedure of a simple GA; the rest of the behavior of the GA remains the same. Also, a simple GA can be easily used as a building block of coarse-grained parallel GAs because it is only necessary to add communication functions to the simple GA to turn it into a deme.

As we mentioned before, in the design of the system we emphasized simplicity, so that the code could be easily modified. One strategy that we adopted to simplify the code was to hide many of the supporting activities in the individual and population classes. This makes the code related with the GA and the parallel GA much more readable and easy to modify.

Another aspect that we were interested in in designing the code was portability. The language that we used (C++) is available for many (if not most) computers today and we handled all the parallel programming aspects using PVM, which is available in commercial aud free versions for many platforms.

PVM implements a Parallel Virtual Machine (hence its name) from a collection of (possible heterogeneous) computers. It can be used to simulate parallel computers, but it can also be used as a programming environment on top of an existing parallel machine. PVM provides a uniform message-passing interface to create and communicate processes making the hardware invisible to the programmer. This last feature enables the same parallel program to be executed efficiently on a multitude of parallel computers without modification.

3.3.1 Simple Genetic Algorithms

The first object that we shall describe in detail is the class GA. It implements a simple genetic algorithm and it is the base class for GlobalGA and Deme, which we shall discuss later. The class GA uses the class Population to store and obtain statistics about its population. Recall that our idea is to use the class GA to implement all the functions of a simple GA and leave all the necessary supporting activities (like allocating and managing dynamic memory and gathering statistics about the population) to other classes.

The GA class can be configured to use different selection methods (currently roulette wheel and tournament selection of any size are implemented) and crossover operators (uniform and multiple-point).

The main loop of the GA is in the method run(), which simply executes the method generate () until the termination condition is met. We normally run a GA until its population has converged completely to a single individual, but the termination condition can be easily changed to a fixed number of generations or evaluations, for example, by modifying GA::done ().

A single generation of the simple GA consists of evaluating the population, generating and reporting statistics on the state of the GA, selecting the parents, and applying crossover and mutation to create the next generation. The code for GA::generate() is in Figure 3.3.

The GA reports some statistics about tile progress of the run every generation and it can be instructive to see how the population diversity, its average, or its variance change over time. To aid in our research, the program also reports statistics about the average and variance of the building blocks in the population. Many times in research we use artificial test functions and we know what building blocks are needed to find the global solution.

Some of the methods of this class are implemented as virtual functions. This is to allow the classes that are derived from GA to extend some of the basic functionality that is provides. For example, there is a ConstrainedCostGA class that interrupts the run of the GA when a certain number of function evaluations have been executed, even if the GA is in the middle of a generation. The purpose of this class is to be used in

time-constrained applications, where there is a fixed amount of time
available to find a solution. Most of the virtual functions are overridden by
the classes that implement the parallel GAs and we will discuss them in
later sections.

How could we make the GA code more efficient? The GA operators are very
simple and there is no need to optimize them and squeeze a one percent
improvement in the performance. The supporting classes (individual and
population) are also very simple and it is unlikely that any change in them
will reduce the execution time. Probably the only optimization of this code
that we did was to avoid unnecessarily copying memory. In particular, the
GA class has a pointer to the current population, so when a new temporary
population of individuals is created only a pointer needs to be updated.

```
// do one generation of the GA
void GA::generate()
{evaluate();statistics();report();
generation++;
(*this.*select)();Population
*temp;currentpop;currentpop = poptemp;poptemp =
temp;
crossover();mutate();
}
```

Figure 3.3 The code for one generation of a simple genetic algorithm. Note
that selection is executed using a pointer to a function providing flexibility
as it is easy to add different selection algorithms to the system and to select
between them. For efficiency, the mating pool created by selection
(poptemp) is not copied directly into the current population; instead just a
pointer needs to be swapped.

3.3.2 Master-Slave Parallel GAs

The class GlobalGA implements a master-slave (or global) GA. Recall
that a master-slave GAs is identical to a simple GA, except that the
evaluation of the individuals is distributed to several slave processes.
GlobalGA is the class that defines the master process and it inherits all the
functionality of a simple genetic algorithm from class GA. The main
differences between these two classes are on the initialization of the class
and on the evaluation of the individuals. In this section we will describe
those differences in detail, but first we briefly look at the slaves processes.

The slaves are implemented by the class Slave, which basically executes a
loop with three operations: (1) it receives some number of strings that
represent individuals, (2) it evaluates them using the user-defined fitness
function, and (3) it returns the fitness values back to the master. When a
GlobalGA is created it launches the slave processes using PVM.
Launching processes require a considerable amount of time and, to avoid

excessive delays, slaves are created only once at the beginning of a series of experiments with the same setup. The GlobalGA keeps track of the number of repetitions and instructs the slaves to terminate only when the series of runs is completed. The code for GlobalGA::run() is in Figure 3.4.

The most interesting part of the GlobalGA is the evaluate() method. It sends the individuals to the slaves and waits for the results to come back. This is a pretty straightforward procedure, but here we had to make an important decision about the size of the messages that we were going to use to send the individuals to the slaves and collect the results. In general, there are two options when one needs to send information across a network: (1) use many small messages, or (2) use a few large messages. In the first case each individual would be sent in a separate message, and in the second case all the individuals that correspond to one slave would be sent in one message. Usually, the best alternative is to use few large messages and only this option is implemented in the current version (the first alternative was implemented in an earlier version, but the tests on different platforms confirmed that it was very inefficient).

```
unsigned long GlobalGA::run(char *outfile)
{int i;
for (i=0; i<repetitions-1; i++) { GA::run(outfile);
   init_GA();}GA::run(outfile);
// tell the slaves to diefor (i=0; i<slaves; i++){
   pvm_initsend(ENCODING); pvm_send(tids[i],
END_OF_RUN);}
}
```

Figure 3.4 The main loop of a GlobalGA. The only difference with the main GA loop is that after repeating an experiment several times, the master instructs the slaves to terminate sending them a specially-tagged message.

To understand why few large messages are usually preferable we have to realize that the time needed to send a message across a network has two components. The first is an overhead from the operating system and the message-passing interface each time a message is sent. This overhead is called latency and is independent of the size of the message. The second component of the communication time depends upon the bandwidth of the network (a measure of the amount of information that the network can transmit in a unit of time) and the size of the message. Overall, the time needed to send a message of certain size across a network is $Tr_{nsg} = Bsize +$ L where B is bandwidth and L is the latency. For example, if we need to communicate x bytes using N messages, the total time would be $Bx + LN$, but if we use only one message the total time is only[1] $Bx + L$.

Avoiding excessive overhead by using fewer messages is not the only way to improve performance of this class of parallel GAs. In the current implementation, the master process sits idle as the slaves evaluate the population. A simple – and potentially effective optimization – is to change the code so that the master process also evaluates a few individuals. Although this involves rather trivial changes in GlobalGA::evaluate (), we decided not to change the code for the sake of clarity. Also, it is possible to instruct PVM to spawn a slave on the same node that the master is running and, in this way, the processor will not remain idle. Our tests confirm that the performance is not negatively affected by having an extra slave perform the work that the master could do, and the code remains remarkably simple (see Figure 3.5 for the complete code of GlobalGA:: evaluate()).

```
void GlobalGA::evaluate()
{unsigned num_strings;double eval_value;int i,j;
assert(size % slaves ++ 0);
num_strings = size / slaves;
// send the strings to the slavesfor (i=0;
i<slaves; i++)[  // send: #strings, length, strings
     pvm_initsend(ENCODING);
     pvm_pkuint(&num_strings, 1, 1);
     pvm_pkuint(&length, 1, 1); for(j=0;
j<num_strings; j++)
     pkbyte((*currentpop)[i*num_strings+j].data,len
gth.,1); pvm_send(tids[i], STRINGS);}
// get the evaluations back from the slavesfor
(i=0; i<slaves; i++){ pvm_recv(tids[i], EVALS);
     for(j=0; j<num_strings; j++){
     pvm_upkdouble(&eval_value,1,1);
     (*currentpop)[i*num_strings+j].set_fitness(eva
l_value);}}// update the evaluations
counterevaluations+=size;
}
```

Figure 3.5 This is the code for evaluating the population in a master-slave GA. The first loop sends the individuals to the slaves, and each slave evaluates the same number of individuals. The second loop receives the evaluations back from the slaves and sets the fitness values in the population.

---

[1] Of course, this is just a coarse example and it may be possible to send the same amount of information faster using N messages in a system with low latency.

3.3.3 Multiple-Deme Parallel GAs

This type of parallel GAs is more complex than the master-slave type and its implementation uses two classes: PGA and Deme. The class PGA creates, initializes, and collects results from a set of Demes. Most of the interesting (and complicated) aspects of the coarse-grained GA are implemented in the class Deme, so this section will focus on it.

The class Deme is derived from the class GA, and it adds the communication capabilities to the simple GA. The most important difference between these two classes is that after each generation a deme checks to see if it is time to migrate. If migration does not occur, then the execution of the next generation proceeds as usual. But, if a migration is due, then the deme sends a predetermined number of individuals to each of its logical neighbors on the communications graph. Then the deme waits to receive the migrants from its neighbors and incorporates them into its population.

The main cycle for a deme is pretty straightforward: execute for a few generations, send some individuals, receive others, incorporate the newly arrived, and execute again (see Figure 3.6). The complications are not at this (high) level, but in the particular workings of each of these steps. In the remainder of this section we will look at each step in detail and explain the options that we face as designers of this type of program.

```
void DEME::generate()
{// do a normal generationGA::generate();
// check to see if it is time to migrateif
(generation % migration_interval ++ 0){
   SendMigrants();  ReceiveMigrants();
   IncorporateMigrants();}
}
```

Figure 3.6 The code for one generation in a deme. The difference with a simple GA is that migration occurs every migration_interval generations.

Let's discuss first the schedule for migrations. How often should we migrate? Nobody has a definitive answer because this is a very difficult question and, therefore, in the current implementation, the user can specify an interval of generations between migrations. This simply requires a modification to the main loop of the simple GA to check every generation to see if it is time to send and receive individuals. Because migration involves expensive communications, it is desirable to avoid it as much as possible. Luckily, our experiments show that it is not necessary to migrate very frequently to find good solutions.

There is also an option to migrate only after the population has completely converged. In our tests the quality of the solutions is similar to the cases when migration is more frequent, with the advantage of using very sparse communication.

We must also decide which individuals migrate. The number of migrants is determined by the user, but how do we choose individuals from the population to be sent away? We could choose migrants at random or we could choose the best individuals of the current generation. But selecting randomly has the advantage of disseminating more diversity and the chances of exploring new regions of the search space may be improved. Selecting the best individuals may help disseminate genetic material that has already been tested and shown to be useful. Because the two options may be beneficial, the program implements both (in our research we mainly choose migrants at random).

The topology used by the demes to communicate is also specified by the user as a parameter to the program and its implementation is probably the most interesting part of the class Deme. In theory, any arbitrary topology can be used, but there are some patterns that are very common and are already supported by our implementation of the coarse-grained GA: linear topologies, rings, hypercubes, grid, star, fully connected, and isolated. The class PGA is responsible for creating and sending to each deme the logical links that specify the topology. The demes receive only the links that correspond to them, so they know where to send migrants and from where to receive them, but they have no knowledge about the global topology. For example, in a ring topology each deme receives only one incoming link (to receive migrants from the deme behind) and one outgoing link (to send migrants to the deme ahead).

Each link could have a different migration rate associated with it, but when we designed the system it was not clear that this feature could be very useful and we decided to implement only a uniform migration rate. However, it would not be terribly difficult to implement this option in the future.

A common problem in the design of concurrent systems is to avoid deadlocks. One form of deadlock occurs when one process is blocked waiting to receive data from another, which is itself blocked waiting to receive data from the first. It is very complicated to detect deadlocks and correct them while the system is running, and so it is much better to avoid them altogether. In the parallel GA there is a possibility of a deadlock during migration, because one deme may be blocked waiting to receive individuals from another deme that is also blocked. One way to avoid deadlocks is to use non-blocking send commands and send all the migrants before attempting to receive any. With non-blocking sends, a process that initiates the communication does not have to wait until the recipient finishes reading data from the communication channel. The data is stored in buffers that can be read at the recipient's convenience.

When migrants arrive, they are stored in a temporary Population object before they are incorporated into the current population. There are several alternatives to incorporate migrants and there are two implemented in the system: random replacement of current members by migrants and also a form of elitist or competitive replacement. In the latter method, the newly arrived compete in tournaments with randomly selected members of the current population and the winner stays in the population. In our research we use only random selection and random replacement of migrants, but the other options could be very useful in real-life optimization problems.

Another interesting part of the implementation is concerned with the termination of the algorithm. When the termination condition becomes true in a deme, it cannot simply send its best individual to the process that collects solutions and disappear. The problem is that there are other demes that might still be running and will attempt to send and to receive individuals to a process that no longer exists. Sending a message to a nonexistent process is not really a problem, as PVM will signal the error and discard the information. The real issue is that a process will be locked forever if it waits to receive a message from a deme that has already terminated. This is another form of deadlock problem and it is also very difficult to detect and solve at run time.

To avoid this form of deadlock each deme notifies all its neighbors when it terminates by sending a message with a special tag. Then the neighbors update a list of active incoming and outgoing links, so when the time to migrate comes they do not send or attempt to receive from a deme that does not exist anymore. This is a simple and effective solution for avoiding deadlock when processes terminate, but there is another complication at the end of a run when there are multiple repetitions of the same experiment.

It is possible that one deme is ready to terminate but its neighbors have already sent migrants to it. The obvious behavior would be to ignore the migrants and terminate as usual, but the incoming migrants will remain in a buffer until the process reads them. The problem is that these migrants will be read at the time of the first migration of the next experiment, and they will probably dominate the population because they come from the end of a previous experiment and likely have a good fitness. A solution to this problem is to empty all the buffers of the incoming links before terminating the execution of an experiment.

The last question we will examine in this section is: How could we make a faster implementation of the multiple-deme GA? Something that can affect the performance greatly is that demes are idle while they wait to receive individuals from their neighbors. To reduce this idle time, we can resort to a common trick in parallel computation and overlap communications and computations. In our case, a simple way to accomplish this would be to evaluate the population after the migrants have been sent and before waiting to receive others. Of course, we could choose only individuals at random because we ignore their fitness values and we would not be able to select the best. With this scheme we are giving time for the migrants to arrive from other demes while the population is being evaluated. So, when the deme attempts to receive migrants, it is more likely that they have already arrived.

A less obscure way to improve the performance of multi-population parallel GAs is to combine them with the master-slave model. This is a good choice when there are many processors available, but only a few demes are used. With this "hybrid" approach, the evaluation of the individuals of one deme can be distributed to several processors and the computation power of the parallel machine would be more efficiently used. To implement this optimization we would need to derive the class Deme from GlobaIGA, instead of from the class GA.

3.4 Test Results

The code described in the previous section has been tested extensively on a variety of parallel computers and this section shows the results of some of those tests. Most of the material here has been used to validate theoretical results about parallel GAs, and it is good evidence of the accuracy of the models and also of the performance of the code.

We show only results of the code executed on a network of workstations, as this is the most difficult environment in terms of achieving good performance. The same code has been executed on a Connection Machine CM-5, a Silicon Graphics Power Challenge, and an IBM SP without any problems.

In general, a fair comparison of parallel and serial programs involves running the two versions and measuring the execution times. There is an implicit condition that the two versions must give the same results, otherwise, the comparison is meaningless. Of course, the same condition has to be enforced in the particular case of GAs, so we compare the execution times of serial and multiple-deme parallel GAs when they find solutions of the same quality.

Most of the fitness functions used in these tests are artificial problems of known difficulty. We created the fitness functions by concatenating several copies of fully deceptive trap functions. Fully deceptive functions are bimodal and the deceptive optimum has a wide attraction basin, while the global optimum is completely isolated. The difficulty of the fitness function depends upon three factors: (1) the relative lengths of the deceptive and global basins, (2) the fitness difference between the global and the deceptive optima, and (3) the number of copies of the trap function that are concatenated.

The first tests are of a master-slave parallel GA. In this case the communications are more frequent and the system can be expected to yield good results only if it takes a considerable time to evaluate the fitness function. We added artificial variable delays to the trap functions so we could simulate the execution time of difficult functions. The results of these tests can be found, along with a theoretical model that predicts the performance of master-slave parallel GAs in Cantú-Paz, 1997a. Note that the efficiency of the code increases as it takes longer to evaluate the fitness, and the speedups can be proportional to the number of slaves for problems that take a very long time to evaluate.

The next set of tests that we will show are of multiple deme GAs. There is also a plot of a theoretical prediction of the speedup in the figures (Cantú-Paz & Goldberg, 1997b), and it is evident that for both functions the speedups are very low. The importance of these results is that they do not depend upon the particular hardware (as there is no communication and the computation time gets normalized when the speedups are computed) or on the particular test functions. The theory shows that, in general, there are no significant gains in performance when the demes are isolated.

The next set of tests are also of multiple-population GAs, but, in this case, each deme is connected to all the others (fully connected topology) and the migration rate is set to the highest possible value. The migration strategy was to migrate only once after the demes had converged completely. There is a maximum speedup for each test function and the theory predicts the actual results very accurately. Notice that in the case of the easier function (composed of 4-bit traps), the speedup is not as significant as for the harder function (composed of 8-bit traps). The reason is that tile savings on computation time brought by the use of multiple demes are not enough to overcome the increase in communications. Recall that these tests are performed on a slow network of (slow) workstations, but even under these circumstances the parallel GA performed significantly better when it was used to optimize the harder problem.

The results from the two algorithms seem to point out that it is more efficient to use parallel GAs with difficult fitness functions. The reason is that in both algorithms there is a tradeoff between savings on computations time and increasing communications. When the fitness functions take a long time to evaluate, the ratio between computation and communication times is much higher making the parallel algorithms more efficient.

3.5. Conclusions and Extensions

It is becoming increasingly important to design faster GAs as they are entering industry and are applied to harder problems. Parallel implementations of GAs are a particularly promising method to make GAs faster as parallel computers are becoming more widely spread. GAs are easy to implement on parallel platforms and they are particularly amenable to parallel computation because the communications-to-computation ratio can be kept very low.

Users of parallel GAs can benefit from the results that research has produced recently. There are some interesting results that can help practitioners to more productively use the resources they have available and take a greater advantage of parallel GAs, but the research is not yet complete. The effect of some of the parameters that control parallel GAs is not completely understood, so it is necessary to continue to systematically experiment and develop new models to predict the performance of parallel GAs using different parameter values. A program like the one presented in this chapter is an indispensable tool to perform research in this area.

In this contribution we have discussed the design and the implementation of a parallel GA code that is both flexible and simple. Our goal was not to produce the ultimate parallel GA code, but to transmit ideas that should help other programmers implement their own parallel GAs. We preferred not to implement every possible variation of codings and operators, but instead we produced a code that is efficient, easy to understand, and very portable.

The code that we presented here is a good starting point for new developments, and it can be extended in many directions. In particular, it is not difficult to add other forms of selection or specialized genetic operators that are suitable to a particular application.

An extension that could significantly improve performance would be to use asynchronous communications. In our system there are times when processors sit idle as they wait for others to finish their work. For master-slave GAs we showed that one way to avoid these idle times is to do an asyncronous implementation. A better implementation of multi-population GAs should also try to overlap communications and computations and, in Section 3.3, we discussed one easy strategy to accomplish this.

An interesting extension is to combine coarse-grained and master-slave GAs and we discussed how this "hybrid" parallel algorithm can be implemented just by changing the base class for Deme. This is an example of the flexibility that results from the object-oriented design.

Of course, there are many other possible extensions to the code, but we have also shown in this contribution that gains in efficiency do not come only from better implementations, but also from more rational utilization of resources. As the theory for parallel GAs advances, better guidelines are emerging to help users exploit the potential of parallel computers.

Acknowledgments
I would like to thank Prof. David E. Goldberg for his encouragement during the development of this project.

I was supported by a Fulbright-Garcia Robles Fellowship.

This study was sponsored by the Air Force Office of Scientific Research, Air Force Materiel Command, USAF, under grants number F49620-94-

10103, F49620-95-1-0338, and F49620-97-1-0050. The U.S. Government is authorized to reproduce and distribute reprints for Governmental purposes notwithstanding any copyright notation thereon.

The views and conclusions contained herein are those of the authors and should not be interpreted as necessarily representing the official policies or endorsements, either expressed or implied, of the Air Force Office of Scientific Research or the U.S. Government.

References

Abramson, D. & Abela, J. (1992). A parallel genetic algorithm for solving the school timetabling problem. *Proceedings of the Fifteenth Australian Computer Science Conference (ACSC-15),* 14, pp 1-11.

Abramson, D., Mills, G., & Perkins, S. (1993). Parallelisation of a genetic algorithm for the computation of efficient train schedules. *Proceedings of the 1993 Parallel Computing and Transputers Conference,* 139-149.

Booch, G. (1994). *Object Oriented Design with Applications* (2 ed.). Reading, MA: Addison-Wesley.

Cantú-Paz, E. (1997a). *Designing efficient master-slave parallel genetic algorithms* (IlliGAL Report No. 97004). Urbana, IL: University of Illinois at Urbana-Champaign.

Cantú-Paz, E. (1997b). *A survey of parallel genetic algorithms* (IlliGAL Report No. 97003). Urbana, IL: University of Illinois at UrbanaChampaign.

Cantú-Paz, E. & Goldberg, D.E. (1997a). Modeling idealized hounding cases of parallel genetic algorithms. In Koza, J., Deb, K., Dorigo, M., Fogel, D., Garzon, M., Iba, H., & Riolo, R. (Eds.), *Genetic Programming 1997: Proceedings of the Second Annual Conference* (pp. 353-361). San Francisco, CA: Morgan Kaufmann Publishers.

Cantú-Paz, E. & Goldberg, D.E. (1995). Predicting speedups of idealized hounding cases of parallel genetic algorithms. In Back, T. (Ed.), *Proceedings of the Seventh International Conference on Genetic Algorithms* (pp. 113-120). San Mateo, CA: Morgan Kaufmann Publishers.

Fogarty, T.C. & Huang, R. (1991). Implementing the genetic algorithm on transputer based parallel processing systems. *Parallel Problem Solving from Nature,* 145-149.

Goldberg, D.E. & Deb, K. (1991). A comparative analysis of selection schemes used in genetic algorithms. *Foundations of Genetic Algorithms, 1,* 69-93. (Also TCGA Report 90007.)

Goldberg, D.E., Deb, K. & Clark, J.H. (1992). Genetic algorithms, noise, and the sizing of populations. *Complex Systems,* 6, 333-362.

Grefenstette, J. J. (1981). *Parallel adaptive algorithms for function optimization* (Tech. Rep. No. CS-81-19). Nashville, TN: Vanderbilt University, Computer Science Department.

Harik, G., Cantu-Paz, E., Goldberg, D. & Miller, B. (1997). The gambler's ruin problem, genetic algorithms, and the sizing of populations. In Back, T. (Ed.), *Proceedings of the Fourth International Conference on Evolutionary Computation* (pp. 7-12). New York: IEEE Press.

Hauser, R. & Manner, R. (1994). Implementation of standard genetic algorithm on MIMD machines. In Davidor, Y., Schwefel, H.-P., & Manner, R. (Eds.), *Parallel Problem Solving from Nature, PPSN 111* (pp. 504-513). Berlin: Springer-Verlag.

Thierens, D. & Goldberg, D.E. (1993). Mixing in genetic algorithms. In Forrest, S. (Ed.), *Proceedings of the Fifth International Conference on Genetic Algorithms* (pp. 38-45). San Mateo, CA: Morgan Kaufmann.

Chapter 4 Pattern Evolver

An Evolutionary Algorithm that Solves the Nonintuitive Problem of Black and White Pixel Distribution to Produce Tiled Patterns that Appear Gray

Keith Wiley
102 Keith Rd.
Carrboro, NC 27510
(919) 929-2783

keithw@wam.umd.edu

http://www.wam.umd.edu/~keithw

Abstract

Pattern Evolver is an evolutionary program that produces black and white bitmapped patterns, which, when tiled, create a smooth unbroken gray appearance to the human eye. Pattern recognition in the human eye and the human visual cortex is highly developed, to the point where humans will reconstruct lines and shapes that do not exist in the actual physical stimulus. As a result, creating smooth gray patterns is not intuitively easy. Pattern Evolver harnesses Darwinian selection to solve this task and thus falls into the category of evolutionary programs. Like biological systems, this program exemplifies evolution in an adaptive landscape with multiple local peaks.

4.1. Introduction

For a program I was recently hired to write, I discovered that I would need to produce degrees of contrast between light and dark areas on a black and white, one-bit deep monitor. Gray patterns had to be produced by varying ratios of black to white pixels. Since I was writing the program for an Apple Macintosh computer, the system was equipped to use 8 x 8 pixel patterns. I decided to use tiles of the same size so that I could easily encode the patterns into my program, and started making the patterns I would need. In a 64 pixel grid there are 65 possible ratios of black to white pixels 0:64 through 64:64. The smoothest possible gray for some ratios is intuitively easy to design. These are the ratios that divide evenly into 64: 0:64, 1:64, 2:64, 4:64, 8:64, 16:64, 32:64, and 64:64, for both black to white and white to black (Figure 4.1).

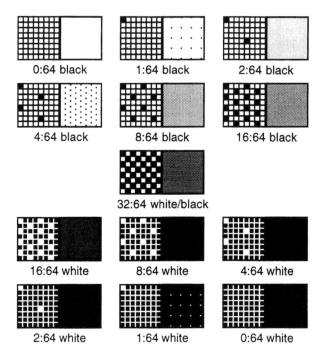

Figure 4.1 The only thirteen perfectly smooth gray patterns on a grid of 64 pixels are those with black to white or white to black ratios that divide evenly into 64

For these ratios (see Figure 4.1) there is an exact and methodical pattern that produces an ideally smooth gray tiling effect. In all figures illustrating patterns, the genotype is enlarged on the left for easy viewing and an example of the phenotype is shown regular size and tiled four across and four down on the right.

A problem arises when one attempts to make all the other numerous patterns corresponding to the leftover ratios look like a smooth gray surface. In many cases, no symmetrical pattern of pixels is possible. Furthermore, when symmetrical patterns are possible, they often do not appear smoothly gray. For example, to make a gray pattern of ratio 15:64, simply removing any one of the pixels in the predetermined 16:64 pattern from Figure 4.1 will result in a symmetrical pattern, but not a smoothly gray pattern. It will have a big white dot on it and consequently will not look evenly gray. Instead it will look like a quilted pattern (Figure 4.2).

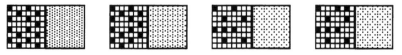

Figure 4.2 Examples of symmetrical but unsmooth patterns

It is possible to make patterns that tile perfectly and yet do not produce the desired smooth gray appearance (see Figure 4.2). Such patterns are abundant, nearly filling the set of remaining patterns that are left after removing the thirteen perfect smooth gray patterns in Figure 4.1. Despite their frequency and ease of use, they defeat the purpose of creating smooth gray patterns and must be discarded as relatively nice, but inadequate, patterns. A better solution can be evolved for the desired ratio.

In order to make all the other gray patterns, I had to work them out one by one, by moving individual pixels to spread the pixels out as evenly as possible, without producing lines and squiggles, or dark and light areas. It was tedious and time consuming, and I had a feeling the results in several cases were not the smoothest, most nearly ideal patterns possible.

After finishing the program, I decided to try another approach to the patterns. The first idea I had was a standard biomorph program, originally written by Richard Dawkins and described in his book *The Blind Watchmaker* (Dawkins, 1986). So, I wrote the first version of Pattern Evolver to produce a family of entirely random patterns of any desired ratio. The user clicked on the pattern that looked the most like what was wanted. In my case I was searching for a smooth gray appearance, but the program could also be used to evolve any pattern the user wanted. When a pattern was chosen by the user, it became the parent of a new family of sixteen patterns, each slightly mutated from the parent. By continuing this process, the user could "push" the family from generation to generation up any fitness peak desired: gray, vertical lines, circles, whatever the user wanted. With this program I found that I could evolve smooth gray patterns of any desired ratio in a matter of minutes, while designing the patterns on my own took at least half an hour each.

This program nevertheless seemed inadequate for two reasons. First, I wasn't sure that I obtained the best possible patterns, because searching all the possible patterns still took a long time. There were too many different kinds of nearly smooth grayness that had to be compared to each other. To look at all seventeen family members (the parent and sixteen offspring), decide upon the best pattern, and click on it takes at least a few seconds and may take considerably longer if the decision is hard to make. This time quickly adds up. Second, smoothness, or grayness, just felt like such a clearly defined state for a pattern of pixels. Why did the user have to make the choices of each generation? The computer could pick the grayest pattern in the family of each generation much faster then a human could possibly do manually.

At this point the idea of using an evolutionary algorithm struck me. If I could simply define a mathematical way for the computer to determine how smooth a pattern was, it could automatically pick the pattern with the highest calculated smoothness. My next version of Pattern Evolver was capable of picking the smoothest pattern in the family and automatically making it the next parent. I left the user-controlled choices as an option in the program. For the computer to pick the grayest pattern, the program evaluates each pattern in a family and assigns it a grayness score. Devising this algorithm was difficult. I used the thirteen perfect patterns as my guides. I was looking for algorithms that would assign a higher score to the perfect patterns than to any other distribution of pixels for those ratios. It took several attempts to find a satisfactory algorithm. The final result was capable of evolving a smooth-looking gray pattern of almost any ratio in less than a minute or two and, in several cases, in less than thirty seconds. A couple of ratios still produce unsmooth gray patterns as the highest scoring, but only very few. Fixing this would be a task of optimizing the grayness calculating algorithm. Overall, the program works well.

4.2. Mutation

Each time one pattern from the family is chosen as the next parent, either because of its grayness score (see below) or because the user manually picks it, a new family is generated in the following manner:

- The chosen pattern is copied into the parent pattern, replacing the old parent.

- Each offspring is cloned from the new parent pattern and immediately mutated.

- The new family is redrawn on the screen (if the "Update Family" switch is on. It can be turned off to speed up the program. Of course, redrawing the screen is not necessary for the computer to pick the grayest pattern, only for the user to see what is happening).

This process is repeated for each generation as one pattern from the family is chosen as the next parent.

Mutation consists of two steps. First, micromutations occur. For every black pixel in the pattern, there is a probability that the pixel will drift one space in a randomly determined empty direction (Figure 4.3). Each pixel has eight possible directions in which to move. Pixels cannot drift onto or over each other. So, if a certain pixel has seven neighbors that are black and the program determines that the pixel will move, it will predictably go to the only empty space available. Pixels that are entirely surrounded by black pixels cannot by moved.

Figure 4.3 An example of a micromutation. A single pixel is moved one space in a randomly assigned available direction

Next, macromutation occurs. There is a set probability that one of two possible macromutations will occur. Each consists of exchanging entire quarters of the pattern (so a four-by-four group of pixels is moved together). One consists of switching the two bottom quarters and one consists of switching the two right quarters (Figure 4.4). Note that macromutation occurs after micromutation.

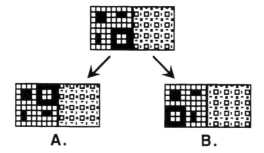

A.   B.

Figure 4.4 Two examples of macromutations

The top pattern (see Figure 4.4) is the original pattern. The bottom patterns are the two possible macromutations. "A" shows that the lower right corner of the pattern has been switched with the top right corner. "B" shows that the lower right corner of the pattern has been switched with the lower left corner.

The probabilities of micro- and macromutations are selected by the user.

4.3. Measuring Grayness

The algorithm I devised for evaluating the grayness of a pattern is rather complicated. To begin, the grayness score is set to zero. Then the pattern accumulates points for having an even distribution of black pixels. For each of the sections of the pattern listed below, the number of black pixels is tallied. If the number is exactly the correct proportion for that section, then the maximum number of points is annexed to the total score. If either more or less than the precise proportion of black pixels are present, then fewer points are awarded down to as few as zero. By proportion I mean that in the sections with 32 possible cells (with half the area of the entire pattern), the number of black pixels must equal half the ratio in order to receive the most points.

The sections checked are as follows:

• Check the 8 horizontal halves (8 pixels horizontal x 4 pixels vertical); wrap vertically but not horizontally (Figure 4.5).

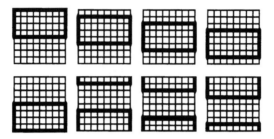

Figure 4.5 The eight horizontal halves that are checked for pixel distriubution

For each of the eight sections (see Figure 4.5), if the number of black pixels equals exactly half the ratio, then the maximum number of points is annexed to the grayness score. Fewer or more black pixels results in fewer points being annexed. The last three quadrants must be wrapped around to the opposite side of the pattern.

- Check the 8 vertical halves (4 x 8 pixels); wrap horizontally but not vertically

- Check the 8 horizontal quarters (8 x 2 pixels); wrap vertically but not horizontally

- Check the 8 vertical quarters (2 x 8 pixels); wrap horizontally but not vertically

- Check the 64 square quarters (4 x 4 pixels); wrap both horizontally and vertically

- Check the 8 horizontal eighths (8 x 1 pixels); one pixel thick, no wrapping is necessary

- Check the 8 vertical eighths (1 x 8 pixels); one pixel thick, no wrapping necessary

- Check the 64 square sixteenths (2 x 2 pixels); wrap both horizontally and vertically

Next, points are deducted from the newly accumulated score for black pixels that are near each other. For every black pixel in the pattern, the overall score is decreased a number of points based on that pixel's position in relation to the pixel being tested (Figure 4.6).

This evaluation (see Figure 4.6) is carried out on all 64 pixels in the pattern, wrapping around the edges for the necessary pixels.

Based on Figure 4.6, the maximum number of points that can be lost due to black pixel neighbors of a single black pixel is 196 (the sum of the deductions in Figure 4.6) if all twenty cells surrounding it have black pixels in them. This does wrap around the edge of the pattern since the pattern will be tiled when it is displayed.

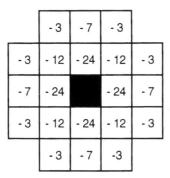

Figure 4.6 A diagram for determining how many points are deducted for the presence of black pixels around the pixel in the center.

It is important to note that so many points are deducted from higher ratio patterns (31:64 for example) that even the best pattern possible will have a low grayness score simply because the pixels are close together. The scores are thus useful only in comparison to the scores of other patterns of the same ratio.

4.4. Adaptive Peaks as a Hindrance and Resilience as a Solution

Pattern Evolver displays the evolutionary phenomenon of adaptive peaks very well. Adaptive peaks in evolving systems were first emphasized by Sewall Wright in his theory of an adaptive landscape (Wright, 1932, 1982); I was amazed when I discovered this effect in Pattern Evolver.

If you set up a random family for Pattern Evolver and let it try to evolve the grayest pattern possible for that ratio, it will eventually reach a state in which it keeps picking the parent as the grayest pattern for endless generations because the parent is perennially awarded a higher grayness score than any of the offspring. It would appear that Pattern Evolver has found the smoothest gray possible. However, when I checked this conclusion, I got a surprise. Remember that I optimized the algorithm for calculating grayness so that the thirteen perfect patterns of the evenly divisible ratios would receive the highest scores for their ratios. By looking at the patterns that emerged for these evenly divisible ratios, I could Figure out whether or not Pattern Evolver was indeed finding the global peak every time. I discovered that sometimes Pattern Evolver was finding persistent parents for the perfect ratios that were not the perfect patterns shown in Figure 4.1, that indeed did not have higher scores than the perfect patterns from Figure 4.1, and yet that could not be beaten by any of their offspring even if the program ran for an extended period of time.

I soon decided that local adaptive peaks must be my problem. Pattern Evolver picks the single pattern in a family that has the highest grayness score, and the next generation is based upon that pattern. As a result, Pattern

Evolver will not move downward in fitness (grayness score) from a local maximum so that it can climb up a different, possibly higher peak.

My solution was to create what I called "resilience." I would take any parent that persisted a specified number of generations, add it to a list of likely best patterns, reseed a completely new, randomly chosen family, and try again. In this way a collection of good patterns could be compiled by taking one resilient parent after another and saving them all for the user to choose from.

The necessary resilience steadily increases throughout any single run of the program so that the present parent must be chosen as the next parent in an increasingly higher number of generations in a row in order to be added to the list of best grays. In essence, it must have a higher grayness score than an increasingly larger number of offspring to be dubbed a successful gray pattern. Consequently, the best grays list first fills up with likely, but not well tested candidates. These best grays are eventually replaced by newer, better grays that have had to remain parents for a longer time (withstanding the grayness scores of a higher number of varying offspring) and thus have better overall scores. Once the best grays list is full (the capacity is nine), any new parent that withstands the resilience test will be added to the best grays list only if its score is higher than the lowest scoring pattern in the best grays list. If this is the case, the previous lowest scoring pattern is removed and the new pattern is inserted into the best grays list in order of grayness score. If the parent withstands the resilience test but its score isn't good enough to get into the best grays list, it is discarded and a counter is annexed.

Every fifth time a parent is discarded, the resilience is increased by five generations, on the assumption that the resilience is too low to evolve patterns good enough to get into the best grays list. By raising the resilience, only parents with better scores will pass the resilience test and thus get into the best grays list.

In this fashion, Pattern Evolver both gives the user access to fairly good grays quickly and keeps getting tougher on the patterns as time goes on so that the user receives better and better patterns. Eventually, the best gray pattern possible for the desired ratio should emerge and not be beaten out of the best grays list, no matter how high the resilience is.

4.5. A Sample Run

The following is an example of what happens in a sample run. Some figures show the entire family for that generation, but most just show the pattern from the previous generation chosen as parent of the present generation. Where full families are shown, the pattern at the top of the family is the parent. The other sixteen patterns are the offspring, all based on the parent (with the exception of generation zero which is randomly seeded). Micromutation rate is 20% and macromutation rate is 20%. For this example, I have chosen the ratio of 11:64, because designing a smooth gray pattern of this ratio manually is exceedingly difficult. Watch as Pattern Evolver effortlessly produces a beautiful smooth gray pattern. First, a random family of the desired ratio is generated. Only at this point is there no relation between the pattern in the parent position and the rest of the patterns. From generation one onwards, the sixteen offspring will always be related to the parent.

The number to the right above each pattern (see Figure 4.7) is the maximum number of points that could be tallied in the first half of the evaluation process, before any points are deducted for black pixels that are close to each other, in other words, the highest score for this ratio. This score is generally not realistically attainable since points will certainly be deducted in even the smoothest pattern at ratios higher than about 4:64 to 6:64. The number to the left above each pattern is the score that Pattern Evolver has assigned to that particular pattern. The pattern chosen from this generation was the one with the highest score: second column, second row. Therefore, it is the parent shown for generation 1 in Figure 4.8.

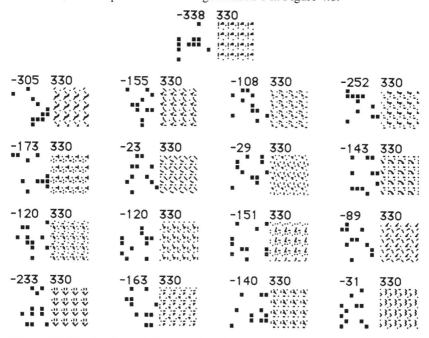

Figure 4.7. Randomly seeded generation 0

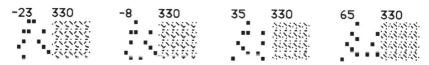

Figure 4.8 Parents for generations 1, 2, 3, and 4

In only 4 generations (see Figure 4.8) (generation 0 is not counted because it is randomly seeded), some improvement in the smoothness is already apparent.

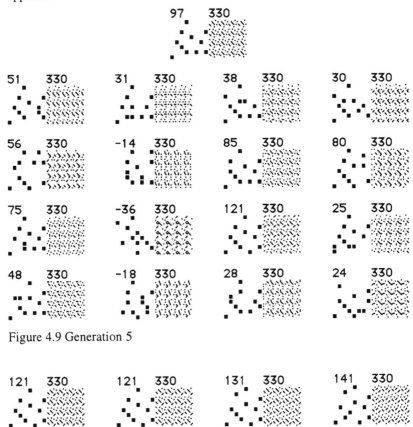

Figure 4.9 Generation 5

Figure 4.10 Parents for generations 6, 7, 8, and 9

The parent pattern (see Figure 4.10) is beginning to look fairly smooth at this point. Notice that one parent was repeated for a second generation. This pattern would have become a best gray pattern if it had lasted a couple of more generations.

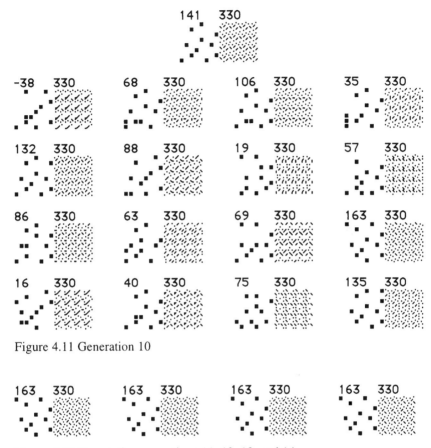

Figure 4.11 Generation 10

Figure 4.12 Parents for generations 11, 12, 13, and 14

The parents (see Figure 4.12) are all identical because the parent of each family had the highest score and was consequently chosen as parent for the next generation.

This pattern (see Figure 4.14) has passed the first resilience test by being chosen for five sequential generations after becoming parent. It was added to the best grays list, a new family was randomized, and the process was repeated.

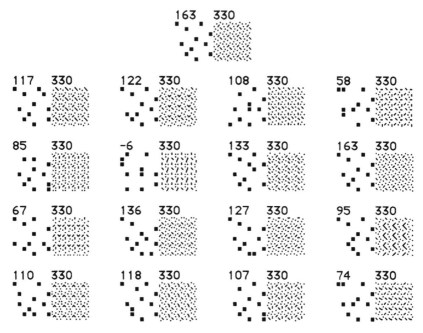

Figure 4.13 Generation 15

Figure 4.14 Parent for generation 16

The score of the first parent chosen from the randomly seeded family was 23. The score went up almost every generation and ended up at 163 for the first pattern added to the best grays list, a much better score, and sure enough, the resulting pattern does look much smoother than any from the preceeding generations.

Figure 4.15 The smoothest gray pattern for a ratio of 11:64 that Pattern Evolver could discover.

Nevertheless, the resulting pattern (Figure 4.14) is not perfectly smooth. This is because the necessary resilience starts very low, only five generations. It slowly increases to as high as 1000 if necessary. From the starting point, it took Pattern Evolver only seven seconds to come up with the best pattern possible for a ratio of 11:64 (Figure 4.15). At this point, the resilience had increased to twenty.

Others with the same score (see Figure 4.15) also exist and it may require a little human discretion to choose between the maximally equal scoring patterns. Nevertheless, it is impossible for a human to design a pattern this smooth in a time equal to that of Pattern Evolver.

A couple of other patterns with scores of 179 appeared soon afterwards, but they all looked similar in grayness (thus the identical scores), and no pattern ever emerged that beat the score of 179, even after the necessary resilience had risen to 45 generations as parent. It is reasonable to conclude that Pattern Evolver has found the smoothest possible gray patterns for a ratio of 11:64 pixels. There is most likely no pattern that would yield a score higher that 179 for this ratio using Pattern Evolver's grayness calculating algorithm, and there may very well be no pattern that would appear smoother to the human eye.

In seven seconds, Pattern Evolver discovered a smooth gray pattern that a human would be lucky to find in thirty minutes or an hour. If you consider that the 65 possible ratios minus the thirteen easy-to-do perfect ratios leaves 52 such patterns, a human could spend a full 24 hours or more completing the entire range of gray patterns and would probably not even do a very good job. Pattern Evolver could produce the entire collection in a few minutes.

4.6. Local Maxima

That Pattern Evolver discovered local adaptive peaks makes it a remarkable program. Local maxima also occur in biological evolution, as first argued by Sewall Wright (Wright, 1932, 1982). In biological systems, the numerous possible combinations of genes presumably have complex patterns of adaptiveness. The combinations of genes that correspond to well-adapted individuals are called local maxima. Evolutionary biologists call these local maxima "adaptive peaks" and the topography of adaptive values an "adaptive landscape." It is probable that such a landscape will have several local peaks which may be of comparable adaptiveness. If a population has combinations of genes that cluster around one peak, then successful migration to another peak requires some strategy for traversing the valleys between the peaks. The easiest such strategy is increased mutation. In Pattern Evolver, mutation is increased either by a higher rate of micromutations or by the presence of macromutations. These two possibilities correspond to two methods of moving to a different peak. In the first possibility, the population spreads out around the first peak, to the point where certain lineages of the population reach a valley and then advance up another peak. The second possibility is that a lineage jumps, as a result of a rather large mutation, onto a different peak. This second method is what macromutations in Pattern Evolver attempted to replicate.

It is interesting to speculate about when genetic patterns will have a complex adaptive landscape with numerous local peaks. Perhaps if the phenotypes of the genes, whether a tile of pixels or a biological organism, are evaluated by complex and overlapping criteria, then the combinations of individual genes that have high fitness will be nearly unpredictable. The interaction of combinations of genes with a particular environment might be so complex that the maxima cannot be easily specified. In this case, Darwinian evolution, coupled with a process for exploring multiple peaks, can provide a good method for discovering the maxima in the landscape.

Pattern Evolver differs from biological evolution in one important way. The latter has no mechanism for systematically comparing many peaks in the landscape. Pattern Evolver picks a new starting location on the landscape every time it randomizes the family and, consequently, searches as many local peaks across the landscape as possible. The result of this process is that Pattern Evolver has a good chance of discovering the global peak simply by trial and error. Biological evolution has no such good fortune. If a population climbs up a local peak, it stands a good chance of being indefinitely stranded on that peak without ever exploring other areas of the landscape. In order to try another peak, a biological population would have to evolve slowly, generation by generation, down the present peak and back up another. Because descending a peak means evolving toward lower adaptiveness, it cannot be realistically accomplished. To jump from peak to peak in a single generation would require both a significant change in the genome and the slight chance that the resulting genome would land, not only on another local slope, but actually higher on the new slope than it is on its current local peak. If local peaks are few and far between in a biological landscape, such a change in the genome in a single generation would essentially never happen.

The way Darwinian selection moves patterns in such a landscape will depend on both the shape of the landscape and the nature of mutation. A flatter landscape, consisting of shallow divides and closely spaced peaks, will more likely make transitions between peaks by chance. A mountainous landscape, with wide, deep valleys, increases the chance that a population will be stranded on a local peak without being able to migrate or jump to another peak.

The nature of mutation will also affect how a population evolves in a landscape. If only minor mutations are possible between generations, then the step size across the landscape is smaller. Consequently, it is easier to get stranded on a peak, unable to step over a valley to another peak. If rather significant mutations are possible, then the possible step size between generations across the landscape is greater, and it is thus easier to cross valleys to other peaks. Macromutations, in this case, might enable a population to more adequately explore the landscape in search of the global peak.

Macromutations, however, are a double-edged sword in evolution. By increasing the chance of jumping from peak to peak, they also increase the chance that a population will jump off the global peak before reaching the summit. In other words, if macromutations were too frequent, a population may not spend the time necessary to perfect its score on one peak before giving up and jumping to another peak.

In Pattern Evolver I included two degrees of mutation, micro- and macromutation. Macromutation was added once I realized that I was dealing with local peaks in an adaptive landscape. I recognized that local peaks would hinder the performance of Pattern Evolver and thought that if I could induce more dramatic change between a parent and its offspring, then some macromutated offspring might launch the lineage in a new direction, in essence, jumping from one local peak onto the slope of another, higher peak.

4.7. Testing the Influence of Macromutations

To see whether or not the method of macrocmutation I devised, by shifting quarters of the pattern, actually increased the chances of finding the global maximum, I ran an experiment. Choosing the perfect ratio of 8:64 black:white pixels, I ran combinations of micro- and macromutations at different resiliences, 1000 times each, and recorded the number of global peaks found for each combination of settings.

The results did not confirm my expectation of significantly better performance with macromutations (Figure 4.16). Nevertheless, macromutations often did improve the performance to some degree, but not in a simple, linear fashion. My particular method of macromutation seemed to hurt Pattern Evolver's performance for certain combinations of micro- and macromutations. Especially when the rate of micromutation was low, zero macromutation generally outperformed low macromutation. Furthermore, the most successful combinations involved both high rates of micro- and macromutation. In these cases, macromutations did seem to play some role in discovering the global peak.

It is interesting that medium rates of macromutation produced poorer results than zero and high rates, especially for low rates of micromutation (exhibited most effectively in Figure 4.16D, as well as 4.16B, 4.16C, 4.16E and 4.16G). As micromutation increases, the span of the U-shaped pattern becomes smaller, perhaps because increased micromutation levels the field for macromutation, negating differences that would otherwise be exhibited across various rates of macromutations. Imagine a population randomly walking around on an adaptive landscape. With no macromutations, it can only take small steps and gradually climb up one peak. With high rates of macromutation, the population would hop all over the landscape from one slope to another, always slowly edging its way higher but not very quickly or methodically, certainly not consistently up any one peak. If a population jumped off the global peak, it would have a good chance of jumping back later. With an intermediate rate of macromutation, there might be enough jumping between peaks to coax a lineage into jumping off a slope, perhaps even off the global peak, but macromutations might be so few that the lineage never jumps back onto the global slope in a finite period of time. In this way, a medium rate of macromutation would perform worse than no macromutation at all and likewise would perform worse than a high rate of macromutation. I had not foreseen this possibility prior to analyzing the data from the experiment.

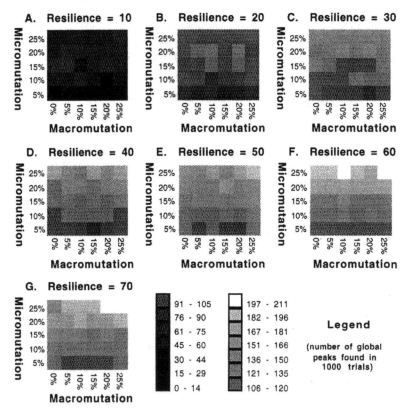

Figure 4.16 Number of runs per 1000 that discovered the global peak, as a function of rates of micro- and macromutation.

It appears that there is little overall improvement in performance associated with increased macromutation. However, there are some interesting patterns present. For low rates of micromutation, it appears that medium rates of macromutation perform worse than both lower and higher rates of macromutation. My explanation is that medium rates of macromutation are just high enough to make lineages jump off of good peaks, but are not high enough to find the global peak. Since such lineages may very well jump off of the global slope but may never jump back, this rate performs worse than both lower and higher rates of macromutation. Lower rates would prevent the lineage from jumping off the global slope, and higher rates would allow the lineage to correct itself if it jumps off the global slope by jumping back.

Another interesting observation is that the number of trials that found global peaks is rather low. The highest number was 211 in 1000 trials (resilience = 70, micromutation = 25% and macromutation = 20% (Figure 4.16G)). It is possible that with a higher resilience, global peaks would be found more often as each lineage has more time to search the landscape (although performance seems to level off as resilience increases). This low overall performance might be interpreted in several ways. My hunch is that the adaptive landscape associated with Pattern Evolver consists of numerous thin-based, spike-shaped peaks with wide valleys between them, as opposed to a few slowly inclining peaks with shallow valleys. Such a scenario would make wandering about on the landscape much more difficult. Once a lineage starts up a local peak, the odds of any small or even a large mutation landing on another slope or peak instead of in a valley are very low. If the landscape has numerous inferior local peaks and only one global peak, or a few inferior peaks with very wide bases that consequently funnel lineages, then it would be hard to find global peaks, pretty much regardless of mutation rates, both micro and macro.

4.8. Closing Remarks

I set out to solve a nonintuitive problem by using evolutionary means, to recursively take a sample of possible solutions, pick the best solution, and breed that solution with mutations. I later introduced an automatic grayness-measuring algorithm into the program so that the computer would solve the problem set forth entirely without human guidance.

The distribution of pixels in gray patterns similar to those discussed in this chapter is truly nonintuitive in nature. There is no other way for a human to design such patterns except to move pixels about one by one in a rather unmethodical fashion, almost entirely by trial and error. Such an approach may yield an acceptable solution, yet, when nothing less than the ideal solution will suffice and a limited amount of time is available in which to find that solution, the human design approach falls desperately short of acceptable performance. As the number of pixels, N, increases, the number of possible patterns increases extremely fast. For a ratio of just 5:64 black to white pixels, there are over 900 billion possible distributions in an 8 x 8 grid, 64 x 63 x 62 x 61 x 60. The human design approach is thus highly inefficient beyond a rather small number of pixels. Coupling a computer's computational speed with Darwinian evolution trivializes the task to the point where I cannot imagine why people have ever had to design such patterns in the first place. Pattern Evolver quickly produces patterns that rival any pattern ever designed by a human, and, in the case of many of the most obscure ratios, they probably surpass those designed by all humans in the past.

4.9. How to Obtain Pattern Evolver

Pattern Evolver is available at several sites on the World Wide Web as well as being registered on the major search engines. Pattern Evolver webpage is located at http://www.wam.umd.edu/~keithw/patternEvolver.html. Please e-mail me if you have any questions.

References

Dawkins, Richard. 1986. *The Blind Watchmaker*. W. W. Norton & Company.

Wright, Sewall. 1932. "The Roles of Mutation, Inbreeding, Crossbreeding, and Selection in Evolution". *Proceedings of the Sixth International Congress of Genetics*. The University of Chicago Press.

Wright, Sewall. 1982. "Character Change, Speciation, and the Higher Taxa." *Evolution*. The University of Chicago Press.

Chapter 5 Matrix-Based GA Representations in a Model of Evolving Animal Communication

Michael Levin
Cell Biology Dept.
Harvard Medical School,
240 Longwood Ave.
Boston, MA 02115

mlevin@fas.harvard.edu

Keywords: matrix, evolution of communication, genetic algorithm, self-organization

Abstract

Much animal communication takes place via symbolic codes, where each symbol's meaning is fixed by convention only and not by intrinsic meaning. It is unclear how understanding can arise among individuals utilizing such arbitrary codes and, specifically, whether evolution unaided by individual learning is sufficient to produce such understanding. Using a genetic algorithm implemented on a computer, I demonstrate that a significant though imperfect level of understanding can be achieved by organisms through evolution alone. The population as a whole settles on one particular scheme of coding/decoding information (there are no separate dialects). Several features of such evolving systems are explored and it is shown that the system as a whole is stable against perturbation along many different kinds of ecological parameters.

5.1. Introduction

An act of communication on the part of one system can be defined as one which changes the probability pattern of behavior of another; it is the functional relationship between signal and response (Wilson, 1975). Communication in animals is a very important part of their ecological profile (e.g., Wilson, 1971 and 1975, Halliday and Slater, 1983, Alcock, 1989). Whether social or solitary, animals often encounter members of their own or another species and exchange information. Ethologists analyze such communication along three aspects: (1) the physical channel for the information transfer (such as emitters and detectors of light, sound, or chemicals), (2) the present-day function of the message (such as alarms, advertising or requesting resources, individual or class recognition, and assembly or recruitment), and (3) the evolutionary or cultural derivation of the communication (plausible models of how the behavior evolved from other simpler behaviors and became established within the population).

Dawkins (1982) expounds one popular and very fruitful view of communication: that sending signals is a way of manipulating one's environment to one's advantage (via the actions of other living things). This suggests that selection should favor those signals which maximally increases the likelihood of a particular behavior by another animal, relative to the amount of effort it takes to signal.[1] This also illustrates another general property of communication (pointed out by J. B. S. Haldane): that it often involves great energetic amplification, because the relatively small amount of energy expended in producing a signal is magnified (at the expense of the perceiver) into potentially great consequences.

Thus, one approach is to study the evolution and ecology of signal emitters – to discover how such signals arise and how they benefit the animal and increase its fitness (Dawkins and Krebs, 1978). The complementary approach focuses on the receiver, and asks what discriminatory faculties enable another organism to perceive signals and act upon them in such a way as to maximize its fitness, and to minimize being detrimentally manipulated by others. This involves issues of classification and thresholds, optimal allocation of resources (such as time and energy) for processing various signals, etc.

A third, and somewhat orthogonal approach, involves the theory of games, and allows analysis of general behavioral strategies with respect to conflicts (Rapaport and Chammah, 1965; Axelrod, 1984). The paradigms of arms races, the Prisoner's Dilemma, ESSs (evolutionarily stable strategies), and other such models can be used to shed light on why certain animals behave as they do.

Communication usually takes the form of energy or chemical emissions, or body postures and displays, and, as with any information exchange, is supervenient upon a coding system. A symbolic or arbitrary code is one in which the symbols have only a contingent relationship to what they represent. Thus, in English (an arbitrary-coded language), the symbol "dog" (whether written or vocalized) has nothing dog-like about it. It is a symbol the meaning of which is fixed by convention only; it could just as easily have been assigned another meaning. This is in contrast to codes like pictographs and hieroglyphics, where the symbol for dog would actually resemble a little dog, and would, of necessity, carry that meaning. These kinds of codes are said to be self-grounded, because they carry their meaning (or some part thereof) within the symbols themselves.

---

[1]Redundancy, conspicuousness, small signal repertoires, memorability, and alerting features are commonly used to achieve this (Wiley and Richards, 1983; Guilford and Dawkins, 1991).

Much animal communication takes place via a symbolic code, since things like "wagging the tail" or elaborate dancing rituals do not in and of themselves mean anything; their meaning (if any) is fixed by mutual understanding. A dog's wagging tail could just as easily mean "I am happy" as "I am very angry" (in contrast to behavior such as displaying sharp teeth or claws, inflating or expanding to seem larger, or spraying with a physiologically noxious chemical, all of which, of necessity, mean "go away or suffer the consequences" because of their obvious and unambiguous physical meanings).

When human beings design communications systems (like computer networks, telegraphs, etc.), the engineers can agree (using a meta-language) on meanings for the various symbols of the language, so everyone can understand each other. Clearly, most animals have no opportunity for such a means of fixing referents to symbols (although Wilson, 1975, pp. 191-193, discusses several interesting examples of primate meta-communication, such as contextualization and play). Thus, the problem arises: how are the meanings for completely arbitrary symbolic gestures fixed in a large population, when no one has the opportunity to discuss their meanings with anyone else (no meta-language is available for discussion, and meanings must be assigned *de novo*)? This is also one feature of the problems which would arise on a successful SETI (Search for Extra-Terrestrial Intelligence).

Clearly, most codes used in nature have some self-grounded components (for example, a lengthy and physically-strenuous display may directly indicate the stamina and agility of a prospective mate). Information-bearing behaviors exist on a continuum between purely arbitrary-coded ones and purely self-grounded ones, and it is often difficult for observers to know where a given behavior might lie on such a scale. Ethologists are often concerned with issues of semanticization and ritualization of behaviors as they evolve into formal communication. Thus, it is interesting to ask how much understanding can be achieved within a population of organisms which is subject only to evolution (no individual learning), and which utilizes only arbitrary codes. Unlike the general field of the biology of animal communication (Sebeok, 1977; Guilford and Dawkins, 1991, etc.), very little work currently exists on this issue (see Seyfarth and Cheney, 1980, for one example).

In order to investigate in a controlled context the idea that meanings are fixed by interaction with others (as opposed to denotational theories of meaning), MacLennan (1991) showed that communication can arise when cooperation is rewarded. Werner and Dyer (1991) likewise investigated the evolution of communication in a population of artificial neural networks. Both of these approaches focused on simulating real-world interactions (i.e., simulating pursuit of mates, etc.), and thus provided some level of ecological detail.

In this study, I abstract from such detail (in the spirit of Kanevsky et al., 1991, Kaneko and Suzuki, 1994, and Balescu, 1975), and simulate a system where agents evolve under a selection which rewards mutual understanding, to study how the members of an initially noncommunicating population can all converge on particular (and identical) meanings for arbitrary symbols. Once defined, this kind of system can be studied under a variety of perturbations to yield information applicable to all classes of evolving communicators. The experiments described below were designed to answer questions such as: what are the dynamics of such a population? Is there any increase of understanding over time? If so, does the population converge on a single "dialect" or do groups form which can understand each other but not members of other groups? How much mutual understanding can be achieved by these means?

5.2. Implementation

In functionalist terms, this problem concerns a population of agents, where each agent has a set of internal states (hunger, anger, closeness to its nest or territory, strength, etc.) and a set of external observables (position of tail, posture of body, display of teeth, manipulation of external objects such as food, etc.). Only these external observables are directly perceived by other creatures. The internal states, while not observable, are what determines the future behavior of the creature.

If chaos is not to ensue (it is here assumed that a population of creatures which all understand each other does better than one where misunderstandings are the norm), each agent is driven (in an evolutionary sense) to attempt to guess or derive the internal states of whatever other agent it interacts with, by observing its external features. There is also usually pressure for each agent to make the mapping of internal states to observables as simple and direct as possible, so that other agents will be able to understand it more easily (although there are cases, such as birds acting as if they were injured to lead predators away from the nest, which exemplify misrepresentation of one's internal states). The fitness of an agent is defined as the average level of understanding of an agent's internal states by other members of the population.

The mapping of a given agent's internal states to observable states, as well as a reverse mapping which the agent uses to guess others' internal states from their observables, is defined by the agent's genome. In this model I neglect individual learning and socialization (i.e., an agent's behavior is assumed to be hardwired from birth), as well as "necessary" codings (the codes are assumed to be truly arbitrary – that all possible mappings between internal states and observables are in themselves indistinguishable with respect to fitness; all that matters in determining fitness is to be understood by other agents).

In interaction, each agent performs a coding on a string of internal state values (the "input" string). The exact nature of the coding is governed by one piece of the agent's genotype, and represents the physiologically defined mapping between a creature's internal states and what it portrays by behavior and body signals. The other agents directly observe this coded string as observables, and decode it (using the complementary piece of their genotype, representing the neural mechanisms by which creatures estimate others' intent from observed data). Thus, each organism's genome consists of two "genes," one governing how it maps its internal states for display to others, and the other which governs how it, in turn, interprets its observations of others. This scheme is illustrated in Figure 5.1.

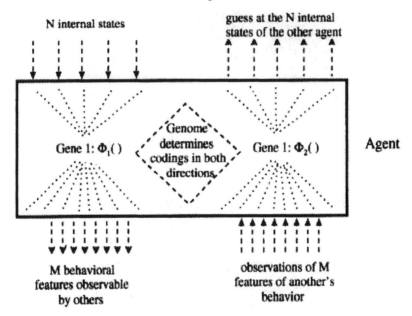

Figure 5.1 Functional diagram of a single agent

The fitness of an organism is highest when others' decodings most closely match the original string (the internal states of the agent). Specifically, the internal states and observables of each individual are represented by vectors of integers; the genomes consist of matrices bearing weights (coefficients) for polynomials which map one vector into another. Thus, for a given vector I representing some set of internal states of agent X (for example, a hungry animal who is moderately strong, and not close to its home territory), the observables vector O is obtained by $O = I\Sigma C$, where C is a matrix whose elements are contained in the genome of agent X. In an interaction, another agent observes the vector O, applies its own matrix D, and arrives at its guess as to what agent X's internal states might be ($I' = O\Sigma D$). The fitness of individual X is given as the average understanding of its internal states by others:

$$fitness(x) = \sum_{x=1, x \neq R}^{g \cdot popsize} U(x, R)$$

where R is a randomly chosen individual, popsize is the size of the population, and g is a number between 0.0 and 1.0 which indicates gregariousness (i.e., how many of the other members of the population each individual interacts with, in determining its fitness). This is important because the level of sociality varies very widely among species. Values of g close to 1.0 make this algorithm very computationally intensive because of the combinatorial nature of the fitness function (each member has to interact with every other member). For larger population sizes, this will be impractical (on a single-processor machine).

U(a,b) determines how well agent A is understood by agent B. It is defined as the average error individuals make in attempting to guess one another's internal states by applying their decoding function to the encoded vector

$$U(a,b) = \sum_{i=1}^{interactions} distance\big(M, G(F(M))\big)$$

In this expression, interactions determines how many interactions with each individual a given agent has (i.e., how many messages they exchange when computing how well they understand each other). This is important because in "dove-like" (nonviolent) species, fitness is determined over a large number of interactions (i.e., no one or few interactions determine fitness because none leads to catastrophic results). In very violent species, a single misunderstanding may lead to death, so fitness needs to be determined over a smaller number of messages. M is a random message over the space of valid internal state sets, and G(F(M)) is agent B's decoding of agent A's coding of that message. The maximum understanding occurs when the distance between them is minimal (i.e., the decoding is maximally similar to an inverse of the coding). The distance (simple Pythagorean hypervolume distance) between two vectors is computed as follows:

$$distance(M_1, M_2) = \sqrt{\sum_{i=1}^{lengthM_1} (M_1[i] - M_2[i])^2}$$

A genetic algorithm (GA, pseudocode is given in Figure 5.2) is used to simulate evolution of this system, with fitness being determined through some number of interactions (on randomly chosen sets of internal states) with some number of other (randomly chosen) members of the population. The numbers used in the algorithm are parameters which may be changed to study various properties of this evolving system.

The vector/matrix representation was chosen for this model instead of other possibilities like finite state automata and neural networks because they provided a computationally nonintensive algorithm for coding and decoding (important because of the combinatorial nature of the fitness measure), covered a large area of possible mappings (because every output element can be a function of every input element), supported mutation and crossover operators which were closed with respect to the space of legal genotypes, and provided an obvious (but not unique) optimal solution (the identity matrix $I_{n,m}$ which corresponds to the simplest mapping between inputs and outputs).

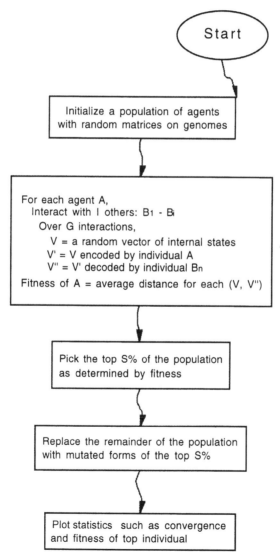

Start

Initialize a population of agents
with random matrices on genomes

For each agent A,
   Interact with I others: B_1 - B_i
   Over G interactions,
      V = a random vector of internal states
      V' = V encoded by individual A
      V'' = V' decoded by individual B_n
Fitness of A = average distance for each (V, V'')

Pick the top S% of the population
as determined by fitness

Replace the remainder of the population
with mutated forms of the top S%

Plot statistics such as convergence
and fitness of top individual

Figure 5.2 Flowchart of the algorithm used for the simulation

Note that there is a fundamental difference between this GA and usual GA applications. In the normal genetic search, each candidate solution has a fitness; this fitness is a measure of how well that solution fits a given problem and is thus independent of any other solutions which may exist at the time. In this GA simulation, however, all fitnesses are relative, since the fitness of an individual is defined by how well others understand it. This has been termed "competitive fitness" (Axelrod 1984, Axelrod, 1987), and has several important consequences: (1) there can be no true elitist selection, since the "best" individuals can easily become poor when others are mutated, and (2) there will be very complex dynamics as the population evolves. Of course, this is much closer to true biological evolution since most characteristics' fitness values are very much dependent upon the other members of the ecology. Thus, this is the logical extension of Hillis (1991) which showed that coevolving two separate populations can be beneficial, since, in this case, every single individual potentially deforms the others' landscapes.

5.3. Results

In order to study the properties of an evolving system of agents seeking to understand each other, several experiments were performed in which various key parameters of the simulation were changed. In all of these experiments, the top fitness (defined as the scaled log10 of the fitness of the most-fit individual) and the population convergence (defined as the scaled average difference of each individual's matrix from the population's matrix average) were plotted as a function of generation number.

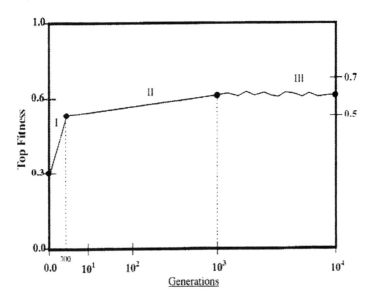

Figure 5.3 Schematic of evolution of base population runs

In the first series, the natural (unperturbed) variability of the system was explored, in order to make meaningful analyses of its behavior under alterations of its parameters. The results for the 50 preliminary runs are summarized in the schematic of Figure 5.3. This will be referred to as the "base population." In general, an evolution of this system consists of three phases, labeled with Roman numerals I through III in Figure 5.3. A sample plot of one actual run appears in Figure 5.4.

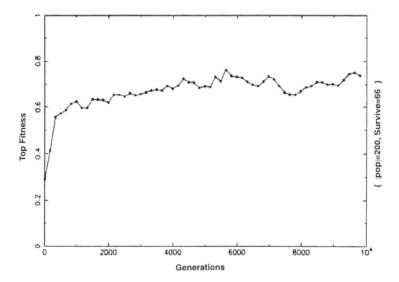

Figure 5.4 Course of evolution of a base population run

All repetitions of this experiment gave approximately the same result. The best individual of a randomly chosen population has a fitness value of about 0.3 ± 0.01. The top fitness rises sharply to a value of about 0.5 until about generation 300 (phase I), then slowly reaches a maximum of 0.6 ± 0.05 by generation 1000 (phase II), and then meanders about that value from then on (phase III). This phase is stable, and no further major increases occur; the population continues to cycle about the value of 0.6 (equivalent to a two orders of magnitude reduction of error in guessing another agent's internal state vector). The population converges quickly (at around generation 100). Thus, it is seen that a population of such agents is able to arrive at a significant although imperfect level of understanding by virtue of evolution alone. Interestingly, the understanding level is not perfect, and never becomes so, even if the evolution is carried out to 106 generations. It was also found (data not shown) that the population converges upon one coding, not sub-populations which each utilize a different "language." In all of the experiments described below, the results given represent the average of 5 runs with the same parameters (this represents a balance between getting statistics which are truly representative of the system and having a computationally feasible set of experiments).

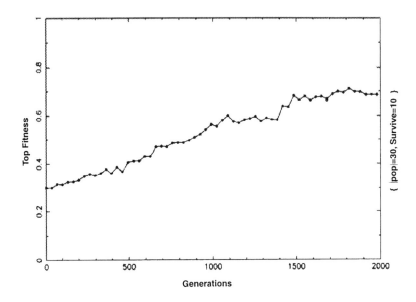

Figure 5.5 Course of evolution of a population of size 30

The next series of experiments was designed to study the affect that population size has upon the population dynamics as it evolves understanding. It is difficult to make a hypothesis as to what size is optimal, because while larger populations in GAs tend to locate solutions quicker than small ones, it may well be that it is more difficult for a large population to achieve mutual understanding (due to the larger range of individual codings available). For population sizes of 100 or more (up to 5000, data not shown), the results were all like those of the base trials summarized in Figure 5.3 (data not shown). For a population of size 30, the improvement in fitness from 0.3 to 0.6 occurred in 1500 generations, in an almost linear fashion (Figure 5.5). For a population of size 10 there was no net improvement whatsoever in 2000 generations (Figure 5.6). Thus, there is a critical population size (somewhere between 10 and 30) such that smaller populations are unable to achieve effective communication. There is also a critical population size (between 30 and 100) at which the manner in which the population converges on the maximum attainable level of understanding changes.

The next experiments were designed to test the effects of various survival rates upon the rate of the evolution of understanding. The percentage of top individuals which were allowed to survive between generations varied from 5% to 95%. It was found (data not shown) that survival rates of 5% to 60% are all equivalent in terms of the behavior of the population, and are very similar to that of the base population described in Figure 5.3. For survival rates of more than 60%, the initial rise in top fitness was slow and, on average, the population reached a fitness of only 0.5 in 3000 generations.

This implies that the evolution of understanding in animal populations is not very sensitive to the fraction of individuals which survive to breed at each generation, as long as that fraction is small enough to allow effective selection to take place. This transition appears to lie at around 60%.

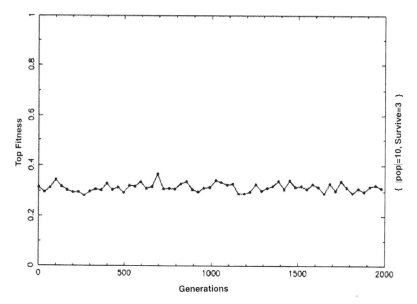

Figure 5.6 Course of evolution of a population of size 10

The next variable to be tested was mutation rate. The mutation rate is defined as the number of times a given individual is mutated (values less than 1.0 indicate a probability of being mutated). For mutation rates of 0.1 to 32, the population's behavior is not significantly different from the base population. For rates above 32, the initial rise is slow and the population requires 1500 generations to reach a fitness value of 0.6.

It was also interesting to determine the effect that crossover (rather than pure mutation) had on the population's behavior. Crossover was seen to achieve the maximum at around generation 100, and the maximum fitness achieved was somewhat higher (0.65). This is as expected, since crossover tends to lead to more rapid convergence, which here (unlike in most GA applications) is a benefit.

The next series of experiments studied the population behavior under various numbers of internal states and observables. In all experiments, the number of internal states and external observables (referred to as N) was equal. It was seen that, as expected, for smaller values of N understanding was achieved more easily than for large values. For N = 3, the population was able to achieve top fitness values of 0.7, whereas when N = 5 there was a very slow rise to a fitness value of 0.6. For N = 6, the rise was also very slow and achieved a fitness of only about 0.53 (shown in Figure 5.7).

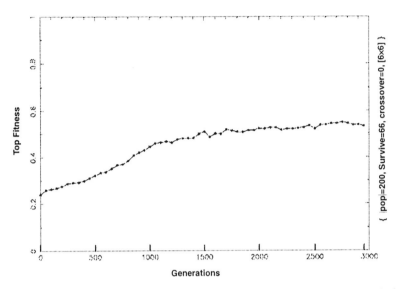

Figure 5.7 Course of evolution of agents with 6 internal states and 6 observables

The next series of experiments studied the importance of the values for gregariousness (i.e., what percent of the population each individual gets to interact with in determining its fitness) and interaction duration (i.e., how many messages are exchanged in an interaction). It was found that when all other parameters are held respectively, gregariousness levels of 0.05 – 0.9, and interaction levels of 1 – 10 all result in the same behavior as the base population. There seems to be an optimal level of gregariousness around 0.4 which gives a maximal fitness of 0.7.

The final series of experiments was designed to explore the stability of an optimal genotype (the $I_{m,n}$ matrix which corresponds to a null coding transformation) under an influx of randomly coding individuals, and the effect of an in-migration of $I_{m,n}$ individuals on a normally evolving population. In the first experiment, a population was allowed to evolve normally, and then various numbers (percentages of the population size from 0 to 90) of individuals bearing the $I_{m,n}$ genotype were artificially inserted into the population. It was seen that injection of 2% or more of $I_{m,n}$ individuals causes the entire population to achieve a fitness level of 0.7 within 50 generations of when the insertion was performed. The state of the population at the time does not matter. Likewise, a population of $I_{m,n}$ individuals is stable against an influx of 99% or fewer random individuals.

5.4. Discussion

The major finding of this series of experiments is that a significant level of understanding among units utilizing purely symbolic codes can be achieved through evolution alone. The evolution profile consists of three stages, and is very consistent between runs, suggesting that it is a real feature of such systems. Furthermore, the fitness profile of such a population as a function of time is very stable against perturbations of various parameters. Surprisingly, gregariousness level and interaction duration do not seem to have a large effect on the evolution of understanding. The same is true for a fairly wide range of selection stringencies and population size. Use of the crossover operator is seen to accelerate convergence on understanding.

The major factors influencing the rate of evolution are the number of internal states and external observables involved in the communication. The results suggest that misunderstandings should be more common in species which utilize larger numbers of signals to represent larger numbers of internal states. It is also seen that once achieved, a good genotype is very stable and as few as 2% of such individuals are able to catalyze optimal understanding among the whole population within 50 generations.

5.5. Future Directions

This chapter presented only preliminary data on this complex system. Work is currently in progress to investigate several important features of such a model. First, it is important to determine what can be said about the characteristics of the codes upon which such populations converge (such as their complexity and other measures from information theory). It would also be interesting to determine the effects of the following modifications on the rate of convergence: (1) making some portion of the code non-arbitrary (i.e., some mappings of inputs to outputs are physiologically constrained), (2) rewarding simplicity of genome (parsimony) along with understanding, (3) providing loci on the chromosome (meta-GA) which control GA parameters (such as locations of mutation hotspots, whether an individual uses crossover or mutation, the value for gregariousness, etc.), and (4) keeping a constant ratio of observables to internal states.

Likewise, it is possible to determine whether the system's self-organizing behavior is robust enough to be able to handle additional uncontrollable or very noisy outputs (which simulate external environmental factors unrelated to the internal state of an agent). Finally, a more complex form of this model is also planned which will utilize steady state GAs, as opposed to discrete generational GAs, and include nongenetic (cultural) information storage, as well as the ability to misrepresent one's internal state in certain circumstances (lying, as in Dawkins and Krebs, 1978; Dawkins, 1982), and other complexities such as eavesdropping and withholding information.

Acknowledgments

I would like to thank Ginger and Ivan Levin for many helpful ideas and discussions.

References

Alcock, J., 1989, Animal Behavior (Sinauer Associates, Massachusetts)

Axelrod, R., 1984, The Evolution of Cooperation (Basic Books, New York)

Axelrod, R., 1987, Evolution of Strategies in the Iterated Prisoner's Dilemma, in: Genetic Algorithms and Simulated Annealing, L. Davis (Ed.) (Morgan Kaufmann, New York)

Balescu, R., 1975, Equilibrium and Nonequilibrium Statistical Mechanics, (Wiley, New York)

Dawkins, R. and Krebs, J. R., 1978, Animal signals: information or manipulation, in: *Behavioral Ecology*, J. R. Krebs and N. B. Davies, (eds.), (Blackwell Scientific Publications, Oxford) pp. 282-309

Dawkins, R., 1982, The Extended Phenotype, (Oxford University Press, New York) pp. 58-66

Guilford, and R. Dawkins, 1991, Receiver psychology and the evolution of animal signals. *Animal Behaviour*, 42: 1-14

Halliday, T. R. and Slater, P. J. B., (Eds.), 1983, Animal Behaviour, vol. 2, (W. H. Freeman & Co., New York)

Hillis, W. D., 1991, Co-evolving parasites improve simulated evolution as an optimization procedure, in: Artificial Life II, C. G. Langton, C. Taylor, J. D. Farmer, and S. Rasmussen, (Eds.), (Addison-Wesley, Massachusetts) pp. 313-324

Kaneko, K.. and Junji, S., 1994, Evolution to the edge of chaos in an imitation game, in: Artificial Life III, C. G. Langton, (Ed.), (Addison-Wesley, Massachusetts)

Kanevsky, V., Garcia, A. and Naroditsky, V., 1991, Consensus in small and large audiences, in: Modeling Complex Phenomena, Lam, L. and Naroditsky, V. (Eds.), (Springer-Verlag, New York)

MacLennan, B., 1991, Synthetic Ethology, in: Artificial Life II, C. G. Langton, C. Taylor, J. D. Farmer and S. Rasmussen, (Eds.), (Addison-Wesley, Massachusetts)

Rapoport, A. and Chammah, A. M., 1965, Prisoner's Dilemma: a Study in Conflict and Cooperation (University of Michigan Press, Ann Arbor)

Sebeok, T. A., 1977, How Animals Communicate, (Indiana University Press, Bloomington)

Seyfarth, R. M. and Cheney, D. L., 1980, The ontogeny of velvet monkey alarm calling behavior. Zeitschrift fur Tierpsychologie 54: 37-56

Werner, G. M. and Dyer, M. G., 1991, Evolution of Communication in Artificial Organisms, in: Artificial Life II, C. G. Langton, C. Taylor, J. D. Farmer, and S. Rasmussen, (Eds.), (Addison-Wesley, Massachusetts)

Wiley, R. H. and Richards, D. G., 1983, Adaptations for acoustic communication in birds, in: Ecology and Evolution of Acoustic Communication in Birds, Kroodsma, D. E. and Miller, E. H., (Eds.), (Academic Press, New York)

Wilson, E. O., 1971, The Insect Societies, (Harvard University Press, Massachusetts)

Wilson, E. O., 1975, Sociobiology: the New Synthesis, (Harvard University Press, Massachusetts)

Chapter 6 Algorithms to Improve the Convergence of a Genetic Algorithm with a Finite State Machine Genome

Natalie Hammerman and Robert Goldberg
Computer Science Department
Graduate Center and Queens College of CUNY

email: hammer@pacevm.dac.pace.edu,

goldberg@qcunix1.acc.qc.edu

Abstract: Strategies for solving different types of problems can be represented as a finite state machine (FSM). For such problems, finite state machines are being used as the genotype (operand) for genetic algorithms (GAs) in artificial life and artificial intelligence research. Algorithms which are FSM-specific and which were designed to improve the convergence of the genetic algorithm for FSM genomes are presented. Because a single finite state machine has different representations (simply by changing state names), two reorganization operators (named SFS and MTF) were developed so identical machines would appear the same and not have to compete against each other for their share of the next generation. The algorithms were designed with the intent of enhancing schemata growth for an FSM genome by reorganizing a population of these machines during run time. Experiments were performed with these new operators in order to determine how they would affect the convergence of genetic algorithms.

Keywords: Ant Trail, Competition, Finite State Machines, Genetic Algorithms, Reorganization

6.1. Introduction to the Genetic Algorithm

Genetic Algorithms (GAs) are modeled after Darwinian evolution. They are used for two purposes: 1) to study the evolutionary process, and 2) as a search procedure. In the former case, scientists focus on the dynamics of the process and the properties of the evolving population(s). In the latter case, researchers simulate, utilize, and modify a process which has been operating in the natural world.

The operand for a genetic algorithm (GA) is generally a bitstring. The bitstring, void of meaning, is called the *genotype* or *genome*. The *phenotype* is the functionality or expression of the bitstring. When researchers decide to use a genetic algorithm to find a solution to a problem, they must determine the information which is needed in the phenotype. Then the phenotype-genotype mapping must be designed.

Each functional piece of information, referred to as a field in the genome, is represented as a bitstring. The bits of each field are then assigned fixed locations along the genome. A *fitness function* must then be defined to evaluate the ability of each phenotype to perform the required task. The mapping of the bits of each field into the genome can greatly affect the convergence of the genetic algorithm.

This section introduces a simple genetic algorithm (Section 6.1.1) and reviews variations of this outline which have appeared in the literature (Section 6.1.2); this is summarized in Section 6.1.3.

6.1.1 A Simple Genetic Algorithm

The genetic algorithm treats bitstrings (or other data structures) as genomes (chromosomes). A GA is generally started with a population of randomly generated bitstrings (genomes, individuals). Each genotype (bitstring) has an associated phenotype, which is the interpretation or functionality of the bitstring. Using the phenotype, each individual is assigned a fitness value using some predefined function or procedure. Operators, generally biologically based, such as fitness-based selection, reproduction (cloning), crossover (mating), and mutation are applied to a population of bitstrings in order to breed a new population. Selection determines which bitstrings will be copied into and used as the parents of the next generation. Reproduction copies (clones) the selected individuals to parent the next generation. Crossover is a mating operation in which genetic material (parts of two bitstrings) is exchanged between two parents to form the genome of the offspring. Finally, mutation arbitrarily alters the genotype, generally by changing the value of a bit. After mutation, the new generation is in its final form (Goldberg 1989).

The following is an outline of a typical genetic algorithm.

1) Create the initial generation as the present generation.

2) Find the fitness of each member of the present generation using a predefined function or procedure.

3) Form the initial phase of the next generation by selecting mating pairs and copying them into the next generation.

4) Generate the intermediate phase of the next generation from the initial phase by exchanging genetic material (performing a crossover operation) between the individuals of each mating pair.

5) Apply mutation to the intermediate population to get the final phase of the next generation. At this point, this becomes the present generation. New operators will be introduced in Sections 6.3 and 6.4 which will reorganize the genomes before the newly formed population becomes the present generation.

6) Repeat from step 2 until the criteria for termination is satisfied.

In steps 3, 4, and 5, the operations are not deterministically implemented; they are implemented with respect to some probability density function. For example, individuals are selected for reproduction with a probability proportional to their fitness; the more-fit individuals have a correspondingly higher probability of being selected. Mutation of the genomes occurs with some fixed probability. For example, if the mutation rate is 0.5% per bit, and the population contains 100 individuals with 50 bits per individual, one would expect .005 x 50 x 100 = 25 bits to be changed. Because the genetic algorithm is implemented as a stochastic (probabilistic, nondeterministic) model, there may be more or less than 25 bits altered. The following very simple example will clarify this.

Consider a problem with the genotype consisting of six bits. The phenotype views the bitstring as a binary integer, and the value of this integer is its fitness. For example, the genotype 101100 has a phenotype of 101100_2, that is, the same six bits are viewed as a binary integer. Its fitness is 44, the decimal equivalent of 101100_2. The population contains six individuals, and mutation is applied with a probability of 10% per bit. This problem has been chosen to illustrate the operation of a genetic algorithm because its solution is obvious, and because it is easy to observe the GA's progress towards the solution.

To start the genetic algorithm for this problem, randomly generate 36 bits to form the initial population of six individuals consisting of six bits each. A run generated the results shown in Table 6.1. The maximum fitness for this population is 52 and the average fitness is 34.67.

Generation 0

Individual #	0	1	2	3	4	5
Genotype (bitstring)	101100	000011	110100	100101	100010	100110
Phenotype (binary integer)	101100_2	000011_2	110100_2	100101_2	100010_2	100110_2
Fitness	44	3	52	37	34	38
Fitness: partial sums	44	44 + 3 = 47	47 + 52 = 99	99 + 37 = 136	136 + 34 = 170	170 + 38 = 208
Fitness range for selection	1-44	45-47	48-99	100-136	137-170	171- 208

Table 6.1 Initial Population

The creation of the *next* generation occurs in three steps:

a) Using fitness-based selection, choose and copy the parents into the initial phase of the next generation.

b) Form an intermediate population by pairing the parents of the initial phase into mating pairs. The crossover operator exchanges genetic material (bits) between the parents in each mating pair producing two offspring from each mating pair.

c) Apply mutation to the intermediate population to get the final population for the next generation.

To get a fitness-based selection for generation 1 as indicated in step 1 above, determine the partial sums of the fitnesses as indicated in Table 6.1. This is used to define a fitness range for each individual; that is, a unique range of integers is associated with each individual. The individual fitness ranges are contiguous. This defines a total fitness range with each integer in the range corresponding to an individual in the population and with all individuals in the population included in the range for the selection process. The individual fitness ranges reflect the relative fitness of the individuals with more fit individuals having a correspondingly larger part of the total fitness range. If, by chance, the genotype 000000 appears, there is no possibility that it will be selected since it has a fitness of 0.

Generation 1: Initial phase

Individual # in Generation 1	0	1	2	3	4	5
Random # for Selection	71	168	103	25	197	37
# of individual selected from generation 0 of Table 6.1	2	4	3	0	5	0
Corresponding genotype from Table 6.1	110100	100010	100101	101100	100110	101100

Table 6.2 First Generation after Selection

As indicated in Table 6.1, the complete fitness range for generation 0 is 1 to 208. Since six individuals are needed to parent the next generation, generate six random integers between 1 and 208 in order to implement the fitness-based selection. Each of the six numbers is used to select an individual to be duplicated (reproduced) into the first phase of the new generation. For this example, an integer between 1 and 44 selects individual #0, between 45 and 47 chooses individual #1, and similarly across the last row of Table 6.1. The result of this selection process appears in Table 6.2. It is interesting to note that in generating the initial phase of generation 1, the least-fit individual from generation 0 (#1 from Table 6.1) was not selected, and one of the more-fit individuals (#0 from Table 6.1) was selected twice. As a result, the most significant bit of all individuals is now one. This is quite advantageous as this is what the most significant bit will have to be to maximize the fitness; the selection process has started to move the population towards the optimal solution.

To generate the second phase for this population, the individuals from the initial phase of generation 1 are paired for mating in the order in which they were selected; that is individuals 0 and 1, 2 and 3, and 4 and 5 from Table

6.2 are paired, with each pair ultimately producing two children. Three random integers are then generated for crossover points, one number for each mating pair. These integers indicate the number of bits to retain from each individual in the corresponding pair; the remainder of the strings are exchanged. To insure that some genetic material is exchanged, the integers used for crossover must result in the retention of at least the first bit of the parents, and the exchange of at least the last bit of the parents. A 1 will retain the first bit, and a number which is one less than the length of the genome will exchange the last bit. Since the genome in this example consists of six bits, the integers generated must be between 1 and 5 inclusive. The numbers generated for the mating process are 3, 3, and 5. The crossover (mating, exchange of genetic material) is carried out in Table 6.3. The space in each pair of bitstrings indicates the crossover point. The bits after the crossover point are then exchanged. Note that numbers 4 and 5 in Table 6.3 were not altered by the crossover.

Generation 1: Intermediate phase

Pair	Individual #	Before Crossover	After Crossover
1	#0	110 100	110 010
	#1	100 010	100 100
2	#2	100 101	100 100
	#3	101 100	101 101
3	#4	100 110	101 110
	#5	101 100	100 100

Table 6.3 First Generation after Crossover

To apply mutation to this intermediate population, generate 36 random numbers between 0 and 1 with each number corresponding, respectively, to each of the 36 bits in this intermediate population. To implement the mutation rate of 10%, change the value of a bit if the corresponding random number is less than 0.1. With 36 bits and a 10% mutation rate, it is expected that (0.1 x 36) three or four bits would be mutated. Applying mutation produced the results indicated in Table 6.4. Note that four bits are mutated – those in boldface and underlined. The line labeled "After mutation" is the final population for generation 1. It has a maximum fitness of 60 and an average fitness of 44.83. In one generation (Tables 6.1 and 6.4), this simple genetic algorithm has come close to finding the optimal individual and the population as a whole has shown a significant increase in fitness.

Generation 1: Final phase

Individual #	0	1	2	3	4	5
Before mutation	110010	100100	100100	101101	100110	101100
After mutation	110**1**10	100100	100100	1**1**1100	100110	10110**1**
Fitness after mutation	54	36	36	60	38	45

Table 6.4 First Generation after Mutation

While in a single generation the results need not be so impressive (see Figure 6.1a, generations 2 and 3), over many generations the trend is towards increasing maximum and average fitness. The table summarizes the results of a sample short run for this problem.

Generation	Fitness for individual number						Maximum fitness	Average fitness
	0	1	2	3	4	5		
0	39	50	41	50	19	23	50	37.00
1	50	18	19	47	51	42	51	37.83
2	47	19	43	62	55	41	62	44.50
3	23	1	15	31	39	55	55	27.33
4	55	55	55	38	7	55	55	44.17
5	23	55	47	54	50	39	55	44.67
6	51	15	51	46	34	59	59	42.67
7	34	33	59	59	59	59	59	50.50
8	59	3	59	43	58	59	59	46.83
9	63	59	59	59	8	56	63	50.67
10	63	59	59	8	63	56	63	51.33

Figure 6.1a. Table of ten generations of sample run.

Note that while both the maximum fitness and average fitness do decrease at times (Figure 6.1b), the general trend is increasing. This genetic algorithm was executed for 10 generations. It found and retained the optimal individual, but this is not necessarily the case in general.

A genetic algorithm will not necessarily locate an individual with the optimal fitness. In such cases, it will generally find an individual with a fitness which is close to the optimal value and acceptable for the problem at hand. For example, a GA may be used to determine the structure and initial weights of a neural network; then a learning algorithm can be used to finalize the weights (Belew et al. 1992). There are problems (such as the traveling salesman problem) for which the optimal solution cannot be found without an extensive search using an algorithm of unacceptable complexity and for which the fitness value of the optimal individual in a genetic algorithm is not known (Nilsson 1980, Goldberg 1989).

For applications in which a near optimal solution is acceptable, a genetic algorithm can be used. For example, Hillis (1992) used a genetic algorithm to try to find a minimal sorting network to sort a list of 16 items. After several attempts which included modifications of his genetic algorithm, a 15 minute run on his connection machine resulted in a 62 comparison network. A subsequent run reduced the number of comparisons by one. Levy (1992) documents that in 1969 a 60 comparison network was published.

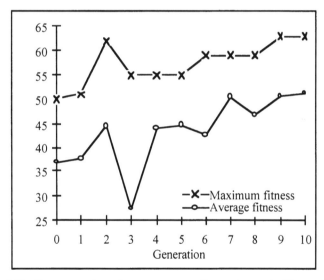

Figure 6.1b. Corresponding chart summary of sample run in Figure 6.1a.

It is obvious that Hillis' modified GA found a suboptimal solution because the 60 comparison network was published, but, in turn, the 60 comparison network could also be suboptimal. Perhaps new techniques can be injected into Hillis' last GA, techniques which might force the GA to explore the genome space in a different way and which can locate a better solution. Some variations on the simple genetic algorithm are presented in the next section. This section focuses on the different stages of the genetic algorithm and surveys from the literature different ways of defining the operation or method at hand. This is important for understanding the theory and adaptation of the simple genetic algorithm to the domain of finite state machines and, in particular, for training robotic ants to follow a trail (Section 6.2).

6.1.2 Variations on the Simple Genetic Algorithm

There are many variations of the simple genetic algorithm. Hillis' (1992) GA, discussed in the previous section, incorporated one possible variation. One of Hillis' modifications was the installation of a second population or species which challenged his sorting networks. This produced a changing environment for the sorting networks to master. The fitness was not aiming for a fixed value, but was determined by how well an individual fared against a changing competitor. This exemplifies an alternate way to define fitness. It is significantly different from the explicitly defined fitness function used in the example developed in the beginning of the previous section. In that example the population was reaching for a fixed target. Different ways of defining fitness are described in the next Section 6.1.2.1.

The example developed in the beginning of the previous section demonstrates one way to implement the biologically-based operators of selection, crossover, and mutation. Various implementations, in fact, appear in the literature. Other operators and numerous variations of the biologically-based operators have also been used. Section 6.1.2.2 presents the alternatives for selection and Section 6.1.2.3 for the rest of the operations. Although the above subsections technically discuss the operations involved in a genetic algorithm, Sections 6.1.2.4 - 6.1.2.7 deal with the practical question of improving the convergence of a genetic algorithm to the desired solution. This requires a brief discussion of the theory of schema (Section 6.1.2.4) with a worked out example (Section 6.1.2.5). This allows for a description of the fitness landscape of the search space of possible solutions (Section 6.1.2.6). Based on this, in Section 6.1.2.7, competition is presented as a technique for preventing premature convergence. This will be utilized in the algorithms of this chapter (Section 6.5). Section 6.1.2 concludes appropriately with the possible criteria of termination for a genetic algorithm. Each section relates the material and concepts introduced to that of the problem considered in this chapter – that of training a robotic ant to follow a trail.

The next section presents different ways to define fitness. In the sections which follow, variations of the basic operators are introduced in addition to other operators, some of which are not biologically based.

6.1.2.1 Variations on Fitness

There are many ways to define a fitness function. In the example presented in the beginning of the previous section, the fitness function was defined explicitly yielding a single numerical value for the fitness, a value which is a constant for an individual. In another example of explicitly defined fitness, the bitstring is interpreted as an algorithm and the fitness function quantifies an individual's ability to perform some task, for example, to traverse a trail (Jefferson et al. 1992, Koza 1992).

Alternately, the fitness can define an individual's relationship to a changing environment. In this case the individual's fitness varies with the environment. One way to implement this is to hold a competition between the individuals in a population. For example, Reynolds (1994) interprets a bitstring as a strategy to catch and evade an opponent when playing the game of tag. Individuals compete against each other and the fitness measures the percent of the time that an individual avoids being the pursuer during the game. Several researchers, for example, Axelrod (1987) and Fogel (1991), experimented with the iterated prisoner's dilemma. In this problem, the fitness is based on different scores for mutual cooperation, mutual defection, and a cooperation-defection situation when different strategies in the evolving population play against each other. In these two different applications, the fitness measures an individual's relationship to its changing environment, where the environment consists of the changing population.

Another way to create a changing environment is by breeding opposing populations. As previously mentioned, Hillis (1992) bred a population of sorting algorithms. After initial testing he installed an opposing population (species) of permutations of numbers which were to be sorted. In this case each species created the environment for the other. Each member of each population faced a fixed number of competitions against individuals of the opposing population. The fitness of each sorting network was the number of competing permutations which it could sort, and the fitness for each permutation was the number of sorting networks which could not sort it. Hillis noted that each population evolved to exploit the weaknesses of, and avoid the strengths of the other. This method, which is similar to a predator-prey situation in nature, made it unnecessary to test each sorting network on the 16! permutations of 16 items in order to determine the fitness of each network.

Another variation for fitness is presented in *Tierra*, in which a genetic algorithm is applied to self-replicating programs. The individuals are placed in *Tierra's* Reaper queue; those individuals at the front of this queue, generally the older ones and those unable to self-replicate, are eliminated when memory becomes too crowded. In this system, fitness is defined by the Reaper. Ray, who is a tropical ecologist, documented the development of an ecological system by the computer similar to that found in nature. (Ray 1992).

Evolutionary computation (including GA) is used for two different types of research: 1) to study the changing properties of an evolving population, and 2) as a search algorithm. The difference between these two forms of research is important because the purpose of the fitness function is different. Ray (1992) investigated the emergent properties of a population during evolution. Jefferson et al. (1992) and Koza (1992) were looking for a solution to a problem. Hillis (1992) was interested in observing the evolutionary process and used the GA as a search procedure to do so (Levy 1992). When a researcher's main purpose is to study the changing properties of an evolving population, an environment defines the individual's fitness; changing the environment during execution changes the optimal value for the fitness function during the run. A changing environment can also be used when a GA is used as a search procedure.

While variations in defining the fitness function are somewhat dependent on the problem, variations in the selection process are not. Some alternatives in implementing the selection process are reviewed in the next section.

6.1.2.2 Variations on the Selection Process

In a genetic algorithm a newly bred population may completely or partially replace the old one when forming each subsequent generation. In the example presented in Section 6.1.1, the fitness-based selection used is called the dart board method. An individual could be selected multiple times, as shown in Table 6.2; in fact, the more-fit individuals generally are selected more frequently (Goldberg 1989). In an alternative to the dart board method, the population is sorted by fitness and only the most fit (for example, the top 5%) are permitted to reproduce. Once this group is chosen, each individual in the group has an equal chance of being selected to parent each child (Jefferson et al. 1992). If an optimal individual is found by a genetic algorithm using either of these selection algorithms, it can be lost since the whole population is replaced by each subsequent generation. If an individual seems to appear in two adjacent generations, it is not due to the survival of the individual, but rather to the development of an identical one.

Other selection methods copy one or more of the most-fit individuals into the next generation and these clones are not subject to further operations. In these selection processes, the bitstrings are ordered by fitness and the least fit is/are replaced. In the *genitor* method, only the single, least fit individual is replaced, one at a time (Paredis 1994). In the *elitist* method, only the most-fit individual is cloned into the next generation (Goldberg 1989). Between these two extremes of replacing only one or all but one individual at a time, the selection process can replace some part, for example, the bottom half, of each generation. With these three partial replacement methods, an optimal individual will not be lost since the most fit from each generation is/are carried over into the new one. This produces *overlapping* generations, that is, generations which share identical genomes because the genomes have been cloned into the next generation and have not been subjected to any other operations. With the retention of the most fit in a population, there is generally a more consistent climb of the average fitness of the population.

This section presented variations in the selection process. In the next section, new operators and variations on the biologically based operators of crossover and mutation are discussed.

6.1.2.3   Crossover   and   Mutation   Variations   and   Other Operators

The literature indicates a wide variety of selection methods in use. It also showcases variations on the other two biologically based operators, in addition to operators which are not biologically based.

Jefferson et al. (1992) breed finite state machines (FSMs) and implement a variation of the crossover operator presented in Section 6.1.1. Instead of selecting a single crossover point, they apply a 1% crossover rate per bit. Two parents are selected with one of the two chosen as the bit donor. After each bit is copied from the genome of the donating parent into the child's genome, a random number is generated. If the random number is less than 0.01, the other parent becomes the donor. When the random number generator again produces a number less than 0.01, the bit donor is again changed. Essentially, each random number less than 0.01 produces a crossover point. With this crossover algorithm, zero to several crossover points can result and only a single child is produced from the mating.

The basic operators used in genetic algorithms are fitness-based selection, reproduction, crossover, and mutation. Human ingenuity and the use of phenotypes with different structures (for example, trees, FSMs) have resulted in redefining the basic operators and the development of new operators. Levinson (1994) experimented with the conversion operator. Gene conversion is "defined as replacement of a random segment of random string *a* by a corresponding segment from random string *b*, without altering string *b*" (Levinson 1994). For example, the middle three bits of 11100 can replace the center three bits of 00110, replacing the latter individual with 01100 – but unlike with crossover the 11100 remains unaltered in the population.

Koza (1992) defined crossover and mutation so as to maintain the integrity of his data structure. In his genetic programming paradigm (GPP), Koza's phenotype is an s-expression, a data structure which is easily destroyed by previously defined crossover and mutation operators. Ignoring the structure of the s-expressions, treating them as linear bitstrings, and arbitrarily exchanging the tails or arbitrary parts of such bitstrings would frequently result in an invalid s-expression; that is, one with unmatched parentheses. Similarly, changing the value of a bit might create an invalid operator or undefined operand within the s-expression. To prevent breeding from yielding invalid s-expressions, in Koza's GPP s-expressions are viewed as trees with crossover exchanging subtrees between the parents, and mutation replacing a subtree with a new, randomly generated one.

Angeline and Pollack (1993) define a compression operator which freezes part of a genome and prevents that part from being altered by mutation. Their expand operator "releases a portion of the compressed components so they can once again be" subject to alteration. These operators are defined for, and applied to, finite state machines and s-expressions. The researchers approach several problems and report that the runs that utilize their compress and expand operators resulted in the breeding and evaluation of fewer generations to find solutions to problems than comparable runs which did not incorporate their innovative operators. The faster convergence to a solution, which resulted when part of the genomes were protected from alteration, indicates the importance of retaining those parts of a genome which cause an individual to be more fit. Angeline and Pollack created their new operators with the expectation of enhancing the performance of the evolutionary algorithm based on the underlying theoretical concepts explaining why evolutionary computation succeeds as a search procedure. Similarly, the reorganization operators, which are presented in this document and which are fully described in sections 6.3 and 6.4, were designed with the same goals in mind. The underlying concepts explaining the success of the genetic algorithm as a search procedure are presented in the next section.

6.1.2.4 Schema: Definition and Theory

The success of the genetic algorithm as a search procedure is based on the formation and proliferation of useful building blocks called *schemata*. This section covers general schema theory. The next section presents an example which clearly demonstrates the theory.

A *schema* is a template which indicates the similarities in strings. The alphabet for schemata for bitstrings is {0, 1, *}. A bitstring matches a schema if it contains a zero in each position that the schema has a zero, and a one in each position that the schema has a one. An asterisk in a schema acts as a place holder and does not identify the bit required in the given position of a matching bitstring; that is, a bitstring will match a schema regardless of whether it contains a one or zero in the schema position containing an asterisk. For example, 10011 and 11010 have a maximum match represented by the schema 1*01*. This schema matches four bitstrings – 10010 and 11011 in addition to the two already given. Replacing any zero or one in this schema with an asterisk will result in a schema which still matches all four individuals, but since each of the asterisks can match a zero or one in a bitstring, the schema with three asterisks will match a total of 2 x 2 x 2 = 8 bitstrings.

The *fixed* positions of a schema are those positions containing a zero or one. The *defining length* of a schema is the difference between the first and last positions containing a symbol other than an asterisk, or, alternatively the number of crossover points between the first and last fixed bits of the schema. It is the building of useful schemata with increasing length within a population that moves the population towards a solution (Goldberg 1989). Looking at Figure 6.1, in generation 0, two individuals match 11**** and none match 111***. By generation 2, one individual matches 111***.

The defining length of a schema affects the chance of its loss due to crossover. Let crossover exchange the tails (consisting of a randomly chosen number of bits) of two bitstrings. If the schema ab*** (a,b ∈ {0,1}) matches individuals with high fitness, there is a one in four chance that a crossover point will come between these two fixed bits of an individual bitstring which matches it. On the other hand, if a***b (a,b ∈ {0,1}) matches highly fit individuals, any crossover on an individual matching this schema will separate these two fixed bits. When two individuals matching a schema are crossed, their two children will match the schema. For example, 111010 and 110111 match 11**1*; any crossover point on these two bitstrings will create two individuals which still match this schema.

Useful schemata are those which match highly fit individuals and which give rise to the solution to the problem at hand. It is the growth of both the number and defining length of useful schemata within a population which drives the population as a whole toward higher fitness. Looking at Figure 6.1, four individuals in generation 0 match 1*****, and in generation 2, five individuals match it. 11**** is another useful schema for this problem. In generation 0, two individuals match it. Only one bitstring matches it in generation 3 when both the maximum and average fitness decrease. This increases to four matches in generation 4 when the maximum fitness retains its value and the average fitness surges back almost to its prior level. Comparing the populations in Tables 1 and 4, five individuals in generation 0 match 1*****; by generation 1, all six match it. As pointed out earlier, this is beneficial as the final solution must match 1*****.

The theory explaining the success of the genetic algorithm as a search procedure focuses on schemata growth – not on the individual bitstrings which are the operands of the GA. A genetic algorithm, while breeding individuals, processes many more schemata (Goldberg 1989). For example for a seven-bit genome the genome space consists of 2^7 individuals which match 3^7 schemata. A single seven-bit genome matches 2^7 schemata as each position in a schema can contain either the bit which fills the position in the genome or an asterisk. The theory underlying the progression of a population towards a solution to a problem is based on determining the expected growth in the number of useful schemata during the execution of a genetic algorithm.

In general, let f_S = the fitness of an individual which matches schema S, $E(S,g+1)$ = the expected number of matches to schema S in generation g+1, $N(S,g)$ = the number of matches to schema S in generation g, and f_i = the fitness of the ith individual in generation g. Let $f_{S:AV}$ = the average fitness of members of generation g which match S $= \dfrac{\sum f_{Sj}}{N(S,g)}$, where f_{Sj} = the sum of the fitnesses of those individuals in the population which match S.

Let f_{AV} = the average fitness of generation g = $\dfrac{\sum f_i}{n}$, where and f_i = the sum of the fitnesses over the n members of the population. Let d = the number of crossover points between the first and last fixed positions in schema S, and c = the number of crossover points in the genome; c is equal to one less than the length of the genome. Let b = the number of fixed positions in S and p_m = the probability of mutation of each bit. P_m is generally very small; usually $p_m \prec\prec 1$; frequently it is less than 0.01. Then, the expected number of matches to schema S in generation g+1 can be modeled as follows:

$$E(S, g+1) \succeq N(S,g)\, \frac{f_{S:AV}}{f_{AV}}\left(1 - \frac{d}{c} - bp_m\right) \qquad \text{(Equation 1)}$$

Therefore, it is expected that the number of matches to S should increase when S has above average fitness, short defining length, and a small number of fixed positions (Goldberg 1989). Note that when every member of a population matches S, there will be no losses due to crossover.

Based on this theory, it is the schemata, rather than the individuals in the population, which are responsible for moving the population toward a solution. It is the schemata which actually do the work in the genetic algorithm. The reorganization algorithms which are described in sections 6.3 and 6.4 were designed to enhance schemata growth. One of these was designed to reduce the defining length of useful schemata. Because a short defining length is desirable, a GA using a genome in which the most significant bits are in close proximity should do better than one in which the most significant bits are not adjacent to each other. In the next section an example is presented to demonstrate this point. This will be the basis of the Move To Front reorganization operator described in Section 6.4.

6.1.2.5 Schemata at Work: an Example

A simple example demonstrates the effect of the organization of the genome on a genetic algorithm in light of schema theory. Moving each run towards a solution is the growth of the useful schemata in each successive population. In this experiment the phenotype consists of two non-negative integers, x and y. x contains three bits; y has four. Therefore, the genotype needs seven bits. The fitness function is $f(x,y) = x^2 - y + 17$. Clearly the maximum value f can attain is 66 when $x = 111_2 = 7_{10}$ and $y = 0000_2 = 0_{10}$. One way this differs from the example developed in the beginning of Section 6.1.1 is that the optimum individual contains a mixture of zeroes and ones. As in that previous example, the population contains six individuals and the mutation rate is 10%. For this example, the maximum and average fitnesses for each generation are averaged over ten runs.

This experiment was run with two different phenotypes. Let $x = x_2x_1x_0$ and $y = y_3y_2y_1y_0$ where $x_i,y_i \in \{0,1\}$. In Case I, the seven bits of the genotype represent $y_3x_2y_2x_1y_1x_0y_0$. In Case II, they represent $y_3y_2y_1y_0x_0x_1x_2$. In the first case, the most significant bits are close to each other; in the latter case, they are as far apart as the genome permits. Because a short defining length is desirable, a GA using a genome in which the most significant bits are in close proximity should do better than one in which the most significant bits are not adjacent to each other. For Case I, $d/_c = 1/_6$ for schema $01*****$; that is, when the most significant bits of x and y contain the bits needed for the solution. For Case II, the schema for the same condition is $0*****1$, and $d/_c = 1$ indicating that these two important bits are certain to be separated by the crossover operation.

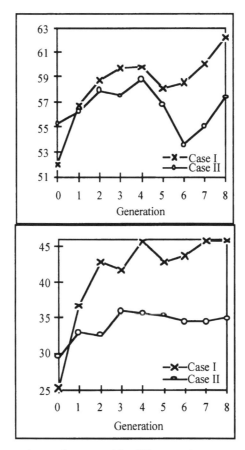

Figure 6.2 Comparison of runs with different phenotypes. Case I uses phenotype $y_3x_2y_2x_1y_1x_0y_0$. Case II uses phenotype $y_3y_2y_1y_0x_0x_1x_2$. Fitnesses are averaged over 10 runs.

Looking at Figure 6.2, note that even though the first case starts with a lower maximum and average fitness than the second case, its performance surpasses that of the second case. Variations will occur, most notably in the maximum, but the average is used to determine the progress of the system and the difference there is clearly noticeable. The proximity of the significant bits in Case I aids the growth of useful schemata, moving the population as a whole more quickly towards the solution. When the bits which were more important to the solution were placed together in the genome, useful schemata had a shorter defining length, were less likely to be lost due to crossover, and were therefore able to overtake the population faster as indicated by the big difference in the average fitness.

But, this is considered a simple problem for a genetic algorithm. If the fitness of each individual in the genome space for this problem were graphed outward from a seven-dimensional hypercube, the optimum fitness would be at the peak of a single hill. The GA used as a search procedure tries to locate this single peak. The significance of this is pursued in the next section.

6.1.2.6 Fitness Landscape

The fitness landscape for a bitstring genome space graphically shows the fitness of that space. This is useful in determining the difficulty of a problem for a GA. The landscape is formed by plotting the fitness of each n-bit individual in the space outward or upward from an n-dimensional hypercube, cube, or square. Each corner of these n-dimensional figures is adjacent to n other corners. The corners are labeled so that the labels for adjacent corners differ by one bit.

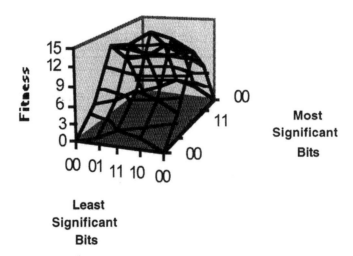

Figure 6.3 Fitness landscape for a four-bit genome with phenotype x = a binary integer and fitness f(x) = x.

Fitness landscape examples appear in Figures 6.3 and 6.4. For Figure 6.3 refer back to the example developed in Section 6.1.1. Reduce the genotype to four bits. Retain the phenotype as the binary integer and the fitness as its decimal equivalent. Figure 6.3 contains the graphical representation of the fitness landscape for this smaller problem. Similarly, reducing the problem developed in Section 6.1.2.5 to a four-bit genome, let $x = Xx$ and $y = Yy$, $(x,y,X,Y \in \{0, 1\})$, and retain the fitness function $f(x,y) = x^2 - y + 17$. Figure 6.4 contains the fitness landscapes for the two different phenotypes utilized in the example in Section 6.1.2.5.

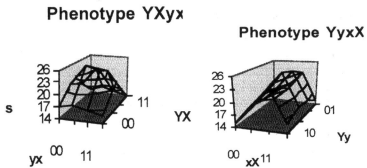

Phenotype YXyx

Phenotype YyxX

Figure 6.4 Fitness landscape for a four-bit genotype with two different phenotypes. $x = Xx$, $y = Yy$ where $x,y,X,Y \in \{0,1\}$, and $f(x,y) = x^2 - y + 17$

Utilizing this imagery, the most GA-friendly landscape consists of a single peak or plateau with hills sloping down to a single valley. The examples developed in Sections 6.1.1 and 6.1.2.5 are instances of this as Figures 6.3 and 6.4 indicate. At the other extreme are the GA-hard landscapes. These are rough and rugged, and have many peaks at different heights. Whereas in a GA-friendly landscape the evolving population of genetic algorithm can climb the hill to the single peak, in the GA-hard case, the population must be able to hop from one hill to another to avoid being stuck on a local maximum that is not an absolute maximum. The undesirable situation of being stuck on a suboptimal peak is called *premature convergence* (Goldberg 1989). Several strategies have been developed to reduce the chance of this occurring. One such strategy is used in the evolutionary algorithm (EA). In the next section the EA is described and then compared to the genetic algorithm.

6.1.2.7   One Strategy to Reduce the Chance of Premature Convergence

There are different philosophical emphases when simulating an evolutionary process. In this section some of these differences are presented along with a strategy to reduce the chance of premature convergence. Genetic algorithms (GAs) and evolutionary algorithms (EAs) comprise the field of *evolutionary computation* (Fogel 1994). "Evolutionary algorithms emphasize phenotypic adaptation, while genetic algorithms emphasize genotypic transformations"

(Fogel 1993). The GA views the evolutionary process as the development and combining of useful schemata, while the EA views the same process as the ability of the individual and the species to adapt to and survive a changing environment (Fogel 1993).

The genetic algorithm focuses on the genotype and is typified by the use of crossover as the primary operator to explore the genome space. Point (bit) mutation plays a secondary role. Even a small change in a genotype can result in a big difference in performance.

While genomes are bitstrings in the computer, the evolutionary algorithm focuses on the phenotype and its behavior. Mutation of the structure is the primary operator used to explore the genome space. For example, a real number is mutated by adding a Gaussian random variable with a mean of zero (Fogel 1993). Mutation of a finite state machine (FSM) consists of adding or deleting a state or changing an output, a transition, or the start state (Fogel 1994). The evolutionary algorithm is based upon mutations which aim for small behavioral changes, and has parents and children competing against each other for a place in the following generation (Fogel 1993).

Within the evolutionary algorithm, a *feature* consists of the functional parts of a genome contributing to an individual's behavior. When focusing on the phenotype, a feature is more relevant than the individual bits which more successful individuals have in common. Since crossover is generally not used in an evolutionary algorithm, the physical location of a feature in the genome is not relevant. Let F represent a given feature, and E(F,g) and N(F,g) represent the expected number and actual number, respectively, of individuals which exhibit feature F in generation g. Then, for features, the equivalent of Equation 1 is the following:

$$E(F,g+1) = n \ (1\text{-}p_d) \ N(F,g) \ p_{AV} \qquad \qquad \text{(Equation 2)}$$

where n = the population size, p_d = the probability that feature F will be disrupted (lost due to mutation), and p_{AV} = the average probability that an individual with feature F will be selected to reproduce into the next generation. When $n > \dfrac{1}{p_{AV}(1-p_d)}$, the number of representatives of feature F can be expected to increase in the next generation (Angeline and Pollack 1993).

As indicated earlier, a GA is viewed as manipulating schemata, while an EA aims to modify features. Angeline and Pollack (1993) created a new operator which is based on the theoretical view of features as the most relevant contributing factors in the success of the EA as a search procedure. Angeline and Pollack's compress operator "freezes" a physical part of a genome. It does not freeze bits within a field (functional unit) of the phenotype; only a complete field of a genome is frozen such as a future state or output of an

FSM genome. Theoretically, a genetic algorithm retains relevant physical units in the form of schemata. But schemata consist of bits, and the functional units of the phenotype are ignored by both schemata and the GA's operators. Mutation alters a single bit within a field, and there is nothing to prevent a crossover point from falling within a field.

Another philosophical difference between the GA and EA relates to fitness definition. When a genetic algorithm uses a fitness-based selection process and the fitness is an explicitly defined function, the environment consists of a static fitness landscape of valleys and hills for the population to overcome; that is, the fitness landscape remains unchanged throughout a run. The population could gravitate towards a local rather than a global optimum when the fitness landscape consists of multiple peaks.

In an evolutionary algorithm, however, fitness is determined by competition between the individuals in a population, even when the fitness can be explicitly defined. Each individual competes against a fixed number of randomly chosen members of the population. The population is then ranked based on the number of wins. Generally, the selection process retains the top half of the population for the next generation. The remaining half of the next generation is filled by applying mutation to each retained individual, with each individual producing a single child. Consequently, for each succeeding generation parents and children compete against each other for a place in the following generation. This provides a changing environment (fitness landscape) and is less likely to prematurely converge (Fogel 1994).

The two different philosophies have relative strengths and weaknesses. Each is more appropriate for different types of problems (Angeline and Pollack 1993), but a competition can be easily integrated into the fitness procedure of a genetic algorithm to reduce the chance of premature convergence. Since the newly designed reorganization operators reduce innovation, they might make the GA more prone to premature convergence. Competition offers a way to counteract this.

Angeline and Pollack's (1993) success with their innovative operators and the use of competition to reduce the chance of premature convergence prompted consideration of creating new operators to enhance schemata formation in a GA, and incorporation of competition to reduce the chance of premature convergence. These two seeds of thought formed the basis for the algorithms which are presented in Sections 6.3 and 6.4.

This section has introduced competition as a method for avoiding premature convergence for GA and EA. This will be implemented in Section 6.5. Coupled with the concern of premature convergence is the identification of a peak. The next section investigates common criteria for termination used in both GA and EA.

6.1.2.8 Variations on Termination

There are different ways to terminate a genetic algorithm. The GA can loop until an individual is found which successfully completes a task. This could result in an infinite loop if the algorithm prematurely converges to a suboptimal peak of the fitness landscape. The genetic algorithm can be stopped after a preselected number of generations have been bred as was the case in the examples developed in sections 6.1.1 and 6.1.2.5, or after a preselected number of children have been "born" (Koza 1992). This could halt the evolutionary process just short of a solution. Similarly, when a researcher is observing the evolutionary process and how the individuals adapt to the environment, as in *Tierra* (Ray 1992), the researcher may not want to chance stopping the process during an interesting evolutionary occurrence. In this case the genetic algorithm can be run as an infinite loop which must be externally terminated. Regardless of the purpose of the experimentation, the researcher must choose a looping structure to cycle through the generations, and decide upon the manner of termination which will best serve the project at hand.

6.1.3 Reflections on the Genetic Algorithm

When designing a genetic algorithm for a particular application, the first step is to define the phenotype along with the phenotype-genotype correlation. Then the fitness function and operators must be determined, and the parameters (like population size and mutation rate) given values. After a few runs, the system may be modified to alter its performance, for example, by changing the mutation rate or population size.

If a genetic algorithm seems to be progressing nicely but is unable to find a solution in the allotted number of generations, the termination criteria can be changed by increasing the number of generations which are bred. When this situation occurs, this simple modification can greatly alter the ability of the GA to succeed. For more complex convergence problems, a new operator may be defined and added to the program, or an old operator may be differently defined. The fitness function may be redefined. This is sometimes done when a researcher is trying to determine how differences in an environment can affect the evolutionary process. These are just a few of the ways that a genetic algorithm can be altered to permit it to explore the genome space when looking for a solution to a given problem or when studying the evolutionary process.

A wide variety of strategies have been reviewed in this section. For the reorganization algorithms which will be presented in sections 6.3 and 6.4, the benchmark case will be achieved by implementing the work of Jefferson et al. (1992). Their work is fully described in the next section. This implementation will be altered in steps, initially by separately adding two newly designed operators which reorganize FSM genomes during run time. Based on schema theory, each of these algorithms could result in either faster convergence to a solution or premature convergence. Competition will then be added to the three programs. In the next section, the benchmark is developed.

6.2. The Benchmark

The importance of the organization of the genome to the convergence of a genetic algorithm can be utilized to improve performance of a GA. This can be particularly useful when a common data structure, such as a finite state machine, is used to model the solution to a problem, and, subsequently, as the operand of a genetic algorithm. The next Section 6.2.1 discusses the finite state machine as a genome and explores, in Section 6.2.2, its use within the context of the genome used by Jefferson et al. (1992). In Section 6.2.3, the discussion of code implementing the benchmark will be presented.

6.2.1 The Finite State Machine as a Genome

Before considering the finite state machine (FSM) as a genome, the FSM is defined. A finite state machine (FSM) is a transducer. It is defined as an ordered septuple $(Q, s, F, I, O, \delta, \lambda)$, where Q, F, and I are finite sets; Q is a set of states; $F \subseteq Q$ is the set of final states; I is a set of input symbols; O is a set of output symbols; $s \in Q$ is the start state; $\delta: Q \times I \rightarrow Q$ is a transition function; and $\lambda: Q \times I \rightarrow O$ is an output function. A finite state machine is initially in state s. It receives a string of symbols as input. An FSM which is in state $q \in Q$ and receiving input symbol a I will move to state $q_{next} \in Q$ and output $b \in O$ based on transmission rule $\delta(q,a) = q_{next}$ and output rule $\delta(q,a) = b$. This data structure has been used to study diverse problems in conjunction with a simulated evolutionary process. Angeline (1994), Fogel (1991), and Stanley et al. (1994) used FSMs to explore the iterated prisoner's dilemma. MacLennan (1992) represented his *simorgs* (simulated organisms) as FSMs to study communication development. (MacLennan's work is interesting in that he incorporates learning into his model. Generations overlap and learning is passed on from generation to generation.) And, Jefferson et al. (1992), and Angeline and Pollack (1993) used a finite state machine genome to breed an artificial ant capable of following an evaporating pheromone trail.

Jefferson et al. (1992) defined an artificial ant as an FSM and bred them to be able to follow a given trail representing a dissipating pheromone trail. Ants lay down a chemical (pheromone) trail when returning from a food source in order to direct other ants in the colony to the foodstore. Since the trail is put down as an ant returns from the food, the part of the trail farthest from the home base has partially dissipated and tends to be weakest and more likely to contain unmarked sections. The trail designed by Jefferson et al. mimicked this. The genetic algorithm they used did find FSM ants which were able to complete the trail in a time frame they imposed on the ants, but it leaves open the question of what modifications might enhance the search process.

The genome used by Jefferson et al. (1992) had 453 bits and therefore defined a space of 2^{453} bitstrings. This space can be divided into equivalence classes with each class containing equivalent FSMs. For example, the two machines in Figure 6.5 are the same machine; q_1 and q_2 have simply exchanged names. The GA used by Jefferson et al. did not utilize this fact to build and retain useful schemata. Since there are many equivalent FSMs, a number of variations of a nearly successful FSM may exist in a population, but the similarity may not be evident due to different representations. Useful schemata are competing with each other during the selection process, each hampering the growth of the other.

Theoretically, the genetic algorithm moves a population towards a solution to a problem as useful schemata grow in number and length. Because a single finite state machine has different representations (simply by changing state names), identical machines with different representations do not necessarily share a useful schema; that is, a set of identical transitions/outputs which provide a successful strategy could reside in different parts of the genome and appear very different due to having different state names. Consequently, these machines hamper schema growth as the machines compete against each other for their share of the next generation. The reorganization algorithms presented in this chapter compensate for this. The new operators (algorithms), called SFS and MTF, will reorganize finite state machines during run time so identical machines will appear the same. The purpose of these new operators is to enhance the growth of useful schemata and thereby hasten the convergence of the genetic algorithm. The strategies these operators employ are FSM-specific, and, theoretically, should improve the convergence of the genetic algorithm for FSM genomes. Experiments incorporating these new operators into a genetic algorithm indicate that they reduce the number of generations for a genetic algorithm to converge to a solution.

Present State	Future State/Output for		Present State	Future State/Output for	
	input=0	input=1		input=0	input=1
q0	q1/1	q2/0	q0	q2/1	q1/0
q1	q2/0	q0/0	q1	q1/1	q1/0
q2	q2/1	q2/0	q2	q1/0	q0/0

Figure 6.5 Equivalent FSMs

The benchmark or starting point is based on the work of Jefferson et al. (1992) and presented in the next section. In sections 6.3 and 6.4, the reorganization algorithms will be added to this benchmark. The benchmark and its implementation are developed in the next section.

6.2.2 The Benchmark: Description and Theory

In this section, the benchmark is described. Equations 1 and 2 are updated to apply to the variations on the simple genetic algorithm which are implemented by Jefferson et al. (1992). The benchmark for this study was developed by implementing the work done by Jefferson et al. It was designed to be easily modifiable. Jefferson et al. defined an artificial ant as a finite state machine, and bred them to be able to follow a given trail representing a dissipating pheromone trail. The trail used by Jefferson et al. has "a series of turns, gaps, and jumps that get more difficult as [they] progress." (Jefferson et al. 1992).

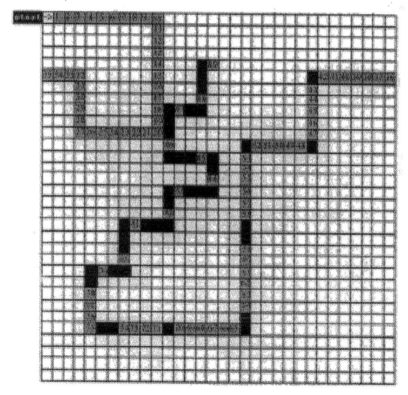

Figure 6.6 The John Muir Trail

Their John Muir Trail (Figure 6.6) starts off with three straight way paths, each followed by a right turn. Jefferson et al. (1992) used this trail to see if the ants would develop a bias for right turns. In Figure 6.6 the gray, numbered squares represent the marked portions of the trail, that is, the part of the trail which still has pheromone. The black squares represent the part of the trail from which the pheromone has evaporated and appear the same as the white squares to the ant. (See Figure 6.7.) The trail lies on a 32 x 32 toroidal grid and consists of 89 marked steps. In addition to the 89 marked steps, the John Muir trail has 38 steps which have lost the pheromone and must be interpolated by the ant. There are 42 marked steps on the trail before an ant encounters an unmarked step (i.e. evaporated pheromone). This first unmarked step is at a corner.

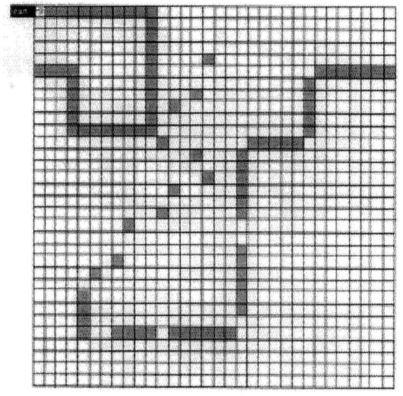

Figure 6.7 The John Muir Trail as it appears to the ant.

An ant's progress along the trail is timed where each time step represents the application of a transition rule of the ant's finite state machine. At each time step an ant can sense the presence (1) or absence (0) of pheromone in the box directly ahead; this is the input to the FSM. The ant can either turn 90^0 to the left or right, move 1 step ahead, or do nothing; the latter permits the ant to change its state without taking any action. Jefferson et al. (1992) found that doing nothing to allow a state change seemed to be bred out of the best FSMs; that is, some FSM minimization was occurring because the fitness function encouraged this. The output of the FSM directed the ant's action for the given time step. Ants were permitted 200 time steps to traverse the trail. The fitness consisted of the number of marked trail steps an ant was able to cover in 200 time steps; consequently, doing nothing just wasted a time step. To prevent retracing part of the trail or back-tracking on the trail, the part of the trail which was traversed was erased. No additional credit was given for completing the trail in fewer than 200 time steps.

It seems that a good trail-following strategy should move the ant forward when there is a marked step directly in front. When there is no trail ahead, look to the right and to the left to see if the trail has turned; if the trail is gone, continue in the same direction as before. Since the John Muir Trail contains 12 right turns and only 8 left turns, it seems that, when there is no trail ahead, a right turn would be a good first step. If, after the turn, the trail is still not in evidence, an about face (2 left turns or 2 right turns) is appropriate. If there is still no trail ahead, turn back to the original direction (a right turn) and then step forward. This strategy makes the solution to the problem seem simple, but the imposition of a limit of 200 time steps makes it difficult. The FSM in Figure 6.8 (similar to one presented by Jefferson et al. (1992)) permits the ant to follow the trail, but it covers only 81 trail steps in 200 time steps. The remainder of the trail takes a significant number of additional time steps since there are many unmarked boxes on the remainder of the trail, and it takes 5 time steps to move from the present position to an unmarked box which immediately follows. According to Jefferson et al., this FSM ant needs 314 time steps to get a perfect score of 89.

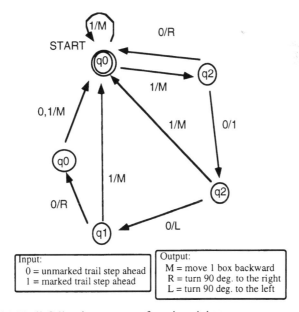

Figure 6.8 Trail-following strategy favoring right turns

The phenotype for this problem is the finite state machine. The genome used by Jefferson et al. (1992) allows for a maximum of 32 states. Each FSM consists of a table with 32 lines for states q_0 - q_{31}, and has two columns for inputs: 0 (no pheromone on the step immediately in front of the ant) and 1 (pheromone present on that next step). The genome contains the next state/output recorded along each row and the sequential listing of the rows of the FSM table. In addition the initial state can be any one of the 32 states. Since there are 32 states, 5 bits are needed for the state number. Four possible outputs (turn left or right, move ahead, do nothing) require 2 bits.

Bit #	0 4	5 9	10-11	12 16	17-18	state	state
Contents	start state	next output *for q_0 with input 0*		next output *for q_0 with input 1*		state	state

Bit #	19 23	24-25	26 30	31-32		state	state
Contents	next output *for q_1 with input 0*		next output *for q_1 with input 1*			state	state

Bit #	5+14i+7j	...	11+14i+7j				
Contents	... next output *for q_i with input j, j$\in$ {0,1}*		state				

Bit #	425 - 429	430-431	432 - 436	437-438		state	state
Contents	next output *for q_{30} with input 0*		next output *for q_{30} with input 1*			state	state

Bit #	439 - 443	444-445	446 - 450	451-452		state	state
Contents	next output *for q_{31} with input 0*		next output *for q_{31} with input 1*			state	state

Figure 6.9 32-state/453-bit FSM genome map

Each ant (FSM) is represented by a genome consisting of 453 (Figure 6.9) bits which was initialized by randomly setting all 453 bits. The first 5 bits of the genome indicate the start state. The next $64*(5+2) = 448$ bits represent the next state (5 bits) and output (2 bits) for q_0 with input 0, q_0 with input 1, q_1 with input 0, q_1 with input 1, and similarly down the table (Jefferson et al. 1992). Consequently, the position of the future state/output for a given present state/input is fixed on the bitstring. While the genome allows for 32 states, the actual FSM which any one genome represents may have fewer than 32 states. For example, the FSM with start state q_{13} and state transition Table 6.5 uses only 4 states. Other parts of the genome may eventually be used as a result of crossover or mutation. For example by inverting a bit within the state portion of the genome, mutation could change the next state for q_{13} with input 0 from q_5 ($5_{10} = 00101_2$) to q_4 ($4_{10} = 00100_2$) or q_7 ($7_{10} = 00111_2$), amongst others.

The population of Jefferson et al. (1992) consisted of 65,536 (64k) FSMs. The bottom 95% (least fit) of each generation was discarded. 65,536 mating pairs were selected from the remaining 5% of the population without regard to fitness. Crossover was applied with a probability of 1% per bit; that is, a random number between 0 and 1 was generated for each bit along the genome and, when the number was under 0.01, the subsequent genetic material was taken from the other parent. (Note that in GAs, genetic material is swapped between two parents producing two offspring; however Jefferson et al. retain only one of the children.) A 1% per bit mutation rate was used, with mutation inverting a bit. Using these parameters and operations, Jefferson et al. got a perfect scoring ant in generation 52.

Start state q_{13}

Present state	Next state/output for input = 0	Next state/output for input = 1
q_{13}	$q_5/0$	$q_9/2$
q_5	$q_{13}/2$	$q_5/2$
q_9	$q_5/1$	$q_{24}/2$
q_{24}	$q_{13}/3$	$q_{24}/2$

Table 6.5 Four state FSM with start state q_{13}.

To find the equivalent of equations 1 and 2 for this scenario, let $E(S,g+1) =$ the expected number of individuals which match schema S in generation $g+1$ and $T(S, g) =$ the number of individuals which match schema S in the top $100t\%$ of generation g. Note that t is in decimal rather than percent form. Let $p_S =$ the probability that an individual with schema S is selected as a parent $= \frac{T(S,g)}{nt}$ where n is the population size. Since a match to S can occur with this probability when each individual is selected to fill the n members of the next generation

$$E(S, g+1) = n\frac{T(S, g)}{nt} = \frac{T(S,g)}{t} \qquad \text{(Equation 6.3)}$$

But this does not take into consideration the losses of schema S due to crossover and mutation.

Let $p_X =$ the per bit probability of crossover and $p_M =$ the per bit probability of mutation. Let $c =$ the number of crossover points in schema S (defining length) and $b =$ the number of fixed bits in the schema. Then the probability that a schema survives crossover = the probability that there is no crossover point within the schema = $(1-p_X)^c$. Similarly, the probability that S is not lost due to mutation = $(1-p_M)^b$. Therefore, the probability that schema S survives both crossover and mutation = $(1-p_X)^c(1-p_M)^b \geq (1-cp_X)^c$ $(1-bp_M) \geq (1-bcp_Xp_M)$. Incorporating this into equation 3 yields equation 4:

$$E(S,g+1) \geq \frac{T(S,g)}{t}(1-bcp_xp_m) \qquad \text{(Equation 6.4)}$$

Thus schema S is most likely to carry into the next generation when it matches a large number of genomes in the parent pool, has a short defining length (c), and fewer fixed bits (b). If every member of the parent pool matches S there is no loss due to crossover. For this case equation 6.4 reduces to equation 6.5.

$$E(S,g+1) \geq \frac{T(S,g)}{t}(1-bp_m) \qquad \text{(Equation 6.5)}$$

indicating that schema S will be disrupted only by mutation, and that it is less likely to be lost when there are fewer fixed bits.

Jefferson et al.'s ant problem (1992) has been chosen as the benchmark method because the problem has been widely explored. In addition, Angeline and Pollack (1993) used it to test the effect of their innovative operators on the convergence of a simulated evolutionary process. This is the aim of the research with the two reorganization algorithms (described in sections 6.3 and 6.4). It is important to note that Jefferson et al. were interested in simulating a natural evolutionary process (although differences can be found when a GA is compared to the biological process (Fogel 1993)), while the present research is concerned with producing a search procedure which converges to a solution in fewer generations. Relevant to this goal is the fact that in nature, a gene that codes for a specific purpose has a set location in a genome, but the nature of Jefferson et al.'s FSM genome defies this natural phenomenon. The start state alone can be in any one of 32 different positions. As indicated earlier, identical finite state machines need not appear the same. For example, the two machines in Figure 6.5 consist of states q_0, q_1, and q_2. This machine put into the 32 state genome shown in Figure 6.9 can have any one of 32 x 31 x 30 = 29,760 different representations. Many of these machines don't even have state names in common, nor need they have the same start state when they do have the same set of state names. Schema growth is not encouraged by this genome since equivalent states of equivalent machines (differing only in state names) do not necessarily reside in the same location of the genome. It is this realization which prompted the design of the two reorganization algorithms which are the focus of this chapter, and which are incorporated into the benchmark (sections 6.3 and 6.4). In the next section, the benchmark program is detailed.

6.2.3 The Coding

The benchmark program retains the selection process, crossover and mutation rates and algorithms, population size, and phenotype-genotype mapping, utilized by Jefferson et al. (1992) as described in the previous section. The initial population is created using Goldberg's (1989) random number generator to create 57 bytes (with values of 0 to 255) per genome, as opposed to 453 bits per genome. Since Jefferson et al. reported that their GA found a successful ant in generation 52, our GA was permitted to breed a maximum of 70 generations.

As already indicated, the programs are coded in C and executed on a Sun Sparc Station 20, as opposed to Goldberg (1989) who used Turbo-Pascal. The program was designed so that planned modifications could be easily incorporated. To permit additional future testing on other types of problems, parameters which are constant for this problem are given values with #define statements.

First, the run number, seed for the random number generator, and files to receive the output are defined as follows:

```
/*      (r)un number (1) for (b)enchmark   */

#define run_num 1          /* run number   */

#define printed_output "r1b.st" /* statistics
output */

#define perfect_score "r1b.ps" /* perfect score
file */

#define seed .69343845 /* seed for random number
generator */
```

The file with the st extension receives the statistics for the run, and the file with the ps extension receives the phenotype of successful ants which appear. In both file names the **r** precedes the run number and the **b** indicates that the run is for the benchmark. These lines are to be appropriately changed for each run, or the program can be modified to read in these values.

Then the FSM and GA parameters are defined.

```
#define bits_per_state_num 5

#define max_num_states 32 /* 2^bits_per_state_num */

#define num_of_inputs 2

#define bits_per_output
/**********************************/
```

```
/*                                            */
/* Genome consists of start state followed by
sequential listing */
/* of the rows of the transition/output table */
/*                                              */
/*     bits_per_genome = bits_per_state_num + */
/*     num_of_inputs * max_num_states *         */
/*     (bits_per_state_num + bits_per_output);*/
/*                                              */
/*****************************************/
#define bits_per_genome 453
#define bytes_per_genome 57 /* contains
bits_per_genome bits */


/***************************************************/
/*                                               */
/*      Genetic Algorithm Constants         */
/*                                               */
/***************************************************/


#define popsize 65536       /* population size */
#define select 3277       /* choose parents from top
members of population    */
#define maxfitness 89       /* maximum fitness which
can be attained      */
#define last_generation 70
#define mutation_rate .01
#define xover_rate .01
```

The data types needed for the program are defined as follows:

```
typedef int table
[max_num_states][num_of_inputs];/* transition table
*/
```

```
typedef unsigned char byte;

typedef byte unpacked_genome [bits_per_genome];

typedef byte packed_genome [bytes_per_genome];

typedef packed_genome population [popsize];
```

Two tables are used to store the phenotype for an ant. One table will hold the next states for each present state/input pair, and the other similarly stores the outputs. The first index into the table is the present state number. The second is the value of the input.

With a population of 64k, machines' efficient storage of the population is a concern, so a generation of FSMs are stored in a bytestring format which is the data type packed_genome population holds a complete generation of ants. When converting a phenotype to a genotype, or vice versa, bits of the genome are manipulated. Consequently, for these conversions an intermediate form of 453 bits is employed. Hence, the genome is stored as a bitstring, that is as data type unpacked_genome, for the conversion.

Finally, the variables are declared.

```
population pop[2];         /* holds present and newly
evolving generations */

table next_state, output; /* holds a phenotype */

unpacked_genome genome;

int generation=0,

   old_pop=0, new_pop=1,    /*the two generations'
indices in the pop array*/

   start_state,


fitness[popsize],fitness_count[maxfitness+1]={0},/*
histogram fitness */

   sorted[3][popsize]={0},/* for merge-sort:
indices of population    */

   row_of_sorted=1;   /* row of sorted array that
has result of the sort */

FILE *fp_stats, *fp_perfect;
```

The declaration for the pop array doesn't reveal the complexity of the structure. This array actually requires three indices. The first index can be a 0 or 1. One value, a 0 or a 1, is stored in old_pop and indicates the population of parents. The other value is stored in new_pop and indicates the row into which the newborns will be placed. The values of old_pop and new_pop are exchanged as each new generation becomes the parents for the subsequent generation (double buffering). But pop is of type population, which is itself an array. Therefore, the second index of pop is the number of the individual in the population. The data type population is a packed_genome, so the third index of pop permits accessing the bytes (and hence the bits) of the genome for the individual pointed to by the first two indices of pop.

With these parameters and variables defined, the program can be coded. Before discussing the GA proper, there are several utility procedures which are needed. The most complex utilities required are a random number generator and a sort procedure. The random number generator coded in Turbo-Pascal in Goldberg's book (1989) has been converted to C. A stable merge-sort is implemented (Knuth 1973), but any sorting procedure can be used. The population is sorted in order of decreasing fitness. The results of the sort appear in the sorted array. The first index of the array indicates the row of the array which holds the results of the sort and is given by row_of_sorted. The second index of the pop array is stored in sorted. So, after the sort, the most-fit individual is in pop[new_pop] [sorted[row_of_sorted] [0]], and the least-fit individual is in pop[new_pop] [sorted[row_of_sorted] [popsize-1]].

power_of_2(n) (Figure 6.10) calculates 2^n for a non-negative n. If n<0, pow_of_2 returns a 0. This is used by main to test the validity of the #defined values.

```
/* ------------------------------------     */
    /*          Power base 2        */
/* ------------------------------------   */

int power_of_2 (int exponent)
/* this will only return the proper answer when the
exponent is non-neg */

{

  int answer;
  if (exponent < 0) answer = 0;
  else
    for (answer = 1; exponent; --exponent)
      answer *= 2;

  return answer;

} /* power_of_2 */
```

Figure 6.10 Calculate 2^n

As earlier stated, the genotype is stored as 57 bytes. The phenotype is stored as a table with base 10 values. The conversion from genotype to phenotype (bytestring_to_FSM, Figure 6.15), and vice versa (FSM_to_bytestring, Figure 6.16), requires conversions between binary and base 10 numbers (bin_to_dec in Figure 6.13, and dec_to_bin in Figure 6.14). The conversions between genotype and phenotype utilizes an intermediate form of the genome, a 453-bitstring. Therefore, routines are needed to unpack the 57 bytes into 453 bits (unpack_genome, Figure 6.12) and pack the 453 bits into 57 bytes (pack_genome, Figure 6.11).

```
/* ----------------------------------- */
/*          Pack Genome          */
/* ----------------------------------- */
void pack_genome(unpacked_genome bitstring,
packed_genome bytestring)
{
line #
1: int i, j, k;

2: for (i=0, j=0; j<bits_per_genome; ++i) {

3:     bytestring[i] = 0;
4:     for (k=0; k<8 && j<bits_per_genome; ++k, ++j)
5:       bytestring[i] = bytestring[i] << 1 |
bitstring[j];

6: }

7: if (k < 8)                   /* fill in extra bits in
last byte */
8:     bytestring[bytes_per_genome-1] <<= (8 - k);
} /* pack_genome */
```

Figure 6.11 Bitstring to Bytestring Conversion

```
/* ----------------------------------- */
/*          Unpack Genome          */
/* ----------------------------------- */
void unpack_genome (packed_genome bytestring,
unpacked_genome bitstring)
{
line #
1: int i, j, k;
2: byte mask;

3: for (i=0, j=0; j < bits_per_genome; ++i)
```

```
4:     for (k=0, mask=0x80; k<8 &&
j<bits_per_genome; ++k, ++j, mask >>= 1)
5:        if (bytestring[i] & mask)
6:bitstring[j] = 1;
7:        else
8:bitstring[j] = 0;
} /* unpack_genome */
```

Figure 6.12 Bytestring to Bitstring Conversion

pack_genome (Figure 6.11) packs 8 bits into a byte. i is the index into the array of 57 bytes, j is the index into the bitstring, and k limits each byte to 8 bits. Indices i, j, and k serve the same purpose in unpack_genome (Figure 6.12). pack_genome shifts a byte one bit left, and ORs it with the next bit to fill the eight bits of a byte (lines 4-5, Figure 6.11). Since 57 bytes = 456 bits, the last 3 bits of the last byte are left blank (lines 7-8, Figure 6.11). In unpack_genome (Figure 6.12), the single one in mask is used to isolate the individual bits of a byte. As the mask is shifted and ANDed with a byte (lines 4-8, Figure 6.12) the individual bits of the byte are sequentially isolated.

bin_to_dec (Figure 6.13) converts a binary integer to its decimal equivalent. The loop (lines 2-3) is based on the binary expansion

$$2^n b_n + 2^{n-1} b_{n-1} + 2^{n-2} b_{n-2} + +2^i b_i + +2^2 b_2 + 2^1 b_1 + 2^0 b_0 =$$

$$b_0 + 2(b_1 + 2(b_2 + 2(+2(b_i + +2(b_{n-2} + 2(b_{n-1} + 2b_n))))))$$

or,

$$((((((b_n x2) + b_{n-1})x2 + b_{n-2})x2 + +b_i)x2 + +b_2)x2 + b_1)x2 + b_0$$

of the binary integer $b_n b_{n-1} b_{n-2}$ b_i $b_2 b_1 b_0$. The start_position is advanced (line 2) as each bit is processed. num_of_bits gives the number of bits in the binary number being converted. When bin_to_dec returns control to the calling program (line 4), start_position indicates the start of the next field in the genome.

```
/* ---------------------------------------- */
/*        Binary to Decimal      */
/* ---------------------------------------- */
int bin_to_dec (int *start_position, int
num_of_bits, unpacked_genome bitstring)
/* when this function terminates, start_position is
the start position  of the next field to be
processed */
{
line #
1: int i, dec=0;
```

```
2:  for (i=0; i<num_of_bits; ++i,
++*start_position)
3:     dec = 2*dec + bitstring[*start_position];
4:  return dec;
} /* bin_to_dec */
```

Figure 6.13 Binary to Decimal Conversion

dec_to_bin (Figure 6.14) utilizes the fact that by dividing a decimal integer d by two to get an integer quotient and a remainder generates the least significant bit of its binary equivalent. Repeatedly divide sequentially derived quotients by two and retain the remainders. The remainders are the bits of the binary equivalent. The bits of the binary equivalent of i are generated from least significant to most significant. More formally,

$d = d/2 + b_0$. Let $q_1 = d/2$.

Then, $q_1 = q_1/2 + b_1$. Let $q_2 = q_1/2$.

Then, $q_2 = q_2/2 + b_2$. Let $q_3 = q_2/2$.

$\vdots$

Then, $q_i = q_i/2 + b_i$. Let $q_{i+1} = q_i/2$.

This iterative process continues until $q_{n+1} = 0$ for some n. Then the binary equivalent of d is $b_n b_{n-1} \ldots b_2 b_1 b_0$. In dec_to_bin (Figure 6.14), the index i controls the storage of bits into the bitstring. Since i is counted down by the for loop (line 2), the most significant bit is stored when i = 0. Since the 453 bits of the bitstring must be appropriately converted into state numbers (0 - 31) and outputs (0 - 3), the start_position is changed (line 4) to indicate the beginning of the next field to be processed. As in bin_to_dec, when dec_to_bin returns control to the calling program, start_position indicates the start of the next field in the genome.

```
/* ------------------------------------------ */
/*        Decimal to Binary       */
/* ------------------------------------------ */

      void dec_to_bin (int *start_position, int
num_of_bits, int number,
            unpacked_genome bitstring)
/* when this function terminates, start_position is
the start position  of the next field to be
processed */
{
```

```
line #
1:  int i;

2:  for (i=num_of_bits-1; i>=0; --i, number /= 2)
3:    bitstring[*start_position + i] = number % 2;

4:  *start_position += num_of_bits;
} /* dec_to_bin */
```

Figure 6.14 Decimal to Binary Conversion

bytestring_to_FSM (Figure 6.15) and FSM_to_bytestring
(Figure 6.16) invoke the four previously described subroutines. For
genotype to phenotype conversion, bytestring_to_FSM first unpacks
the 453 bits out from the 57 bytes (line 3, Figure 6.15). The start state is
then decoded (lines 4-5). Then the for loops (lines 6-7) march through the
32 states of the FSM and the two inputs for each state to fill in the
next_state and output tables of the phenotype (lines 8-9).

```
/* ------------------------------------- */
/*      Store byte string        */
/*      information as a         */
/*      finite state machine       */
/* ------------------------------------- */
void bytestring_to_FSM(int *start_state,
packed_genome bytestring)
{
line #
1:  int i, state, input;
2:  unpacked_genome bitstring;

3:  unpack_genome(bytestring, bitstring);
4:  i = 0; /* present position in bitstring */
5:  *start_state = bin_to_dec(&i,
bits_per_state_num, bitstring);

6:  for (state=0; i<bits_per_genome; ++state)
7:    for (input=0; input<num_of_inputs; ++input) {
8:      next_state [state] [input] =
                    bin_to_dec(&i,
            bits_per_state_num, bitstring);
9:        output [state] [input] = bin_to_dec(&i,
bits_per_output, bitstring);
10:     }
} /* bytestring_to_FSM */
```

Figure 6.15 Genotype to Phenotype Conversion

```
/* -------------------------------------- */
/*      Store finite state     */
/*      machine information       */
/*      as a byte string       */
/* -------------------------------------- */
void FSM_to_bytestring (int start_state,
packed_genome bytestring)
{
line #
1: int i, j, k;
2: unpacked_genome bitstring;

3: k = 0; /* present position in bitstring */
4: dec_to_bin(&k, bits_per_state_num, start_state,
bitstring);

5: for (i=0; i<max_num_states; ++i)
6:    for (j=0; j<num_of_inputs; ++j) {
7:       dec_to_bin(&k, bits_per_state_num,
next_state[i][j], bitstring);
8:       dec_to_bin(&k, bits_per_output,
output[i][j], bitstring);
9:    }

10:   pack_genome(bitstring, bytestring);
} /* FSM_to_bytestring */
```

Figure 6.16 Phenotype to Genotype Conversion

FSM_to_bytestring (Figure 6.16) is needed to reverse the process
when the phenotype has been altered. First each functional part of the FSM
is converted to its binary form using dec_to_bin (line 4). The bits are
placed in their proper position in the bitstring as shown in Figure 6.9. As
in bytestring_to_ FSM, conversion starts with the start_state
(lines 3-4, Figure 6.16). The for loops (lines 5-6) move through the 32
states and the two inputs for each state. Lastly the 453 bits are packed into
57 bytes for more efficient storage (line 10).

```
/**********************************************/
/*            Trail Related Constants         */
/**********************************************/


#define maxtime 200    /* maximum number of steps
FSM allowed on trail */
```

```
#define trail_width 32
#define trail_length 32
```

Figure 6.17 Trail related parameters

With the utilities complete, the program for the GA can be written. The fitness function, `calc_fitness` (Figure 6.18), is problem specific. To reduce run-time overhead, the `trail` array could be instantiated as a global variable. (Then, it would appear in Figure 6.17, instead of Figure 6.18. However, this would affect other parts of the program and, hence, was avoided in this implementation.) The `trail` array contains the trail, listed row by row from Figures 6.6 and 6.7. The ones in the array indicate the marked trail steps. When an ant covers a marked step, the marking is erased to prevent the ant from backtracking on the trail. To prevent loss of the `trail`, the array is local to `calc-fitness`, and is therefore set up each time the subroutine is invoked. `maxtime` is the number of time steps an ant is allotted to traverse the trail (Figure 6.17). `trail_width` and `trail_length` are the number of rows and columns, respectively, in the trail lattice as shown in Figures 6.6 and 6.7.

When `calc_fitness` (Figure 6.18) is invoked, the ant is positioned in the upper left box of the trail grid (line 1) and is facing the first step of the trail (line 3). `i` is the row number and `j` the column number of the grid of Figures 6.6 and 6.7. `move_dir_i` and `move_dir_j` indicate the change in `i` and `j` needed to move the ant one step straight ahead. Consequently `move_dir_i` and `move_dir_j` also indicate the direction the ant is facing. To simplify visualization of the trail and an ant's progress along the trail, the common directions of north, east, south, and west are imposed on the trail. An ant's initial direction is considered to be easterly.

```
/* ------------------------------------ */
/* Calculate fitness of an individual   */
/* ------------------------------------ */

int calc_fitness (int present_state)
/* argument must be start_state when function
called */

/* fitness = the number of trail steps the ant can
cover in maxtime steps
Note: trail step is erased when an ant steps on it
in order to prevent the retracing of trail steps.

  FSM output = 0: turn 90 degrees to the right
            1: turn 90 degrees to the left
            2: move forward one step
            3: do nothing--permits change of state  */
```

```
/* The ordered pair (i, j) is the ant's location
(row, column) in the trail array. The ordered pair
(ii, jj) is the location of the step in front of
the ant. The ordered pair (move_dir_i, move_dir_j)
indicates how i and j change when the ant moves
ahead 1 step. It also indicates the direction the
ant is facing:
  when (move_dir_i, move_dir_j) =
  (0, 1)        ant is facing E
  (0,-1)        ant is facing W
  (1, 0)        ant is facing S
  (-1,0)        ant is facing N              */
{
line #
1: int i = 0, j = 0,  /* ant starts in the upper
left of the trail array */
2:    ii, jj,
3:    move_dir_i = 0, move_dir_j = 1, /* ant starts
facing E */
4:    fit = 0,                      /* fitness      */
5:    time = 0, future_state,
6:    input /*FSA input:pheromone content of trail
step immediately ahead */

7: short int trail [trail_width] [trail_length] =
{
          /* the trail used by Jefferson et al */
{0,1,1,1,1,1,1,1,1,1,1,0,0,0,0,0,0,0,0,0,0,0,0,0,0,0,0,0,0,0,0
,0},
{0,0,0,0,0,0,0,0,0,0,1,0,0,0,0,0,0,0,0,0,0,0,0,0,0,0,0,0,0,0,0
,0},
{0,0,0,0,0,0,0,0,0,0,1,0,0,0,0,0,0,0,0,0,0,0,0,0,0,0,0,0,0,0,0
,0},
{0,0,0,0,0,0,0,0,0,0,1,0,0,0,0,0,0,0,0,0,0,0,0,0,0,0,0,0,0,0,0
,0},
{0,0,0,0,0,0,0,0,0,0,1,0,0,0,0,1,0,0,0,0,0,0,0,0,0,0,0,0,0,0,0
,0},
{1,1,1,1,0,0,0,0,0,0,1,0,0,0,0,0,0,0,0,0,0,0,0,0,0,0,1,1,1,1,1,1
,1},
{0,0,0,1,0,0,0,0,0,0,1,0,0,0,0,0,0,0,0,0,0,0,0,0,0,1,0,0,0,0,0,0
,0},
{0,0,0,1,0,0,0,0,0,0,1,0,0,0,1,0,0,0,0,0,0,0,0,0,1,0,0,0,0,0,0
,0},
{0,0,0,1,0,0,0,0,0,0,1,0,1,0,0,0,0,0,0,0,0,0,0,0,1,0,0,0,0,0,0
,0},
{0,0,0,1,0,0,0,0,0,0,1,0,0,0,0,0,0,0,0,0,0,0,0,0,1,0,0,0,0,0,0
,0},
{0,0,0,1,1,1,1,1,1,1,1,0,0,0,0,0,0,0,0,0,0,0,0,0,1,0,0,0,0,0,0
,0},
{0,0,0,0,0,0,0,0,0,0,0,1,0,0,0,0,0,0,0,1,1,1,1,1,0,0,0,0,0,0,0
,0},
{0,0,0,0,0,0,0,0,0,0,0,0,0,0,1,0,0,0,1,0,0,0,0,0,0,0,0,0,0,0,0
,0},
{0,0,0,0,0,0,0,0,0,0,0,0,0,0,0,0,0,0,0,1,0,0,0,0,0,0,0,0,0,0,0,0
```

```
,0},
{0,0,0,0,0,0,0,0,0,0,0,0,0,0,0,1,0,0,1,0,0,0,0,0,0,0,0,0,0,0,0
,0},
{0,0,0,0,0,0,0,0,0,0,0,0,1,0,0,0,0,0,1,0,0,0,0,0,0,0,0,0,0,0,0
,0},
{0,0,0,0,0,0,0,0,0,0,0,0,0,0,0,0,0,0,1,0,0,0,0,0,0,0,0,0,0,0,0
,0},
{0,0,0,0,0,0,0,0,0,0,0,1,0,0,0,0,0,0,1,0,0,0,0,0,0,0,0,0,0,0,0
,0},
{0,0,0,0,0,0,0,0,1,0,0,0,0,0,0,0,0,0,0,0,0,0,0,0,0,0,0,0,0,0,0
,0},
{0,0,0,0,0,0,0,0,0,0,0,0,0,0,0,0,0,0,0,0,0,0,0,0,0,0,0,0,0,0,0
,0},
{0,0,0,0,0,0,0,0,0,0,0,0,0,0,0,0,0,0,1,0,0,0,0,0,0,0,0,0,0,0,0
,0},
{0,0,0,0,0,0,0,1,0,0,0,0,0,0,0,0,0,0,1,0,0,0,0,0,0,0,0,0,0,0,0
,0},
{0,0,0,0,0,1,0,0,0,0,0,0,0,0,0,0,0,0,1,0,0,0,0,0,0,0,0,0,0,0,0
,0},
{0,0,0,0,0,0,0,0,0,0,0,0,0,0,0,0,0,0,1,0,0,0,0,0,0,0,0,0,0,0,0
,0},
{0,0,0,0,1,0,0,0,0,0,0,0,0,0,0,0,0,0,1,0,0,0,0,0,0,0,0,0,0,0,0
,0},
{0,0,0,0,1,0,0,0,0,0,0,0,0,0,0,0,0,0,1,0,0,0,0,0,0,0,0,0,0,0,0
,0},
{0,0,0,0,1,0,0,0,0,0,0,0,0,0,0,0,0,0,0,0,0,0,0,0,0,0,0,0,0,0,0
,0},
{0,0,0,0,1,0,0,1,1,1,1,0,1,1,1,1,1,1,0,0,0,0,0,0,0,0,0,0,0,0,0
,0}


}; /* trail array */

8: do { /* traverse trail if have not completed
trail and time permits */

/* ii and jj indicate row and column position in
front of ant */
9:    ii = (i + move_dir_i) % trail_width;
10:      if (ii < 0) ii += trail_width;
11:      jj = (j + move_dir_j) % trail_length;
12:      if (jj < 0) jj += trail_length;

13:      input = trail[ii][jj];
14:      future_state = next_state [present_state]
[input];

15:      switch (output [present_state] [input]) {

16:         case 0: /* turn 90 degrees to the right
*/    /* ant facing E or W */
17: if (move_dir_i == 0) {
                    /* change E to S, or W to N */
18:    move_dir_i = move_dir_j;
19:        move_dir_j = 0;
20:        break;
```

```
21: }

 /* ant facing N or S */
22: else {
 /* change N to E, or S to W */
23:   move_dir_j = -move_dir_i;
24:   move_dir_i = 0;
25:   break;
26: }

27:        case 1: /* turn 90 degrees to left */

 /* ant facing E or W */
28: if (move_dir_i == 0) {
 /* change E to N, or W to S */
29:   move_dir_i = -move_dir_j;
30:   move_dir_j = 0;
31:   break;
32: }

 /* ant facing N or S */
33: else {
 /* change N to W, or S to E */
34:   move_dir_j = move_dir_i;
35:   move_dir_i = 0;
36:   break;
37: }

38:        case 2: /* move 1 step ahead */ {

39: i = ii;
40: j = jj;

41: if (trail [i][j]) {
 /* ant on marked trail step */
42:   ++fit;         /* increase fitness */
43:   trail [i][j] = 0; /* erase trail step */
44: }

45:      }
    /* case 3: do nothing--permits change of state
*/

46:    }/* switch */

47:    present_state = future_state;
```

```
48:      ++time;

49:    } while (time < maxtime && fit < maxfitness);

50:    if (fit == maxfitness) {
51:        fprintf(fp_perfect,
52:    "Generation %d; Start State=%d;  ," generation,
start_state);
53:        fprintf(fp_perfect,
54:    "Fitness of 89 reached in %d time steps.\n,"
time);
55:        save_FSM(start_state); /* save a perfect
score */
56:    }
57:    ++fitness_count[fit];       /* histogram
fitnesses */
58:    return fit;
} /* calc_fitness*/
```

Figure 6.18 Determine an ant's fitness

A do-while (lines 8-49) loop moves the ant along the trail. The ant's response to the input (1 for marked step, 0 for unmarked step) is controlled by a switch statement (line 15). A left or right turn requires changing the values of move_dir_i and move_dir_j (lines 16-37). To move onto the next step, the ant's location, indicated by i and j, is changed (lines 9-12 and 39-40). When an ant moves onto a marked step (due to line 13), the ant's fitness is incremented (line 42) and the marking is removed from the trail step (line 43) to prevent the ant from backtracking on the trail.

An ant's fitness is the number of pheromone-marked trail steps which the ant traverses. Since the trail has 89 marked trail steps, the optimal fitness is 89. An ant is moved along the trail until it attains a fitness of 89, or until it has spent 200 time steps following the trail. Hence, this is the criteria for terminating the loop (line 49). An ant has achieved a perfect score if it attains a fitness of 89; the FSM representing a perfect scoring ant models a successful solution to the problem. Perfect scoring FSMs are output to the file with the ps extension (lines 50-56). The ant's fitness is reflected in the fitness histogram (line 57) before returning the fitness to the calling program (line 58). The calc_fitness procedure can be easily rewritten for other types of problems.

calc_fitness (line 55, Figure 6.18) calls save_FSM (Figure 6.19) to record the perfect scoring FSMs in a *.ps file. Only those states which can be reached via a transition are printed out. For example, for the FSM in Table 6.5, save_FSM will print out only the four states which appear in the table. The reachable array indicates whether a state is reachable. reachable[i]=1 if state i is reachable (lines 4, 9 and 14);

otherwise, it is 0 (line 1). As each state is reached for the first time, it is placed on the queue (lines 3, 7-8, and 12-13) so that its next states can be tagged as reachable. The reachable states are output to the *.ps file (lines 18-29).

```
/* ------------------------------- */
/*     Save Finite State Machine */
/* ------------------------------- */
void save_FSM(int start_state)/* print out state
transitions and output in table form only (5 per
row)  for those states which are reachable from the
start state */{
line #
1:  int i,j, reachable[max_num_states]={0},
2:    first=0, last=0, queue[max_num_states];
3:  queue[0] = start_state;
4:  reachable[start_state]=1;
/* breadth first search approach to finding which
states are reachable */
5:  while (first <= last) {

6:    if (!reachable [next_state
[queue[first]][0]]) {
7:        ++last;
8:        queue [last] = next_state
[queue[first]][0];
9:        reachable [queue[last]] =1;
10:       }

11:      if (!reachable
[next_state[queue[first]][1]]) {
12:          ++last;
13:          queue[last] =
next_state[queue[first]][1];
14:          reachable[queue[last]] =1;
15:      }

16:      ++first;

17:  } /* while */
/* print out state transition table for those
states reachable from the  start state, 5 in a row
*/
18:    for (i=0; i<5; ++i) fprintf(fp_perfect, "ps
fs/output  ");
```

```
19:    fprintf(fp_perfect, "\n");
20:    for (i=0; i<5; ++i) fprintf(fp_perfect, "
in=0:in=1   ");
21:    for (i=0, j=0; i<max_num_states; ++i)
22:      if (reachable[i]) {
23:          if (j==0) fprintf(fp_perfect, "\n");
24:          fprintf(fp_perfect,
"%2d%3d/%1d:%2d/%-4d,"
25:    i, next_state[i][0], output[i][0],
26:    next_state[i][1], output[i][1]);
27:          j = ++j % 5;
28:       }

29:    fprintf(fp_perfect, "\n");
} /* save_FSM */
```

Figure 6.19 Output Successful FSMs

The main procedure (Figure 6.20) of the GA consists of four parts. First validate_data (Figure 6.21, invoked on line 2 of Figure 6.20) checks to make sure that the values assigned to the constants do not contradict each other. For example,

```
#define bits_per_genome 50
```

```
#define bytes_per_genome 6
```

would cause an error message (lines 18-23 of Figure 6.21, invoked on line 2 of Figure 6.20) because seven bytes are needed to store a 50-bit genome. Then print_header (Figure 6.22, invoked on line 3 of Figure 6.20) writes the headings to the two output files. Prologue (Figure 6.23, invoked on line 13 of Figure 6.20) closes out those files when the program is finished executing. main (Figure 6.20) starts the GA and carries out the iterative evolutionary process.

```
/***********************************************/
/*                                             */
/* This is the basic driver for the GAs        */
/* The initial data validation tests are FSM   */
/* specific                                    */
/*                                             */
/***********************************************/
void main(void)
{
line #
1: int i, j;
2: if (!validate_data()) return;
3: print_header(&fp_stats, &fp_perfect);
```

```
/* the actual GA starts here */
4: initialize(); /* generation 0 */
5: save_stats();
6: for (generation=1; /* maximum fitness has not
been reached and more generations available */
7:     (generation <= last_generation) &&
8:     (fitness [sorted[row_of_sorted][0]] <
maxfitness);
9:     ++generation) {

10:        create_next_generation();
11:        save_stats();

12:    }

13:    prologue();
} /* main */
```

Figure 6.20 The main Part of the GA

```
/* ----------------------------------- */
/*    Validate Value of Constants */
/* ----------------------------------- */
int validate_data(void)/* tests to validate values
assigned to constants as not contradictory */
{
line #
1: int calculated;
/* TEST #1: Are there enough bits for storing the
number of each state? */

2: calculated = power_of_2(bits_per_state_num); /*
number of bits necessary */
3: if (max_num_states != calculatcd) {
4:    printf("(%d) max_num_states not = (%d)
2^bits_per_state_num \n,"
5:"make corrections and try again\n,"
max_num_states, calculated);
6:    return 0;
7: }
/* TEST #2: Are there enough bits for storing the
entire genome?        */

8: calculated =                  /* number of bits
necessary */
9: bits_per_state_num + num_of_inputs *
```

```
max_num_states
10: * (bits_per_state_num + bits_per_output);
11:   if (bits_per_genome != calculated) {
12:      printf("(%d) bits_per_genome not =\n, "
"(%d) bits_per_state_num +\n, "
13:"   num_of_inputs * max_num_state\n, "
14:"      * (bits_per_state_num +
bits_per_output)\n, "
15:"make corrections and try again\n, "
bits_per_genome,calculated);
16:      return 0;
17:   }
/* TEST #3: Are there enough bytes for storing the
entire genome?    */

18:   calculated = (bits_per_genome + 7)/8;   /*
number of bits necessary */
19:   if (bytes_per_genome != calculated) {
20:      printf("(%d) bytes_per_genome not = (%d)
(bits_per_genome + 7)/8\n, "
21:"make corrections and try
again\n, "bytes_per_genome,calculated);
22:      return 0;
23:   }

24:   return 1;
} /* validate_data */
```

Figure 6.21 Check for Contradictory Constant Values


```
/* ---------------------------------- */
/*    Print Header on Output Files    */
/* ---------------------------------- */
void print_header(void)/* print out header
information for the GA
         in the statistics and perfect score files
*/
{
line #
1: fp_stats =  fopen(printed_output, "w");
2: fp_perfect = fopen(perfect_score, "w");

3: fprintf(fp_stats,  "\nBenchmark\nStatistics for
run #%d on file %s\n, "
```

```
4:      run_num, printed_output);
5: fprintf(fp_perfect, "\nBenchmark\nPerfect
scorers for run #%d on file %s\n,"
6:      run_num, perfect_score);
7: fprintf(fp_stats,   "Seed=%.8f\n," seed);
8: fprintf(fp_perfect, "Seed=%.8f\n," seed);
9: fprintf(fp_stats,   "Crossover rate=%.2f;
Mutation rate=%.2f\n,"
10:      xover_rate, mutation_rate);
11:      fprintf(fp_perfect, "Crossover rate=%.2f;
Mutation rate=%.2f\n,"
12:      xover_rate, mutation_rate);
13:      fprintf(fp_stats,   "Population size=%d\n,"
popsize);
14:      fprintf(fp_perfect, "Population size=%d\n,"
popsize);
15:      fprintf(fp_stats,   "Select the top %d
individuals or %.2f%%,"
16:      select, 100. * (double)select /
(double)popsize );
17:      fprintf(fp_perfect, "Select the top %d
individuals or %.2f%%,"
18:      select, 100. * (double)select /
(double)popsize );
19:      fprintf(fp_stats,   " of each generation for
reproduction.\n\n");
20:      fprintf(fp_perfect, " of each generation for
reproduction.\n\n");
21:      fprintf(fp_stats,   " Gen  Max/# with Max  ");
22:      fprintf(fp_stats,   "Average/Median/Mode/#
with Mode Standard dev.\n");
23:   } /* print header information */
```

Figure 6.22 Prints Headings on Output Files

```
/* ------------------------------- */
/*      Close Output Files      */
/* ------------------------------- */
void prologue(void)/* close the statistics and
perfect score files */{
 fprintf(fp_stats,"\n"); fprintf(fp_perfect,"\n");
 fclose(fp_stats); fclose(fp_perfect);
} /* prologue */
```

Figure 6.23 Close Output Files

Typically, the initialize procedure (Figure 6.24, invoked on line 4 of Figure

6.20) starts the random number generator (line 2) and uses it to create the initial population. The random number generator coded in Turbo-Pascal in Goldberg's book (1989) has been converted to C. Generation 0 is created (lines 4-5, Figure 6.24) by generating bytes_per_genome = 57 bytes (with values of 0 through 255) as opposed to bits_per_genome = 453 bits. The extra bits which result are discarded when each child is bred for generation 1. The initialize procedure also prepares generation 0 to parent generation 1. Invoking procedure bytestring_to_FSM (line 6, Figure 6.24) converts a single genotype to its corresponding phenotype. The phenotype is used by calc_fitness (line 7, Figure 6.24) to determine the fitness for a single individual. The sorted array is initialized to prepare for the sorting of the population (line 8). After the population has been processed in this way, the population is sorted in decreasing order of fitness (line 10). As earlier stated, a stable merge-sort is implemented (Knuth 1973), but any sorting procedure can be used. To avoid moving the large genomes around in memory, the indices, which indicate the location of each individual in the pop array, is sorted. After the population is sorted, it is ready to parent the next generation.

```
/* ------------------------------------ */
/*      Initialize population     */
/* ------------------------------------ */
void initialize(void)
{
line #
1:  int i, j;
 /* Randomly generate initial population */
2:  Randomize(); /* initialize random number
generator */

3:  for (i=0; i < popsize; ++i) {

4:    for (j=0; j < bytes_per_genome; ++j)
5:      pop [new_pop] [i] [j] = Rnd(0, 255);
6:
bytestring_to_FSM(&start_state,&pop[new_pop][i][0])
;
7:    fitness[i] = calc_fitness(start_state);
8:    sorted[0][i]=i;
9:  }
10:   row_of_sorted = sort(0, popsize - 1, 0); /*
in order of fitness */
} /* initialize */
```

Figure 6.24 Initialization

Control is then returned to main (Figure 6.20) which then calls
save_stats (Figure 6.25), invoked on line 11 of Figure 6.20.
save_stats computes and outputs the statistics for a generation. First
the maximum fitness is set (line 3, Figure 6.25) and the number individuals
with that fitness is determined (lines 4-6). The median is located at the
center of the sorted array (lines 7-12). While popsize for this program
is an even number (lines 8-10), the coding allows for it to be odd (lines 11-
12). The calculations for the average (lines 13-15) and standard deviation
(lines 16-18) are straightforward. The fitnesses are added together and divided
by popsize to calculate the average (lines 13-15). The standard deviation
is similarly coded utilizing the built-in functions of pow and sqrt. (lines
16-18). fitness_count[i] is a count of the number of FSMs in the
population which have a fitness of i. This is used to find the mode (lines
19-26). The code to find the mode must allow for a multimodal possibility.
There will be at least a single mode (lines 20-23). Multimode counts the
number of additional modes (lines 24-25). After the statistics for the
generation are recorded on the *.st file (lines 27-29), multimode permits the
recording of all the modes (lines 30-37). If multimode = 0, there was
only one mode and the single line which was output (lines 27-29) fully
describes the statistics for the generation.

```
/* ------------------------------------   */
/*      Save Statistics        */
/* ------------------------------------   */
void save_stats(void)
{
line #
1:  int max, num_of_max=1, mode=maxfitness,
multimode=-1, i, j, k;
2:  double avg=0, median, standard_dev=0;


/* ------------------------------------   */
/*      Determine maximum      */
/* ------------------------------------   */

3:  max = fitness[ sorted[row_of_sorted][0] ];
4:  while (num_of_max<popsize
5:      && max==fitness[
sorted[row_of_sorted][num_of_max] ])
6:    ++num_of_max;

/* ------------------------------------   */
/*      Determine median       */
/* ------------------------------------   */
```

```
7:  i = popsize/2;
8:  if (2*i == popsize)
9:     median = (double) (fitness[
sorted[row_of_sorted][i-1] ] +
10:       fitness[ sorted[row_of_sorted][i] ] ) /
2.0;
11:   else
12:      median = (double) fitness[
sorted[row_of_sorted][i] ];

/* ------------------------------------  */
/*      Determine average       */
/* ------------------------------------  */
13:   for (i=0; i<popsize; ++i)
14:      avg += (double) fitness[i];
15:   avg /= (double) popsize;

/* ------------------------------------  */
/*      Determine standard dev.   */
/* ------------------------------------  */

16:   for (i=0; i<popsize; ++i)
17:      standard_dev += pow( (double) fitness[i] -
avg, 2.);
18:   standard_dev = sqrt(standard_dev/ (double)
popsize);

/* ------------------------------------  */
/*      Determine mode         */
/* ------------------------------------  */

19:   for (i=maxfitness, j=5; i>=0; --i) {
20:      if (fitness_count[mode] < fitness_count[i])
{
21:         mode=i;
22:         multimode=0;
23:      }
24:      else
25:         if (fitness_count[mode] ==
fitness_count[i]) ++multimode;
26:   }

/* ------------------------------------  */
/*      Print Statistics        */
/* ------------------------------------  */
```

```
27:    fprintf(fp_stats, "%4d%6d/%-10d%10.4f/%5.1f
/%3d /%-11d%11.4f\n,"
28:        generation, max, num_of_max, avg, median,
29:        mode, fitness_count[mode], standard_dev);
30:    if (multimode)
31:      for (j=0, i=mode; j<multimode; ++j) {
32:        --i;
33:        while (i >= 0 && fitness_count[mode] !=
fitness_count[i]) --i;
34:        if (i >= 0)
35:          fprintf(fp_stats, "%31.4f/%5.1f /%3d
/%-11d\n,"
36:      avg, median, i, fitness_count[i]);
37:      }
} /* save_stats */
```

Figure 6.25 Calculating and Outputting the Statistics for Each Generation

Each time through the for loop of main (lines 6-12, Figure 6.20) yields a new generation. The iterative process halts when at least one perfect scoring ant has evolved (line 8) or when last_generation generations have been bred (line 7). create_next_ generation (Figure 6.26, called on line 10 of Figure 6.20) exchanges the values of old_pop and new_pop (line 3, Figure 6.26) to label the previous generation as the parents and indicate where the newborns are to be placed, and initializes the fitness histogram (line 2).

```
/* ----------------------------------   */
/*      Create next generation      */
/* ----------------------------------   */
void create_next_generation(void){
line #
1:  int fit, i;

2:  for (i=0; i<=maxfitness; ++i)
fitness_count[i]=0;
3:  old_pop = new_pop; new_pop = ++new_pop % 2;

4:  for (i=0; i<popsize; ++i) {

5:    xover(&pop[new_pop][i][0]);
6:    mutate(&pop[new_pop][i][0]);
7:    bytestring_to_FSM(&start_state,
&pop[new_pop][i][0]);
8:    fitness[i] = calc fitness(start_state);
9:    sorted[0][i] = i;
```

```
10:    }
```

```
11:    row_of_sorted = sort(0, popsize - 1, 0); /*
in order of fitness */
} /* create_next_generation */
```

Figure 6.26 Breeding the Next Generation

The `xover` procedure (Figure 6.27, invoked by line 5 of Figure 6.26)
selects a mating pair from the top `select` members of the population
(lines 3-4, Figure 6.27) to produce a single child. One parent is the bit
donor indicated by `present_parent` (line 1). Each time a bit is copied
to the child (line 10), a random number between 0 and 1 is generated (line
8). Whenever this number falls below `xover_rate` (line 8), a crossover
point is created by changing the donating parent (line 9). The `mask`
permits a single bit to be copied from the `present_parent` to the child
(lines 6 and 10). As in `unpack_genome` (Figure 6.12), the `mask`
contains a single 1 (line 6, Figure 6.27) which is shifted through the eight
bits of the byte (line 7, Figure 6.27). ANDing `mask` with a byte from the
parent isolates the desired bit (line 10, Figure 6.27). ORing this with the
corresponding byte of the child (line 10), which was initialized to zeroes
(line 6), copies the bit from the parent to the child. `i` is the index into the
bytestrings of the parent and child, `j` is the bit count, and `k` counts the bits
in the ith byte (lines 5-11). The for loops (lines 5-11) move along the
genome processing eight bits per byte for 453 bits.

```
/* ------------------------------------    */
/*           Crossover        */
/* ------------------------------------    */
void xover(packed_genome child)/* selects a pair of
parents and mates them to produce a single child */
{

line #
1:  int i, j, k, parent[2], present_parent=0;
2:  byte mask;
/* i = byte position in genome   j = bit position in
genome */

3:  k = Rnd(0, select - 1); parent[0] =
sorted[row_of_sorted][k];
4:  k = Rnd(0, select - 1); parent[1] =
sorted[row_of_sorted][k];

5:  for (i=0, j=0; j < bits_per_genome; ++i)
```

```
6:     for (k=0, mask=0x80, child[i]=0;
7:       k<8 && j<bits_per_genome; ++k, ++j, mask
>>= 1) {
8:       if (Random() <= xover_rate) /* change
present parent */
9:present_parent = (++present_parent) % 2;
10:        child[i] |= mask & pop [old_pop]
[parent[present_parent]] [i];
11:      }
} /* xover */
```

Figure 6.27 Mating

Mutation (Figure 6.28) is then applied (on line 6 of create_next_ generation, Figure 6.26) at mutation_rate per bit. In the mutation procedure i, j, and k, and their loops (lines 3-4) serve the same purposes as they do in xover (lines 5-7, Figure 6.27). In mutate (Figure 6.28), a random number is generated for each bit (line 5). To implement the #defined mutation_rate, a bit is inverted when the corresponding random number falls below mutation_rate (lines 5-6). The EXOR can either retain a value or invert (NOT) it. Zero is the identity element for EXOR, therefore, when one operand of an EXOR is a zero, the other operand is returned as the answer. When one operand of an EXOR is a 1, the value of the other operand is inverted. Here again the mask contains a single 1 (line 4) which is used to invert a bit (line 6). When a bit is to be mutated, the mask is EXORed with the present byte to invert the indicated bit and retain the seven other bits. The rest of create_next_generation (lines 7-11, Figure 6.26) prepares these children to parent the next generation. The process is identical to the one carried out by initialize (lines 6-10, Figure 6.24). When control is returned to main, save_stats (Figure 6.25, called on line 11 of Figure 6.20) calculates the statistics for each generation and outputs it to the file with the st extension.

```
/* -------------------------------------  */
/*          Mutate              */
/* -------------------------------------  */
void mutate(packed_genome bytestring)
/* ^ is EXOR operator:   0^x=x , i.e. 0 preserves
value; 0 is the identity element.   1^x=x', i.e. 1
inverts value.
   Therefore, mask^byte inverts the single position
in byte
where mask has a 1.                     */
{
line #
```

```
1: int i, j, k;
2: byte mask;
/* i = byte position in genome   j = bit position in
genome */

3: for (i=0, j=0; j < bits_per_genome; ++i)
4:    for (k=0, mask=0x80; k<8 &&
j<bits_per_genome; ++k, ++j, mask >>= 1)
5:       if (Random() <= mutation_rate)
6:bytestring[i] = mask ^ bytestring[i];} /* mutate
*/
```

Figure 6.28 Mutation

Figure 6.29 contains a listing of an **st** file, and Figure 6.30 is the corresponding **ps** file. All numbers in Figure 6.29, except for those in the first column, are related to fitness. The column labeled "Max/# with Max" is the maximum fitness appearing in the generation and the number of individuals in the generation which have that fitness. Similarly for "Mode" and "# with Mode."

Benchmark

Statistics for run #1 on file r1b.st

Seed = 0.69343845

Crossover rate = 0.01;   Mutation rate = 0.01

Population size = 65536

Select the top 3277 individuals or 5.00% of each generation for reproduction.

Gen	Max/# with Max	Average	/Median	/Mode	/# with Mode	Standard dev
0	64/1	3.7148	/1.0	/0	/26956	7.2664
1	79/1	7.8863	/2.0	/0	/18636	11.6169
2	67/1	11.7417	/3.0	/0	/16454	15.2043
3	65/1	14.1120	/5.0	/0	/15110	16.6865
4	75/1	15.9189	/10.0	/42	/15638	17.4709
5	79/1	17.4440	/10.0	/42	/18097	18.0109
6	79/1	18.4360	/10.0	/42	/18899	18.1688
7	80/1	19.0267	/10.0	/42	/18411	18.2153
8	81/1	18.5346	/10.0	/42	/13883	17.9810
9	81/1	18.3342	/10.0	/0	/10563	18.0576
10	81/1	19.8597	/11.0	/0	/9674	18.6749
11	81/1	21.6483	/14.0	/0	/8610	19.2626
12	81/2	23.7276	/21.0	/42	/8057	20.0531
13	81/5	25.1715	/22.0	/42	/8555	20.5392
14	81/12	27.8097	/32.0	/42	/8688	21.2958
15	82/1	31.9812	/35.0	/42	/9499	22.5329
16	83/1	42.3225	/42.0	/74	/17242	26.0879
17	83/1	41.4549	/42.0	/74	/15167	26.9132
18	83/4	37.0724	/35.0	/42	/8977	26.4808
19	83/12	38.0038	/37.0	/80	/11992	27.4448
20	83/27	40.4289	/42.0	/80	/10893	28.2652
21	83/48	43.4303	/42.0	/81	/14017	29.0688
22	83/93	43.6623	/42.0	/81	/14858	29.2561
23	84/3	43.3132	/42.0	/81	/13922	29.2306

24	84/3	44.5001	/42.0	/81	/9594	29.1818
25	84/20	44.8606	/42.0	/82	/9478	29.2679
26	84/58	50.9981	/47.0	/83	/23495	29.8806
27	84/146	55.1226	/58.0	/83	/27333	29.0099
28	85/2	56.2042	/58.0	/83	/28122	28.7117
29	85/5	54.1236	/52.0	/83	/25421	29.1874
30	86/2	48.1696	/42.0	/83	/16582	29.7283
31	86/5	54.6107	/52.0	/84	/23666	29.6827
32	86/6	57.0353	/79.0	/84	/26154	29.2422
33	86/10	57.2668	/79.0	/84	/25964	29.2187
34	86/29	55.3385	/57.0	/84	/22706	29.5564
35	86/78	50.7199	/43.0	/84	/12132	30.0218
36	86/217	51.5211	/44.0	/85	/16674	29.8535
37	87/2	52.2516	/46.0	/85	/17704	29.6856
38	88/1	52.0490	/46.0	/85	/16301	29.6341
39	88/2	50.4423	/43.0	/85	/8220	29.2398
40	88/3	53.4057	/47.0	/86	/18050	28.7444
41	88/13	54.2980	/52.0	/86	/18943	28.7117
42	88/37	54.8068	/52.0	/86	/19261	28.6745
43	88/136	54.8835	/52.0	/86	/18390	28.5455
44	88/477	53.7232	/50.0	/86	/15006	28.4901
45	89/1	49.9201	/43.0	/87	/8672	28.0389

Figure 6.29 Listing of file with st extension from sample run.

Benchmark

Perfect scorers for run #1 on file r1b.ps

Seed = 0.69343845

Crossover rate = 0.01;   Mutation rate = 0.01

Population size = 65536

Select the top 3277 individuals or 5.00% of each generation for reproduction.

Generation 45; Start State=4; Fitness of 89 reached in 197 time steps.

ps fs/output ps fs/output   psfs/output   ps fs/output   ps fs/output

in=0:in=1   in=0:in=1   in=0:in=1   in=0:in=1   in=0:in=1

0 22/2:4/2   1 10/0:9/2   2   8/3:1/2   4 10/0:4/2   6   1/2:6/2

7 6/0:8/2   8 6/2:6/2   9 7/0:22/2   10 9/0:0/2   22 25/2:22/2

25 7/1:2/0

Figure 6.30 Listing of file with ps extension from sample run.

With the benchmark program now complete, additional strategies can be added. Two of the new strategies are mutually exclusive. Each of these algorithms reorganizes each genome during run time in order to enhance schema development. The next two sections describe these two algorithms.

6.3. The SFS Algorithm

The SFS algorithm reorganizes a finite state machine by $\underline{\underline{S}}$tandardizing the transitions to the $\underline{\underline{F}}$uture or next $\underline{\underline{S}}$tates according to a mathematical function. The purpose of standardizing transitions is to enhance schema growth. In the next two sections, the SFS algorithm and its coding are described and two SFS reorganizations are presented.

6.3.1 SFS: Description and Coding

In order to standardize transitions, SFS has to rename the states. Therefore, it has to move the next state and output for a renamed state to the proper location in the genome based on its new state name. It also has to change all references to the renamed state in the next state part of the FSM table. Since there is limited space in the genome, a state, which is assigned a new name, has its name and location exchanged with the state which has the desired name. For example, if SFS reassigns state q_{20} to state q_4, the twentieth and fourth rows of the FSM table must be interchanged, and their names in the next state part of the table must also be exchanged. Procedures xchange_states and switch_rows perform all necessary exchanges. switch_rows (Figure 6.31) exchanges the data in two rows of a table. It is called by xchange_states (Figure 6.32) which not only exchanges the rows of data (lines 3-4, Figure 6.32) but also appropriately exchanges the next state names in the next_state table (lines 7-11, Figure 6.32) The for loops (lines 5-6) move through the states, and inputs for each state.

```
/* ------------------------------------------    */
/*          Switch Rows              */
/* ------------------------------------------    */
void switch_rows(int row1, int row2, table tbl)
{
line #
1:  int i, temp;

2:  for (i=0; i<num_of_inputs; ++i) {
3:    temp = tbl[row1][i];
4:    tbl[row1][i] = tbl[row2][i];
5:    tbl[row2][i] = temp;
6:  }
} /* switch rows */
```

Figure 6.31 Interchange Rows of a Table

```
/* ------------------------------------------ */
/*          Exchange States           */
/* ------------------------------------------ */
void xchange_states(int state1, int state2)
```

```
{
line #
1:  int i,j;

2:  if (state1 != state2) {

3:    switch_rows(state1, state2, next_state);
4:    switch_rows(state1, state2, output);

5:    for (i=0; i<max_num_states; ++i)
6:      for (j=0; j<num_of_inputs; ++j)
7:if (next_state[i][j] == state1)
8:  next_state[i][j] = state2;
9:else
10:  if (next_state[i][j] == state2)
11:    next_state[i][j] = state1;
12:   }
} /* xchange_states */
```

Figure 6.32 Interchange States Based on Name Change

The SFS procedure, called by reorganizeFSM (Figure 6.33), utilizes the procedures in Figures 6.31 and 6.32 to reorganize the FSM. reorganizeFSM loops through the top select individuals (line 2) for their reorganization. There is no need to reorganize those FSMs which cannot be selected as a parent. Since SFS operates on the phenotype, the genome must be converted to the phenotype (lines 3-4) before SFS is invoked (line 5), and then returned to the bytestring form for storage (lines 6-7).

```
/* ---------------------------------- */
/*        Reorganization        */
/* ---------------------------------- */
void reorganizeFSM(void){
line #
1:  int i;

2:  for (i=0; i<select; ++i) {
3:    bytestring_to_FSM (&start_state,
4:      &pop[new_pop][sorted[row_of_sorted][i]][0]);
5:    SFS (&start_state);
6:    FSM_to_bytestring (start_state,
7:      &pop[new_pop][sorted[row_of_sorted][i]][0]);
8:    }
} /* reorganizeFSM */
```

Figure 6.33 Reorganize parent pool

SFS was designed with the intent of enhancing schema growth by standardizing transitions to the next state. A mathematical function determines the desired next state for each present state/input pair. A finite state machine with 32 states and two inputs will have 64 transitions to next states. Since the genome has only 32 states, each state becomes the desired future state for two present state/input pairs. In addition the start state is standardized to q_0. Consequently, SFS consists of three steps. First the initial state is exchanged with state q_0, and q_0 is placed on a queue so its future states can be reassigned.

The second step (Figure 6.34) of SFS (completed in Figure 6.35) processes the state at the front of the queue to_be_processed2 (lines 3-4, Figure 6.34)) by trying to assign its next state names based on a mathematical function. Most of the state reassignments, sometimes all of them, are carried out by step2. This is the most complex part of the SFS algorithm. This algorithm was designed for input $j \in \{0,1\}$ and is not applicable for other values of j. Next states are assigned by the following multi-part mathematical function. The argument cut_off = max_num_states/2 (line 12, Figure 6.35). Then state i with input j has desired next state =

1) $2i+j+1$ when $i \le$ cut_off - 2 (line 7, Figure 6.34),

2) max_num_states - 1 when $i =$ cut_off - 1

   (line 8, Figure 6.34), and

3) $2(\text{max\_num\_states}-2-i)+j+1$ for

cut_off - 1 < i < max_num_states - 1   (line 10, Figure 6.34).

Note that state max_num_states - 1 is not included in this function and, for state cut_off - 1 the desired next state is the same for the two inputs. As implemented in the program, the desired next state for state max_num_states - 1 is set to 0 (line 9). This case will be covered shortly.

```
/* ----------------------------------  */
/*       Step 2 of SFS           */
/* ----------------------------------  */
void step2 (int cut_off, int *next_step2, int
*last_step2, int *last_step3,
  int used[], int to_be_processed2[],
  int to_be_processed3[2*max_num_states][2])
{
line #
1: int i, j, new_next_state;
```

```
2:  while (*next_step2 <= *last_step2) {

3:    i = to_be_processed2[*next_step2];
4:    ++*next_step2;
5:    for (j = 0; j < num_of_inputs; ++j)

6:       if (!used[next_state[i][j] + 1]) {
      /* next state has not been reassigned */
7:if (i < cut_off-1) new_next_state = 2*i + j + 1;
8:else if (i == cut_off-1) new_next_state =
max_num_states-1;
9:else if (i == max_num_states-1) new_next_state =
0;
/* since state 0 is always used, the last state
  will always be processed by step 3 */
10:else new_next_state = 2*(max_num_states-2-i) + j
+ 1;

11:if (used[new_next_state + 1]) {
  /* save state for processing by step 3 */
12:  ++*last_step3;
13:  to_be_processed3[*last_step3][0] = i;
14:  to_be_processed3[*last_step3][1] = j;
15:}
16:else {
17:  xchange_states(next_state[i][j],
new_next_state);
18:  used[new_next_state+1] = 1;
19:  ++*last_step2;
20:  to_be_processed2[*last_step2] =
new_next_state;
21:}
22:      }
23:   } /* while */} /* step2 */
```

Figure 6.34 Second step of the SFS algorithm.

The while loop (lines 2-23) in step2 processes the states which have been put on the queue to_be_processed2. This queue contains the states which have been reassigned new state names (numbers) and which have to be processed by step2 for next state reassignments. next_step2 and last_step2 (line 2) are, respectively, the positions of the first and last elements on the queue. The array to_be_processed2 is sufficiently large so that a cyclic queue will not be needed; that is, to_be_processed2 will never have to store an element after the last position in the array is filled.

The for loop of `step2` on line 5 (Figure 6.34) makes sure next states are reassigned for both inputs. When a state is reassigned, it is tagged by setting `used[i]=1` (line 18) to indicate that reassignment has occurred. This prevents a state from again being reassigned if it is the next state for more than one present state-input pair. It also prevents a second present state-input pair from mapping into an already assigned position, since the reassignment function is two-to-one.

Looking at Figure 6.34, if the next state for a given present state-input pair has been reassigned, there is nothing to do. The test for the need to reassign the next state is on line 6. When reassignment is required, three levels of cascading if-then-else (lines 7-10) determine the desired `new_next_state` base on the mathematical function. Then an if (line 11) determines whether `new_next_state` is available. When `new_next_state` is available (lines 16-21), it is exchanged with the `next_state` (line 17). `new_next_state` is then tagged as `used` (line 18), and added to the queue for processing by `step2` (lines 19-20). On the other hand, when `new_next_state` is used (lines 11-15), present state i with input j are stored on queue `to_be_processed3` for next state reassignment by step 3 of SFS.

`step2` is a separate subroutine because it is invoked from two different places of `SFS` (lines 13 and 32, Figure 6.35). The `SFS` algorithm (Figure 6.35) consists of three steps. The first step standardizes the `start_state` as state 0. `start_state` and state 0 are interchanged (lines 6 and 8). State 0 is then tagged as reassigned (line 7) and placed on the queue for processing by the second step (lines 9-10). The queue for the third step of `SFS` is initialized (line 11). Then `step2` is called (line 13). It executes until its queue is empty (line 2, Figure 6.34). Then step 3 (lines 15-35, Figure 6.35) processes the present state-input pairs on its queue. `next_step3` and `last_step3` (line 15, Figure 6.35) are, respectively, the positions of the first and last elements on the queue. The array `to_be_processed3` is sufficiently large so that a cyclic queue will not be needed; that is, `to_be_processed3` will never have to store an element after the last position in the array is filled. Between the time a present state-input pair is placed on `to_be_processed3` and the time they are taken off the queue for processing by step 3 (lines 16-18), the `next_state` could have been reassigned. Consequently, the `next_state` must be checked to see if it has been reassigned (line 19). When it has not been, step 3 looks for the closest available state in the genome (lines 20-27). This will be the `new_next_state`. k (lines 20-21) indicates the first available state to the left of the present state, and n (lines 22-23) indicates the closest available state to its right. Whichever is closer to the present state is the `new_next_state` (lines 24-27). `next_state` and `new_next_state` are interchanged (line 28). `new_next_state` is then tagged as `used` (line 29), and added to the queue for processing by `step2` (lines 30-31). SFS requires that reassignment of next states is always first attempted by step2; a state is reassigned by step 3 only when step2 has failed. Therefore step2 is called (line 32) to process this newly assigned state and any states which are reassigned during this execution of step2. Step 3 continues processing the states on its queue when step2's queue is again empty.

```
/* ------------------------------------- */
/*  Standardize Next (Future) States   */
/* ------------------------------------- */
void SFS (int *start_state)
/* step 1: Make state 0 the start state.  step 2:
Standardize some present state/next state pairs on
the genome.      Let cut_off=max_num_states/2.
Then, for state i with input j = 0,1, next state k
is      k = 2i+j+1          for i <= cut_off-2
k = max_num_states-1      for i = cut_off-1      k
= 2(max_num_states-2- i)+j+1 for 31> i>= cut_off
step 3: Reassign the remaining next states by
placing them in the genome      as close as
possible to the present state. Newly assigned
states      are first processed through step 2.
*/
{
line #
1: int i, j, k, n, cut_off, new_next_state,
2:    next_step2, last_step2, next_step3,
last_step3,
3:    used [max_num_states+2] = {0},    /* The
to_be_processed arrays are queues which hold the
states      which must be processed by steps 2 & 3.
*/
4:    to_be_processed2 [max_num_states],
5:    to_be_processed3 [2*max_num_states] [2];
 /* STEP 1 */
6: xchange_states(0, *start_state);
7: used[1] = 1; /* state 0 is used */
8: *start_state = 0;
9: last_step2 = next_step2 = 0;
10:    to_be_processed2 [next_step2] = 0;
11:    next_step3 = 0; last_step3 = -1;
 /* STEP 2 */
12:    cut_off = max_num_states/2;
13:    step2(cut_off, &next_step2, &last_step2,
&last step3,
14:      &used[0], &to_be_processed2[0],
```

```
to_be_processed3);
 /* STEP 3 */
15:   while (next_step3 <= last_step3) { /* States
appear on this list because the the next state had
not been reassigned and the desired next state had
already   been used. */
16:      i = to_be_processed3[next_step3] [0];
17:      j = to_be_processed3[next_step3] [1];
18:      ++next_step3;
19:      if (!used[next_state[i][j] + 1]) {
   /* the next state may have been reassigned since
this      state/output pair was put on this list */
/* Look for closest available state.   */
20:      k = i - 1;
21:      while (used[k + 1]) --k; /* at worst will
stop at used[0] */
22:      n = i + 1;
23:      while (used[n + 1]) ++n; /* at worst will
stop at
            used[max_num_states + 1]   */
24:      if (((i-k <= n-i) && (k > 0)) || (n ==
max_num_states))
25:new_next_state = k;
26:      else
27:new_next_state = n;
28:      xchange_states(next_state [i] [j],
new_next_state);
29:      used[new_next_state+1] = 1;
30:      ++last_step2;
31:      to_be_processed2[last_step2] =
new_next_state;
32:      step2(cut_off, &next_step2, &last_step2,
&last_step3,
33: &used[0], &to_be_processed2[0],
to_be_processed3);
34:      }
35:   } /* while */
} /* SFS */
```

Figure 6.35 The SFS Algorithm

The benchmark program can be easily altered to include a SFS reorganization. In addition to adding the procedures in Figures 31 through 35, in the #define statements alter the file names for printed_output (r1sfs.st is suggested) and perfect_score (r1sfs.ps) to indicate that the SFS algorithm is implemented. Add

#define reorganization 1 /* use 1 for reorganization; 0 otherwise */

to the #define statements at the beginning of the program. Change the beginning of the print_header subroutine of Figure 6.21 to the beginning indicated in Figure 6.36.

```
/* --------------------------------   */
/*   Print Header on Output Files   */
/*     Modified to include SFS    */
/* --------------------------------   */
void print_header(void)/* print out header
information for the GA in  the statistics and
perfect score files */
{
line #
1: fp_stats =  fopen(printed_output, "w");
2: fp_perfect = fopen(perfect_score, "w");

3: if (reorganization) {
4:    fprintf(fp_stats,  "\nSFS\nStatistics for run
#%d on file %s\n,"
5:       run_num, printed_output);
6:    fprintf(fp_perfect, "\nSFS\nPerfect scorers
for run #%d on file %s\n,"
7:       run_num, perfect_score);
8: }

9: else {
10:     fprintf(fp_stats,  "\nBenchmark\nStatistics
for run #%d on file %s\n,"
11:      run_num, printed_output);
12:     fprintf(fp_perfect, "\nBenchmark\nPerfect
scorers for run #%d on file %s\n,"
13:      run_num, perfect_score);
14:   } fprintf(fp_stats,  "Seed=%.8f\n," seed);
fprintf(fp_perfect, "Seed=%.8f\n," seed);    !   }
/* print header information */
```

Figure 6.36 Prints Headings on Output Files–Modified to Include SFS

The FSMs are reorganized before parenting the next generation. Since a generation is prepared for parenting in initialize (Figure 6.24) and create_next_generation (Figure 6.26), reorganizeFSM is called in these two procedures. The line

```
if (reorganization) reorganizeFSM();
```

must be inserted into initialize (Figure 6.24) and create_next_ generation (Figure 6.26) immediately after the common line (line 10 of Figure 6.24, line 11 of Figure 6.26)

```
row_of_sorted = sort(0, popsize - 1, 0);
```

for the reorganization of the parent pool to be carried out.

There is one other change which is not necessary for proper execution, but can be installed for esthetic purposes. Perfect scoring individuals are printed out before they are reorganized. If it is desirable to see these machines in their reorganized format, SFS must be called just before a perfect scoring machine is printed out. To do this add the line

```
if (reorganization) SFS(&start_state);
```

between lines 54 and 55 of Figure 6.18, the calc_fitness procedure. This reorganizes the machine strictly for the printout, but it does not alter its stored format.

Except for the changed title, the output for an SFS run appears the same as for the benchmark. If perfect scoring machines are reorganized for the printout, an observer familiar with the algorithm might recognize a machine as having been reorganized by the SFS algorithm. In the next section two FSMs are walked through an SFS reorganization.

6.3.2 Examples of SFS Reorganization

Because the SFS algorithm is rather complex, two examples will be presented. In each successive table the changes from the previous table are indicated in boldface. For the first example, SFS is applied to the FSM in Table 6.5. The first step (lines 6-11, Figure 6.35) changes the start state, q_{13}, to q_0 changing Table 6.5 to Table 6.6.

Start state: q_0

Present state	Next state/output for input = 0	Next state/output for input = 1
q_0	$q_5/0$	$q_9/2$
q_5	$q_0/2$	$q_5/2$
q_9	$q_5/1$	$q_{24}/2$
q_{24}	$q_0/3$	$q_{24}/2$

Table 6.6 FSM of Table 6.5 after Step 1 of SFS

q_0 is then processed by step2 (Figure 6.34). According to the function, the next state for q_0 with input 0 should be 2 x 0 + 0 + 1 q_1, and for q_0 with input 1 the next state should be 2 x 0 + 1 + 1 q_2 (line 7, Figure 6.34).

Since q_5 is the next state for q_0 with input 0 (Table 6.6), q_5 must be changed to q_1. Similarly, because q_9 is the next state for q_0 with input 1 (Table 6.6), q_9 must be changed to q_2. Making these changes yields Table 6.7. Start state: q_0

Present state	Next state/output for input = 0	Next state/output for input = 1
q_0	$q_1/0$	$q_2/2$
q_1	$q_0/2$	$q_2/2$
q_2	$q_1/1$	$q_{24}/2$
q_{24}	$q_0/3$	$q_{24}/2$

Table 6.7 FSM of Table 6.6 after Next States for q_0 Reassigned

At this point states q_0, q_1, and q_2 have been reassigned and tagged as used, and any transitions to these states will not be changed as SFS continues. q_1 and q_2 must now be processed by step2. Since the next states for q_1 (q_0 and q_2) have been tagged, no further processing is needed for q_1. Similarly, for the next state for q_2 with input 0. But the next state for q_2 with input 1 has to be reassigned. Using the formula (line 7, Figure 6.34), q_{24} should become 2 x 1 + 1 + 1 q_6. Since q_6 has not been used, the change is made, yielding the FSM in Table 6.8. q_6 is then tagged and processed by step2. Because the next states of q_6 have been tagged, the SFS reorganization is complete.

Start state: q_0

Present state	Next state/output for input = 0	Next state/output for input = 1
q_0	$q_1/0$	$q_2/2$
q_1	$q_0/2$	$q_1/2$
q_2	$q_1/1$	$q_6/2$
q_6	$q_0/3$	$q_6/2$

Table 6.8 FSM of Table 6.5 after SFS

A large FSM is needed to demonstrate more of the SFS algorithm. The reorganization of the FSM in Table 6.9 offers more insight to SFS. In the following discussion to_bc_processed$i$ and queue$i$, $i \in \{2,3\}$, are used interchangeably.

A simple summary of step2 will clarify the SFS reorganization which follows.

When a state is removed from queue2 for processing, for that state and a given input, check the next state in the next_state table.

1) If the next state is used, it has already been reassigned. An attempt to reassign it at this point will undo the previous reassignment for that state. A previous reassignment is never changed by SFS. Therefore, when the next state is used, no further processing is needed for the present state-input being processed.

2) When the next state is not used, calculate the desired next state (lines 7-10, Figure 6.34). If the desired next state is used, there is a conflict of state assignment. An attempt to reassign it at this point will undo the previous reassignment for that state. Since step2 cannot reassign the next state, the present state-input pair is put on queue3 (lines 11-15, Figure 6.34) for reassignment by step 3.

3) Only when both the next state and the desired next state are not used will they be interchanged (lines 16-21, Figure 6.34).

It is suggested that the reader duplicate Table 6.9 and append a list of the states that do not appear on the table. Changing the states as indicated in the following description will make it easier to follow the walk-through. It is very difficult to follow the progress of the algorithm without a complete updated table available.

Start state: q_8

Line #	Present state	Next state/output for input = 0	Next state/output for input = 1
1	q_8	$q_{19}/3$	$q_{16}/3$
2	q_0	$q_{13}/2$	$q_{21}/2$
3	q_2	$q_{21}/3$	$q_{18}/3$
4	q_3	$q_{19}/0$	$q_{11}/1$
5	q_5	$q_{11}/0$	$q_{20}/1$
6	q_6	$q_5/0$	$q_{15}/1$
7	q_9	$q_2/3$	$q_{23}/0$
8	q_{10}	$q_2/1$	$q_0/2$
9	q_{11}	$q_{25}/1$	$q_{14}/0$
10	q_{13}	$q_{16}/3$	$q_8/0$

11	q_{14}	$q_9/1$	$q_{16}/1$
12	q_{15}	$q_{28}/3$	$q_{20}/2$
13	q_{16}	$q_0/0$	$q_{10}/1$
14	q_{18}	$q_{11}/3$	$q_9/1$
15	q_{19}	$q_{13}/0$	$q_{18}/2$
16	q_{20}	$q_{13}/2$	$q_3/0$
17	q_{21}	$q_0/2$	$q_3/0$
18	q_{23}	$q_{10}/0$	$q_6/2$
19	q_{25}	$q_{30}/0$	$q_{25}/1$
20	q_{26}	$q_{20}/3$	$q_{15}/0$
21	q_{28}	$q_{18}/1$	$q_{26}/0$
22	q_{30}	$q_3/1$	$q_{16}/2$

Table 6.9 FSM before SFS

In the following walk-through, all statements which start with "As a result of this exchange, the next states for ..." refer to the updated next state table. These next states are the result of interchanging state names in the "present state" column of the most recent version of Table 6.9. All other state changes are made by the program and the program lines which carry out the changes are given.

Updated tables are periodically provided indicating only those lines of the table which have been changed with the changes on each line in boldface.

When SFS is applied to Table 6.9, first step 1 interchanges q_0 and q_8 to make q_0 the start state (lines 6-8, Figure 6.35). As a result of this exchange, the next states for q_0 become q_{19} and q_{16}, and for q_8 they become q_{13} and q_{21}.

Also, this exchange changes the next state (lines 5-11, Figure 6.32) of q_{10} with input 1 to q_8, of q_{13} with input 1 to q_0, of q_{16} with input 0 to q_8, and of q_{21} with input 0 to q_8. q_0 is tagged as used and put on queue2 for processing by step2 (lines 7 and 9-10, Figure 6.35). When this is completed (Table 6.10), used: q_0, and to_be_processed2: q_0.

Start state: q_0

Line # from Table 6.9	Present state	Next state/output for input = 0	Next state/output for input = 1
1	q_0	$q_1/3$	$q_2/3$
2	q_8	$q_{13}/2$	$q_{21}/2$
8	q_{10}	$q_2/1$	$q_8/2$
10	q_{13}	$q_{16}/3$	$q_0/0$
13	q_{16}	$q_8/0$	$q_{10}/1$
17	q_{21}	$q_8/2$	$q_3/0$

Table 6.10

step2 is then invoked (line 13, Figure 6.35). q_0 is removed from queue2 for processing by step2 (lines 3-4, Figure 6.34). The next state for q_0 with input 0 is q_{19}. q_{19} is not used and should be 2x0+0+1 q_1 (line 7, Figure 6.34). Since q_1 is not used, q_1 and q_{19} are interchanged (line 17, Figure 6.34). As a result of this exchange, the next states for q_1 become q_{13} and q_{18}. The next states for q_{19} are irrelevant because, after the exchange is complete, q_{19} is not reachable. In actuality the irrelevant states are retained in the table; they are just omitted from this discussion.

As was planned, the next state of q_0 with input 0 is changed to q_1 (lines 5-11, Figure 6.32). Also, this exchange changes the next state (lines 5-11, Figure 6.32) of q_3 with input 0 to q_1. q_1 is tagged as used and put on queue2 for processing by step2 (lines 18-20, Figure 6.34).

Moving on to q_0 with input 1, the next state for q_0 with input 1 is q_{16}. q_{16} is not used and should be 2 x 0 + 1 + 1 q_2 (line 7, Figure 6.34). Since q_2 is not used, q_2 and q_{16} are interchanged (line 17, Figure 6.34). As a result of this exchange, the next states for q_2 become q_8 and q_{10}, and for q_{16} they become q_{21} and q_{18}. As was planned, the next state of q_0 with input 1 is changed to q_2 (lines 5-11, Figure 6.32).

Also, this exchange changes the next state (lines 5-11, Figure 6.32) of q_9 with input 0 to q_{16}, of q_{10} with input 0 to q_{16}, of q_{13} with input 0 to q_2, of q_{14} with input 1 to q_2, and of q_{30} with input 1 to q_2. q_2 is tagged as used and put on queue2 for processing by step2 (lines 18-20, Figure 6.34). When this is completed (Table 6.11), used: q_0, q_1, and q_2; and to_be_processed2: q_1 and q_2.

Then q_1 is removed from queue2 for processing by step2 (lines 3-4, Figure 6.34). The next state for q_1 with input 0 is q_{13}. q_{13} is not used and should be 2x1+0+1 q_3 (line 7, Figure 6.34). Since q_3 is not used, q_3 and q_{13} are interchanged (line 17, Figure 6.34). As a result of this exchange the next states for q_3 become q_2 and q_0, and for q_{13} they become q_1 and q_{11}.

Line # from Table 6.9	Present state	Next state/output for input = 0	Next state/output for input = 1
1	q_0	$q_1/3$	$q_2/3$
3	q_{16}	$q_{21}/3$	$q_{18}/3$
4	q_3	$q_1/0$	$q_{11}/1$
7	q_9	$q_{16}/3$	$q_{23}/0$
8	q_{10}	$q_{16}/1$	$q_0/2$
10	q_{13}	$q_2/3$	$q_8/0$
11	q_{14}	$q_9/1$	$q_2/1$
13	q_2	$q_8/0$	$q_{10}/1$
15	q_1	$q_{13}/0$	$q_{18}/2$
22	q_{30}	$q_3/1$	$q_2/2$
–	q_{19}	–	–

Table 6.11

As was planned, the next state of q_1 with input 0 is changed to q_3 (lines 5-11, Figure 6.32). Also this exchange changes the next state (lines 5-11, Figure 6.32) of q_8 with input 0 to q_3, of q_{21} with input 1 to q_{13}, and of q_{30} with input 0 to q_{13}. In addition, the next states (lines 5-11, Figure 6.32) of q_{20} are changed to q_3 and q_{13}. q_3 is tagged as used and put on queue2 for processing by step2 (lines 18-20, Figure 6.34).

Moving on to q_1 with input 1, the next state for q_1 with input 1 is q_{18}. q_{18} is not used and should be 2 x 1 + 1 + 1 q_4 (line 7, Figure 6.34). Since q_4 is not used, q_4 and q_{18} are interchanged (line 17, Figure 6.34). As a result of this exchange, the next states for q_4 become q_{11} and q_9. The next states for q_{18} are irrelevant, because after the exchange is complete, q_{18} is not reachable.

As was planned, the next state of q_1 with input 1 is changed to q_4 (lines 5-11, Figure 6.32). Also, this exchange changes the next state (lines 5-11, Figure 6.32) of q_{16} with input 1 to q_4, and of q_{28} with input 0 to q_4. q_4 is tagged as used and put on queue2 for processing by step2 (lines 18-20, Figure 6.34). When this is completed (Table 6.12), used: q_0, q_1, q_2, q_3, and q_4; and to_be_processed2: q_2, q_3, and q_4

q_2 is then removed from queue2 for processing by step2 (lines 3-4, Figure 6.34). The next state for q_2 with input 0 is q_8. q_8 is not used and should be $2 \times 2 + 0 + 1$ q_5 (line 7, Figure 6.34). Since q_5 is not used, q_5 and q_8 are interchanged (line 17, Figure 6.34). As a result of this exchange, the next states for q_5 become q_3 and q_{21}, and for q_8 they become q_{11} and q_{20}.

As was planned, the next state of q_2 with input 0 is changed to q_5 (lines 5-11, Figure 6.32). Also this exchange changes the next state (lines 5-11, Figure 6.32) of q_6 with input 0 to q_8, of q_{10} with input 1 to q_5, and of q_{21} with input 0 to q_5. q_5 is tagged as used and put on queue2 for processing by step2 (lines 18-20, Figure 6.34). (See Table 6.13.)

Moving on to q_2 with input 1, the next state for q_2 with input 1 is q_{10}. q_{10} is not used and should be $2 \times 2 + 1 + 1$ q_6 (line 7, Figure 6.34). Since q_6 is not used, q_6 and q_{10} are interchanged (line 17, Figure 6.34). As a result of this exchange, the next states for q_6 become q_{16} and q_5, and for q_{10} they become q_8 and q_{15}.

As was planned, the next state of q_2 with input 1 is changed to q_6 (lines 5-11, Figure 6.32). In addition, the next states (lines 5-11, Figure 6.32) of q_{23} are changed to q_6 and q_{10}. q_6 is tagged as used and put on queue2 for processing by step2 (lines 18-20, Figure 6.34). When this is completed (Table 6.14)

Line # from Table 6.9	Present state	Next state/output for input = 0	Next state/output for input = 1
2	q_8	$\mathbf{q_3}/2$	$q_{21}/2$
3	q_{16}	$q_{21}/3$	$\mathbf{q_4}/3$
4	$\mathbf{q_{13}}$	$q_1/0$	$q_{11}/1$
10	$\mathbf{q_3}$	$q_2/3$	$q_0/0$
14	$\mathbf{q_4}$	$q_{11}/3$	$q_0/1$
15	q_1	$\mathbf{q_3}/0$	$\mathbf{q_4}/2$

16	q_{20}	$q_3/2$	$q_{13}/0$
17	q_{21}	$q_5/2$	$q_{13}/0$
21	q_{28}	$q_4/1$	$q_{26}/0$
22	q_{30}	$q_{13}/1$	$q_2/2$
–	q_{18}	–	–

Table 6.12

used: q_0, q_1, q_2, q_3, q_4, q_5, and q_6; and to_be_processed2: q_3, q_4, q_5, and q_6.

Line # from Table 6.9	Present state	Next state/output for input = 0	Next state/output for input = 1
2	q_5	$q_3/2$	$q_{21}/2$
5	q_8	$q_{11}/0$	$q_{20}/1$
6	q_6	$q_8/0$	$q_{15}/1$
8	q_{10}	$q_{16}/1$	$q_5/2$
13	q_2	$q_5/0$	$q_{10}/1$
17	q_{21}	$q_5/0$	$q_{10}/2$

Table 6.13

Line # from Table 6.9	Present state	Next state/output for input = 0	Next state/output for input = 1
6	q_{10}	$q_8/0$	$q_{15}/1$
8	q_6	$q_{16}/1$	$q_5/2$
13	q_2	$q_5/0$	$q_6/1$
18	q_{23}	$q_6/0$	$q_{10}/2$

Table 6.14

Then q_3 is removed from queue2 for processing by step2 (lines 3-4, Figure 6.34). The next state for q_3 with input 0 is q_2. Since q_2 is used, no further processing is needed for q_3 with input 0 (line 6, Figure 6.34). Moving on to q_3 with input 1, the next state for q_3 with input 1 is q_0. Since

q_0 is used, no further processing is needed for q_3 with input 1 (line 6, Figure 6.34). At this point (no changes to table), used: q_0, q_1, q_2, q_3, q_4, q_5, and q_6; and to_be_processed2: q_4, q_5, and q_6.

q_4 is then removed from queue2 for processing by step2 (lines 3-4, Figure 6.34). The next state for q_4 with input 0 is q_{11}. q_{11} is not used and should be $2 \times 4 + 0 + 1$ q_9 (line 7, Figure 6.34). Since q_9 is not used, q_9 and q_{11} are interchanged (line 17, Figure 6.34). As a result of this exchange, the next states for q_9 become q_{25} and q_{14}, and for q_{11} they become q_{16} and q_{23}.

As was planned, the next state of q_4 with input 0 is changed to q_9 (lines 5-11, Figure 6.32). Also this exchange changes the next state (lines 5-11, Figure 6.32) of q_4 with input 1 to q_{11}, of q_8 with input 0 to q_9, of q_{13} with input 1 to q_9, and of q_{14} with input 0 to q_{11}. q_9 is tagged as used and put on queue2 for processing by step2 (lines 18-20, Figure 6.34). (See Table 6.15.)

Moving on to q_4 with input 1, the next state for q_4 with input 1 is q_{11}. q_{11} is not used and should be $2 \times 4 + 1 + 1$ q_{10} (line 7, Figure 6.34). Since q_{10} is not used, q_{10} and q_{11} are interchanged (line 17, Figure 6.34). As a result of this exchange, the next states for q_{10} become q_{16} and q_{23}, and for q_{11} they become q_8 and q_{15}.

As was planned, the next state of q_4 with input 1 is changed to q_{10} (lines 5-11, Figure 6.32). Also this exchange changes the next state (lines 5-11, Figure 6.32) of q_{14} with input 0 to q_{10}, and of q_{23} with input 1 to q_{11}. q_0 is tagged as used and put on queue2 for processing by step2 (lines 18-20, Figure 6.34). When this is completed (Table 6.16), used: q_0, q_1, q_2, q_3, q_4, q_5, q_6, q_9, and q_{10}; and to_be_processed2: q_5, q_6, q_9, and q_{10}.

Line # from Table 6.9	Present state	Next state/output for input = 0	Next state/output for input = 1
4	q_{13}	$q_1/0$	$q_9/1$
5	q_8	$q_9/0$	$q_{20}/1$
7	q_{11}	$q_{16}/3$	$q_{23}/0$
9	q_9	$q_{25}/1$	$q_{14}/0$
11	q_{14}	$q_{11}/1$	$q_2/1$
14	q_4	$q_9/3$	$q_{11}/1$

Table 6.15

Then q_5 is removed from queue2 for processing by step2 (lines 3-4, Figure 6.34). The next state for q_5 with input 0 is q_3. Since q_3 is used, no further processing is needed for q_5 with input 0 (line 6, Figure 6.34). Moving on to q_5 with input 1, the next state for q_5 with input 1 is q_{21}. q_{21} is not used and should be 2x5+1+1 q_{12} (line 7, Figure 6.34). Since q_{12} is not used, q_{12} and q_{21} are interchanged (line 17, Figure 6.34). As a result of this exchange the next states for q_{12} become q_5 and q_{13}. The next states for q_{21} are irrelevant because after the exchange is complete q_{21} is not reachable.

As was planned, the next state of q_5 with input 1 is changed to q_{12} (lines 5-11, Figure 6.32). Also this exchange changes the next state (lines 5-11, Figure 6.32) of q_{16} with input 0 to q_{12}. q_{12} is tagged as used and put on queue2 for processing by step2 (lines 18-20, Figure 6.34). When this is completed (Table 6.17), used: q_0, q_1, q_2, q_3, q_4, q_5, q_6, q_9, q_{10}, and q_{12}; and to_be_processed2: q_6, q_9, q_{10}, and q_{12}.

Line # from Table 6.9	Present state	Next state/output for input = 0	Next state/output for input = 1
6	q_{11}	$q_8/0$	$q_{15}/1$
7	q_{10}	$q_{16}/3$	$q_{23}/0$
11	q_{14}	$q_{10}/1$	$q_2/1$
14	q_4	$q_9/3$	$q_{10}/1$
18	q_{23}	$q_6/0$	$q_{11}/2$

Table 6.16

Line # from Table 6.9	present state	next state/output for input=0	next state/output for input=1
2	q_5	$q_3/2$	$q_{12}/2$
3	q_{16}	$q_{12}/3$	$q_4/3$
17	q_{12}	$q_5/2$	$q_{13}/0$
–	q_{21}	–	–

Table 6.17

Then q_6 is removed from queue2 for processing by step2 (lines 3-4, Figure 6.34). The next state for q_6 with input 0 is q_{16}. q_{16} is not used and should be 2 x 6 + 0 + 1 q_{13} (line 7, Figure 6.34). Since q_{13} is not used,

q_{13} and q_{16} are interchanged (lines 17, Figure 6.34). As a result of this exchange, the next states for q_{13} become q_{12} and q_4, and for q_{16} they become q_1 and q_9.

As was planned, the next state of q_6 with input 0 is changed to q_{13} (lines 5-11, Figure 6.32). Also this exchange changes the next state (lines 5-11, Figure 6.32) of q_{10} with input 0 to q_{13}, of q_{12} with input 1 to q_{16}, of q_{20} with input 1 to q_{16}, and of q_{30} with input 0 to q_{16}. q_{13} is tagged as used and put on queue2 for processing by step2 (lines 18-20, Figure 6.34). Moving on to q_6 with input 1, the next state for q_6 with input 1 is q_5. Since q_5 is used, no further processing is needed for q_6 with input 1 (line 6, Figure 6.34). At this point (Table 6.18) used: q_0, q_1, q_2, q_3, q_4, q_5, q_6, q_9, q_{10}, q_{12}, and q_{13}; and to_be_processed2: q_9, q_{10}, q_{12}, and q_{13}.

Line # from Table 6.9	Present state	Next state/output for input = 0	Next state/output for input = 1
3	q_{13}	$q_{12}/3$	$q_4/3$
4	q_{16}	$q_1/0$	$q_9/1$
7	q_{10}	$q_{13}/3$	$q_{23}/0$
8	q_6	$q_{13}/1$	$q_5/2$
16	q_{20}	$q_3/2$	$q_{16}/0$
17	q_{12}	$q_8/2$	$q_{16}/0$
22	q_{30}	$q_{16}/1$	$q_2/2$

Table 6.18

q_9 is then removed from queue2 for processing by step2 (lines 3-4, Figure 6.34). The next state for q_9 with input 0 is q_{25}. q_{25} is not used and should be $2 \times 9 + 0 + 1$ q_{19} (line 7, Figure 6.34). Since q_{19} is not used, q_{19} and q_{25} are interchanged (line 17, Figure 6.34). As a result of this exchange, the next states for q_{19} become q_{30} and q_{25}. The next states for q_{25} are irrelevant because, after the exchange is complete, q_{25} is not reachable.

As was planned, the next state of q_9 with input 0 is changed to q_{19} (lines 5-11, Figure 6.32). Also, this exchange changes the next state (lines 5-11, Figure 6.32) of q_{19} with input 1 to q_{19}. q_{19} is tagged as used and put on queue2 for processing by step2 (lines 18-20, Figure 6.34). Moving on to q_9 with input 1, the next state for q_9 with input 1 is q_{14}. q_{14} is not used and should be $2 \times 9 + 1 + 1$ q_{20} (line 7, Figure 6.34). Since q_{20} is not used, q_{20} and q_{14} are interchanged (line 17, Figure 6.34). As a result of this exchange, the next states for q_{14} become q_3 and q_{16}, and for q_{20} they become q_{10} and q_2.

As was planned, the next state of q_9 with input 1 is changed to q_{20} (lines 5-11, Figure 6.32). Also this exchange changes the next state (lines 5-11, Figure 6.32) of q_8 with input 1 to q_{14}, of q_{15} with input 1 to q_{14}, and of q_{26} with input 0 to q_{14}. q_{20} is tagged as used and put on queue2 for processing by step2 (lines 18-20, Figure 6.34). When this is completed (Table 6.19), used: q_0, q_1, q_2, q_3, q_4, q_5, q_6, q_9, q_{10}, q_{12}, q_{13}, q_{19}, and q_{20}; and to_be_processed2: q_{10}, q_{12}, q_{13}, q_{19}, and q_{20}.

Then q_{10} is removed from queue2 for processing by step2 (lines 3-4, Figure 6.34). The next state for q_{10} with input 0 is q_{13}. Since q_{13} is used, no further processing is needed for q_{10} with input 0 (line 6, Figure 6.34). Moving on to q_{10} with input 1, the next state for q_{10} with input 1 is q_{23}. q_{23} is not used and should be 2x10+1+1 q_{22} (line 7, Figure 6.34). Since q_{22} is not used, q_{22} and q_{23} are interchanged (line 17, Figure 6.34). As a result of this exchange, the next states for q_{22} become q_6 and q_{11}. The next states for q_{23} are irrelevant because, after the exchange is complete, q_{23} is not reachable.

As was planned, the next state of q_{10} with input 1 is changed to q_{22} (lines 5-11, Figure 6.32). q_{22} is tagged as used and put on queue2 for processing by step2 (lines 18-20, Figure 6.34). When this is completed (Table 6.20), used: q_0, q_1, q_2, q_3, q_4, q_5, q_6, q_9, q_{10}, q_{12}, q_{13}, q_{19}, q_{20}, and q_{22}; and to_be_processed2: q_{12}, q_{13}, q_{19}, q_{20}, and q_{22}.

Line # from Table 6.9	Present state	Next state/output for input = 0	Next state/output for input = 1
5	q_8	$q_9/1$	$q_{14}/1$
9	q_9	$q_{19}/1$	$q_{20}/0$
11	q_{20}	$q_{10}/1$	$q_2/1$
12	q_{15}	$q_{28}/3$	$q_{14}/2$
16	q_{14}	$q_3/2$	$q_{16}/0$
19	q_{19}	$q_{30}/0$	$q_{19}/1$
20	q_{26}	$q_{14}/3$	$q_{15}/0$

–	q_{25}	–	–

Table 6.19

Line # from Table 6.9	Present state	Next state/output for input = 0	Next state/output for input = 1
7	q_{10}	$q_{13}/3$	$q_{22}/0$
18	q_{22}	$q_6/0$	$q_{11}/2$
–	q_{23}	–	–

Table 6.20

q_{12} is then removed from queue2 for processing by step2 (lines 3-4, Figure 6.34). The next state for q_{12} with input 0 is q_5. Since q_5 is used, no further processing is needed for q_{12} with input 0 (line 6, Figure 6.34). Moving on to q_{12} with input 1, the next state for q_{12} with input 1 is q_{16}. q_{16} is not used and should be 2 x 12 + 1 + 1 q_{26} (line 7, Figure 6.34). Since q_{26} is not used, q_{26} and q_{16} are interchanged (line 17, Figure 6.34). As a result of this exchange, the next states for q_{16} become q_{14} and q_{15}, and for q_{26} they become q_1 and q_9.

As was planned, the next state of q_{12} with input 0 is changed to q_{26} (lines 5-11, Figure 6.32). Also this exchange changes the next state (lines 5-11, Figure 6.32) of q_{14} with input 1 to q_{26}, of q_{28} with input 1 to q_{16}, and of q_{30} with input 0 to q_{26}. q_{26} is tagged as used and put on queue2 for processing by step2 (lines 18-20, Figure 6.34). When this is completed (Table 6.21), used: $q_0, q_1, q_2, q_3, q_4, q_5, q_6, q_9, q_{10}, q_{12}, q_{13}, q_{19}, q_{20}, q_{22}$, and q_{26}; and to_be_processed2: $q_{13}, q_{19}, q_{20}, q_{22}$, and q_{26}.

Then q_{13} is removed from queue2 for processing by step2 (lines 3-4, Figure 6.34). The next state for q_{13} with input 0 is q_{12}. Since q_{12} is used, no further processing is needed for q_{13} with input 0 (line 6, Figure 6.34). Moving on to q_{13} with input 1, the next state for q_{13} with input 1 is q_4. Since q_4 is used, no further processing is needed for q_{13} with input 1 (line 6, Figure 6.34). At this point (no change to the table), used: $q_0, q_1, q_2, q_3, q_4, q_5, q_6, q_9, q_{10}, q_{12}, q_{13}, q_{19}, q_{20}, q_{22}$, and q_{26}; and to_be_processed2: q_{19}, q_{20}, q_{22}, and q_{26}.

Line # from Table 6.9	Present state	Next state/output for input = 0	Next state/output for input = 1
4	q_{26}	$q_1/0$	$q_9/1$
16	q_{14}	$q_3/2$	$q_{26}/0$
17	q_{12}	$q_5/2$	$q_{26}/0$
20	q_{16}	$q_{14}/3$	$q_{15}/0$
21	q_{28}	$q_4/1$	$q_{16}/0$
22	q_{30}	$q_{26}/1$	$q_2/2$

Table 6.21

q_{19} is then removed from queue2 for processing by step2 (lines 3-4, Figure 6.34). The next state for q_{19} with input 0 is q_{30}. q_{30} is not used and should be 2x(32 2 19)+0+1 q_{23} (line 10, Figure 6.34). Note that a different part of the function is being used to determine the next state. Since q_{23} is not used, q_{23} and q_{30} are interchanged (line 17, Figure 6.34). As a result of this exchange, the next states for q_{23} become q_{26} and q_2. The next states for q_{30} are irrelevant because, after the exchange is complete, q_{30} is not reachable.

As was planned, the next state of q_{19} with input 0 is changed to q_{23} (lines 5-11, Figure 6.32). q_{23} is tagged as used and put on queue2 for processing by step2 (lines 18-20, Figure 6.34). Moving on to q_{19} with input 1, the next state for q_{19} with input 1 is q_{19}. Since the next state is used, no further processing is needed for q_{19} with input 1 (line 6, Figure 6.34). At this point (Table 6.22), used: q_0, q_1, q_2, q_3, q_4, q_5, q_6, q_9, q_{10}, q_{12}, q_{13}, q_{19}, q_{20}, q_{22}, q_{23}, and q_{26}; and to_be_processed2: q_{20}, q_{22}, q_{26}, and q_{23}.

Then q_{20} is removed from queue2 for processing by step2 (lines 3-4, Figure 6.34). The next state for q_{20} with input 0 is q_{10}. Since q_{10} is used, no further processing is needed for q_{20} with input 0 (line 6, Figure 6.34). Moving on to q_{20} with input 1, the next state for q_{20} with input 1 is q_2. Since q_2 is used, no further processing is needed for q_{20} with input 1 (line 6, Figure 6.34). At this point (no changes to the table), used: q_0, q_1, q_2, q_3, q_4, q_5, q_6, q_9, q_{10}, q_{12}, q_{13}, q_{19}, q_{20}, q_{22}, q_{23}, and q_{26}; and to_be_processed2: q_{22}, q_{26}, and q_{23}.

Line # from Table 6.9	Present state	Next state/output for input = 0	Next state/output for input = 1
19	q_{19}	$q_{23}/0$	$q_{19}/1$
22	q_{23}	$q_{26}/1$	$q_2/1$
–	q_{30}	–	–

Table 6.22

q_{22} is then removed from queue2 for processing by step2 (lines 3-4, Figure 6.34). The next state for q_{22} with input 0 is q_6. Since q_6 is used, no further processing is needed for q_{22} with input 0 (line 6, Figure 6.34). Moving on to q_{22} with input 1, the next state for q_{22} with input 1 is q_{11}. q_{11} is not used and should be 2x(32 2 22)+1+1 q_{18} (line 10, Figure 6.34). Since q_{18} is not used, q_{18} and q_{11} are interchanged (line 17, Figure 6.34). As a result of this exchange, the next states for q_{18} become q_8 and q_{15}. The next states for q_{11} are irrelevant because, after the exchange is complete, q_{11} is not reachable.

As was planned, the next state of q_{22} with input 1 is changed to q_{18} (line 5-11, Figure 6.32). q_{18} is tagged as used and put on queue2 for processing by step2 (lines 18-20, Figure 6.34). When this is completed (Table 6.23), used: q_0, q_1, q_2, q_3, q_4, q_5, q_6, q_9, q_{10}, q_{12}, q_{13}, q_{18}, q_{19}, q_{20}, q_{22}, q_{23}, and q_{26}; and to_be_processed2: q_{26}, q_{23}, and q_{18}.

Line # from Table 6.9	Present state	Next state/output for input=0	Next state/output for input=1
6	q_{18}	$q_8/0$	$q_{15}/1$
18	q_{22}	$q_6/0$	$q_{18}/2$
–	q_{11}	–	–

Table 6.23

Then q_{26} is removed from queue2 for processing by step2 (lines 3-4, Figure 6.34). The next state for q_{26} with input 0 is q_1. Since q_1 is used, no further processing is needed for q_{26} with input 0 (line 6, Figure 6.34). Moving on to q_{26} with input 1, the next state for q_{26} with input 1 is q_9. Since q_9 is used, no further processing is needed for q_{26} with input 1 (line 6, Figure 6.34). At this point (no change to the table), used: q_0, q_1, q_2, q_3, q_4, q_5, q_6, q_9, q_{10}, q_{12}, q_{13}, q_{18}, q_{19}, q_{20}, q_{22}, q_{23}, and q_{26}; and to_be_processed2: q_{23}, and q_{18}.

q_{23} is then removed from queue2 for processing by step2 (lines 3-4, Figure 6.34). The next state for q_{23} with input 0 is q_{26}. Since q_{26} is used, no further processing is needed for q_{23} with input 0 (line 6, Figure 6.34). Moving on to q_{23} with input 1, the next state for q_{23} with input 1 is q_2. Since q_2 is used, no further processing is needed for q_{23} with input 1 (line 6, Figure 6.34). At this point (no change to the table), used: q_0, q_1, q_2, q_3, q_4, q_5, q_6, q_9, q_{10}, q_{12}, q_{13}, q_{18}, q_{19}, q_{20}, q_{22}, q_{23}, and q_{26}; and to_be_processed2: q_{18}.

Then q_{18} is removed from queue2 for processing by step2 (lines 3-4, Figure 6.34). The next state for q_{18} with input 0 is q_8. q_8 is not used and should be 2x(32 2 18)+0+1 q_{25} (line 10, Figure 6.34). Since q_{25} is not used, q_{25} and q_8 are interchanged (line 17, Figure 6.34). As a result of this exchange, the next states for q_{25} become q_9 and q_{14}. The next states for q_8 are irrelevant because, after the exchange is complete, q_8 is not reachable.

As was planned, the next state of q_{18} with input 0 is changed to q_{25} (lines 5-11, Figure 6.32). q_{25} is tagged as used and put on queue2 for processing by step2 (lines 18-20, Figure 6.34). Moving on to q_{25} with input 1, the next state for q_{25} with input 1 is q_{14}. q_{14} is not used and should be 2x0(32 2 18)+1+1 q_{26} (line 10, Figure 6.34). Since q_{26} is used, it cannot be assigned as the next state for q_{18} with an input of 1. Therefore, (q_{18},1) is put on queue3 for processing by step 3 (lines 11-15, Figure 6.34). At this point (Table 6.24), used: q_0, q_1, q_2, q_3, q_4, q_5, q_6, q_9, q_{10}, q_{12}, q_{13}, q_{18}, q_{19}, q_{20}, q_{22}, q_{23}, q_{25}, and q_{26}; to_be_processed2: q_{25}; and to_be_processed3: (q_{18},1).

Line # from Table 6.9	Present state	Next state/output for input = 0	Next state/output for input = 1
5	q_{25}	$q_9/0$	$q_{14}/1$
6	q_{18}	$q_{25}/0$	$q_{15}/1$
–	q_8	–	–

Table 6.24

q_{25} is then removed from queue2 for processing by step2 (lines 3-4, Figure 6.34). The next state for q_{25} with input 0 is q_9. Since q_9 is used, no further processing is needed for q_3 with input 0 (line 6, Figure 6.34).

Moving on to q_{25} with input 1, the next state for q_{25} with input 1 is q_{14}. q_{14} is not used and should be 2x(32 2 25)+1+1 q_{12} (line 10, Figure 6.34). Since q_{12} is used, it cannot be assigned as the next state for q_{25} with an input of 1. Therefore (q_{25},1) is put on queue3 for processing by step 3 (lines 11-15, Figure 6.34). At this point (no change to the table), used: q_0, q_1, q_2, q_3, q_4, q_5, q_6, q_9, q_{10}, q_{12}, q_{13}, q_{18}, q_{19}, q_{20}, q_{22}, q_{23}, q_{25}, and q_{26}; to_be_processed2: empty; and to_be_processed3: (q_{18},1) and (q_{25},1).

Since queue2 is empty, control is returned to SFS (lines 13-15, Figure 6.35), and step 3 is executed. Remove q_{18} with input 1 from queue3 (lines 16-18, Figure 6.35) for processing by step 3. The next state for q_{18} with input 1 is q_{15} which is still not used. Looking at the used states (in the previous paragraph), the state, which is closest to q_{18} and which is not used, is q_{17} (lines 20-27, Figure 6.35). Therefore q_{17} will be the new_next_state for q_{18} with input 1 (lines 24-27, Figure 6.35). q_{17} has been selected because it is not used, therefore q_{17} and q_{15} can be interchanged (line 28, Figure 6.35). As a result of this exchange, the next states for q_{17} become q_{28} and q_{14}. The next states for q_{15} are irrelevant because, after the exchange is complete, q_{15} is not reachable.

As was planned, the next state of q_{18} with input 1 is changed to q_{17} (lines 5-11, Figure 6.32). Also this exchange changes the next state (lines 5-11, Figure 6.32) of q_{16} with input 1 to q_{17}. q_{17} is tagged as used and put on queue2 for processing by step2 (lines 29-31, Figure 6.34). (Whenever possible, next states are assigned by step2.) When this is complete (Table 6.25), used: q_0, q_1, q_2, q_3, q_4, q_5, q_6, q_9, q_{10}, q_{12}, q_{13}, q_{17}, q_{18}, q_{19}, q_{20}, q_{22}, q_{23}, q_{25}, and q_{26}; to_be_processed2: q_{17}; and to_be_processed3: (q_{25},1).

Line # from Table 6.9	Present state	Next state/output for input = 0	Next state/output for input = 1
6	q_{18}	q_{25}/0	$\mathbf{q_{17}}$/1
12	$\mathbf{q_{17}}$	q_{28}/3	q_{14}/2
20	q_{16}	q_{14}/3	$\mathbf{q_{17}}$/0
–	$\mathbf{q_{15}}$	–	–

Table 6.25

step2 takes precedence over step 3 and queue2 is not empty, therefore, step2 is called (lines 32-33, Figure 6.35). q_{17} is removed from queue2 for processing by step2 (lines 3-4, Figure 6.34). The next state for q_{17} with input 0 is q_{28}. q_{28} is not used and should be 2x(32 2 17)+0+1 q_{27} (line 10, Figure 6.34). Since q_{27} is not used, q_{27} and q_{28} are interchanged (line 17, Figure 6.34). As a result of this exchange, the next states for q_{27} become q_4 and q_{16}. The next states for q_{28} are irrelevant because, after the exchange is complete, q_{28} is not reachable.

As was planned, the next state of q_{17} with input 0 is changed to q_{27} (lines 5-11, Figure 6.32). q_{27} is tagged as used and put on queue2 for processing by step2 (lines 18-20, Figure 6.34). (See Table 6.26.) Moving on to q_{17} with input 1, the next state for q_{17} with input 1 is q_{14}. q_{14} is not used and should be 2x(32 2 17)+1+1 q_{28} (line 10, Figure 6.34). Since q_{28} is not used, q_{28} and q_{14} are interchanged (line 17, Figure 6.34). As a result of this exchange, the next states for q_{28} become q_3 and q_{26}. The next states for q_{14} are irrelevant because, after the exchange is complete, q_{14} is not reachable.

Line # from Table 6.9	Present state	Next state/output for input = 0	Next state/output for input = 1
12	q_{17}	$q_{27}/3$	$q_{14}/2$
21	q_{27}	$q_4/1$	$q_{16}/2$
–	q_{28}	–	–

Table 6.26

Line # from Table 6.9	Present state	Next state/output for input = 0	Next state/output for input = 1
5	q_{25}	$q_9/0$	$q_{28}/1$
12	q_{17}	$q_{27}/3$	$q_{28}/2$
16	q_{28}	$q_3/2$	$q_{26}/0$
20	q_{16}	$q_{28}/3$	$q_{17}/0$
–	q_{14}	–	–

Table 6.27

As was planned, the next state of q_{17} with input 1 is changed to q_{28} (line 5-

11, Figure 6.32). Also this exchange changes the next state (lines 5-11, Figure 6.32) of q_{16} with input 0 to q28, and of q25 with input 1 to q28. q28 is tagged as used and put on queue2 for processing by step2 (lines 18-20, Figure 6.34). When this is completed, (Table 6.27) used: q0, q1, q2, q3, q4, q5, q6, q9, q10, q12, q13, q17, q18, q19, q20, q22, q23, q25, q26, q27, and q28; to_be_processed2: q27 and q28; and to_be_processed3: (q25,1).

Then q_{27} is removed from queue2 for processing by step2 (lines 3-4, Figure 6.34). The next state for q_{27} with input 0 is q_4. Since q_4 is used, no further processing is needed for q_4 with input 0 (line 6, Figure 6.34). Moving on to q_{27} with input 1, the next state for q_{27} with input 1 is q_{16}. q_{16} is not used and should be 2x(32 2 27)+1+1 q_8 (line 10, Figure 6.34). Since q_8 is not used, q_8 and q_{16} are interchanged (line 17, Figure 6.34). As a result of this exchange, the next states for q_8 become q_{28} and q_{17}. The next states for q_{16} are irrelevant because, after the exchange is complete, q_{16} is not reachable.

As was planned, the next state of q_{27} with input 1 is changed to q_8 (lines 5-11, Figure 6.32). q_8 is tagged as used and put on queue2 for processing by step2 (lines 18-20, Figure 6.34). When this is completed, (Table 6.28) used: q_0, q_1, q_2, q_3, q_4, q_5, q_6, q_8, q_9, q_{10}, q_{12}, q_{13}, q_{17}, q_{18}, q_{19}, q_{20}, q_{22}, q_{23}, q_{25}, q_{26}, q_{27}, and q_{28}; to_be_processed2: q_{28} and q_8; and to_be_processed3: (q_{25},1).

q_{28} is then removed from queue2 for processing by step2 (lines 3-4, Figure 6.34). The next state for q_{28} with input 0 is q_3. Since q_3 is used, no further processing is needed for q_{28} with input 0 (line 6, Figure 6.34). Moving on to q_{28} with input 1, the next state for q_{28} with input 1 is q_{26}. Since q_{26} is used, no further processing is needed for q_{28} with input 1 (line 6, Figure 6.34). At this point (no changes to the table), used: q_0, q_1, q_2, q_3, q_4, q_5, q_6, q_8, q_9, q_{10}, q_{12}, q_{13}, q_{17}, q_{18}, q_{19}, q_{20}, q_{22}, q_{23}, q_{25}, q_{26}, q_{27}, and q_{28}; to_be_processed2: q_8; and to_be_processed3: (q_{25},1).

Line # from Table 6.9	Present state	Next state/output for input = 0	Next state/output for input = 1
20	q_8	$q_{28}/3$	$q_{17}/0$
21	q_{27}	$q_4/1$	$q_8/0$
–	q_{16}	–	–

Table 6.28

Then q_8 is removed from queue2 for processing by `step2` (line 3-4, Figure 6.34). The next state for q_8 with input 0 is q_{28}. Since q_{28} is `used`, no further processing is needed for q_8 with input 0 (line 6, Figure 6.34). Moving on to q_8 with input 1, the next state for q_8 with input 1 is q_{17}. Since q_{17} is `used`, no further processing is needed for q_8 with input 1 (line 6, Figure 6.34). At this point (no changes to the table), `used`: q_0, q_1, q_2, q_3, q_4, q_5, q_6, q_8, q_9, q_{10}, q_{12}, q_{13}, q_{17}, q_{18}, q_{19}, q_{20}, q_{22}, q_{23}, q_{25}, q_{26}, q_{27}, and q_{28}; `to_be_processed2`: empty; and `to_be_processed3`: $(q_{25},1)$.

Since queue2 is empty, control is returned to SFS (Figure 6.35), and step 3 is continued since the last call to `step2` was from within step 3 (lines 32-33, Figure 6.35). Remove q_{25} with input 1 from queue3 (lines 16-18, Figure 6.35) for processing by step 3. The next state for q_{25} with input 1 is q_{28} which is now `used`. Therefore no further processing is needed for q_{25} with input 1 (line 19, Figure 6.35). The two queues are empty, hence the SFS reorganization of Table 6.9 is complete.

Table 6.29 contains the FSM of Table 6.9 after the SFS reorganization. Table 6.30 correlates the states between Tables 6.9 and 6.10.

SFS reorganizes the genome to enhance schemata growth by standardizing transitions to some extent. Referring back to equation 6.4, another way to enhance schemata growth is to reduce the chance that a useful schema will be disrupted by crossover by reducing the schema's defining length. The MTF algorithm, which is presented in the next section, was designed with this in mind.

Start state: q_0

Line # from Table 6.9	Present state	Next state/output for input = 0	Next state/output for input = 1
1	q_0	$q_1/3$	$q_7/3$
15	q_1	$q_3/0$	$q_4/2$
13	q_2	$q_5/0$	$q_{61}/1$
10	q_3	$q_2/3$	$q_0/0$
14	q_4	$q_9/3$	$q_{10}/1$
2	q_5	$q_3/2$	$q_{12}/2$
8	q_6	$q_{13}/1$	$q_5/2$

20	q_8	$q_{28}/3$	$q_{17}/0$
9	q_9	$q_{19}/1$	$q_{20}/0$
7	q_{10}	$q_{13}/3$	$q_{22}/0$
17	q_{12}	$q_5/2$	$q_{26}/0$
3	q_{13}	$q_{12}/3$	$q_4/3$
12	q_{17}	$q_{27}/3$	$q_{28}/2$
6	q_{18}	$q_{25}/0$	$q_{17}/1$
19	q_{19}	$q_{23}/0$	$q_{19}/1$
11	q_{20}	$q_{10}/1$	$q_2/1$
18	q_{22}	$q_6/0$	$q_{18}/2$
22	q_{28}	$q_{26}/1$	$q_2/2$
5	q_{25}	$q_9/0$	$q_{28}/1$
4	q_{26}	$q_1/0$	$q_9/1$
21	q_{27}	$q_4/1$	$q_8/0$
16	q_{28}	$q_3/2$	$q_{26}/0$

Table 6.29 FSM of Table 6.9 after SFS

State # from Table 6.9	0	2	3	5	6	8	9	10	11	13	14
State # from Table 6.10	5	13	26	25	18	0	10	6	9	3	20

State # from Table 6.9	15	16	18	19	20	21	23	25	26	28	30
State # from Table 6.10	17	2	4	1	28	12	22	19	8	27	23

Table 6.30 Correlation of States between the FSMs in Tables 6.9 and 6.29

6.4. The MTF Algorithm

The MTF algorithm systematically M̲oves the reachable states of a finite state machine T̲o the F̲ront of the genome. As indicated, the phenotype allows for 32 states, but not all machines that are created by the GA utilize all 32 states. By reorganizing an FSM to use states zero through one less than the number of states which can be reached, the relevant data is more closely positioned in the genome. For example, the machine in Table 6.5 is spread out across the genome. MTF would reorganize this FSM to Table 6.31. Like SFS, MTF standardizes q_0 as the start state.

Start state: q_0

present state	next state/output for input=0	next state/output for input=1
q_0	$q_1/0$	$q_2/2$
q_1	$q_0/2$	$q_1/2$
q_2	$q_1/1$	$q_3/2$
q_3	$q_0/3$	$q_3/2$

Table 6.31 FSM of Table 6.5 after MTF

Recall that the start state appears in the first 5 bits of the genome and each state needs $2 \times (5+2) = 14$ bits to code for its next states (5 bits per state) and outputs (2 bits per output). Therefore, the machine in Figure 6.5 is spread out across the first $5+25 \times 14 = 355$ bits of the genome. The equivalent machine in Table 6.31 uses only the first $5+4 \times 14 = 61$ bits of the genome for the relevant data. The FSM in Table 6.31 will obviously have shorter useful schemata. The next section describes the algorithm and its coding. The following section presents a simple example.

6.4.1 MTF: Description and Coding

The code and description of the MTF reorganization algorithm appear in Figure 6.37. The algorithm is very simple. Step 1 is the same as for SFS – rename the start state as state 0 (lines 2-3, Figure 6.37). The `next_available state` along the genome is state 1 (line 4). MTF moves along the genome, starting at state 0. The for loops move the processing through the states and through the inputs for each state (lines 5-6). When state i with input j is processed, if the next state is past the end of the reorganized part of the genome, the next state is reassigned to the `next_available_state` (lines 7-8), and the `next_state` variable is advanced (line 9). MTF does not need a separate queue to retain the states which have to be processed because the phenotype contains that queue. Since each state is moved into the next available position and because this position sequentially moves through the states, the `next_state` table actually forms the processing queue. A `used` array is also not needed because all states before the `next_available_state` have been reassigned (`used`). The looping terminates when all `reachable` states have been moved to the front of the table (line 5), and hence the front of the genome.

```
/* ------------------------------------  */
/*        Move To Front          */
/* ------------------------------------  */
void MTF (int *start_state)
/* move all reachable states to the front of the
genome.   step 1: make 0 the start state.   step 2:
for each state processed     a) make its next state
on input=0 the next available state       on the
genome, unless it is already a reassigned state.
b) do the same for input=1.                        */
{
line #
1: int i, j, next_available_state;
 /* STEP 1 */
2: xchange_states(0, *start_state);
3: *start_state = 0;
4: next_available_state = 1;
 /* STEP 2 */
5: for (i=0; i < next_available_state; ++i)
6:    for (j=0; j<2; ++j)
7:       if (next_state[i][j] >=
next_available_state) {
8:xchange_states(next_state[i][j],
next_available_state);
9:++next_available_state;   /* must be performed
even if equal */
10:         }
} /* MTF */
```

Figure 6.37 The MTF Algorithm

To add MTF to the program, first add the MTF procedure. Then add

#define method 1 /* 0 for SFS; 1 for MTF (Irrelevant if reorganization is 0) /*

to the #define statements at the beginning of the program. Also change the lines which #define printed_output (r1mtf.st) and perfect_ score (r1mtf.ps). reorganizeFSM has to be modified to allow for either an SFS or MTF reorganization. The new version of reorganizeFSM appears in Figure 6.38. An if/else (lines 5-8) controls which reorganization algorithm is called. In addition, print_header (Figures 6.21 and 6.36) must be altered to be able to identify an MTF run. The beginning of the updated version of print_header appears in Figure 6.39.

```
/* ------------------------------------     */
/*      Reorganize (Final)        */
/* ------------------------------------     */
void reorganizeFSM(void){
line #
1: int i;

2: for (i=0; i<select; ++i) {
3:    bytestring_to_FSM (&start_state,
4:       &pop[new_pop][sorted[row_of_sorted][i]][0]);
5:    if (method == 1)
6:       MTF (&start_state);
7:    else
8:       SFS (&start_state);
9:    FSM_to_bytestring (start_state,
10:
&pop[new_pop][sorted[row_of_sorted][i]][0]);
11:    }
} /* reorganizeFSM */
```

Figure 6.38 Reorganization Modified to Permit SFS or MTF

```
/* ---------------------------------------
*/
/*  Print Header on Output Files     */
/*  Modified to include SFS and MTF        */
/* ---------------------------------------
*/
void print_header(void)/* print out header
information for the GA in  the statistics and
perfect score files */
{
line #
1: fp_stats =  fopen(printed_output, "w");
2: fp_perfect = fopen(perfect_score, "w");

3: if (reorganization) {

4:    if (method ==1) {
5:       fprintf(fp_stats,  "\nMTF\nStatistics for
run #%d on file %s\n,"
6:          run_num, printed_output);
7:       fprintf(fp_perfect, "\nMTF\nPerfect scorers
for run #%d on file %s\n,"
8:          run_num, perfect_score);
```

```
9:    }
10:    else {
11:        fprintf(fp_stats,   "\nSFS\nStatistics for
run #%d on file %s\n,"
12:         run_num, printed_output);
13:        fprintf(fp_perfect, "\nSFS\nPerfect
scorers for run #%d on file %s\n,"
14:         run_num, perfect_score);
15:    }
16:    }
17:    else {
18:      fprintf(fp_stats,   "\nBenchmark\nStatistics
for run #%d on file %s\n,"
19:         run_num, printed_output);
20:        fprintf(fp_perfect, "\nBenchmark\nPerfect
scorers for run #%d on file %s\n,"
21:         run_num, perfect_score);
22:    } fprintf(fp_stats,   "Seed=%.8f\n," seed);
fprintf(fp_perfect, "Seed=%.8f\n," seed);        !}
/* print header information */
```

Figure 6.39 Prints Headings on Output Files – Modified to Include SFS and MTF

If the output was modified to print out a perfect scoring FSM in its reorganized form, change in calc_fitness (added between lines 54 and 55 of Figure 6.18) to

```
if (reorganization)
    if (method == 1) MTF(&start_state);
    else SFS(&start_state);
```

This reorganizes the FSM strictly for printout, but does not alter its stored format. In the next section a simple example of an MTF reorganization is presented.

6.4.2 Example of an MTF Reorganization

Because MTF is so simple, a single short example will suffice to demonstrate it. Start with the FSM in Table 6.5. The first step of MTF (Figure 6.37) makes q_0 the start state; that is, interchanges states q_0 and q_1 (lines 2-3). This changes Table 6.5 to Table 6.6. This is exactly the same thing SFS does for its first step. The next_available_state (line 4) is the state after q_0, which is q_1. Starting with q_0, assign the next state for inputs 0 and 1 to the next two available states (lines 5-10), unless the next states have already been reassigned (line 7).

The next_state for q_0 with input 0 is q_5 and by MTF should be the next_available_state, which is q_1. Therefore, q_5 and q_1 are interchanged (line 8) and the next_available_state is incremented to 2 (line 9). Moving along to q_0 with input 1, next_state for q_0 with input 1 is q_9 and by MTF should be the next_available_state, which is now q_2. Therefore, q_9 and q_2 are interchanged (line 8) and the next_available_state is incremented to 3 (line 9). While this process is different from that of SFS, the result at this point is the same (Table 6.7).

q_1 is next processed. The next states for q_1 with inputs 0 and 1 are q_0 and q_2, respectively. These states have already been reassigned, indicated by the fact that they are both before next_available_state (line 7). Hence no further processing is needed for q_1. q_2 is the next state to be processed. The next state for q_2 with input 0 is q_1, which has already been reassigned (line 7). Therefore, no further processing is needed for q_2 with input 0. But the next state for q_2 with input 1 is q_{24} and that is beyond the next_available_state, which is now q_3. Consequently, q_3 and q_{24} are exchanged (line 8) and the next_available_state is advanced to q_4 (line 9).

The next state to be processed is q_3. Both next states, q_0 and q_3, have been reassigned. Moving on to q_4 for processing halts the looping (line 5). q_4 is not before next_available_state (line 5) indicating that all reachable states have been moved to the front of the table, and hence the genome. The final result appears in Table 6.31.

Both SFS and MTF could reduce innovation. With more machines appearing the same, crossover is less likely to create something new. Utilizing these algorithms could make the GA more prone to premature convergence. One way to compensate for this is to add a competition between individuals in order to change the static landscape of the fitness function to a landscape which changes with each generation. The next section explains competition and provides the coding to implement it.

6.5. Competition

Competition is one way to compensate for the higher probability of *premature convergence* when SFS or MTF are implemented. Premature convergence occurs when the parent pool is overtaken by a suboptimal genome. This occurred in the run which produced the output in Figure 6.40. The evolving population quickly attains a maximum fitness of 85. Similar individuals quickly take over a larger part of the population and fill up the parent pool.

Benchmark
Statistics for run #2 on file r2b.st
Seed = 0.48296342
Crossover rate = 0.01;   Mutation rate = 0.01
Population size = 65536

Select the top 3277 individuals or 5.00% of each generation for reproduction.

Gen	Max/# with Max	Average	/Median	/Mode	/# with Mode	Standard dev.
0	78/1	3.7161	/1.0	/0	/27008	7.2989
1	70/1	7.8577	/2.0	/0	/18385	11.6240
2	70/2	12.1766	/3.0	/0	/16026	15.5308
3	70/1	14.7429	/6.0	/0	/14862	17.0850
4	74/1	16.7762	/10.0	/42	/17745	17.8928
5	80/1	18.3567	/10.0	/42	/20133	18.3020
6	80/1	19.2623	/10.0	/42	/21130	18.4483
7	79/1	19.8498	/11.0	/42	/20784	18.4541
8	79/2	19.4379	/11.0	/42	/17181	18.2266
9	80/1	18.6725	/10.0	/0	/11011	18.0874
10	80/1	20.6723	/11.0	/0	/10427	19.0944
11	81/2	21.8529	/13.0	/0	/9569	19.9821
12	81/10	29.9752	/28.0	/61	/9611	24.5029
13	82/5	39.2769	/42.0	/79	/9849	28.0425
14	82/35	48.3407	/47.0	/79	/26073	31.7393
15	83/1	49.7960	/58.0	/80	/30219	32.0161
16	83/1	49.7920	/53.0	/80	/26615	31.9080
17	84/1	48.9350	/43.0	/82	/14444	31.4711
18	84/4	54.1667	/79.0	/82	/28899	31.2277
19	84/16	54.7868	/79.0	/82	/29726	31.1520
20	84/66	54.9068	/79.0	/82	/29851	31.0771
21	84/227	54.4014	/79.0	/82	/29269	31.2023
22	84/812	52.8376	/78.0	/82	/26664	31.6384
23	85/8	47.8541	/42.0	/82	/16200	32.2410
24	86/1	45.6152	/42.0	/84	/18487	32.3465
25	85/106	46.9562	/42.0	/84	/19711	32.4502
26	85/403	48.0203	/42.0	/84	/20296	32.4963
27	85/1860	49.4446	/42.0	/84	/19561	32.4576
28	85/9925	49.1650	/42.0	/85	/9925	32.4568
29	85/20745	49.8664	/42.0	/85	/20745	32.7179
30	85/21926	50.9184	/44.0	/85	/21926	32.6939
31	85/23690	52.2127	/50.0	/85	/23690	32.6795
32	85/25027	53.2120	/59.0	/85	/25027	32.5720
33	85/25847	53.8965	/78.0	/85	/25847	32.5748
34	85/26472	54.3810	/78.0	/85	/26472	32.5290
35	85/26555	54.4535	/78.0	/85	/26555	32.5426
36	85/26796	54.4999	/78.0	/85	/26796	32.5788
37	85/26954	54.6478	/78.0	/85	/26954	32.5519
38	85/26900	54.5289	/78.0	/85	/26900	32.5846
39	85/27154	54.8597	/79.0	/85	/27154	32.5391
40	85/27285	54.8983	/79.0	/85	/27285	32.5821
41	85/27066	54.6912	/78.0	/85	/27066	32.5471
42	85/26842	54.4200	/78.0	/85	/26842	32.7018
43	85/26992	54.4673	/78.0	/85	/26992	32.6669
44	85/27320	54.7612	/79.0	/85	/27320	32.6243
45	85/27251	54.7477	/79.0	/85	/27251	32.6546
46	85/27266	54.7601	/79.0	/85	/27266	32.6276
47	85/26963	54.2763	/78.0	/85	/26963	32.7590
48	85/26881	54.3092	/78.0	/85	/26881	32.7520
49	85/27326	54.9054	/79.0	/85	/27326	32.6446
50	85/27151	54.5582	/78.0	/85	/27151	32.7468
51	85/27275	54.5559	/79.0	/85	/27275	32.7504
52	85/27039	54.4343	/78.0	/85	/27039	32.8020
53	85/26778	54.2419	/78.0	/85	/26778	32.8035
54	85/27161	54.6802	/79.0	/85	/27161	32.7173
55	85/27032	54.4643	/78.0	/85	/27032	32.7540

56	85/27240	54.3823	/78.0	/85	/27240	32.8372
57	85/27065	54.5332	/79.0	/85	/27065	32.7871
58	85/27053	54.3053	/78.0	/85	/27053	32.7971
59	85/27065	54.3079	/78.0	/85	/27065	32.8426
60	85/26866	54.1505	/78.0	/85	/26866	32.9019
61	85/27027	54.2435	/78.0	/85	/27027	32.8454
62	85/27061	54.1236	/78.0	/85	/27061	32.9368
63	85/26978	54.0425	/78.0	/85	/26978	32.9256
64	85/26922	54.1561	/78.0	/85	/26922	32.9152
65	85/27270	54.3465	/79.0	/85	/27270	32.9389
66	85/26993	54.0707	/78.0	/85	/26993	32.9375
67	85/27078	54.1169	/78.0	/85	/27078	32.9898
68	85/27173	54.1822	/78.0	/85	/27173	33.0029
69	85/27134	54.3050	/78.0	/85	/27134	32.9180
70	85/27175	54.3922	/79.0	/85	/27175	32.9406

Figure 6.40 An example of premature convergence.

There are several situations that can prevent the GA from moving the population ahead. Recall that the parents are selected from the top `select` = 3277 members of a generation. If these machines have the same or very similar phenotypes, crossover has little chance of creating something new. If these machines are small, say 10 states, only $5+10 \times 14 = 145$ bits of the genome carry relevant data. With a mutation rate of 1%, one or two of these bits (0.01×145) can be expected to be changed by mutation. Most of these changes will reduce the fitness of the individual. For example, one of our runs which was permitted to breed 70 generations, even though the first perfect scoring individual appeared in generation 32, produced a set of machines whose similarities are represented in Table 6.32. The table represents a schema shared by the ants. For one FSM, *** is $q_0/0$ yielding an ant which takes 195 time steps to traverse the trail. When *** is $q_7/2$, $q_0/2$, or $q_4/2$, the ant needs 193 time steps. Other relatives which appeared were $q_5/2$ for *** requiring 197 time steps and $q_6/3$ needing the full 200 time steps to traverse the trail. The success of this set of ants, differing in only a single transition/output, might suggest that, for some inexplicable reason, the GA is altering only this part of the genome. In fact, this indicates that only this set of changes does not destroy the fitness; the numerous other machines which had other changes in the phenotype had a lower fitness.

Another run, which was also permitted to breed 70 generations, even though a perfect score appeared in generation 38, produced the set of relatives whose similarities are represented in Table 6.33. When ### is replaced with $q_1/0$ and @@@ is replaced with $q_7/2$, it takes the machine 187 time steps to cover the 89 trail steps. When @@@ in the previous machine was changed to $q_3/2$, probably as the result of mutation of a single bit, the machine still scores perfectly, but this machine needs 191 time steps to cover the trail. When ### in the second machine was changed to $q_0/3$, the fitness and time remain unchanged. The changing of $q_1/0$ (which appears as 0000100 in the genome) to $q_0/3$ (which appears as 0000011 in the genome) could have occurred as the result of mutation 3 adjacent bits, but it is more likely to have occurred as the result of a crossover. The first of these three machines must have produced many progeny, but only a few different ones attained the optimal fitness.

Present state	Next state/output for input = 0	Next state/output for input = 1
q_0	$q_1/1$	$q_2/2$
q_1	$q_3/1$	$q_4/2$
q_2	$q_5/2$	$q_1/2$
q_3	$q_5/0$	$q_2/2$
q_4	$q_5/2$	$q_6/2$
q_5	$q_7/2$	***
q_6	$q_6/0$	$q_1/2$
q_7	$q_0/0$	$q_6/2$

Table 6.32 Template for a family of ants with differences in one transition/output

Present state	Next state/output for input = 0	Next state/output for input = 1
q_0	$q_1/0$	$q_2/2$
q_1	$q_3/1$	$q_4/2$
q_2	###	$q_3/2$
q_3	$q_5/1$	$q_6/2$
q_4	$q_6/2$	$q_3/2$
q_5	$q_7/0$	$q_8/2$
q_6	$q_3/2$	$q_2/2$
q_7	$q_0/2$	$q_7/2$
q_8	$q_7/2$	@@@

Table 6.33 Template for a few perfect scoring ants with differences in two transitions/outputs.

When the parent pool consists of similar individuals, crossover is not likely to be innovative. Recall that crossing two parents which share a schema does not destroy the schema. In addition, when relevant data does not fill every bit in the genome, fewer bits of relevant data will be mutated. When crossover and/or mutation create a new phenotype, the fitness of the newborn can be lower than that of the parents. Note that in Figure 6.29, the average fitness levels off around generation 26. Also look at the mode and number of individuals with that fitness. The parent pool (consisting of the top 3277 ants) does not have individuals with a fitness less than the mode. Therefore, starting at generation 19 (Figure 6.29), parents have a fitness of at least 80. A high fitness fills up a large portion of the population (between 12.5% and 43%) as indicated by the number of individuals with mode. Also there are a significantly lower average and median for the same population. This indicates that parents with a high fitness are producing low scoring children. In addition there is probably a reduction of innovation due to a parent pool consisting of similar phenotypes. Consequently, the high scoring children which will form the next parent pool may not be significantly different from their parents. Another one of our runs (Figure 6.41) shows another pattern. Note that FSMs with a fitness of 87 intermittently appear, but are unable to maintain a continuous presence. Apparently there is not enough schema support in the parent pool to create enough children with a more useful schema, which, in turn, would move the population to a new area of the genome space. The run in Figure 6.41 might move ahead if it were permitted to continue beyond generation 70, but the run in Figure 6.40 appears to be stuck. Adding a few individuals with a lower fitness into the pool can dramatically alter the results by creating more innovation. Introducing a competition to determine an individual's fitness can let some less-fit individuals, based on trail-following ability, into the parent pool. This, in turn, can push the population into another portion of the genome space. Competition has been widely used in evolutionary simulations. (For example, Angeline 1994, Angeline and Pollack 1993, Fogel 1991, Fogel 1993, and Reynolds 1994, to name a few.)

Benchmark
Statistics for run #3 on file r3b.st
Seed = 0.85823880
Crossover rate = 0.01;   Mutation rate = 0.01
Population size = 65536
Select the top 3277 individuals or 5.00% of each generation for reproduction.

Gen	Max/# with Max	Average	/Median	/Mode	/# with Mode	Standard dev.
0	58/7	3.7190	/ 1.0	/ 0	/27247	7.3241
1	70/1	7.8323	/ 2.0	/ 0	/18569	11.6077
2	78/1	11.7236	/ 3.0	/ 0	/16270	15.2258
3	78/2	14.1280	/ 5.0	/ 0	/15171	16.7544
4	78/4	16.1601	/ 10.0	/ 42	/16003	17.5561
5	81/1	17.7928	/ 10.0	/ 42	/18489	18.0866

6	81/1	18.3326	/ 10.0	/ 42	/18494	18.1056
7	80/2	18.3824	/ 10.0	/ 42	/16674	17.9642
8	81/2	17.8925	/ 10.0	/ 42	/11839	17.7116
9	81/1	17.9904	/ 10.0	/ 0	/10848	17.9229
10	81/1	18.1175	/ 10.0	/ 0	/9891	17.7954
11	81/1	19.5357	/ 11.0	/ 0	/9016	18.3897
12	81/3	20.8326	/ 12.0	/ 0	8508	19.2128
13	81/7	23.0388	/ 16.0	/ 0	/8349	20.6642
14	81/14	25.3135	/ 21.0	/ 58	/9872	21.9316
15	81/39	27.7204	/ 27.0	/ 58	/11878	22.9996
16	82/1	28.8656	/ 27.0	/ 42	/9288	24.5009
17	82/1	41.4281	/ 42.0	/ 80	/18567	30.4038
18	82/5	44.9957	/ 42.0	/ 80	/17054	30.8681
19	82/2	49.6899	/ 46.0	/ 81	/24185	31.0441
20	82/1	53.5471	/ 80.0	/ 81	/29200	30.4957
21	82/4	55.3736	/ 80.0	/ 81	/31783	30.1042
22	82/20	56.3548	/ 81.0	/ 81	/33363	29.9466
23	82/57	56.3084	/ 81.0	/ 81	/33504	29.8992
24	82/187	56.5181	/ 81.0	/ 81	/33705	29.9683
25	84/1	56.1573	/ 81.0	/ 81	/ 32553	29.9602
26	84/2	55.3858	/ 80.0	/ 81	/28704	29.8535
27	84/6	52.9048	/ 56.0	/ 81	/13161	29.5627
28	84/7	52.2641	/ 52.0	/ 82	/19673	29.3857
29	84/22	52.4103	/ 52.0	/ 82	/19895	29.5209
30	84/111	51.5901	/ 49.0	/ 82	/17933	30.0348
31	84/531	50.6058	/ 47.0	/ 83	/12593	31.0139
32	84/2495	55.2657	/ 79.0	/ 83	/25126	30.0359
33	85/1	53.1192	/ 52.0	/ 84	/15929	29.7098
34	85/3	54.3094	/ 56.0	/ 84	/23657	29.7017
35	85/10	54.6090	/ 57.0	/ 84	/23808	29.6042
36	85/41	54.8912	/ 58.0	/ 84	/24131	29.5418
37	85/169	55.0085	/ 58.0	/ 84	/24292	29.5680
38	85/730	54.9841	/ 58.0	/ 84	/23356	29.5169
39	86/1	54.4352	/ 55.0	/ 84	/19191	29.4439
40	86/5	54.1050	/ 53.0	/ 85	/20799	29.6215
41	86/20	54.3600	/ 53.0	/ 85	/21567	29.6883
42	86/52	54.4796	/ 53.0	/ 85	/21663	29.7553
43	86/159	54.5285	/ 53.0	/ 85	/21619	29.7139
44	86/503	53.9719	/ 53.0	/ 85	/20668	29.7487
45	86/1697	52.9792	/ 52.0	/ 85	/18922	29.9071
46	86/7237	51.1180	/ 52.0	/ 85	/11591	29.9374
47	86/17496	52.2747	/ 52.0	/ 86	/17496	29.8381
48	86/17475	52.1763	/ 52.0	/ 86	/17475	29.8532
50	86/17677	52.5339	/ 52.0	/ 86	/17677	29.7720
51	87/1	52.6176	/ 52.0	/ 86	/17941	29.8655
52	86/18037	52.8100	/ 52.0	/ 86	/18037	29.8821
53	86/18037	52.6003	/ 52.0	/ 86	/18037	29.9441
54	87/1	52.4585	/ 52.0	/ 86	/17989	29.9091
55	87/1	52.6242	/ 52.0	/ 86	/18033	29.9459
56	86/18323	52.7689	/ 52.0	/ 86	/18323	29.9270
57	87/1	52.3235	/ 52.0	/ 86	/17822	29.9750
58	86/18090	52.5240	/ 52.0	/ 86	/18090	29.9793
59	86/17965	52.5231	/ 52.0	/ 86	/17965	29.9541
60	86/18146	52.6640	/ 52.0	/ 86	/18146	29.9816
61	87/1	52.5207	/ 52.0	/ 86	/18011	29.9798
62	87/1	52.5978	/ 52.0	/ 86	/18110	29.9897
63	87/1	52.5226	/ 52.0	/ 86	/18047	30.0488
64	86/18158	52.5684	/ 52.0	/ 86	/18158	30.0247
65	86/18203	52.5325	/ 52.0	/ 86	/18203	30.0613
66	86/17987	52.5670	/ 52.0	/ 86	/17987	29.9707
67	86/18248	52.6566	/ 52.0	/ 86	/18248	30.0245
68	86/18110	52.4253	/ 52.0	/ 86	/18110	30.0879
69	87/1	52.0291	/ 52.0	/ 86	/17652	30.0739
70	87/1	52.3869	/ 52.0	/ 86	/18035	30.0552

Figure 6.41 Example of an unsuccessful run.

To add competition to the program, add

```
#define competition 1 /* 0 for no competition; 1
for competition*/

#define num_competitions 10 /* irrelevant if
competition is 0      */
```

to the #define statements at the beginning of the program. In this implementation, each individual faces 10 randomly selected competitors. num_competitions is the number of competitions that each individual will face. This will be irrelevant if competition is 0. The files to receive the output should also be changed (perhaps by adding _c to the file names) to indicate that fitness is determined using competition. The fitness will be determined by the results of the competition. The compete procedure (Figure 6.42) carries out the competition and alters the fitness to include the results of the competition. Each individual faces num_competitions (line 3) randomly chosen competitors (line 4). If the competitor covered fewer trail steps, the individual is credited with two points for the competition (lines 5-6). If the competitor traversed more trail steps, the individual receives no credit for the competition. When there is a tie, the individual receives a single point for the competition (lines 7-9). The points for the ten competitions are accumulated (lines 6 and 9). The new fitness is the sum of the number of trail steps traversed and 100 times the points earned from the competitions (lines 11-12). Consequently the competition becomes the primary determination of fitness. The competition cannot be carried out until the trail-following ability (calculated in calc_fitness, Figure 6.18) has been determined for the entire population. Therefore, compete must be invoked after the trail-following ability is determined and before the population is sorted according to fitness. compete must be invoked by create_next_generation (Figure 6.26) and initialize (Figure 6.24) as part of the process of preparing a population for parenting. Therefore, add the line

```
if (competition) compete();
```

just before the line

```
row_of_sorted = sort(0, popsize - 1, 0);
```

which appears at the end of both create_next_generation (line 11, Figure 6.26) and initialize (line 10, Figure 6.24).

```
/* ------------------------------ */
/*         Competition       */
/* ------------------------------ */
void compete(void)/* holds a competition between an
individual and a random opponent.   the altered
fitness favors those with most wins. */
```

```
{
line #
1:  int i, j, k, wins[popsize]={0};

2:  for (i=0; i<popsize; ++i)
3:    for (j=0; j<num_competitions; ++j) {
4:      k = Rnd(0, popsize-1);
5:      if (fitness[k] < fitness[i])
6:wins[i] += 2;
7:      else
8:if (fitness[k] == fitness [i])
9:  ++wins[i];
10:     }

11:   for (i=0; i<popsize; ++i)
12:     fitness[i] += 100*wins[i]; /* bonus points
for winning */
} /* compete */
```

Figure 6.42 Competition Procedure

The procedure which calculates the statistics for the file with the st extension must also be changed. This is necessary because the statistics are based upon the ants' ability to traverse the trail, but a population whose fitness is determined by competition is not ordered by trail-following ability. Since the original save_stats subroutine (Figure 6.25) assumes the data is ordered by trail-following ability, it has to be modified. Also each ant's fitness includes the result of the competition as the primary determination of fitness. This must be removed for the calculations of average and standard deviation which are based upon trail-following ability. The new save_stats subroutine is listed in Figure 6.43. Since the primary determination of fitness is the competition, the FSM which covered the most trail steps is not necessarily the first on the sorted list. In Figure 6.25 (line 3) the maximum number of trail steps covered by an individual in the population is determined by the fitness of the first individual on the sorted list. fitness_count[i] indicates the number of FSMs that covered i trail steps. Therefore, the maximum number of trail steps covered by an individual in the population is the largest i such that fitness_count[i] 0 (line 3, Figure 6.43). The median is no longer at the center of the sorted list (lines 7-12, Figure 6.25). Here also fitness_count[i] is used. The while loop (lines 7-10, Figure 6.43) permits j to accumulate the number of individuals until it has counted more than halfway through the population. i indicates the fitness of the jth individual. If popsize is odd or if j is past the midpoint of the population (lines 11-12, Figure 6.43), i is the median. Otherwise find the next higher fitness and average the two (lines 13-17, Figure 6.43). The computation of the average (lines 18-23, Figure 6.43) and standard deviation (lines 24-29, Figure 6.43) uses the fitness when there is no competition (lines 21-22 and 27-28), but it must isolate the trail-following ability (the least two significant digits of the fitness) when there is a competition (lines 19-20 and 25-26). The location of the mode(s) is identical in the two subroutines (Figures 25 and 43), as is the printing out of the data. save_stats of Figure 6.43 replaces the one in Figure 6.25.

```
/* ------------------------------- */
/*      Save Statistics     */
/* ------------------------------- */
void save_stats(void)
{
line #
1:  int max = maxfitness, num_of_max, mode =
maxfitness, multimode = -1, i, j, k,
      fitness_sum=0;
2:  double avg=0, median, standard_dev=0;

  /* ------------------------------- */
  /*      Determine maximum     */
  /* ------------------------------- */

3:  while (!fitness_count[max]) --max; num_of_max =
fitness_count[max];

  /* ------------------------------- */
  /*      Determine median     */
  /* ------------------------------- */

4:  k = popsize/2;
5:  if (2*k != popsize) ++k;
6:  i = -1; j = 0;
7:  while (j < k) {
8:      ++i;
9:      j += fitness_count[i];
10:   }
11:   if ( j>k || 2*k>popsize)
12:       median = (double) i;
13:   else {
14:       j = i+1;
15:       while (!fitness_count[j]) ++j;
16:       median = (double) (i+j)/2.0;
17:   }
  /* ------------------------------- */
  /*      Determine average     */
  /* ------------------------------- */
```

```
18:    for (i=0; i<popsize; ++i)
19:       if (competition)
20:           avg += (double) (fitness[i] % 100);
21:       else
22:           avg += (double) fitness[i];
23:    avg /= (double) popsize;
```

```
/* ------------------------------ */
/*       Determine standard dev.   */
/* ------------------------------ */
```

```
24:    for (i=0; i<popsize; ++i)
25:       if (competition)
26:          standard_dev += pow( (double)(fitness[i]
% 100) - avg, 2.);
27:       else
28:          standard_dev += pow( (double) fitness[i]
- avg, 2.);
29:    standard_dev = sqrt(standard_dev/ (double)
popsize);
    /* ------------------------------ */
    /*       Determine mode        */
    /* ------------------------------ */
                !
} /* save_stats */
```

Figure 6.43 The procedures to calculate and output the statistics for each generation.

In addition, because the population is not ordered by trail-following ability, the termination test for the for loop in the main procedure must be changed. The loop in the main procedure of the program which does not contain competition (lines 6-9, Figure 6.20) is

```
for (generation=1;
    (generation <= last_generation)
        && (fitness[sorted[row_of_sorted][0]] <
maxfitness);
    ++generation)
```

This will change to

```
for (generation=1;
    (generation <= last_generation)
        && (fitness_count[maxfitness] == 0;
    ++generation)
```

in the program which includes competition. Note that the only change is in the third line. Instead of checking to see if the first individual on the sorted list has the maxfitness based on trail-following ability, the test checks to see if any individual has that maxfitness. Also, the beginning of print_header (Figure 6.39) must be modified so that the title indicates that competition was used to generate the results. The beginning of the new print_header procedure appears in Figure 6.44.

```
/* ------------------------------- */
/* Print Header on Output Files    */
/* Includes SFS, MTF, and competition */
/* ------------------------------- */
void print_header(void)/* print out header
information for the GA in  the statistics and
perfect score files */
{
line #
1: fp_stats =  fopen(printed_output, "w");
2: fp_perfect = fopen(perfect_score, "w");

3: if (reorganization) {
4:   if (method ==1) {
5:      fprintf(fp_stats,   "\nMTF");
6:      fprintf(fp_perfect, "\nMTF");
7:   }
8:   else {
9:      fprintf(fp_stats,   "\nSFS");
10:        fprintf(fp_perfect, "\nSFS");
15:      }
16:   }
17:   else {
18:     fprintf(fp_stats,   "\nBenchmark");
19:     fprintf(fp_perfect, "\nBenchmark");
20:   }

21:   if (competition) {
22:       fprintf(fp_stats,   " with competition");
23:       fprintf(fp_perfect, " with competition");
24:     }

25:   fprintf(fp_stats,   "\nStatistics for run #%d
on file %s\n,"
26:     run_num, printed_output);
27:   fprintf(fp_perfect, "\nPerfect scorers for
run #%d on file %s\n,"
```

```
28:     run_num, perfect_score);
 fprintf(fp_stats,   "Seed=%.8f\n," seed);
 fprintf(fp_perfect, "Seed=%.8f\n," seed);        !
    } /* print header information */
```

Figure 6.44 Prints Headings on Output Files – Modified to Include SFS, MTF, and Competition

Figures 6.45 through 6.50 contain a set of six runs – one for each of the six methods. Note that all six runs use the same seed for the random number generator; therefore, all six runs start with the identical initial population.

6.6. Summary

The organization of data along a genome which is used as the operand of a genetic algorithm can significantly affect the GA's convergence. This is supported by schema theory and suggests reorganizing FSM genomes during execution of a GA to improve convergence. Two reorganization algorithms were designed. The SFS algorithm enhances schema formation by standardizing transitions. The MTF algorithm reduces schema length. To reduce the possibility of premature convergence, the fitness was modified from an explicitly defined function to one which incorporates competition. With competition, the evolving population faces a changing fitness landscape.

Preliminary testing indicated that reorganizing the FSMs during run time improved GA convergence, with reduced schema length showing more promise than standardizing transitions. The advantage of incorporating competition is not as clear, but the addition of both MTF and competition could dramatically improve performance.

As of this writing, the runs are being timed to see how the additional processing which was performed to improve convergence affects run time. Another area presently being investigated is whether it is necessary to reorganize genomes every generation, or whether the effects of reorganization will carry over a few generations.

It is also important to determine whether the convergence patterns are problem-specific or whether they will be similar when these methods are applied to other problems. Also suggested for future research is to see how sensitive these methods are to different parameter values. Testing is now being done to determine the sensitivity to population size and the maximum number of iterations permitted for a run. This research was FSM-specific, but the ideas should be considered for other standard phenotypes, and for problem-specific phenotypes as well.

Benchmark
Statistics for run #4 on file r4b.st
Seed = 0.41974869

Crossover rate = 0.01;   Mutation rate = 0.01
Population size = 65536
Select the top 3277 individuals or 5.00% of each generation for reproduction.

Gen	Max/# with Max	Average	/Median	/Mode	/# with Mode	Standard dev.
0	58/10	3.7280	/1.0	/0	/26879	7.2818
1	69/1	7.7590	/2.0	/0	/18716	11.5410
2	74/3	11.9915	/3.0	/0	/16287	15.2990
3	74/3	14.7576	/8.0	/0	/14528	16.9324
4	78/1	17.0205	/10.0	/42	/17361	17.8306
5	75/2	18.7800	/10.0	/42	/19962	18.2666
6	79/2	19.4777	/10.0	/42	/20317	18.3386
7	79/5	19.8506	/11.0	/42	/19182	18.3038
8	79/8	18.6168	/10.0	/42	/13422	17.8933
9	79/7	17.8272	/10.0	/0	/10411	17.7072
10	79/6	18.7734	/10.0	/0	/10112	18.3726
11	79/11	20.0063	/11.0	/0	/10305	19.7186
12	80/2	22.4649	/12.0	/58	/10168	21.5989
13	80/4	24.6806	/20.0	/58	/13275	22.6235
14	80/8	26.3638	/21.0	/58	/15244	23.1291
15	81/2	27.8293	/28.0	/58	/16224	23.3392
16	81/12	28.6604	/32.0	/58	/15887	23.4982
17	81/71	28.433	/29.0	/58	/12463	23.8299
18	81/253	30.0690	/27.0	/42	/8664	26.3270
19	81/881	41.1425	/42.0	/42	/8977	28.7242
20	82/1	42.6135	/42.0	/42	/9588	29.4298
21	82/7	50.9608	/47.0	/81	/26528	30.5739
22	82/26	54.8194	/80.0	/81	/31304	29.8236
23	82/99	56.2657	/81.0	/81	/33299	29.5352
24	82/450	56.3027	/81.0	/81	/33001	29.4298
25	83/2	55.7378	/80.0	/81	/30259	29.2594
26	83/5	53.8321	/52.0	/81	/14789	28.9199
27	83/12	54.4393	/53.0	/82	/26623	28.8767
28	83/37	54.5782	/58.0	/82	/27095	28.9468
29	83/119	54.6098	/58.0	/82	/27483	29.1305
30	83/460	54.7584	/59.0	/82	/27486	29.1773
31	84/3	54.0505	/53.0	/82	/24844	29.2547
32	85/12	52.4595	/52.0	/82	/15036	29.3201
33	85/38	51.3959	/49.0	/83	/17682	29.5341
34	85/98	51.4716	/51.0	/83	/18120	29.6539
35	85/282	50.7956	/47.0	/83	/17446	29.8910
36	85/1138	48.7928	/43.0	/83	/12495	30.0468
37	85/7192	49.1184	/44.0	/84	/11522	30.5455
38	85/24441	54.3148	/52.0	/85	/24441	30.2781
39	85/24526	54.5387	/52.0	/85	/24526	30.2703
40	85/24872	54.8865	/52.0	/85	/24872	30.1879
41	85/24677	54.6404	/52.0	/85	/24677	30.2351
42	86/1	54.5604	/52.0	/85	/24521	30.2632
43	86/3	54.8971	/52.0	/85	/24959	30.2195
44	86/4	54.7020	/52.0	/85	/24835	30.3057
45	86/7	54.8248	/52.0	/85	/24899	30.2265
46	86/14	54.8295	/52.0	/85	/24966	30.2300
47	86/36	54.7278	/52.0	/85	/24902	30.2146
48	86/103	54.6983	/52.0	/85	/24802	30.2380
49	86/302	54.4777	/52.0	/85	/24105	30.1766
50	86/971	53.9646	/52.0	/85	/22710	30.1696
51	86/3853	53.0190	/52.0	/85	/17935	29.9081
52	86/20196	53.7197	/52.0	/86	/20196	30.2311
53	86/21150	54.4987	/52.0	/86	/21150	30.1254
54	86/21251	54.4297	/52.0	/86	/21251	30.0805
55	86/21178	54.2505	/52.0	/86	/21178	30.1298
56	86/21391	54.1986	/52.0	/86	/21391	30.1132
57	86/21529	54.3324	/52.0	/86	/21529	30.1919
58	86/21533	54.2803	/52.0	/86	/21533	30.1737
59	86/21840	54.4099	/52.0	/86	/21840	30.2477
60	86/21805	54.5126	/52.0	/86	/21805	30.1309

61	86/21761	54.2267	/52.0	/86	/21761	30.2420
62	86/21888	54.4391	/52.0	/86	/21888	30.1982
63	86/21963	54.5552	/52.0	/86	/21963	30.2247
64	86/21514	54.1948	/52.0	/86	/21514	30.1887
65	86/21675	54.3253	/52.0	/86	/21675	30.1334
66	86/21731	54.1884	/52.0	/86	/21731	30.2053
67	86/21678	54.2560	/52.0	/86	/21678	30.1655
68	87/1	54.4397	/52.0	/86	/21800	30.1347
69	86/21914	54.2779	/52.0	/86	/21914	30.1824
70	86/22014	54.5685	/52.0	/86	/22014	30.1294

Figure 6.45. Seed = 0.41974869 for benchmark case.

SFS

Statistics for run #5 on file r5sfs.st

Seed = 0.41974869

Crossover rate = 0.01;   Mutation rate = 0.01

Population size = 65536

Select the top 3277 individuals or 5.00% of each generation for reproduction.

Gen	Max/# with Max	Average	/Median	/Mode	/# with Mode	Standard dev.
0	58/10	3.7280	1.0	/0	/26879	7.2818
1	64/2	9.4931	3.0	/0	/15065	12.5063
2	79/1	16.7633	/10.0	/42	/15400	17.1080
3	77/1	21.2328	/13.0	/42	/22957	18.2337
4	79/1	23.5792	/27.0	/42	/25001	18.1880
5	79/2	22.5542	/21.0	/42	/16137	17.8502
6	81/1	23.6796	/23.0	/42	/8483	18.1358
7	81/9	25.8023	/32.0	/42	/9296	18.5205
8	81/44	28.3749	/32.0	/42	/9846	19.8366
9	81/199	29.9216	/32.0	/42	/9713	21.6730
10	82/1	35.0613	/33.0	/74	/11269	27.2817
11	82/6	36.7620	/37.0	/74	/11559	27.4644
12	82/33	46.8080	/42.0	/81	/21939	28.2858
13	83/2	49.9388	/42.0	/81	/25682	28.6222
14	83/2	51.4632	/43.0	/81	/27267	28.7879
15	83/11	50.7088	/42.0	/81	/24921	28.9342
16	84/3	45.3295	/42.0	/81	/10948	28.4854
17	84/10	44.6686	/42.0	/82	/17264	28.1975
18	84/36	44.9738	/42.0	/82	/17346	28.1933
19	84/162	44.9743	/42.0	/82	/16924	28.2754
20	84/731	43.9175	/42.0	/82	/14297	28.4634
21	84/4006	41.3651	/34.0	/32	/7625	29.1865
22	85/1	56.8664	/79.0	/84	/29059	30.2307
23	85/5	57.2281	/79.0	/84	/29378	30.0881
24	85/12	57.4824	/80.0	/84	/29621	30.0399
25	85/35	57.4722	/79.0	/84	/29471	30.0686
26	85/107	57.3028	/79.0	/84	/29274	30.1173
27	85/303	56.3052	/79.0	/84	/28283	30.4505
28	85/844	54.3921	/58.0	/84	/26077	31.0252
29	85/2895	50.0596	/43.0	/84	/20348	31.7350
30	86/2	43.8951	/42.0	/85	/13911	31.8159
31	86/6	44.7617	/42.0	/85	/16705	31.6947
32	87/1	45.3253	/42.0	/85	/17217	31.7614
33	86/12	45.8809	/42.0	/85	/17639	31.8019
34	86/29	46.2325	/42.0	/85	/17736	31.7316
35	86/47	46.1827	/42.0	/85	/17728	31.7913
36	86/87	46.0375	/42.0	/85	/17590	31.7975
37	86/193	46.2776	/42.0	/85	/17574	31.6771
38	86/433	46.0887	/42.0	/85	/17045	31.6061
39	86/1128	45.0469	/42.0	/85	/15247	31.4144
40	86/3460	42.8525	/42.0	/85	/10915	30.8928
41	86/16021	46.6987	/42.0	/86	/16021	32.1949
42	86/16048	47.0449	/42.0	/86	/16048	32.1498
43	87/1	47.1566	/42.0	/86	/16256	32.1281

44	87/4	47.3444	/42.0	/86	/16238	32.1031
45	87/13	47.2872	/42.0	/86	/16259	32.1767
46	87/26	47.3780	/42.0	/86	/16419	32.2318
47	87/46	47.3228	/42.0	/86	/16004	32.0642
48	87/98	46.8675	/42.0	/86	/15732	32.0816
49	87/198	46.5328	/42.0	/86	/15462	32.0798
50	87/423	45.9071	/42.0	/86	/14673	31.9786
51	87/980	45.5287	/42.0	/86	/14003	31.9481
52	89/1	43.2632	/42.0	/86	/11151	31.5850

Figure 6.46a Seed = 0.41974869 for SFS

SFS
Perfect scorers for run #5 on file r5sfs.ps
Seed = 0.41974869
Crossover rate = 0.01;   Mutation rate = 0.01
Population size = 65536
Select the top 3277 individuals or 5.00% of each generation for reproduction.

Generation 52; Start State = 0; Fitness of 89 reached in 197 time steps.

```
ps  fs/output  ps  fs/output  ps  fs/output  ps  fs/output
    in=0:in=1      in=0:in=1      in=0:in=1      in=0:in=1
0    1/0:2/2    1   3/0:4/3   2   5/2:6/2   3   7/0:8/2
4    9/3:2/2    5  11/2:12/2  6   4/1:8/2   7  15/2:16/3
8    7/2:16/2   9   0/0:9/2  11   3/1:9/2  12   8/1:0/2
15   0/0:5/2   16   1/0:11/2
```

Figure 6.46b Perfect score machine for seed = 0.41974869 in SFS.

MTF
Statistics for run #6 on file r6mtf.st
Seed = 0.41974869
Crossover rate = 0.01;   Mutation rate = 0.01
Population size = 65536
Select the top 3277 individuals or 5.00% of each generation for reproduction.

Gen	Max/# with Max	Average	/Median	/Mode	/# with Mode	Standard dev.
0	58/10	3.7280	/1.0	/0	/26879	7.2818
1	74/1	10.6899	/4.0	/0	/14263	13.2674
2	78/2	18.4981	/10.0	/42	/17353	17.6218
3	79/1	21.6968	/19.0	/42	/20210	18.2004
4	80/2	21.7650	/19.0	/10	/8876	18.0427
5	81/3	24.8135	/21.0	/42	/8620	19.2507
6	81/19	29.5378	/32.0	/42	/9499	20.4956
7	81/124	30.6649	/32.0	/10	/9227	22.0310
8	81/811	34.9615	/34.0	/42	/10516	26.1281
9	82/2	44.0251	/42.0	/42	/11478	27.8943
10	83/1	51.3692	/47.0	/81	/25142	28.4255
11	83/5	51.9636	/47.0	/81	/26131	28.6096
12	83/31	52.8090	/47.0	/81	/27155	28.6107
13	83/198	52.6682	/47.0	/81	/25964	28.6689
14	84/2	51.1153	/47.0	/81	/18211	28.4507
15	84/27	49.4771	/42.0	/42	/11310	28.0274
16	84/134	52.0174	/47.0	/83	/24117	30.0812
17	84/698	52.0406	/47.0	/83	/23469	29.9702
18	85/4	50.3621	/43.0	/83	/18539	29.8418
19	85/18	48.5201	/42.0	/84	/18916	29.4776
20	85/102	49.2816	/42.0	/84	/20026	29.7747
21	85/357	50.3421	/43.0	/84	/21027	30.0432

22	85/1265	50.7561	/43.0	/84	/20690	30.1043
23	85/4798	47.9555	/42.0	/84	/13764	29.6501
24	86/1	44.4820	/42.0	/85	/12635	27.8554
25	86/8	44.7047	/42.0	/85	/12693	27.7549
26	86/34	44.7867	/42.0	/85	/12872	27.9735
27	86/140	44.9085	/42.0	/85	/12922	27.8762
28	86/563	45.0175	/42.0	/85	/12697	27.9083
29	86/2249	45.6832	/42.0	/85	/11818	28.1813
30	86/10096	45.9176	/42.0	/86	/10096	28.1798
31	86/16342	46.9234	/42.0	/86	/16342	28.6926
32	86/17533	47.9438	/42.0	/86	/17533	29.0834
33	86/18941	49.3522	/42.0	/86	/18941	29.3331
34	86/20634	51.2908	/43.0	/86	/20634	29.5176
35	86/21924	52.6308	/45.0	/86	/21924	29.5520
36	86/22241	52.9017	/45.0	/86	/22241	29.5098
37	86/22286	53.0096	/45.0	/86	/22286	29.6017
38	86/22701	53.2843	/45.0	/86	/22701	29.6988
39	87/1	53.4767	/45.0	/86	/22781	29.5384
40	87/3	53.6538	/45.0	/86	/22950	29.5295
41	87/11	53.5298	/45.0	/86	/22914	29.5320
42	87/32	53.1087	/45.0	/86	/22563	29.6992
43	87/110	53.3006	/45.0	/86	/22623	29.6559
44	88/2	53.1126	/45.0	/86	/22145	29.6783
45	88/35	52.2466	/45.0	/86	/19820	29.7569
46	88/305	49.7986	/43.0	/86	/9544	29.7668
47	88/1870	50.2104	/43.0	/87	/16404	29.7635
48	88/10633	50.9468	/43.0	/88	/10633	29.9665
49	88/18939	51.7462	/44.0	/88	/18939	30.1365
50	89/1	52.1405	/44.0	/88	/19329	30.1445

Figure 6.47a Seed = 0.41974869 for MTF.

MTF
Perfect scorers for run #6 on file r6mtf.ps
Seed = 0.41974869
Crossover rate = 0.01;   Mutation rate = 0.01
Population size = 65536
Select the top 3277 individuals or 5.00% of each generation for reproduction.

Generation 50; Start State = 0; Fitness of 89 reached in 199 time steps.

```
ps fs/output ps  fs/output ps  fs/output ps  fs/output
   in=0:in=1      in=0:in=1      in=0:in=1      in=0:in=1
0   1/0:2/2  1    3/0:4/2  2    4/2:4/2  3     4/0:2/2
4   5/2:6/2  5    0/0:1/2  6    4/2:7/2  7     0/0:8/2
8   0/0:6/2
```

Figure 6.47b Perfect score machine for seed = 0.41974869 in MTF.

Benchmark with competition
Statistics for run #7 on file r7b_c.st
Seed = 0.41974869
Crossover rate = 0.01;   Mutation rate = 0.01
Population size = 65536
Select the top 3277 individuals or 5.00% of each generation for reproduction.

Gen	Max/# with Max	Average	/Median	/Mode	/# with Mode	Standard dev.
0	58/10	3.7280	/1.0	/0	/26879	7.2818
1	64/1	7.4415	/2.0	/0	/18879	11.1687
2	78/1	10.7848	/3.0	/0	/16434	14.3049

3	78/2	13.3718	/5.0	/0	/14610	15.9486
4	78/1	14.6616	/10.0	/0	/13296	16.4654
5	78/2	16.0459	/10.0	/42	/12904	17.0432
6	78/5	17.0100	/10.0	/42	/13035	17.2615
7	80/1	17.7767	/10.0	/42	/11731	17.4142
8	80/2	18.6165	/10.0	/42	/9871	17.5932
9	80/3	19.4613	/11.0	/0	/9232	17.8781
10	80/7	20.4615	/13.0	/0	/8711	18.2588
11	80/8	21.7628	/19.0	/0	/8040	18.7908
12	81/2	23.6749	/21.0	/0	/7529	19.9519
13	81/2	28.3628	/32.0	/57	/8277	21.7817
14	81/9	28.9930	/32.0	/57	/9706	21.7499
15	81/15	29.2647	/32.0	/57	/6754	21.5700
16	81/37	31.0760	/32.0	/59	/8362	21.4737
17	82/1	33.6069	/34.0	/59	/11455	21.9847
18	82/5	35.4726	/36.0	/42	/7605	23.1570
19	83/1	40.8318	/42.0	/74	/10363	25.8944
20	83/3	41.6259	/42.0	/80	/9502	27.8089
21	83/6	44.9665	/42.0	/81	/11509	28.4470
22	83/14	48.4539	/42.0	/81	/18446	28.5255
23	83/58	52.4289	/47.0	/81	/19629	28.0524
24	84/1	56.8100	/74.0	/82	/21606	26.9351
25	85/1	57.6122	/79.0	/82	/22423	26.7799
26	84/28	56.3918	/58.0	/83	/22347	27.2813
27	84/89	57.0762	/58.0	/83	/25672	27.3905
28	85/2	57.1154	/59.0	/83	/26582	27.5665
29	85/6	56.6614	/58.0	/83	/25426	27.7676
30	85/45	54.4220	/52.0	/83	/20330	28.6322
31	86/2	50.5783	/47.0	/84	/13454	29.6566
32	86/14	50.7700	/47.0	/84	/13976	29.6266
33	86/44	50.5063	/46.0	/84	/12798	29.7581
34	87/4	49.5480	/44.0	/85	/8535	29.8544
35	87/21	50.0268	/45.0	/85	/11818	29.9229
36	87/71	50.5455	45.0	/85	/12837	29.9228
37	87/271	50.8333	/45.0	/85	/10020	29.9259
38	87/1224	52.7166	/47.0	/86	/11803	30.0705
39	87/6829	55.9179	/54.0	/86	/12269	29.6038
40	88/1	57.3744	/58.0	/87	/18639	29.5516
41	88/1	57.5300	/58.0	/87	/22082	29.6658
42	88/2	57.6128	/58.0	/87	/22471	29.6940
43	88/5	57.3963	/58.0	/87	/22387	29.7864
44	88/7	57.6078	/58.0	/87	/22596	29.7117
45	88/9	57.6693	/59.0	/87	/22733	29.7991
46	88/16	57.8033	/63.0	/87	/22948	29.7903
47	88/36	57.5606	/58.0	/87	/22732	29.8030
48	88/101	57.5585	/58.0	/87	/22714	29.8124
49	88/235	56.9625	/58.0	/87	/22055	30.0146
50	88/594	56.3016	/54.0	/87	/21134	30.1627
51	89/1	54.6802	/52.0	/87	/18646	30.4508

Figure 6.48a Sccd = 0.41974869 for benchmark with competition

Benchmark with competition
Perfect scorers for run #7 on file r7b_c.ps
Seed = 0.41974869
Crossover rate = 0.01;   Mutation rate = 0.01
Population size = 65536
Select the top 3277 individuals or 5.00% of each generation for reproduction.

Generation 51; Start State = 7; Fitness of 89 reached in 200 time steps.

```
ps fs/output ps fs/output ps fs/output ps  fs/output
   in=0:in=1    in=0:in=1     in=0:in=1     in=0;in=1
0   7/0: 2/2 2   1/1: 5/2 3   4/1:17/0 4   6/0:23/2
5  13/1:12/2 6   0/2: 6/2 7   3/1:10/2 8   7/0: 8/2
```

```
 9   23/3:23/2 10  12/2:  8/2 11  29/0:  3/0 12   3/2:  9/2
13    8/1:11/1 14   3/3:22/1 15  13/0:14/1 16   3/3:30/1
17  30/3:  8/1 18  15/1:13/0 22   6/1:  8/3 23   6/2:  6/2
28  31/0:  3/3 29  28/2:18/2 30  16/3:  7/2 31   7/2:18/1
```

Figure 6.48b Perfect score machine for seed = 0.41974869 for benchmark with competition

SFS with competition
Statistics for run #8 on file r8sfs_c.st
Seed = 0.41974869
Crossover rate = 0.01;   Mutation rate = 0.01
Population size = 65536
Select the top 3277 individuals or 5.00% of each generation for reproduction.

Gen	Max/# with Max	Average	/Median	/Mode	/# with Mode	Standard dev.
0	58/10	3.7280	/1.0	/0	/26879	7.2818
1	72/1	9.0284	/3.0	/0	/15254	12.1615
2	77/1	15.0237	/10.0	/0	/11547	16.1218
3	79/1	19.1306	/11.0	/42	/17396	17.4896
4	79/1	22.8893	/21.0	/42	/23341	18.0789
5	79/2	23.3478	/22.0	/42	/18375	18.0045
6	80/1	22.8701	/20.0	/10	/8425	18.4156
7	81/1	25.9128	/27.0	/10	/7561	20.1181
8	81/3	35.5536	/35.0	/42	/9132	22.7704
9	81/18	45.9860	/42.0	/42	/8835	24.1247
10	81/142	48.7653	/46.0	/74	/13690	24.8134
11	82/2	47.2839	/42.0	/75	/12374	26.3217
12	83/1	45.9576	/42.0	/80	/9044	29.0170
13	83/5	49.5052	/42.0	/81	/19573	28.5774
14	83/37	51.6628	/44.0	/81	/24171	27.8746
15	83/195	51.8560	/44.0	/81	/24300	27.9203
16	84/2	48.6786	/42.0	/81	/16958	28.9824
17	84/10	47.1964	/42.0	/82	/12742	30.1493
18	84/65	50.8436	/43.0	/83	/21672	30.2633
19	84/531	53.3024	/47.0	/83	/26307	30.3748
20	84/4447	54.2857	/55.0	/83	/23485	30.2954
21	84/28526	56.5837	/79.0	/84	/28526	30.4192
22	86/1	57.0052	/80.0	/84	/29557	30.2833
23	86/3	56.9180	/79.0	/84	/28911	30.3190
24	86/5	56.3080	/79.0	/84	/28622	30.4840
25	86/6	56.2051	/79.0	/84	/28830	30.5407
26	86/14	56.1910	/79.0	/84	/28535	30.6189
27	86/29	56.3466	/79.0	/84	/28873	30.5591
28	86/62	55.3745	/79.0	/84	/27777	30.9539
29	86/143	53.5541	/49.0	/84	/26087	31.4318
30	86/333	48.8129	/42.0	/84	/20955	32.3894
31	86/890	42.1069	/36.0	/84	/9146	32.4042
32	86/2206	39.0927	/32.0	/85	/8827	32.0811
33	86/5770	36.6393	/28.0	/10	/8079	31.7700
34	86/9975	37.8831	/32.0	/86	/9975	32.1718
35	86/11841	39.4905	/32.0	/86	/11841	32.2608
36	87/1	42.5386	/34.0	/86	/14169	32.7261
37	87/3	44.8340	/42.0	/86	/16062	32.8553
38	87/7	45.8881	/42.0	/86	/17094	33.0009
39	87/11	46.4956	/42.0	/86	/17901	33.0580
40	87/11	46.8366	/42.0	/86	/18248	33.0388
41	87/16	47.2164	/42.0	/86	/18465	32.9956
42	87/23	47.0280	/42.0	/86	/18620	33.0722
43	87/29	47.5056	/42.0	/86	/19000	33.0500
44	87/30	47.6999	/42.0	/86	/19057	33.0201
45	87/38	47.1041	/42.0	/86	/18593	33.0517
46	87/48	47.5201	/42.0	/86	/18907	33.0353
47	88/1	47.6421	/42.0	/86	/18906	32.9693

48	88/4	47.7644	/42.0	/86	/18915	33.0194
49	88/3	47.0699	/42.0	/86	/18244	32.8558
50	88/3	47.1079	/42.0	/86	/18353	32.9362
51	88/3	47.2968	/42.0	/86	/18266	32.8582
52	88/3	46.7418	/42.0	/86	/17544	32.7058
53	88/7	46.2410	/42.0	/86	/16362	32.3921
54	88/12	45.9250	/42.0	/86	/14719	31.8856
55	88/20	45.2344	/42.0	/86	/10566	30.9934
56	88/48	45.9375	/42.0	/42	/7567	30.1948
57	88/146	47.3744	/42.0	/87	/8516	29.9943
58	88/538	47.3186	/42.0	/87	/8596	29.8563
59	88/1932	47.4791	/42.0	/87	/7719	29.8543
60	88/6257	49.5363	/42.0	/42	/7122	30.0529
61	88/10200	49.7405	/42.0	/88	/10200	30.2543
62	88/11363	49.2929	/42.0	/88	/11363	30.3142
63	88/11537	48.6758	/42.0	/88	/11537	30.3291
64	88/11610	48.9829	/42.0	/88	/11610	30.2850
65	88/11752	48.8280	/42.0	/88	/11752	30.1998
66	88/11598	48.3188	/42.0	/88	/11598	30.2948
67	88/11739	48.8638	/42.0	/88	/11739	30.1981
68	88/12073	48.9224	/42.0	/88	/12073	30.3068
69	88/11766	48.2541	/42.0	/88	/11766	30.2316
70	88/11629	48.2657	/42.0	/88	/11629	30.2536

Figure 6.49 Seed = 0.41974869 for SFS with competition.

MTF with competition
Statistics for run #9 on file r9mtf_c.st
Seed = 0.41974869
Crossover rate = 0.01;   Mutation rate = 0.01
Population size = 65536
Select the top 3277 individuals or 5.00% of each generation for reproduction.

Gen	Max/# with Max	Average	/Median	/Mode	/# with Mode	Standard dev.
0	58/10	3.7280	/1.0	/0	/26879	7.2818
1	70/1	9.9999	/3.0	/0	/14592	12.8337
2	75/1	17.1035	/10.0	/42	/13134	16.9880
3	79/1	20.8065	/14.0	/42	/17812	17.8732
4	81/1	22.3551	/20.0	/42	/13740	18.0776
5	81/6	24.3652	/21.0	/42	/8953	18.9684
6	81/19	28.5779	/32.0	/42	/9546	20.6806
7	82/3	35.5118	/42.0	/42	/10330	22.1069
8	83/1	42.1950	/42.0	/42	/11045	22.7813
9	84/1	41.8004	/42.0	/42	/12246	24.7928
10	84/5	40.4713	/42.0	/42	/13298	26.5387
11	84/20	45.4988	/42.0	/42	/12771	27.3285
12	84/77	46.0054	/42.0	/42	/10802	27.6969
13	84/351	48.2076	/42.0	/83	/10696	27.8459
14	84/1732	52.6362	/47.0	/83	/19445	27.6965
15	85/1	51.5933	/47.0	/83	/11760	28.7898
16	85/4	53.2215	/52.0	/84	/22317	29.4939
17	85/11	54.7674	/54.0	/84	/24961	29.4480
18	86/1	55.5945	/58.0	/84	/25856	29.2958
19	86/6	55.6414	/58.0	/84	/25895	29.4441
20	87/1	55.4093	/58.0	/84	/25270	29.3581
21	87/4	54.1301	/53.0	/84	/22321	29.6183
22	87/20	51.7877	/47.0	/85	/11905	29.2420
23	87/66	52.9535	/48.0	/85	/18213	28.4662
24	87/234	52.0408	/47.0	/85	/16677	28.3756
25	87/1005	49.4060	/42.0	/85	/8467	28.4656
26	88/2	51.7654	/44.0	/86	/9794	28.6345
27	88/9	54.5997	/52.0	/87	/12698	28.5412
28	88/16	55.3470	/52.0	/87	/11516	28.7137
29	88/38	55.6882	/54.0	/87	/18670	28.7142
30	88/95	55.4890	/52.0	/87	/18634	28.6981

| 31 | 88/268 | 55.2525 | /52.0 | /87 | /18287 | 28.7210 |
| 32 | 89/1 | 54.4330 | /52.0 | /87 | /17305 | 28.9533 |

Figure 6.50a Seed = 0.41974869 for MTF with competition.

MTF with competition
Perfect scorers for run #9 on file r9mtf_c.ps
Seed = 0.41974869
Crossover rate = 0.01;   Mutation rate = 0.01
Population size = 65536
Select the top 3277 individuals or 5.00% of each generation for reproduction.

Generation 32; Start State=0; Fitness of 89 reached in 196 time steps.

```
ps fs/output ps fs/output ps fs/output ps fs/output
   in=0:in=1    in=0:in=1    in=0:in=1    in=0:in=1
0    1/0: 2/3 1   3/0: 4/2 2   0/0: 5/2 3   6/2: 7/2
4    3/2: 7/2 5   7/2: 8/2 6   2/0: 1/2 7   9/2: 3/2
8    2/0: 8/2 9   1/1:10/0 10  8/3:11/3 11 12/1: 0/0
12 11/1:13/2 13   5/2: 5/0
```

Figure 6.50b Perfect score machine for seed = 0.41974869 in MTF with competition.

The next section discusses ants found during similar experimentation with different seeds for the above algorithms. The focus of this section is on trail traversal strategies that are generated during the GA in the form of finite state machines. While GAs are generally used to solve a search problem along the fitness landscape, the strategies developed by partially fit individuals bred along the way can be equally important.

6.7. Interesting Ants Bred during Experimentation

In the previous section, we demonstrated families of successful FSMs. During further testing of the algorithms, some interesting FSM were observed. In this section, two families are presented that are interesting from an evolutionary standpoint. The focus here is the relevance of the inherent strategies employed by the different finite-state machines bred during the genetic algorithm to solve the trail-following problem used for this study.

Jefferson et al. (1992) presented Champ-100. Champ-100 is a 13-state FSM which takes the full 200 time steps to complete the trail. FSMs which take less time and have fewer states are considered interesting. Not all successful FSMs were recorded during the testing – only the first ten successful ants in a generation. Consequently, the observations do not necessarily indicate the smallest FSM nor the minimum number of time steps needed for this problem.

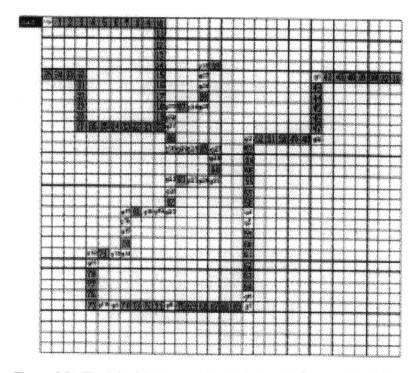

Figure 6.51 The John Muir Trail with Marked and Unmarked Trail Steps Labeled

The examples are presented with decreasing number of states and generally (but not always) with increasing time steps. All of the FSMs presented come from some form of MTF run. This is not unexpected since theoretically the MTF operator should encourage smaller FSMs. In addition, because all the FSMs are the result of MTF, all have a start state of 0. There is no pressure in the fitness function or schema theory to encourage minimization of the time steps needed to traverse the trail. The FSMs are presented exactly as they are printed out by the GA. Recall that an output of 0 indicates a right turn; 1, a left turn; 2, take one step straight ahead; and 3, do nothing. An input of 1 indicates that the next trail step ahead is marked. Note that in all the examples, an output of 2 is the prevalent response; at times the only response, when the input is 1. Also note that an output of 3 is rare in the examples presented. The rendition of the John Muir trail in Figure 6.51 includes labels of gi for the unmarked trail steps. This is done to facilitate the trail traversals presented in the section. In several of the examples, a sequence of five states is presented during which time an ant is turning in place to look around when the trail is not in evidence. Note that right and left turns are sometimes navigated as a subsequence of the five-state sequence. Hence, the five-state sequence is not completed when a marked step is in front of an ant after one of the turns in the sequence.

6.7.1 Best Observed Time: 181 Time Steps

A family of 1 2-state FSMs covered the 89 trail steps in 181 time steps.

Start state: 0

present state	next state/output for input = 0	next state/output for input = 1
0	1/0	2/2
1	3/0	###
2	4/2	3/2
3	5/0	6/2
4	7/2	8/2
5	8/2	9/1
6	5/2	9/2
7	3/1	10/2
8	0/0	11/2
9	@@@	7/2
10	9/2	8/2
11	2/2	4/2

Table 6.34 Template for a 12-state Perfect Scoring Ants with Time = 181

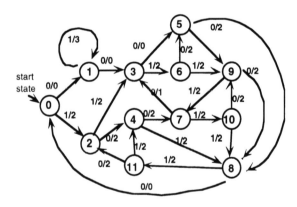

Figure 6.52 12-state Perfect Scoring Ant with Time = 181

The family is represented by the template in Table 6.34. Figure 6.52 represents the first ant and is obtained by substituting the first row of Table 6.35 into Table 6.34. Table 6.35 indicates four of the successful variations. When this FSM is applied to the John Muir trail, the transitions for states 1 and 5 with input 1 are never used. When these two transitions and their corresponding outputs are eliminated, this FSM can then be reduced to an 11-state equivalent by combining states 5 and 9 into a single state.

Referring to Figure 6.51 (the Muir trail), when an ant is on any of the marked steps 10, 20, 27, 75, or unmarked steps g2, g7, g12, g17, g22, g30, g33, g38, the ant is initially in state 8. An input of 0 results in a right turn, thus putting the ant immediately back on track. An ant in state 7 will make a left turn when the trail disappears. This takes an ant through the left turns at marked step 32, and unmarked steps g1, g14, and g35. The sequence of states 8-0-1-3-5 yields a 360^0 clockwise turn when there is no marked step in any of the adjacent boxes. This strategy, or parts of it are used at marked step 47, and unmarked steps g3, g16, g20, g25, g27, g29, and g37. Note that the first two states of this sequence are used for right turns. The state sequences 10-9-8 and 6-5-8 can take an ant across two unmarked steps but only the 6-5-8 sequence is used by the ant. It takes the ant from marked step 78 to g12, from 80 to g16, from 82 to g22, from 84 to g27, from 85 to g29, and from 88 to g37. The sequence of states 11-2-4-7 can take an ant across three unmarked steps, but it is never utilized on the John Muir trail. The shorter subsequences 11-2-4 and 2-4-7 are used to take an ant from marked step 58 to g5 (11-2-4), from 74 to g10 (11-2-4), from 79 to g14 (2-4-7), from 81 to g19 (2-4-7), from 83 to g24 (2-4-7), from 86 to g32 (2-4-7), and from 87 to g35 (2-4-7).

The state sequence 7-3-5 (used at steps 52, g6, g19, g24, and g32) has an ant looking only to the left for the missing trail before stepping ahead in the original direction. This saves time, but, if it is utilized at the wrong location it could take the ant off the trail and result in aimless wandering to relocate the trail. Since the latter does not occur when this FSM is applied to the John Muir trail, it helps keep the time down. It might seem plausible to stay in a single state for a marked straight sectional path, but this is not the case. For example, to get from the start position to marked step 10, an ant goes through the state sequence 0-2-3-6-9-7-10-8-11-4-8. The sequence 8-11-4 cycles on a marked straight sectional path.

###	@@@
1/3	8/2
4/2	8/1
0/1	6/2
4/2	4/2

Table 6.35 Substitutions for ### and @@@ in Table 6.34.

In the next section some smaller FSMs with higher traversal times are presented to analyze strategies that did not perform as well.

6.7.2 Intermediate FSMs

The FSMs presented in this section have neither the best time nor the fewest states, but the combination of time and number of states are notable. The 11-state FSM in Figure 6.53 takes 188 time steps to traverse the trail.

The blank positions in the table are never used which is why they have been omitted.

Start State : 0

Present state	Next state/ output for input = 0	Next state/ output for input = 1
0	1/0	2/2
1	3/0	3/2
2	4/2	5/2
3	6/0	7/2
4	8/2	-
5	0/0	5/2
6	5/2	1/2
7	6/2	9/2
8	3/1	—
9	10/2	9/2
10	1/2	1/3

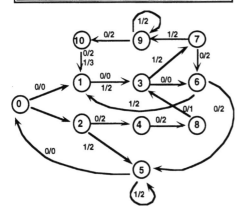

Figure 6.53 11-state Perfect Scoring FSM with Time = 188

This FSM is of interest because at one point it takes the ant off the trail. Instead of wandering aimlessly after that, the ant backtracks onto an unmarked step and continues on to complete the trail before its time is up. When the ant is on step 42, it is in state 9 facing g1, and receiving an input of 0 from g1. This brings the ant to state 10 on g1 with an input of 0 from the square to the left of g1. The ant moves onto the square to the left of g1 and into state 1. The ant is no longer on the trail. During the next two time steps with inputs of 0, the ant makes two right turns and moves into state 3, followed by state 6. The ant is now facing g1 standing on the square to the left of g1, and in state 6 receiving an input of 0. This brings the ant back onto g1, and into state 5. Step 42 is now directly in front of the ant, but since the ant was previously on step 42, the marking was erased to prevent backtracking on the trail. Consequently, the ant on g1 facing step 42 receives an input of 0. The ant makes a right turn and moves into state 0. At this point the ant is on g1, facing marked step 43 and, therefore, receiving an input of 1. This takes the ant onto step 43 and back onto the marked trail. It is this recovery which makes the FSM of Figure 6.53 so interesting.

With both fewer states and better time, the 10-state FSM in Figure 6.54 requires 185 time steps to traverse the 89 marked trail steps. The blank position for state 9 with input 1 contained different values during the run, but these values are irrelevant since an ant on the trail in state 9 never receives an input of 1. Note that the only response to a marked trail step directly ahead (that is, to an input of 1) is to move ahead onto that marked step. Also note that there is no output of 3. That is, an ant always takes some action; a time step is never used strictly to change state.

Start state: 0

Present state	Next state/ output for input = 0	Next state/ output for input = 1
0	1/2	1/2
1	2/0	3/2
2	4/0	5/2
3	6/2	0/2
4	7/0	8/2
5	9/2	7/2
6	0/2	4/2
7	0/0	6/2
8	7/1	0/2
9	8/2	—

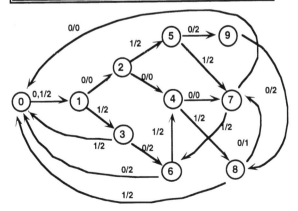

Figure 6.54 10-state Perfect Scoring FSM with Time = 185

To walk from step 10 to step 20, the state sequence is 2-5-7-6-4-8-0-1-3-0-

1. Subsequences of this and the cycle 3-0-1 are utilized along marked straight sectional paths. All right turns are efficiently handled by state sequence 1-2. The left turns at steps 39, g14, and g35 are accomplished with state sequence 8-7. The sequence of states 1-2-4-7-0 takes the ant through a 360^0 clockwise turn when no marked trail steps are in sight. All or part of this sequence is invoked at steps g1, 47, g3, g6, g16, g20, g25, g27, g29, and g37. While the sequence 3-6-0-1 can take an ant from a marked step across three successive unmarked steps, this full sequence is not used in the trail traversal. The subsequences 3-6-0 and 6-0-1 along with 5-9-8 are used to cover the two unmarked steps following steps 58 (3-6-0), 74 (3-6-0), and 78 (6-0-1), 79 (5-9-8), 80 (6-0-1), 81 (5-9-8), 82 (6-0-1), 83 (5-9-8), 84 (6-0-1), 85 (6-0-1), 86 (5-9-8). 87 (5-9-8), and 88 (6-0-1).

The FSM in Figure 6.55 takes 187 time steps to complete the trail. This ant has two interesting relatives. Both relatives take 4 time steps longer to cover the trail. For one relative, state 2 with input 0 moved the FSM to state 0 and output a 3. This simply wastes a time step whenever the FSM is in state 2 and receiving an input of 0. This occurs on steps 10, 20, 27, and 75 explaining the additional 4 time steps that this ant uses. For the other relative, the only part of Figure 6.55 which is affected is the next state for state 8 with an input of 1. For this relative the next state for state 8 with input 1 is 3 instead of 7. This causes this relative to use four additional time steps on step 42 and hence the traversal ends later than it could have.

Unlike the FSMs in Figures 6.53 and 6.54, the FSM of Figure 6.55 does not have the capability to make a 360^0 turn when there are no trail steps in sight. Like the FSM in Figure 6.54, the only response of the FSM in Figure 6.55 facing a marked trail step is to move onto it. When there are no marked trail steps around the ant, the state sequence 0-1-3-5-7 first turns the ant to the right, then follows with two left turns, and ends with a right turn restoring the ant to its original position. This strategy (or parts of it) which tries to locate the next marked trail step is applied at steps g1, 52, g3, g4, *g6*, *g9*, *g11*, g16, *g19*, g20, *g24*, g25, g27, g29, *g32* and g37. At the italicized steps, the subsequence 3-5-7 is invoked. Thus the ant looks only to the left for the missing trail before continuing on in its original direction. Sequences 0-2-3-6 and 1-4-3-6-2 along with the cycle 23-6 are used on several marked straight sectional paths, but state 7 alone is also utilized to walk along some of the marked straight sectional paths. The sequences 4-6-3 and 8-7-0 are used to move from a marked step across two successive unmarked steps. These are invoked at steps 79 (4-6-3), 80 (8-7-0), 81 (4-6-3), 82 (8-7-0), 83(4-6-3), 84 (8-7-0), 85 (8-7-0), 86 (4-6-3), 87 (4-63), and 88 (8-7-0). Right turns are navigated by either state 2 or 0 followed by state 1. The left turn at steps 32, g14, and g35 are handled by state 3 followed by state 5. The FSM of Figure 6.55 does not have the ability to take an ant across three unmarked steps in a row, but this does not seem to hinder its performance based on the FSMs which were examined and those which have that capability do not use it.

Start state: 0

Present state	next state/ output for input = 0	Next state/ output for input = 1
0	1/0	2/2
1	3/1	1/2
2	1/0	3/2
3	5/1	6/2
4	6/2	3/2
5	7/0	8/2
6	3/2	2/2
7	0/2	7/2
8	7/2	7/2

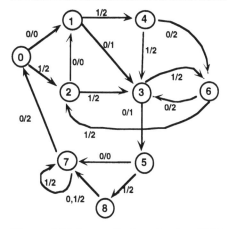

Figure 6.55 9-state Perfect Scoring FSM with Time = 187

The FSMs in the next section contain only 8 states. These are of interest because the memory-restricted storing of FSMs during the runs did not contain FSMs with fewer than eight states.

6.7.3 8-State Perfect Scoring FSMs

The smallest FSMs which were seen during examination of successful FSMs during the runs for the research have eight states. At this point it is not known whether a smaller successful FSM exists for this problem. The ant in Figure 6.56 takes 193 time steps to traverse the trail. This ant, like the two previous ones, always moves onto a marked trail step immediately in front of it. This FSM does not contain any strategies which have not already been discussed. It is of interest because no 8-state machine was seen which needed less than 193 time steps to traverse the trail.

This FSM is also of interest not only because it is one of a family of successful FSMs, but more so because parts of its family appeared in two very different runs. The two runs started with different initial populations. One run applied MTF to generation 0, and then only to every fourth generation after that. This run did not incorporate competition. The other run applied MTF to every generation and used competition during each generation to determine the fitness.

Start state: 0

Present state	Next state/ output for input = 0	Next state/ output for input = 1
0	1/1	2/2
1	3/1	4/2
2	5/2	1/2
3	5/0	2/2
4	5/2	6/2
5	7/2	7/2
6	0/0	1/2
7	0/0	6/2

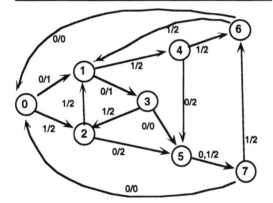

Figure 6.56 8-state Perfect Scoring FSM with Time = 193

For all except two of the successful relatives, the only change in the state transition table is in the next state of state 5 with input 1. Both runs found the FSM in Figure 6.56, and FSMs in which the next state for state 5 with input 1 is either 0 or 4. In addition, a FSM with this next state changed to 3 appeared in one of the runs. It might also have been bred by the other run and not been seen. Recall that only the first 10 successful FSMs from each generation were recorded, and not all were carefully scrutinized. One other variation which was seen replaces the next state for state 2 with input 1 from state 1 to 5. All of these ants required 193 time steps to traverse the trail. Other relatives of the FSM in Figure 6.56 have changes in the next state, or next state and output for state 5 with input 1, but these relatives have times greater than 193. Changing the next state for state 5 with input 1 to state 5 yields an ant which has a time of 197. Replacing the next state with 6 and output with 3 for state 5 with input 1 results in an ant which takes the full allotment of 200 time steps to complete the trail.

The last FSM which is presented appears in Figure 6.57. This FSM requires 199 time steps to cover the trail. As opposed to some of the other FSMs presented, the response of this FSM to a marked trail step immediately ahead is to move onto it. This machine differs from all of those previously presented in that it does not make any left turns. Consequently, a left turn can be navigated only with three right turns. The state sequence 5-7-0-1 (used at g1, g3, g14, g20, g25, g27, and g35), 2-0-1-3 (invoked at step 32) and 7-0-1-3 (not used) take an ant through a 270 right turn as long as the input remains 0. Thus, it takes three time steps to make a left turn. The only other point of interest is that the FSM of Figure 6.57 uses a single state to walk along straight sectional paths. For the first three straight trail sections, the ant is in state 2. For other straight sectional paths the ant is primarily in state 6.

Start state: 0

Present state	Next state/ output for input = 0	Next state/ output for input = 1
0	1/0	2/2
1	3/0	4/2
2	0/0	2/2
3	5/2	5/2
4	6/2	6/2
5	7/0	1/2
6	5/2	6/2
7	0/0	4/2

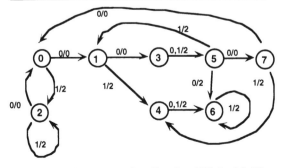

Figure 6.57 8-state Perfect Scoring FSM with Time = 199

The successful ants presented share some common strategies. One strategy which saves time is that anywhere along the trail after successfully navigating the corner at g3, an ant can move ahead onto two successive unmarked trail steps immediately following a marked trail step without losing the trail. This strategy is effective for the John Muir trail, but it is not a good general trail-following strategy. Other FSMs found always check to the left and right before moving ahead when the trail step immediately ahead is not marked. This consumes a great deal of time, but is a better generalized strategy for all trails.

This section, while interesting with respect to the trail problem implemented, stresses the point that GAs develop strategies to solve problems besides obtaining solutions per se. The main goal of this chapter was to introduce changes made to the benchmark GA (Section 6.4) for purposes of the GA efficiency. Extensive further research is necessary to exactly determine the effects of these changes.

6.8. Summary

The organization of data along a genome which is used as the operand of a genetic algorithm can significantly affect the GA's convergence. This is supported by schema theory and suggests reorganizing FSM genomes during execution of a GA to improve convergence. Two reorganization algorithms were designed. The SFS algorithm enhances schema formation by standardizing transitions. The MTF algorithm reduces schema length. To reduce the possibility of premature convergence, the fitness was modified from an explicitly defined function, to one which incorporates competition. With competition, the evolving population faces a changing fitness landscape.

Preliminary testing indicated that reorganizing the FSMs during run time improved GA convergence, with reduced schema length showing more promise than standardizing transitions. The advantage of incorporating competition is not as clear, but the addition of both MTF and competition could dramatically improve performance.

As of this writing, the runs are being timed to see how the additional processing which was performed to improve convergence affects run time. Another area presently being investigated is whether it is necessary to reorganize genomes every generation, or whether the effects of reorganization will carry over a few generations.

It is also important to determine whether the convergence patterns are problem-specific or whether they will be similar when these methods are applied to other problems. Also suggested for future research is to see how sensitive these methods are to different parameter values. Testing is now being done to determine the sensitivity to population size and the maximum number of iterations permitted for a run. This research was FSM-specific, but the ideas should be considered for other standard phenotypes, and for problem-specific phenotypes as well.

References

Angeline, Peter J. (1994) "An Alternate Interpretation of the Iterated Prisoner's Dilemma and the Evolution of Non-Mutual Cooperation." In *Artificial Life IV*, pp. 353-358, edited by Rodney A. Brooks and Pattie Maes. Cambridge, MA: MIT Press.

Angeline, Peter J. and Jordan B. Pollack. (1993) "Evolutionary Module Acquisition." In *Proceedings of the Second Annual Conference on Evolutionary Programming*, edited by D.B. Fogel and W. Atmar. Palo Alto, CA: Morgan Kaufman.

Axelrod, R. (1987) "The Evolution of Strategies in the Iterated Prisoner's Dilemma." In *Genetic Algorithms and Simulated Annealing*, edited by L. Davis. London:Pitman.

Belew, Richard K., John McInerney, and Nichol N. Schraudolph. (1992) "Evolving Networks: Using Genetic Algorithm with Connectionist Learning." In *Artificial Life II*, pp. 511-547, edited by Christopher G. Langton, Charles Taylor, J. Doyne Farmer, and Steen Rasmussen. Reading, MA: Addison-Wesley.

Fogel, D. B. (1991) "The Evolution of Intelligent Decision Making in Gaming." *Cybernetics and Systems*, pp. 223-226, vol. 22.

Fogel, D. B. (1993) "On the Philosophical Differences between Evolutionary Algorithms and Genetic Algorithms." In *Proceedings of the Second Annual Conference on Evolutionary Programming*, edited by D.B. Fogel and W. Atmar. Palo Alto, CA: Morgan Kauffman.

Fogel, D. B. (1994) "An Introduction to Simulated Evolutionary Optimization." *IEEE Transactions on Neural Networks*, pp. 3-14, vol. 5, no. 1, Jan. 1994.

Goldberg, David E. (1989) *Genetic Algorithms in Search, Optimization and Machine Learning*. Reading, MA: Addison-Wesley.

Hillis, W. Daniel. (1992) "Co-Evolving Parasites Improve Simulated Evolution as an Optimization Procedure." In *Artificial Life II*, pp. 313-324, edited by Christopher G. Langton, Charles Taylor, J. Doyne Farmer, and Steen Rasmussen. Reading, MA: Addison-Wesley.

Jefferson, David, Robert Collins, Claus Cooper, Michael Dyer, Margot Flowers, Richard Korf, Charles Taylor, and Alan Wang. (1992) "Evolution as a Theme in Artificial Life: The Genesys/Tracker System " In *Artificial Life II*, pp. 549-578, edited by Christopher G. Langton, Charles Taylor, J. Doyne Farmer, and Steen Rasmussen. Reading, MA: Addison-Wesley.

Knuth, Donald E. (1973) *The Art of Computer Programming: Sorting and Searching*. Reading, MA: Addison-Wesley.

Koza, John R. (1992) "Genetic Evolution and Co-Evolution of Computer Programs." In *Artificial Life II*, pp. 603-629, edited by Christopher G. Langton, Charles Taylor, J. Doyne Farmer, and Steen Rasmussen. Reading, MA: Addison-Wesley.

Levinson, Gene. (1994) "Crossovers Generate Non-Random Recombinants under Darwinian Selection." In *Artificial Life IV*, pp. 90-101, edited by Rodney A. Brooks and Pattie Maes. Cambridge, MA: MIT Press.

Levy, Stephen. (1992) *Artificial Life: The Quest for a New Creation*. New York: Pantheon.

MacLennan, Bruce. (1992) "Synthetic Ethology: An Approach to the Study of Communication." In *Artificial Life II*, pp. 631-658, edited by Christopher G. Langton, Charles Taylor, J. Doyne Farmer, and Steen Rasmussen. Reading, MA: Addison-Wesley.

Nilsson, Nils J. (1980) *Principles of Artificial Intelligence*. Los Altos, CA: Morgan Kaufmann.

Paredis, Jan. (1994) "Steps Towards Co-Evolutionary Classification Neural Networks." In *Artificial Life IV*, pp. 102-108, edited by Rodney A. Brooks and Pattie Maes. Cambridge, MA: MIT Press.

Ray, Thomas S. (1992) "An Approach to the Synthesis of Life." In *Artificial Life II*, pp. 371-408, edited by Christopher G. Langton, Charles Taylor, J. Doyne Farmer, and Steen Rasmussen. Reading, MA: Addison-Wesley.

Reynolds, Craig W. (1994) "Competition, Coevolution and the Game of Tag." In *Artificial Life IV*, pp. 59-69, edited by Rodney A. Brooks and Pattie Maes. Cambridge, MA: MIT Press.

Stanley, E. Ann, Dan Ashlock, and Leigh Tesfatsion. (1994) "Iterated Prisoner's Dilemma with Choice and Refusal of Partners." In *Artificial Life III*, pp. 131-175, edited by Christopher G. Langton. Reading, MA: Addison-Wesley.

Chapter 7 A Synthesizable VHDL Coding of a Genetic Algorithm

Stephen D. Scott, Sharad Seth, and Ashok Samal

Abstract

This chapter presents the HGA, a genetic algorithm written in VHDL and intended for a hardware implementation. Due to pipelining, parallelization, and no function call overhead, a hardware GA yields a significant speedup over a software GA, which is especially useful when the GA is used for real-time applications, e.g., disk scheduling and image registration. Since a general-purpose GA requires that the fitness function be easily changed, the hardware implementation must exploit the reprogrammability of certain types of field-programmable gate arrays (FPGAs), which are programmed via a bit pattern stored in a static RAM and are thus easily reconfigured.

After presenting some background on VHDL, this chapter takes the reader through the HGA's code. We then describe some applications of the HGA that are feasible given the state-of-the-art in FPGA technology and summarize some possible extensions of the design. Finally, we review some other work in hardware-based GAs.

7.1. Introduction

This chapter presents the HGA, a genetic algorithm written in VHDL[1] and intended for a hardware implementation. Due to pipelining, parallelization, and no function call overhead, a hardware GA yields a significant speedup over a software GA, which is especially useful when the GA is used for real-time applications, e.g., disk scheduling [51] and image registration [52]. Since a general-purpose GA requires that the fitness function be easily changed, the hardware implementation must exploit the reprogrammability of certain types of field-programmable gate arrays (FPGAs) [13], e.g., those from Xilinx [57]. Xilinx's FPGAs are programmed via a bit pattern stored in a static RAM and are thus easily reconfigured. While FPGAs are not as fast as typical application-specific integrated circuits (ASICs), they still hold a great speed advantage over functions executing in software. In fact, speedups of 1--2 orders of magnitude have been observed when frequently used software routines were implemented with FPGAs [8], [9], [12], [15], [19], [54]. Characteristically, these systems identify simple operations that are bottlenecks in software and map them to hardware. This is the approach used in this chapter, except that the entire GA is intended for a hardware implementation.

---

[1]VHDL stands for VHSIC (Very High Speed Integrated Circuit) Hardware Description Language.

The remainder of this chapter is organized as follows. Section 7.2 gives a brief primer on VHDL. Section 7.3 describes the HGA design. Then, in Section 7.4 we describe some applications of the HGA which are feasible given current FPGA technology. Possible design improvements and extensions are summarized in Section 7.5. Finally, in Section 7.6 we summarize and review other hardware-based GAs.

Also available are simulation results, a theoretical analysis of this design, and a description of a proof-of-concept prototype. They are beyond the scope of this chapter and are instead given in an associated technical report [44]. As they become available, updates to the code in this chapter will be made available at `ftp://ftp.cse.unl.edu/pub/HGA`.

Finally, it is important to distinguish hardware-based GAs from *evolvable hardware*. The former (e.g., the design presented in this chapter) is an implementation of a genetic algorithm in hardware. The latter involves using GAs and other evolution-based strategies to generate hardware designs that implement specific functions. In this case, each population member represents a hardware design, and the goal is to find an optimal design with respect to an objective function, e.g., how well the design performs a specific task. There are many examples of evolvable hardware in the literature [1], [17], [26], [43].

7.2. A VHDL Primer

In this section we briefly review some of the VHDL fundamentals employed in this chapter. Much more detail is available at the VIUF and VHDL^{UK} WWW sites [2], [3], which include lists of several VHDL books. The material in this section should be sufficient to allow the reader to comprehend the code of the design.

7.2.1. Entities, Architectures and Processes

Each module in the HGA consists of an ENTITY and an ARCHITECTURE. The entity portion of the module defines the PORT, which gives the data type, size and direction of all the inputs and outputs. For example, in the file `fitness.hdl`, ain is an input (due to the keyword IN), is of type qsim_state_vector (see below) and is n bits wide,[1] with the $(n-1)$th bit the most significant and the 0th bit the least significant (due to the keyword DOWNTO).

The architecture portion of a module defines its structure (specifying the interconnections between lower-level modules) or functionality (specifying what a module outputs in response to its inputs). In most of the VHDL files in this design, the architecture is specified behaviorally, and each

---

[1] n and other quantities are defined in Section 7.3.2.

architecture is composed of one or more PROCESSes. Typically the only process in each file is called main, but sometimes other processes exist, e.g., fitness.hdl also contains a process called adder which adds two values. Each process has a list of signals (signals are defined later) to which it is sensitive. When the value of any signal in the list changes, that process is activated. For example in fitness.hdl, if addval1 or addval2 changes, then process adder is activated and adds its inputs. If clk or init changes, then process main is activated.

7.2.2. Finite State Machines

Within the main process of each HGA module, a *finite state machine with asynchronous reset* is defined. Revisiting the example of fitness.hdl, the process main is activated by a change in clk (the system clock) and init (an asynchronous reset signal). The statement IF init='0' is true if the state machine has been reset, so in this part of the code the module shuts itself down. If instead init is 1, then regular state machine execution should proceed. The line

```
ELSIF (clk'EVENT AND clk='1' AND
clk'LAST_VALUE='0') THEN
```

is true only if init is 1, the clock has just changed (clk'EVENT), and the change was a low-to-high transition, i.e., the state machine responds only to positive clock edges, thus the machine's registers are all positive-edge triggered. If this ELSIF passes, we enter a CASE statement, in which we determine what state the machine is in and take the actions specified in that state. The names of the states are specified in the TYPE states IS ... statement and a signal of type states is defined to hold the current state.

7.2.3. Data Types

Designs in VHDL manipulate SIGNALs and VARIABLEs. Signals are used to store states of state machines, communicate between modules, and communicate between different processes within the same module. Hence they are defined globally for an entire architecture or specified in a port. Variables can be locally defined for a single process. Signals can be thought of as sending information to other modules or processes, and variables can be thought of as storing information in a register. Another difference between signals and variables is that signals can have delays associated with them (e.g., propagation delay of a wire), to better resemble reality, whereas variables cannot. Hence it is possible to examine the waveform of a signal but not of a variable. Assignment to a signal uses the <= operator and assignment to a variable uses the := operator.

The data types that variables and signals can assume include integer, bit, bit_vector, and user-defined data types. Type bit is a simple bit and can take on the values 0 and 1, so a signal of type bit is like a wire and a variable of type bit is like a flip-flop. Type bit_vector is an array of bits with a user-specified width, so a signal of type bit_vectoris like a bus and a variable of type bit_vector is like a register. In this design, we use the types qsim_state and qsim_state_vector, which are specific to the simulators and compilers of Mentor Graphics [33]. These types are nearly identical to bit and bit_vector, so a direct substitution should work to map the code of this chapter to a code compatible with other VHDL compilers.

The functions to_qsim_state and to_integer frequently appear in the code. The function to_integer(q) converts q, a qsim_state_vector, to an integer.[1] The function to_qsim_state(i,s) converts integer i to a qsim_state_vector that is s bits wide. These functions are used when adding qsim_state_vectors, using a qsim_state_vector as an index into an array, and to resize a qsim_state_vector. If the types qsim_state and qsim_state_vector are changed (e.g., to bit and bit_vector) for compatibility reasons, equivalent functions for to_qsim_state and to_integer must be written unless the equivalents already exist.

7.2.4. Miscellany

The LIBRARY and USE commands at the top of each VHDL file tells the VHDL compiler which libraries are used in that file. This design uses the mgc_portable library, which defines the typesqsim_state and qsim_state_vector as well as the functions to_integer and to_qsim_state. Our code also uses the libs.sizing library that is described in the file libs/sizing.hdl. In that file are several constants that parameterize the HGA design (Section 7.3.2).

Finally, note that all the VHDL code, except that which is in memory.hdl, is *synthesizable*. That is, it uses a subset of VHDL that cannot only be simulated, but can be mapped to actual hardware by using AutoLogic II from Mentor Graphics [34], [35]. Other synthesizers exist and can be used to synthesize the HGA code if the Mentor-specific code is changed.

7.3. The Design

The functional model of the HGA is as follows. A software *front end* running on a PC or workstation interfaces with the user and writes the GA parameters (Section 7.3.1) into a memory that is shared with the *back end*,

---

[1] Since these functions assume that q is in 2's complement form, a 0 bit is prepended to q before passing it to to_integer. This is done by using the & operator, which performs concatenation.

consisting of the HGA hardware. Additionally, the user specifies the fitness function in some programming or other specification language (e.g., C or VHDL). Then software translates the specification into a hardware image and programs the FPGAs that implement the fitness function. Many such software-to-hardware translators exist [27], [35], [54]. Then the front end signals the back end. When the HGA back end detects the signal, it runs the GA based on the parameters in the shared memory. When done, the back end signals the front end. The front end then reads the final population from the shared memory. The population could then be written to a file for the user to view or have other computations performed on it. Currently the only GA termination condition that the user can specify is the number of generations to run. If other termination conditions are desired (e.g., amount of population diversity, minimum average fitness), the user must tell the HGA to run for a fixed number of generations and then check the resultant population to see if it satisfies the termination criteria. If not, then that population can be retained in the HGA's memory for another run. This process repeats until the termination criteria are satisfied.

Note that the front end could, in fact, be any interface between the HGA and a (possibly autonomous) system that occasionally requires the use of a GA. This system would select its own HGA parameters and initial population, give them to the front end for writing to the shared memory, and invoke the HGA. The system could even program the FPGAs containing the fitness function.

7.3.1. The User-Controlled (Run-Time) Parameters

The run-time parameters specified by the HGA's user are as follows. Their addresses in the shared memory are specified in `libs/sizing.hdl` (Section 7.3.2).

1. The number of members in the population. This is stored as `psize` in `sequencer.hdl` and as `psize` in `fitness.hdl`.

2. The initial seed for the pseudorandom number generator. This is stored as `rn` in `rng.hdl`, but is later overwritten with new random bit strings.

3. The mutation probability. This is stored as `mutprob` in `xovermut.hdl`.

4. The crossover probability. This is stored as `xoverprob` in `xovermut.hdl`.

5. The sum of the fitnesses of the members of the initial population. This is stored as `sum` in `fitness.hdl`, but is later overwritten with new sums as the population changes.

6. The number of generations in the HGA run. This is stored as `numgens` in `fitness.hdl`.

The initial population is stored in memory below the above parameters.

In directory makepop is C source code that will take GA parameters on the command line and create a file that includes these parameters plus a randomly generated population. The format of this file is directly readable by memory.hdl (Section 7.3.5.2) and thus can be immediately used as input when simulating the VHDL code of this chapter.

Other GA parameters such as the length of the population members[1] and their encoding scheme are indirectly specified when the user defines the fitness function.

7.3.2. The VHDL Compile-Time Parameters

Designing the HGA using VHDL allowed the design to be specified behaviorally rather than structurally. It also allowed for general (parameter-independent) designs to be created, facilitating scaling. The specific designs implemented from the general designs depend upon parameters provided at VHDL compile time. When the parameters are specified, the design can be simulated or implemented with a VHDL synthesizer such as AutoLogic II from Mentor Graphics [34], [35]. The parameters are set in libs/sizing.hdl and used in the other VHDL files. They are as follows.

1. The width in bits of the crossover and mutation probabilities and the random numbers sent from rng.hdl to xovermut.hdl is denoted p.

2. The maximum width in bits of the population members is denoted n.

3. The maximum width in bits of the fitness values is denoted f.

4. The precision used in scaling down the sum of fitnesses for selection in selection.hdl is denoted r. See Section 7.3.5.5 for more information.

5. The size of the cellular automaton in rng.hdl is denoted casize. See Section 7.3.5.1 for more information.

6. The maximum size of the population is denoted m.

7. The maximum number of generations is denoted maxnumgens.

8. The number of parallel selection modules is denoted nsel. Note: when changing nsel, the actual number of selection modules in top.hdl must also be changed.

The other constants defined in libs/sizing.hdl depend entirely upon

---

[1] The maximum member size is specified in Section 3.2, but the actual length as interpreted by the fitness function could be less.

the above parameters, e.g., `logn` is the base-2 logarithm of n. Thus these constants are not discussed here.

The last portion of `libs/sizing.hdl` gives the locations in memory of the user-specified parameters described in Section 7.3.1.

7.3.3. The Architecture

The HGA (Figure 7.1) was based on Goldberg's simple genetic algorithm (SGA) [20]. The HGA's modules were designed to correlate well with the SGA's operations, be simple and easily scalable, and have interfaces that facilitate parallelization. They were also designed to operate concurrently, yielding a coarse-grained pipeline. The basic functionality of the HGA design is as follows.[1]

1. After loading the parameters into the shared memory, the front end signals the memory interface and control module (MIC, `meminterface.hdl`). The MIC acts as the main control unit of the HGA during start-up and shut-down and is the HGA's sole interface to the outside world. After start-up and before shut-down, control is distributed; all modules operate autonomously and asynchronously.

2. The MIC notifies the fitness module (FM, `fitness.hdl`), crossover/mutation module (CMM, `xovermut.hdl`), the pseudorandom number generator (RNG, `rng.hdl`) and the population sequencer (PS, `sequencer.hdl`) that the HGA is to begin execution. Each of these modules requests its required user-specified parameters (Section 7.3.1) from the MIC, which fetches them from the shared memory.

3. The PS starts the pipeline by requesting population members from the MIC and passing them along to the selection module (SM, `selection.hdl`).

4. The task of the SM is to receive new members from the PS and judge them until a pair of sufficiently fit members is found. At that time, it passes the pair to the CMM, resets itself, and restarts the selection process.

5. When the CMM receives a selected pair of members from the SM, it decides whether to perform crossover and mutation based on random values sent from the RNG. When done, the new members are sent to the FM for evaluation.

6. The FM evaluates the two new members from the CMM and writes the new members and their fitnesses to memory via the MIC.

---

[1] Note that Figure 7.1 shows only the data path; control lines are omitted.

The FM also maintains information about the current state of the HGA that is used by the SM to select new members and by the FM to determine when the HGA is finished.

7. The above steps continue until the FM determines that the current HGA run is finished. It then notifies the MIC of completion which, in turn, shuts down the HGA modules and signals the front end.

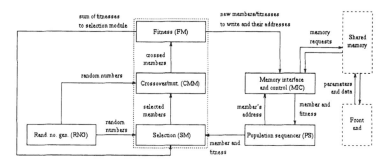

Figure 7.1 The data path of the HGA architecture.

7.3.4. Inter-Module Communication

The modules in Figure 7.1 communicate via a simple asynchronous handshaking protocol similar to asynchronous bus protocols used in computer architectures [24]. When transferring data from the initiating module I to the participating module P, I signals P by raising a request signal to 1 and awaits an acknowledgment. When P agrees to participate in the transfer, it raises an acknowledgment signal. When I receives the acknowledgment, it sends to P the data to be transferred and lowers its request, signaling P that the information was sent. When P receives the data, it no longer needs to interact with I, so P lowers its acknowledgment. This signals I that the information was received. Now the transfer is complete and I and P are free to continue processing. Examples of this kind of transfer occur between the SM and CMM, between the CMM and FM, and between the FM and MIC.

When the HGA first starts up, all the modules except the SM issue requests to the MIC for the values of some of the user-specified parameters of Section 7.3.1. The communication protocol used is similar to that stated above. In this case, P is the MIC. After the MIC raises the acknowledgment, I sends the address of the parameter to read from memory. After the MIC reads the parameter from memory and sends it to I, the MIC lowers its acknowledgment. This same protocol is used when the PS reads population members from memory via the MIC.

7.3.5. Component Modules

In this section we describe in more detail the functionality of each module from Figure 7.1.

7.3.5.1. Pseudorandom Number Generator (RNG)

The output of the pseudorandom number generator (RNG, rng.hdl) is used by two HGA modules. The RNG supplies pseudorandom bit strings (randsel1 and randsel2) to the selection module (SM, Section 7.3.5.5) for scaling down the sum of fitnesses. This scaled sum is used when selecting pairs of members from the population. The RNG also supplies pseudorandom bit strings to the crossover/mutation module (CMM, Section 7.3.5.6) for determining whether to perform crossover and mutation (doxover and domut) and what the crossover and mutation points are (xoverpt and mutpt).

After loading its seed into rn, the RNG uses a linear cellular automaton (CA) to generate a sequence of pseudorandom bit strings. The CA used in the RNG consists of 16 alternating cells which change their states according to rules 90 and 150 as described by Wolfram [56]:

Rule 90: $s_i^+ = s_{i-1} \oplus s_{i+1}$

Rule 150: $s_i^+ = s_{i-1} \oplus s_i \oplus s_{i+1}$.

Here s_i is the current state of site (cell) i in the linear array, s_i^+ is the next state for s_i, and $\oplus$ is the exclusive OR operator. Serra et al. [45] showed that a 16-cell CA whose cells are updated by the rule sequence 150-150-90-150 ... 90-150 produces a maximum-length cycle, i.e., it cycles through all 2^{16} possible bit patterns except the all 0s pattern. It has also been shown that such a rule sequence has more randomness than a linear feedback shift register[1] (LFSR) of corresponding length [28]. This scheme is implemented in state active in rng.hdl.

7.3.5.2. Shared Memory

The shared memory is actually external to the HGA system, but is presented here for completeness. The memory's specifications are known by the memory interface and control module (MIC, Section 7.3.5.3). It is shared by the back end and front end, acting as their primary communication medium. Before the HGA run, the front end writes the GA parameters of Section 7.3.1 into the memory and signals the MIC. When the HGA run is finished, the memory holds the final population which is then read by the front end.

Since an off-the-shelf physical memory would likely be employed in a hardware GA system, the file memory.hdl is intended only for simulation

---

[1] LFSRs are commonly used as pseudorandom number generators in hardware.

purposes. It reads its initial contents from a text file.[1] It is designed to behave like a real, but simple, memory. The memory operates only when the chip select (cs) and memory access (memacc) lines are low. When both these signals are low, if output enable (oe) is low, the memory outputs to dataout the value stored at address. If the read-write (rw) signal is low, then the memory stores into storage at address the value given in datain.

For tracking the progress of the simulation, memory.hdl maintains a variable called pulsecount that counts the total number of clock cycles in the run. It periodically writes this value to a file so the user can observe the rate of simulation. After the HGA shuts down, the memory writes its final contents (including the final population) to a text file, followed by the final value of pulsecount.

7.3.5.3. *Memory Interface and Control Module (MIC)*

The memory interface and control module (MIC, meminterface.hdl) is the only module in the HGA system which has knowledge of the HGA's environment. It provides a transparent interface to the memory for the rest of the system. When the MIC senses that go went low (state start1) and then high (state start2), it takes control of the shared memory[2] by setting memacc to 0, initializes the other modules by setting init to 1, and begins to field requests from the other modules by transitioning to state idle.

The IF-ELSIF structure of state idle imposes priorities over the signals to which the MIC responds. First, it checks if fitdone is 1, indicating that the GA run has completed. If this is true, then it shuts down the other modules by setting init to 0, tells the user which population in memory is the final one by passing NOT toggle to toggleout,[3] relinquishes control of the memory by setting memacc to 1, and tells the front end that the HGA is finished by setting done to 1.

If instead fitdone is 0, then the MIC serves any outstanding requests from the other modules for user-specified parameters (Section 7.3.1). It does this by checking the signals reqrng through reqfit(0) in the order that they appear in meminterface.hdl. If any of these signals is high, then the MIC reads the appropriate parameter from memory by converting the address it receives[4] to an address understood by the memory. After

---

[1] This file can be created by makepop. See Section 7.3.1.

[2] At this point the MIC assumes that the front end has surrendered control of the memory.

[3] See Section 3.5.7.

[4] Which is as specified in libs/sizing.hdl.

reading the parameter from memory using a procedure described below, it (in state read2) sends the value out via valout and informs the module that the parameter was sent by setting all the acknowledgment signals to 0.

The final cases covered by the IF-ELSIF structure of state idle are the ones most frequently encountered during normal HGA operation. If the signal reqfit(1) is 1, then the MIC knows that the fitness module (FM, Section 7.3.5.7) has a new population member to write to memory. The memory holds two populations, the current one and the new one (the new population will be the current one in the next generation). The FM writes to the new one and the population sequencer (PS, Section 7.3.5.4) reads from the current one. In state fit1, the MIC checks toggle, which the FM uses to inform the MIC which population is the new one. The MIC uses toggle to determine which base address (pop0base or pop1base) to add to the address sent from the FM (addrfit) to create the address sent to memory (address). It then takes the value (valfitin) from the FM and writes it to memory.

If reqfit(1) is 0, then the MIC checks reqseq(1) to determine if the PS wants to read a new member from memory. If so, then in state seq1 the MIC checks toggle and builds address as above, using the opposite population as used by the FM. Then it reads the member from memory and passes it along to the PS via valout.

To read from memory, first the MIC transmits address to the memory, then enters state read1. During the state transition time (a single clock cycle), it is expected that the address has reached the input pins of the memory chip. So, in read1, the MIC lowers oe, telling the memory it wants to read the data stored at the given address. After another clock cycle, the MIC enters read2, at which time it is expected that the data has arrived at datain. So the MIC forwards the data via valout, raises oe, and lowers the acknowledgments, indicating that it has forwarded the data.

To write to memory, first the MIC transmits address and dataout to the memory, then enters state write1. During the state transition time (a single clock cycle), it is expected that the address and data have reached the input pins of the memory chip. So in write1 the MIC lowers rw, telling the memory it wants to write the given data to the location given by the address. After another clock cycle, the MIC enters write2, at which time it is expected that the data has been written. So it raises rw, but keeps address and dataout unchanged to avoid corrupting memory. After one more clock cycle (the transition to write3), the MIC assumes the write was successful and informs the FM.[1]

---

[1] The FM is the only one informed here because it is the only module that writes data.

7.3.5.4. Population Sequencer (PS)

The job of the population sequencer (PS, sequencer.hdl) is to cycle through the current population and pass the members on to the SM(s). After loading the population size into psize, the PS enters state getmember1, where it sends the index membaddr of a population member to the MIC via addr. Then it awaits reception of the member (arriving via value) in state getmember2. It then passes this member to the SM(s) via output. If the member sent to the SM(s) is the same as the previous one sent, dup is set to tell the SM(s) to accept it. The index membaddr is then incremented modulo the population size psize so the next population member will be requested from the MIC. This process continues until the GA run is complete and the MIC shuts down all the modules (i.e., init goes low).

7.3.5.5. Selection Module (SM)

The HGA's selection method is similar to the implementation of roulette wheel selection used in Goldberg's SGA [20]. The SGA's selection procedure is as follows.

1. Using a random real number $r \lfloor [0,1]$, scale down the sum of the fitnesses of the current population to get $S_{scale} = r \xi S_{fit}$.

2. Starting at the first population member, examine the members in the order they appear in the population.

3. Each time a new member is examined, accumulate its sum in a running sum of fitnesses S_R. If at that time S_R 3 S_{scale}, then the member under examination is selected. Otherwise, the next population member is examined (Step 2).

Each time a new population member is to be selected, the above process is executed.

The selection module (SM, selection.hdl) implements the roulette wheel selection process used by the SGA, but it selects a pair of population members simultaneously rather than a single member at a time. First, from the FM it receives sum, the sum of the fitnesses of the current population. It then scales down this sum by two random values rand1 and rand2 provided by the RNG. The actual scaling is performed in the process scalefit which multiplies an r-bit random bit string by the sum of fitnesses and then right shifts the result by r bits (Section 7.3.2). This is done to simplify the division operation in the hardware. Thus larger values of r yield more precision in scaling down sum. Since a multiplier that operates in a single clock cycle consumes a significant amount of hardware, only one is instantiated in the SM. So in state init1 the variables serialize and multdone are used to coordinate the two accesses to process scalefit. The two scaled sums are stored in scalea and scaleb. After storing the scaled sums, the SM resets the fitness accumulators accuma and accumb, the members a and b, and the flags donea and doneb. Now selection may begin.

The state getcandidates is where selection is performed. First, the flags donea and doneb are checked. If either of these is 1, then the corresponding member (e.g., a if donea = 1) is not changed because the done flag indicates that it has already been selected. So any member with a done flag = 0 is replaced by input when a new member arrives.[1] After storing a new input into a and/or b, serialize is set to 1 to indicate that the running sums of fitnesses accuma and accumb require updating, since new members were stored. Then the SM checks if accuma > scalea. If so, then a has been selected and the SM sets donea to 1. doneb is set to 1 under similar conditions. When both donea and doneb are 1, the SM moves to state awaitackxover1 and transmits a and b to the CMM via the handshaking protocol described in Section 7.3.4. Then the SM returns to state init1 to select another member. Finally, note that the reset signal comes from the FM to indicate that the current generation has ended and the FM has switched the population it writes to (Sections 3.5.3 and 3.5.7). Thus the PS now reads from a different population. This means that sum is no longer the sum of the fitnesses of the current population. So a reset places the SM into state idle where it stores the new sof into sum and restarts selection.

The HGA is designed to allow multiple SMs to operate in parallel, which is useful when the SM is the bottleneck of the pipeline [44]. The PS sends the same members to all SMs, but this does not pose a problem for selection since each SM uses an independent pair of random bit strings to scale down the sum of fitnesses. Thus, each selection process is independent of the others. All the parallel SMs feed into a single CMM. To add parallel SMs, nsel in libs/sizing.hdl must be changed, and alterations are necessary in top.hdl (Section 7.3.6).

7.3.5.6. Crossover/Mutation Module (CMM)

After loading values into mutprob and xoverprob, the crossover/mutation module (CMM, xovermut.hdl) remains in state findreqsel while polling the SM(s). When SM *i* has selected members, it sends a request. When the index currsel = *i*, the CMM will detect SM *i*'s request and accept the pair of members via the handshaking protocol of Section 7.3.4, storing them in a1 and b1. Then if the random string doxover (from the RNG, Section 7.3.5.1) is smaller than the crossover probability xoverprob, crossover is performed while copying a1 into a2 and b1 into b2, where the crossover point is indicated by

---

[1] A new member's arrival is indicated by a change in input or if dup is 1 (Section 7.3.5.4).

xoverpt. Then a2 is copied into a3 and b2 into b3, where the bit of a2 indexed by mutpt is mutated if domut < mutprob. Then the CMM acknowledges the SM so it may start a new selection process. Finally, the CMM transmits the new members a3 and b3 to the FM via the handshaking protocol of Section 7.3.4 and resumes polling the SMs.

Note that the HGA has only one opportunity per pair of members to perform mutation, but in Goldberg's SGA, the possibility of mutation is explored for *every* bit of *every* member. Thus the SGA's mutation probability is effectively higher than the one used in the HGA. This can be adjusted for by increasing the mutation probability given to the HGA or by a few simple alterations to xovermut.hdl to make the HGA's mutation operator the same as the SGA's.

7.3.5.7. Fitness Module (FM)

When the fitness module (FM, fitness.hdl) starts up, it loads toggleinit into tog, which tells the FM which of the two populations contains the initial one specified by the user. Then the FM stores the population size in psize, stores the sum of fitnesses of the initial population in sum(0) or sum(1), depending upon the value of tog, and stores the number of generations in numgens.

After initializing, the FM enters state newgen to start a new generation.[1] It first checks if the entire HGA run is over (if numgens = 0). If not, it resets psizetmp (the number of population members remaining to fill the current generation) and sof(NOT tog) (the sum of fitnesses accumulator for the new generation). It also toggles tog and sends its value to the MIC via toggle, which tells the MIC which population the FM writes to and which the PS reads from. Finally, it sends a 0 to the SM via reset, which tells the SM that the sum of fitnesses of the population the SM reads from has been updated. Then the FM moves to waitforxovreq, where it awaits a new pair of members from the CMM.

In waitforxovreq, the FM first checks if psizetmp = 0, indicating the end of the current generation. Otherwise the FM awaits a request from the CMM and then receives a and b via ain and bin using the handshaking protocol of Section 7.3.4. For evaluating a and b, the FM can use its default fitness function of $f(x) = 2x$, distributed over states waitforxovreq2 and sumfitb. Alternatively, an optional external fitness evaluator (FE) can be attached to the FM. If it is attached, then offchipfit will be 1, causing the FM to send a and b, one at a time, to

---

[1] The flag cleanxover is used to insert a delay before the FM starts a new generation. This is necessary because of a small technicality: if the CMM was processing members from the old population when the FM switched populations (due to the new generation), then the FM should ignore the CMM's output. The use of ackxover and cleanxover in states newgen and waitforxovreq achieves this goal.

the FE via `offchipfitmemb` and receive each member's fitness via `offchipfitres`. These actions occur in states `offchipfita` and `offchipfitb`. After each member is evaluated (in either the FM or FE), the FM accumulates the sum of their fitnesses in `sum(NOT tog)` by using the process `adder`, which implements a single adder in the FM. After the current generation completes, `sum(NOT tog)` will be sent to the SM.

In the current design, the FM expects the FE to evaluate each member in a single clock cycle. But this restriction can be removed by implementing a slightly more complex communication protocol between the FM and FE. Use of an external FE allows all the other HGA modules (including the FM, which would now perform only bookkeeping) to be implemented in a non-reprogrammable technology such as fabricated chips, to reap a space and time savings over FPGAs. Only the FE need be implemented on reprogrammable FPGAs. Additionally, since the implementation of the FE is independent of the FM's implementation, the FE could be implemented in software if the fitness function is much too complex for an FPGA implementation. All that is required is that the software-based FE adhere to the communication protocol expected by the FM.

After evaluating the new members and accumulating their fitnesses, the FM uses the handshaking protocol of Section 7.3.4 to send the members and their fitnesses to the MIC for writing to memory.

While not in the current design, it is possible to allow for multiple FMs to operate in parallel, much like the parallel SMs of Section 7.3.5.5. This is useful when the FM is the bottleneck of the pipeline [44]. If parallel FMs were to be added to this design, the duty of writing new members to memory and maintaining records of the HGA's state (e.g., maintaining `numgens` and `psizetmp`) would best be shifted from the FM to a new module called the memory writer (MW). This is because maintaining these values in a distributed fashion would be difficult. The CMM would connect to all the FMs, and each FM would connect to the MW.

To add parallel FMs, an MW should be created, the FM should be modified, and a parameter `nfit` in `libs/sizing.hdl` should be added. Also, alterations would be necessary in `top.hdl`.

7.3.6. Combining the Modules

The file `top.hdl` specifies the connections between all the modules and the HGA's interface to its environment. The entity `top` defines the HGA's interface and the COMPONENT definitions describe the interfaces of each HGA module. Below the COMPONENT definitions are the definitions of signals that interconnect the modules. Each signal is explained by comments in the code. The next portion of `top.hdl` defines which VHDL source file corresponds with each component. This is where multiple SMs are instantiated, if desired. Finally, the components are interconnected by specifying which signals connect to each component. Naturally, the size and type of each signal must match the size and type of the port it connects to. Also, multiple driven signals are disallowed unless some arbitration logic is utilized, so, in top.hdl, exactly one port connected to each signal is an OUT port, which can be seen in the COMPONENT definitions.

Note that the instantiation of the mem component (for memory) implies that as written, top.hdl is intended only for simulation. In fact, the file top.hdl might not be used at all in an actual implementation since groups of components will likely be mapped to different FPGAs.

7.4. Example Applications

This section gives a high-level description of several problems to which the HGA is applicable given the current state-of-the-art in FPGA technology.

7.4.1. A Mathematical Function

Our first example application comes from Michalewicz [36]. The problem is to optimize the function

$$f(x_1, x_2) = 21.5 + x_1 \sin(4\pi x_1) + x_2 \sin(20\pi x_2) \qquad (1)$$

where $-3.0 \le x_1 \le 12.1$ and $4.1 \le x_2 \le 5.8$. A plot of $f(x_1, x_2)$ is in Figure 7.2. The spiky nature of the plot indicates that it should be difficult to optimize, given the myriad local minima. To obtain four decimal places of precision for each variable, Michalewicz used 18 bits for x_1 and 15 for x_2. His binary strings were manipulated directly and converted to real values only during fitness evaluation.

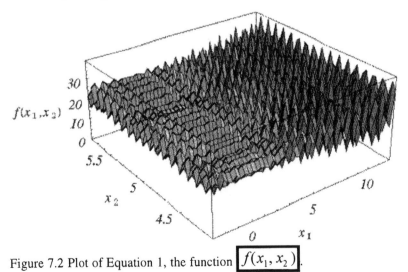

Figure 7.2 Plot of Equation 1, the function $\boxed{f(x_1, x_2)}$.

A hardware implementation of this fitness function is straightforward if the CORDIC algorithm [53] is used to evaluate the sines. To evaluate Equation 1 in hardware, we require a single multiplier,[1] two adder/subtracters, six registers, two shifters, and a lookup table in the form of a 16×18 ROM. All these components easily fit on a Xilinx XC4013 FPGA [57].

In the evaluation process, first x_1 is multiplied by 4π in the multiplier, and then x_2 is multiplied by 20π in the multiplier in the next cycle. Then both sine calculations run concurrently via CORDIC. But since CORDIC works only on arguments between 0 and 2π, before running CORDIC we must subtract from each argument to sine an amount $2i\pi$, where $i = \lfloor x / (2\pi) \rfloor$ for $x \in \{4\pi x_1, 20\pi x_2\}$. These amounts are $2\pi \lfloor 2x_1 \rfloor$ and $2\pi \lfloor 10x_2 \rfloor$, each of which is computable with two more uses of the multiplier.[2] So, after four cycles, one of the two concurrent CORDIC systems is given an argument of $4\pi x_1 - 2\pi \lfloor 2x_1 \rfloor$. After another three cycles, the other CORDIC system receives an argument of $20\pi x_2 - 2\pi \lfloor 10x_2 \rfloor$. Fourteen steps of each CORDIC are required to attain 4 decimal places of precision in the result, which is the precision used in Michalewicz's implementation. Fourteen steps are required because approximately one bit of accuracy is attained in each step, and $2^{-14} < 10^{-4} < 2^{-13}$. After the first CORDIC finishes, multiply x_1 by $\sin(4\pi x_1)$ and add the result to 21.5. One cycle after this, the second CORDIC will finish. So multiply x_2 by $\sin(20\pi x_2)$ and add this result to $x_1 \sin(4\pi x_1) + 21.5$. By overlapping as many operations as possible while sharing the multiplier and adder, the total time to evaluate the member is 23 cycles. Now repeat for the other member, yielding a cumulative delay

---

[1] Since multipliers occupy significant FPGA space and can reduce the maximum possible clock rate, a dedicated multiplier chip, such as the AMD Am29323 multiplier [5], could be used instead. This would increase the maximum possible clock rate and save area on the FPGAs.

[2] For arbitrary $x \geq 0$, $2ip$ can be found with a binary search, repeatedly subtracting $2^j p$ from x for j ranging from $\lfloor \log_2(x / \pi) \rfloor$ down to 1. After each subtraction, if the result ≥ 0, put j in a set S and continue. If the result < 0, then add $2^j p$ back to x and continue. When finished, $2i\pi = \pi \sum_{j \in S} 2^j$. If initially $x < 0$, then perform a similar process, but repeatedly add $2^j p$ to x and test if the result is < 0. When finished, $\pi \sum_{j \in S} 2^j$ yields a quantity between $-2p$ and 0. Now add 2p to this quantity.

of 46 cycles. But note that operations performed in evaluating the first member can partially overlap operations evaluating the second. So, in fact, both members can be evaluated in 43 cycles.

In related work [44], we give simulation results on this problem and contrast the results to those from a software implementation of Goldberg's SGA.

7.4.2. Logic Partitioning

Sitkoff et al. [46] have proposed a scheme to apply GAs to the problem of partitioning logic designs across two FPGAs. A design is comprised of c components and a particular partitioning (population member) is represented by a c-bit string P, where the ith bit is 1 if and only if component i lies in FPGA 1. Accompanying this bit string is a set of c-bit strings, N_j, one per inter-component net in the design, where the ith bit of N_j is 1 if and only if component i is connected to N_j. So net j lies in FPGA 1 of partition P if one of the bits in $P \mid N_j$ is 1, where $\mid$ is the bitwise AND operator. Likewise, net j lies in FPGA 2 of partition P if one of the bits in $\sim P \mid N_j$ is 1, where $\sim P$ is the bitwise not of P. Thus, a net j crosses a chip boundary if and only if some bit from $P \mid N_j$ is 1 and some bit from $\sim P \mid N_j$ is 1. This can be easily determined with combinational logic. A partition's fitness is then the total number of boundary crossings. This fitness function can be easily evaluated in hardware, just like in Sitkoff et al.'s work. The nets N_j used to evaluate each P are the same for each P, so they can be permanently stored in the FM. Since the number of potential nets is exponential in c, the nets N_j might need to be stored in some memory attached to the FM.

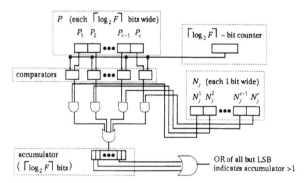

Figure 7.3 Circuit to evaluate a general F-way partition.

This approach can be generalized to an arbitrary (but bounded) number of FPGAs F as follows (Figure 7.3). First store P in c registers, each with $\lceil \log_2 F \rceil$ bits. Each register represents in which of the F FPGAs its corresponding component lies. Then for each net N_j, a counter initializes itself to 0 and cycles through all integer values $v \mathrel{\lfloor} [0, F]$. For each value v, compare it to P_i (the index of the FPGA holding component i) for all i. If they are equal, then component i lies in FPGA v in partition P. Now, logically AND this result with the ith bit of N_j. If this result is 1, then N_j lies, at least in part, on FPGA v. The results for all i are logically ORed yielding a single bit indicating if N_j lies in FPGA v. This result is fed into an accumulator which counts the number of FPGAs in which N_j lies. After looping through all values of $v \lfloor[0, F]$, the accumulator is checked. If it holds a value >1, then N_j crosses a chip boundary. Repeat this process for all N_j. The fitness of P is as defined before.

This scheme will work if F is a constant known a priori. In addition to some control logic, its hardware requirements are as follows. To store P, we need c registers, each of size $\lceil \log_2 F \rceil$. One $(\lceil \log_2 F \rceil)$-bit counter is required to cycle through the values v. The counter output will be fed into c $(\lceil \log_2 F \rceil)$-bit comparators, each comparator taking its other input from one of the registers storing part of P. Each comparator's output is fed into one of c 2-input AND gates along with one value from the c-bit register storing N_j. The outputs of these AND gates feed into one c-input OR gate, whose output enters a $(\lceil \log_2 F \rceil)$-bit accumulator. After cycling through all the vs, all bits except the lowest order one of the accumulator are fed into a $(\lceil \log_2 F \rceil - 1)$-input OR gate to determine if the accumulator's value is >1. This output activates a final accumulator (not shown in Figure 7.3) that counts the number of inter-chip nets in P. The width of this accumulator is, at most, $\lceil \log_2 n \rceil$ where n is the number of nets in the design. Note that $\lceil \log_2 n \rceil \leq c$ since $n \leq 2^c$. Finally, a table of the N_js (not shown in Figure 7.3) is needed either in a bank of registers or in an off-chip memory. In this scheme, the time to evaluate two members is approximately $2n\lceil \log_2 F \rceil$.

There is a potential difficulty of generating invalid partitions if bitwise crossover and mutation are used when $F < 2^{\lceil \log_2 F \rceil}$. This is because a value could appear in P_i that is $> F$, defining an invalid partition. This can be remedied by requiring that the initial population be valid, crossover respects the boundaries between bit groups, and that mutation maps only a bit group into a valid bit group.

Finally, note that given the net specification of a circuit, the set of vectors N_j and the fitness evaluation hardware can be automatically generated by software. So the user's work can be limited to specifying the components and nets of the circuit.

7.4.3. Hypergraph Partitioning

A GA approach to the hypergraph partitioning problem by Chan and Mazumder [16] uses a fitness function similar to that described in Section 7.4.2. After counting the number of nets that span the partition, this value is divided by the product of the sizes of the two partitions. This is known as the *ratio cut cost* and involves only a little extra work (the multiplication and division operations) beyond what is done in Section 7.4.2. Thus the HGA is applicable to this problem. Also, as in Section 7.4.2, the fitness function can be generalized to F-way partitioning where F is arbitrary but known a priori.

7.4.4. Test Vector Generation

When fabricating VLSI chips, the process occasionally introduces faults into the chip, causing incorrect functionality. Thus it is desirable to detect the faulty chips before packaging and shipping them to customers. Since the chip testers typically have access only to the inputs and outputs of the chip (not the internal structure), the only way to detect faults is by applying a *test vector* to the inputs and then contrast the output with the expected output (i.e., the output that would be produced by a fault-free circuit). If they differ, then the chip is faulty and is rejected. If they do not differ, then the other test vectors are applied. If the chip passes all the tests, the chip is accepted. The goal is to generate the smallest set of test vectors that still detects most (or all) possible faults in the chip.

In this section we focus on the *single stuck-at fault model*. In this model any faulty circuit has only a single fault, and the fault is of the following form: some wire in the circuit is permanently stuck at 0 (e.g., the wire is short-circuited to ground) or 1 (e.g., the wire is short-circuited to power). See Abramovici et al. [4] for more information on this and other fault models.

O'Dare and Arslan [38] have described a GA to generate test vectors for stuck-at fault detection in combinational circuits (i.e., circuits with no memory). In their scheme, each population member is a single test vector. The member's fitness is evaluated on the basis of how many new faults it covers. The GA maintains a global table which defines the current set of vectors. A vector is added to the table if it covers a fault not already covered by another vector in the table. Using a software-based fault simulator, each vector is evaluated by first applying it to a fault-free version of the circuit under test (CUT). Then each node within the CUT is, in turn, forced to a logic 1 and a logic 0 to simulate the stuck-at faults. If the circuit's output differs from the fault-free output, then the vector detects the given fault. Each vector gets a fixed number of points for each fault it covers that is not already covered by the test set, and it receives a smaller number of points for each fault it covers that is already covered.

We now propose how to map O'Dare and Arslan's fitness function to hardware for the HGA. Each logic gate in the CUT is mapped to a pair of gates that allow simulation of stuck-at faults. Figure 7.4 gives an example of this for an AND gate. To simulate the output c as stuck at 0, both x and y are set to 0. To simulate c stuck at 1, x is set to 0 and y is set to 1. To simulate fault-free behavior, x is set to 1 and y is set to 0. OR and NOT gates and the original inputs can be modified in a similar fashion. For a circuit with n gates and m inputs, the new circuit has, at most, $2n + 2m$ gates and, at most, $2n + 2m$ extra inputs that are controlled by the fitness module. This hardware-based fault simulation component of our proposed hardware implementation of O'Dare and Arslan's fitness function is similar to hardware accelerators designed for fault simulation [29], [58] and logic simulation [14], [39], [47].

Figure 7.4 An example of mapping a logic gate to a stuck-at fault simulation gate.

Then fitness evaluation simply requires a look-up table of previously selected vectors and the faults that they cover, a counter to cycle through all $2(m+n)$ possible stuck-at faults, an accumulator for the members' scores, and some simple control logic. The time to evaluate a member is approximately the number of faults plus one, so the time to evaluate two members is about $4(m+n)+2$. Finally, the mapping process from the original circuit to the fitness evaluation hardware can be automated as in Section 7.4.2, relieving the user of that responsibility.

7.4.5. Traveling Salesman and Related Problems

Here we consider GA approaches to the traveling salesman problem (TSP), which poses special difficulties. Using a straightforward encoding consisting of a permutation of cities encounters problems if conventional crossover and mutation operators are used. This is because regular (or uniform) crossover, if it changes anything, will create two invalid tours, i.e., some cities will appear more than once and some will not appear at all. Thus, much work in applying GAs to the TSP (e.g., [20], [22]) involve the use of special crossover operators that preserve the validity of tours. This method can be used in the HGA but requires modification of the CMM. In lieu of this, conventional crossover operators can be used in conjunction with a special encoding of the population members. One such encoding is called a *random keys* encoding [11], [37]. In this encoding, each tour is represented by a tuple of random numbers, one number for each city, with each number from [0,1]. After selecting a pair of tours, simple or uniform crossover can be applied, yielding two new tuples. To evaluate these tuples, sort the numbers and visit the cities in ascending order of the sort. For example, in a five-city problem the tuple (0.52,0.93,0.26,0.07,0.78) represents the tour 4 ♦ 3 ♦ 1 ♦ 5♦ 2. This tour can then be evaluated. Note that every tuple maps to a valid tour, so any crossover scheme is applicable.

When employing this scheme, all that is required of the HGA's FM, upon receiving a population member, is to sort the tuple and accumulate the distances between the cities of the tour. Sorting of the numbers can be done with a sorting circuit based on the Odd-Even Merge Sort algorithm [10].[1] For sorting n numbers (i.e., for n-city tours), the depth of the sorting network is $\lceil \log_2 n \rceil (\lceil \log_2 n \rceil + 1)/2$. Each level of the network has n registers and $n/2$ comparators.[2] Also, a single set of n registers and $n/2$ comparators can simulate the sorting network using some finite state logic and $\lceil \log_2 n \rceil (\lceil \log_2 n \rceil + 1)/2$ steps. Thus, the hardware requirements of the FM include some finite state logic, a linear number of registers and comparators, and a lookup table that provides the inter-city distances (with $O(n^2)$ entries). In this case, the time to evaluate two members is approximately $\lceil \log_2 n \rceil (\lceil \log_2 n \rceil + 1)$.

Finally, we note that the scheme just presented can be adapted for application to other problems with similar constraints as the TSP. These include scheduling problems, vehicle routing, and resource allocation (a generalization of the 0-1 knapsack problem and the set partitioning problem) [11], [37]. The HGA can be applied to these problems as well with a slight increase in complexity.

7.4.6. Other NP-Complete Problems

In this section we explore the exploitation of polynomial-time reductions between instances of NP-complete problems. Developing a GA to solve any NP-complete problem (e.g., SAT, the boolean satisfiability problem) yields automatic solutions to all other NP-complete problems via the reductions. All that is required is to (in software) map the instance of any NP-complete problem to SAT, apply the SAT (hardware-based) GA to solve it, and then (in software) map the SAT solution to a solution of the original problem. Of course, the GA must find an optimal solution (i.e., a satisfying assignment for a SAT instance) for this to work. A fit member in the SAT GA may map to a worthless non-solution in another problem unless the SAT member is optimal.

---

[1] Also see an excellent description by Leighton [30].

[2] The size of each register and comparator depends upon the desired precision of the numbers in the tuples, but should be at least $\log_2 n$ bits.

The idea of exploiting the reductions between NP-complete problems was studied extensively by DeJong and Spears [18], who also provided a SAT GA and empirical results on the hamiltonian circuit (HC) problem. Their SAT GA evaluates a population member by quantifying how "close" the bit string is to satisfying the given boolean function f. They do this by assigning a numeric value to each expression in f and combining them. Specifically, given boolean expressions $e_1, \ldots, e_\ell$, the fitness value of e_i, denoted $val(e_i)$, is given as follows for the operators AND ($\wedge$), OR ($\vee$) and NOT ($\sim$):

$$val(e_1 \wedge \cdots \wedge e_\ell) = avg(val(e_1), \ldots, val(e_\ell)),$$

$$val(e_1 \vee \cdots \vee e_\ell) = \max(val(e_1), \ldots, val(e_\ell)),$$

and

$$val(\sim e_i) = 1 - val(e_i),$$

where $avg(x_1, \ldots, x_\ell)$ returns the mean of the values $x_1, \ldots, x_\ell$. An example of these evaluation functions appears in Table 7.1. Notice in the table that an assignment has a fitness of 1.0 if and only if it satisfies f. This is true, in general, if f satisfies some simple conditions.[1] While improvements to these evaluation functions were suggested, empirically those described above performed well in the work of DeJong and Spears and are simple enough for a hardware implementation.

Table 7.1 Evaluating assignments for x_1 and x_2 in the SAT GA when attempting to satisfy $\boxed{f(x_1, x_2) = \sim x_1 \wedge (x_1 \vee x_2)}$.

x_1	x_2	$val(f(x_1, x_2)) = avg(1 - x_1, \max(x_1, x_2))$
0	0	$avg(1 - 0, \max(0,0)) = 0.5$
0	1	$avg(1 - 0, \max(0,1)) = 1.0$
1	0	$avg(1 - 1, \max(1,0)) = 0.5$
1	1	$avg(1 - 1, \max(1,1)) = 0.5$

A hardware GA implementation for the SAT GA could work as follows. Take the given boolean formula f and map it to a circuit consisting of AND, OR and NOT gates where each AND and OR gate has only two inputs. Then replace each inverter in the circuit with a module that subtracts

---

[1] These conditions are not listed here, but any boolean formula can be made to satisfy them with only a linear increase in its size.

its input from the constant 1, replace each OR gate with a module that outputs the maximum of its two inputs, and replace each AND gate with a module that adds its two inputs and right shifts the result by one bit. The result is a circuit that outputs the fitness of the input binary string. The number of gates in the new circuit is more than in the old circuit by a linear factor of only the precision (number of bits) used to represent the fitnesses. Thus, a HGA implementation of the SAT GA is feasible.

7.5. Extensions of the Design

There are many possible ways to extend the design of Section 7.3, many of which require only simple modifications to the VHDL code. First, other genetic algorithm operators could be implemented, including uniform crossover [48], multi-point crossover, and inversion [20]. Permutation-preserving crossover and mutation operators [20], [22] could be implemented for constrained problems such as the TSP. Additionally, the CMM could be parameterized to respect the boundaries of bit groups, i.e., permit crossover only at certain locations. This would be useful in preventing invalid strings in the generalized FPGA partitioning (Section 7.4.2) and hypergraph partitioning (Section 7.4.3) problems. Also, other selection methods [21] could be implemented. When implemented, these methods would be made available to the user via the software front end. The user would select the desired selection and crossover methods as HGA parameters. Also recall from Section 7.3 that if other GA termination conditions are desired besides running for a fixed number of generations (e.g., amount of population diversity, minimum average fitness), the front end can tell the HGA to run for a fixed number of generations and then check the resultant population to see if it satisfies the termination criteria. If not, then that population is retained in the HGA's memory for another run. This process repeats until the termination criteria are satisfied.

Another extension of this design involves allowing parallelization of the fitness modules. This is useful when the pipeline's bottleneck lies in the FM rather than the SM [44]. As mentioned in Section 7.3.5.7, a new module called the memory writer (MW) would be very useful for arbitrating memory writes and performing bookkeeping functions, e.g., maintaining the values numgens and psizetmp.

To extend the parallelization of the system, the SM-CMM-FM pipeline could be parallelized by replicating the highlighted portion (dotted box) of Figure 7.1. As with the parallel FM configuration mentioned above, an MW would be useful in this scheme. Other improvements include merging the PS with the MIC and changing the handshaking protocol of Section 7.3.4 to take fewer clock cycles. These reductions of communication delay should improve performance.

If the population members are extremely large (e.g., hundreds or thousands of bits), then it is unrealistic to send entire members between modules in parallel. Instead, the modules could process data in stream form, where processing begins when the first few bits arrive in the module and output begins while input and processing still occur. That is, the result of processing the first portion of input is sent to the next module before the rest of the input arrives. So the members travel in a "cut-through" fashion through each module rather than the "store-and-forward" fashion of Section 7.3. In the stream model, the SM operates on members' addresses and fitnesses rather than on the members themselves and their fitnesses. Once a pair of members is selected, the SM tells the CMM the addresses of the selected members. The CMM then fetches these members from memory and begins sending them to the FM, performing crossover and mutation while it transmits. If fitness function evaluation can begin before receiving the entire member, then, as it evaluates the member, the FM begins to write it into memory before evaluation or input is complete. If this is not possible, then multiple passes over the members is necessary for evaluation, so on-chip buffers are required.

7.6. Summary and Related Work

This chapter described the HGA, a general-purpose VHDL-based genetic algorithm intended for a hardware implementation. The design is parameterized, facilitating scaling. It is possible to implement the HGA on reprogrammable FPGAs, exploiting the speed of hardware while retaining the flexibility of a software implementation. The result is a general-purpose GA engine which is useful in many applications where software-based GA implementations are too slow, e.g., when real-time constraints apply.

Recently there has been more work on hardware-based GAs. Other VHDL GA implementations include Alander et al. [6] and Graham and Nelson [22]. Salami and Cain applied the design of this chapter to the problems of finding optimal gains for a proportional integral differential (PID) controller [41] and optimization of electricity generation in response to demand [42]. Additionally, Tommiska and Vuori [49] implemented a GA with Altera HDL (AHDL) for implementation on Altera FLEX 10K FPGAs [7].

A subset of the GA operations has been mapped to hardware by Liu [31], who designed and simulated a hardware implementation of the crossover and mutation operators. In similar work, Red'ko et al. [40] developed a GA which implemented crossover and mutation in hardware. Hesser et al. [25] implemented in hardware crossover, mutation, and a simple neighborhood-based selection routine. Hämäläinen et al. [23] designed the genetic algorithm parallel accelerator (GAPA), which is a tree structure of processors for executing a GA. The GAPA is a parallel GA with specialized hardware support to accelerate certain operations. Sitkoff et al. [46] designed a hardware GA for partitioning logic designs across Xilinx FPGAs. After running the GA in software, the bottleneck was determined to be in evaluation of the fitness function. Thus, parallel fitness evaluation modules were implemented on FPGAs and the remainder of the GA ran in software. Megson and Bland [32] present a design for a hardware-based GA that implements all the GA operations except fitness evaluation in a pipeline of seven systolic arrays.

Problem-specific (non-FPGA) implementations include a suite of proprietary GAs in a text compression chip from DCP Research Corporation [55]. Other examples include Turton et al.'s applications to image processing [50], image registration [52], disk scheduling [51], and Chan et al.'s application to hypergraph partitioning [16]. These GAs were designed for implementation on VLSI chips and thus are neither reconfigurable nor general-purpose. They are also expensive to produce in small quantities. However, the intended applications are popular, so a VLSI implementation seems justifiable since the systems can be produced in bulk.

References

[1] Darwin on a chip. *The Economist*, page 85, February 1993.

[2] VHDL International Users Forum (VIUF), 1997. http://www.vhdl.org/viuf.

[3] *VHDL^{UK}*: A communication network for VHDL users, 1997 http://www.vhdluk.org/.

[4] M. Abramovici, M. A. Breuer, and A. D. Friedman. *Digital Systems Testing and Testable Design*. AT&T Bell Laboratories and W. H. Freeman and Company, New York, New York, 1990.

[5] Advanced Micro Devices. *Bipolar Microprocessor and Logic Interface (Am29000 Family) Data Book*, 1985. http://www.amd.com/.

[6] J. T. Alander, M. Nordman, and H. Setälä. Register-level hardware design and simulation of a genetic algorithm using VHDL. In P. Osmera, editor, *Proceedings of the 1st International Mendel Conference on Genetic Algorithms, Optimization, Fuzzy Logic and Neural Networks*, pages 10-14, 1995.

[7] Altera Corporation, San Jose, California. *Flex 10k Embedded Programmable Logic Family*, 1996. http://www.altera.com/.

[8] J. M. Arnold, D. A. Buell, and E. G. Davis. Splash 2. In *Proceedings of the 4th Annual ACM Symposium on Parallel Algorithms and Architectures*, pages 316-324, June 1992.

[9] P. M. Athanas and H. F. Silverman. Processor reconfiguration through instruction-set metamorphosis. *IEEE Computer*, 26(3):11-18, March 1993.

[10] K. Batcher. Sorting networks and their applications. In *Proceedings of the AFIPS Spring Joint Computing Conference*, volume 32, pages 307-314, 1968.

[11] J. Bean. Genetics and random keys for sequencing and optimization. *ORSA Journal on Computing*, 6:154-160, 1994.

[12] P. Bertin, D. Roncin, and J. Vuillemin. Programmable active memories: A performance assessment. In G. Borriello and C. Ebeling, editors, *Research on Integrated Systems: Proceedings of the 1993 Symposium*, pages 88-102, 1993.

[13] S. D. Brown, R. J. Francis, J. Rose, and Z. G. Vranesic. *Field-Programmable Gate Arrays*. Kluwer Academic Publishers, Boston, Massachusetts, 1992.

[14] C. Burns. An architecture for a Verilog hardware accelerator. In *Proceedings of the IEEE International Verilog HDL Conference*, pages 2-11, February 1996. http://www.crl.com/www/users/cb/cburns/

[15] S. Casselman. Virtual computing and the virtual computer. In R. Werner and R. S. Sipple, editors, *Proceedings of the IEEE Workshop on FPGAs for Custom Computing Machines*, pages 43-48. IEEE Computer Society Press, April 1993. http://www.vcc.com/.

[16] H. Chan and P. Mazumder. A systolic architecture for high speed hypergraph partitioning using a genetic algorithm. In X. Yao, editor, *Progress in Evolutionary Computation*, pages 109-126, Berlin, 1995. Springer-Verlag. *Lecture Notes in Computer Science* number 956.

[17] H. de Garis. An artificial brain. *New Generation Computing*, 12:215-221, 1994.

[18] K. A. De Jong and W. M. Spears. Using genetic algorithms to solve NP-complete problems. In J. D. Schaffer, editor, *Proceedings of the Third International Conference on Genetic Algorithms*, pages 124-132. Morgan Kaufmann Publishers, Incorporated, June 1989.

[19] M. Gokhale, W. Holmes, A. Kosper, S. Lucas, R. Minnich, D. Sweely, and D. Lopresti. Building and using a highly parallel programmable logic array. *IEEE Computer*, 24(1):81-89, January 1991.

[20] D. E. Goldberg. *Genetic Algorithms in Search, Optimization, and Machine Learning*. Addison-Wesley, Reading, Massachusetts, 1989.

[21] D. E. Goldberg and K. Deb. A comparative analysis of selection schemes used in genetic algorithms. In G. Rawlings, editor, *Foundations of Genetic Algorithms*, pages 69-93, 1991.

[22] P. Graham and B. Nelson. A hardware genetic algorithm for the traveling salesman problem on Splash 2. In *5th International Workshop on Field-Programmable Logic and its Applications*, pages 352-361, August 1995. http://splish.ee.byu.edu/.

[23] T. Hämäläinen, H. Klapuri, J. Saarinen, P. Ojala, and K. Kaski. Accelerating genetic algorithm computation in tree shaped parallel computer. *Journal of Systems Architecture*, 42(1):19-36, August 1996.

[24] J. P. Hayes. *Computer Architecture and Organization*. McGraw-Hill Book Company, New York, second edition, 1989.

[25] J. Hesser, J. Ludvig, and R. Männer. Real-time optimization by hardware supported genetic algorithms. In P. Osmera, editor, *Proceedings of the 2nd International Mendel Conference on Genetic Algorithms, Optimization, Fuzzy Logic and Neural Networks*, pages 52-59, 1996.

[26] T. Higuchi, H. Iba, and B. Manderick. Evolvable hardware. In H. Kitano and J. A. Hendler, editors, *Massively Parallel Artificial Intelligence*, pages 398-421. MIT Press, 1994.

[27] H. Högl, A. Kugel, J. Ludvig, R. Männer, K.-H. Noffz, and R. Zoz. Enable++: A second generation FPGA processor. In *Proceedings of the IEEE Symposium on FPGAs for Custom Computing Machines*, pages 45-53, April 1995. http://www-mp.informatik.uni-mannheim.de

[28] P. D. Hortensius, H. C. Card, and R. D. McLeod. Parallel random number generation for VLSI using cellular automata. *IEEE Transactions on Computers*, 38:1466-1473, October 1989.

[29] S. Kang, Y. Hur, and S. A. Szygenda. A hardware accelerator for fault simulation utilizing a reconfigurable array architecture. *VLSI Design*, 4(2):119-133, 1996. http://www.ece.utexas.edu/ece/people/profs/Szygenda.html.

[30] F. T. Leighton. *Introduction to Parallel Algorithms and Architectures: Arrays, Trees, Hypercubes*. Morgan Kaufmann Publishers, Incorporated, San Mateo, California, 1992.

[31] J. Liu. A general purpose hardware implementation of genetic algorithms. Master's thesis, University of North Carolina at Charlotte, 1993.

[32] G. M. Megson and I. M. Bland. A generic systolic array for genetic algorithms. Technical report, University of Reading, May 1996. http://www.cs.rdg.ac.uk/cs/research/Publications/reports.html.

[33] Mentor Graphics Corporation, Wilsonville, Oregon. *Mentor Graphics VHDL Reference Manual*, 1994. http://www.mentorg.com/

[34] Mentor Graphics Corporation, Wilsonville, Oregon. *VHDL Style Guide for AutoLogic II*, 1995. http://www.mentorg.com/.

[35] Mentor Graphics Corporation, Wilsonville, Oregon. *Synthesizing with AutoLogic II*, 1996. http://www.mentorg.com/.

[36] Z. Michalewicz. *Genetic Algorithms + Data Structures = Evolution Programs*. Springer-Verlag, Berlin, second edition, 1994.

[37] B. Norman and J. Bean. Random keys genetic algorithm for job shop scheduling. *Engineering Design and Automation*, to appear. http://www-personal.engin.umich.edu/~jbean/.

[38] M. J. O'Dare and T. Arslan. Hierarchical test pattern generation using a genetic algorithm with a dynamic global reference table. In *Proceedings of the First IEE/IEEE International Conference on Genetic Algorithms in Engineering Systems: Innovations and Applications*, pages 517-523, Sept, 1995. http://vlsi2.elsy.cf.ac.uk/group/.

[39] Precedence, Incorporated, Campbell, California. *Product Brief*, 1996. http://www.precedence.com/.

[40] V. G. Red'ko, M. I. Dyabin, V. M. Elagin, N. G. Karpinskii, A. I. Polovyanyuk, V. A. Serechenko, and O. V. Urgant. On microelectronic implementation of an evolutionary optimizer. *Russian Microelectronics*, 24(3):182-185, 1995. Translated from *Mikroelektronika*, Vol. 24, No. 3, pages 207-210, 1995.

[41] M. Salami and G. Cain. An adaptive PID controller based on a genetic algorithm processor. In *Proceedings of the First IEE/IEEE International Conference on Genetic Algorithms in Engineering Systems: Innovations and Applications*, pages 88-93, September 1995.

[42] M. Salami and G. Cain. Multiple genetic algorithm processor for the economic power dispatch problem. In *Proceedings of the First IEE/IEEE International Conference on Genetic Algorithms in Engineering Systems: Innovations and Applications*, pages 188-193, September 1995.

[43] E. Sanchez and M. Tomassini, editors. *Towards Evolvable Hardware: The Evolutionary Engineering Approach*. Springer-Verlag, Berlin, 1996. *Lecture Notes in Computer Science* number 1062.

[44] S. D. Scott, S. Seth, and A. Samal. A hardware engine for genetic algorithms. Technical Report UNL-CSE-97-001, University of Nebraska-Lincoln, July 1997. ftp://ftp.cse.unl.edu/pub/TechReps/UNL-CSE-97-001.ps.gz.

[45] M. Serra, T. Slater, J. C. Muzio, and D. M. Miller. The analysis of one-dimensional linear cellular automata and their aliasing properties. *IEEE Transactions on Computer-Aided Design of Integrated Circuits and Systems*, 9(7):767-778, July 1990.

[46] N. Sitkoff, M. Wazlowski, A. Smith, and H. Silverman. Implementing a genetic algorithm on a parallel custom computing machine. In *IEEE Symposium on FPGAs for Custom Computing Machines*, pages 180-187, April 1995. http://www.lems.brown.edu/arm/.

[47] Synopsys, Incorporated, Mountain View, California. *ARKOS Datasheet*, 1997. http://www.synopsys.com/.

[48] G. Syswerda. Uniform crossover in genetic algorithms. In *Proceedings of the Third International Conference on Genetic Algorithms and their Applications*, pages 2-9, 1989.

[49] M. Tommiska and J. Vuori. Implementation of genetic algorithms with programmable logic devices. In J. T. Alander, editor, *Proceedings of the Second Nordic Workshop on Genetic Algorithms and their Applications (2NWGA)*, pages 71-78, August 1996. http://www.uwasa.fi/cs/publications/2NWGA.html.

[50] B. C. H. Turton and T. Arslan. An architecture for enhancing image processing via parallel genetic algorithms & data compression. In *Proceedings of the First IEE/IEEE International Conference on Genetic Algorithms in Engineering Systems: Innovations and Applications*, pages 337-342, Sept, 1995. http://vlsi2.elsy.cf.ac.uk/group/

[51] B. C. H. Turton and T. Arslan. A parallel genetic VLSI architecture for combinatorial real-time applications---disc scheduling. In *Proceedings of the First IEE/IEEE International Conference on Genetic Algorithms in Engineering Systems: Innovations and Applications*, pages 493-499, September 1995. http://vlsi2.elsy.cf.ac.uk/group/

[52] B. C. H. Turton, T. Arslan, and D. H. Horrocks. A hardware architecture for a parallel genetic algorithm for image registration. In *Proceedings of the IEE Colloquium on Genetic Algorithms in Image Processing and Vision*, pages 11/1-11/6, October 1994. http://vlsi2.elsy.cf.ac.uk/group/.

[53] J. E. Volder. The CORDIC trigonometric computing technique. *IRE Transactions on Electronic Computers*, EC-8:330--334, September 1959.

[54] M. Wazlowski, A. Smith, R. Citro, and H. F. Silverman. Armstrong III: A loosely-coupled parallel processor with reconfigurable computing capabilities. Technical report, Brown University, 1996. http://www.lems.brown.edu/arm/.

[55] L. Wirbel. Compression chip is first to use genetic algorithms. *Electronic Engineering Times*, page 17, December 1992.

[56] S. Wolfram. Universality and complexity in cellular automata. *Physica*, 10D:1-35, 1984.

[57] Xilinx, Incorporated, San Jose, California. *The Programmable Logic Data Book*, 1996. http://www.xilinx.com/.

[58] Zycad Corporation, Fremont, California. *Paradigm XP Product News*, 1996. http://www.zycad.com/.

Chapter 8 Genetic Algorithm Model Fitting

Matthew Lybanon

Naval Research Laboratory, Stennis Space Center, MS

and

Kenneth C. Messa, Jr., Loyola University, New Orleans, LA

8.1 Introduction

Techniques to fit mathematical models ("curves," which is what they are typically) to data are plentiful. However, many of these techniques often fail for large data sets or when complicated functions are used in the fit. The most frequent criterion used in fitting is minimization of a (possibly weighted) sum of squares of "residuals" or deviations between data points and the curve. This idea, that we obtain the "best" fit by minimizing an error measure, casts curve fitting as an optimization problem.

There are two general classes of techniques for solving optimization problems: derivative-based methods and search methods. The latter class is subdivided into exhaustive search and guided search methods. When the "objective function" (function to minimize) is a complicated function of several variables and the equations involved cannot be solved in closed form, it is not difficult to see why derivative-based techniques often fail. Whether the technique involves Taylor-series approximation or proceeding up or down a gradient, it is generally necessary to have a good initial solution estimate. Otherwise, being trapped in a local minimum or having a runaway solution is likely. Exhaustive search guarantees finding the global optimum, but is impractical for problems of any real complexity. This leaves the guided search methods, one example of which is genetic algorithms (GAs).

Many researchers have used GAs as a tool for solving optimization problems (Gallagher and Sambridge, 1994), although De Jong (1992) argues that "Genetic Algorithms Are NOT Function Optimizers." When GAs are applied as optimizers (the distinction De Jong makes), the organisms represent solutions to the problem. In the curve-fitting case, solutions are real values assigned to each fit parameter. These values must be encoded in some way for the GA to operate on them. (Details of the coding, and other implementation details, are left for a later part of this chapter. However, we point out here that the range in which each parameter is assumed to lie is explicitly involved.) The fitness function is a measure of how well the candidate solution fits the data. It is the entity that focuses the GA toward the solution.

This curve-fitting process exhibits an interesting property. If the initial guess for the range in which a parameter lies is wrong, and the true value lies outside the interval, the GA will often "discover" that fact, by producing organisms with values of that parameter clustered near the appropriate end of the interval. This is a signal to move the interval and try again, a procedure that is easy to automate. This is in contrast to the behavior of many other curve-fitting techniques when the "first guess" is bad, as described above. The GA model-fitting technique is highly robust.

8.2 *Program Description*

This chapter describes a C program for the model-fitting problem. In this section we describe some details of the genetic algorithm and how the GA is used to solve the problem. The GA details include data representation, operators, and fitness functions. The GA itself is fairly conventional, but it is one part of a structure tailored to the problem. The structure's design takes advantage of some of the problem's characteristics to efficiently find an accurate solution.

8.2.1 The Genetic Algorithm

Data Representation

The organisms in a genetic algorithm represent solutions to the problem. In the GA used here, an organism is a single chromosome (the solution). For convenience, identifiers for the organism's parents and its fitness values, are appended. The population is a collection of organisms, with statistical information appended.

Solutions to the fitting problem are real values assigned to each adjustable parameter in the model. Thus, if one fits a curve $f(a_1, a_2, ..., a_n; x)$ to data, the solution consists of the real-number values for the parameters a_i. Thus vectors $<r_1, r_2, ..., r_n>$, where a_i is replaced by real numbers r_i, $i = 1, 2, ...,$ n, are candidate solutions. These vectors are a high-level view of the representation. However, the genetic algorithm works at a lower level, the level of bits or alleles. Given upper and lower bounds for each r_i, u_i and l_i, respectively, r_i is represented in the chromosome as a (double-precision) real number d_i in the interval $[0, 1]$ such that

$$r_i = d_i (u_i - l_i) + l_i. \tag{1}$$

Each number d_i corresponds to a real number r_i in the interval $[l_i, u_i]$. Thus, a candidate solution is constructed:

$$\theta = <d_1, d_2, ..., d_n>. \tag{2}$$

In addition to this "real representation," we also used a similar scheme in which each d_i is replaced by $b_i/2^m$ in Eq. (1), where b_i is an m-bit binary integer. The analog to Eq. (2) simply replaces each d_i by b_i. The two approaches work about equally well, but certain obvious changes must be made to the crossover and mutation operators, depending, on which representation is used.

Fitness Functions

This application's fitness function is a measure of how well the candidate solution fits the data. This fitness function is built in several stages. We call the first stage the "prefitness." The usual measure of goodness of fit in a model-fitting problem is a function of the "residuals," or deviations between the measurements and the model. Experimental error contaminates the measurements, and the model is a statistical estimate of an improved (we hope) version of the measurements. The most common criterion for fitting is the "least-squares" criterion; the measure of goodness of fit is the sum of the squares of the residuals. This measure has the virtues of being positive definite, large when the residuals are large, and small when the residuals are small. If the model is the "best" (according to the least-squares criterion) estimate of the process the measurements describe, then the residuals are effectively the errors in the measurements, and the fitting process consists of adjusting the model's adjustable parameters to minimize the error measure.

If the measurements consist of pairs of values (x_i, y_i), i = 1, 2, ..., N, where x_i are measurements of the "independent variable" and y_i are measurements of the "dependent variable," as often happens, then fitting according to the least-squares criterion consists of adjusting the parameters $a_1, a_2, ..., a_n$ to minimize S,

$$S = \sum_{i=1}^{N} \left[f(a_1, a_2, ..., a_n; x_i) - y_i \right]^2 \tag{3}$$

It is important to note that n and N are not the same value. If N < n, there is not enough information to find the parameter values and, if N = n, one obtains an algebraic solution that exactly matches the model to the measured values, error included. A model-fitting problem normally means that N > n, i.e., the problem is overdetermined.

In the genetic algorithm, a candidate solution is represented by a vector θ, where each element d_i of the vector θ is related to a real number r_i by Eq. (1). So, Eq. (1) converts the candidate solution θ into a vector $\mathbf{r} = <r_1, r_2, ..., r_n>$. Then the *prefitness* value of θ is given by

$$prefitness(\theta) = \sum_{i-1}^{N} \left[f(r; x) - y_i \right]^2 \tag{4}$$

Sometimes criteria other than least-squares are used in fitting. The genetic algorithm approach to fitting has the advantage that it can be changed to accommodate a different fitting criterion simply by changing the (pre-) fitness function; the remainder of the program stays the same.

Since genetic algorithms work by searching for maximum values and since the prefitness function has its minimum for the best fit, the roles of max and min must be reversed. For each population p, let max_p represent the largest prefitness value (i.e., the worst fit). We compute the *raw fitness* as the difference.

$$\text{raw fitness } (\theta) = max_p - \text{prefitness } (\theta). \tag{5}$$

The raw fitness is not precisely the fitness function used in the genetic algorithm. Many genetic algorithms require a scaling of these fitness values. Scaling amplifies the distinction between good and very good organisms. Our genetic algorithm employs linear scaling,

$$\text{fitness } (\theta) = m_p \times \text{raw fitness } (\theta) + b_p, \tag{6}$$

where m_p and b_p are constants calculated for each population p. The constants are chosen to scale the maximum raw fitness to a scale factor (which depends upon the GA generation; more details are given later in the text, and in the code) times the average raw fitness, except that when the resulting minimum fitness is negative, all the fitness values are shifted upward.

Clearly, Eqs. (5) and (6) can be combined to form a linear relationship between the fitness and prefitness functions,

$$\text{fitness } (\theta) = \alpha_1 \times \text{prefitness } (\theta) + \alpha_2, \tag{7}$$

where α_1 and α_2 are derived constants for each population. Thus, the final fitness of an organism θ is linearly related to the sum of the squares of the residuals (or other error measure if the fit criterion is different from least-squares).

Genetic Algorithm Operators

The selection operator that our genetic algorithm uses is stochastic sampling without replacement, called "expected value" by Goldberg (1989). Given an organism θ with fitness f_i, the fitness is weighted with respect to the population's average f^*. The truncated division of f_i by f^* yields the expected number of offspring formed by the organism θ; an organism selected (randomly) this number of times is subsequently no longer available for selection. The program calculates the expected number of offspring of each organism, and fills any remaining unfilled slots in the new population based on the fractional parts of f_i/f^* (De Jong, 1975). Besides this selection method, our genetic algorithm uses De Jong's (1975) elitist strategy, by which the single best organism is placed unchanged into the next generation. This strategy gives a little more weight to the best organism than might be achieved from selection alone and prevents the possibility that the best organism might be lost early from the population through crossover or mutation.

Simple crossover works effectively. For problems with many fit parameters or where many bits are needed for greater precision, two-point crossover yields better results than single-point crossover. Typically, 90% of the selected pairings are crossed, and the remaining 10% are added unchanged to the next generation (90% crossover rate). Our program implements crossover with the data representation described by Eq. (2) by *arithmetic crossover in a single gene*. Let $c_1 = <r_1, r_2, ..., r_n>$ and $c_2 = <s_1, s_2, ..., s_n>$ be two chromosomes. Let i be a random integer between 1 and n and let p be a random real between 0 and 1. Then the crossover of c_1 with c_2 is given by

$$c = <r_1, r_2, ..., r_{i-1}, t, s_{i+1}, s_{i+2}, ..., s_n>, \tag{8a}$$

(the other member of the pair is defined analogously), where

$$t = p \cdot r_i + (1 - p) s_i. \tag{8b}$$

The mutation method most commonly used in GAs is changing the value of a bit at a gene position with a frequency equal to the mutation rate. There are various problems with that strategy here, all related to the fact that a gene is a string of bits for this problem, not just one bit. "Flipping" a single bit produces highly variable results depending upon the significance of that particular bit. The real representation we used also requires a different mutation mechanism. See Michalewicz (1992) for a detailed explanation of crossover and mutation for real representation of chromosomes.

The present application uses several alternative mutation techniques. The first two are specific to the real-representation case, while the latter two can be applied to either that case or the binary-integer-representation case.

1. Uniform mutation: Randomly change the value of a gene to a real number between 0 and 1.

2. Boundary mutation: Change the value of a gene to 0 or 1.

3. Non-uniform mutation: Add (or subtract) a small value to (from) the gene while maintaining the range of values between 0 and 1. Decay (decrease) the effect as the generation number increases.

4. Non uniform mutation without decay: This is the same as 3, but without decay.

The mutation probabilities differ for the different techniques. For example, a reasonable value for the probability of boundary mutation is 0.01, while for uniform mutation a reasonable value is 0.1.

One difficulty for genetic algorithms is finding optimal solutions when there are many local optima that are near in value to the global optimum. In circumstances like this (or when there are multiple global optima), one can use the "crowding model" for the genetic algorithm (Goldberg, 1989). With this model, only a certain percentage (the *generation gap*), denoted genGap, of the original population is replaced each generation. (If genGap = 1.0, then this is the standard GA model.) For example, only 10% (genGap = 0.10) of one generation is replaced to form the next generation. The method of replacement follows. Once an offspring is formed, several members of the current generation are inspected. The number inspected is the *crowding factor*. Of those inspected, the one that is most like the offspring is replaced by the offspring. In this way, the number of individuals that are similar (perhaps they are near a local optimum) is limited. With the crowding model, no one group of individuals is allowed to dominate the population. Convergence takes longer, but there is less chance that the GA will converge to a suboptimal solution.

Seeding is the operator by which some organisms used to form the original population are generated, not at random, but from some set. For example, since one cycle produces, say, five organisms, each of which should be "close" to the ideal one, one can place the best of these into the initial generation of the next cycle. This could give the next cycle a "head start." (The following description of the program's organization will clarify the meaning of "cycle.")

Program Organization

Begin
Initialize
Read parameters and data
Parameter initial intervals loop: loop if parameter intervals were provisionally shifted (or if this is the first pass), but no more than MAX_PASS times
    Set initial parameter intervals and trends
    Parameter interval sizes loop: loop while largest parameter interval > EPS, but nomore than maxCycles times
        GA runs loop: loop timesRepeat times
            Generate initial GA population
            GA loop
        End GA runs loop
        Print final values
        Shrink parameter intervals
        Find largest parameter interval
    End parameter interval sizes loop
    Provisionally shift interval(s) if necessary
    Print report
End parameter initial intervals loop
Print elapsed time
End

Figure 8.1. Top-Level Structure of SSHGA Program

Figure 8.1 illustrates the program organization, whose main part consists of loops nested four deep. The innermost loop (GA loop) is the genetic algorithm, which executes several generations. The next most deeply nested loop (GA runs loop) executes the genetic algorithm loop several times, with different randomly chosen initial populations. The loop outside this one (parameter interval sizes loop) reduces the size of the intervals thought to contain the fit parameter values (i.e., decreases the distance between l_i and u_i, $i = 1, 2, ..., n$) and begins the more deeply nested loops again. The outermost loop (parameter initial intervals loop) moves the intervals (defines sets of l_i, u_i that encompass different ranges, when that is appropriate) and executes the three inner loops again. Details follow.

The summary given above starts with the innermost loop and works outward. The following discussion gives an idea of what the program does in the order it does it, beginning with the outermost loop. This discussion complements Figure 8.1.

First, the program prepares to use the initial boundaries specified by user input. The number of parameters, and the ranges within which they are assumed to lie, are problem specific. The more accurate the initial guesses, the better. The GA can probably find a good solution even if the initial estimates are bad, but it will take longer.

The genetic algorithm runs several times and saves the best solution from each run in a table. If convergence is not sufficient (based on whether all GA runs yield best results that are sufficiently near each other, where "sufficiently near" is specified by user input, referred to as *EPS* in Figure 8.1), the original interval boundaries are revised (narrowed) and the program repeats another *cycle*. The program may execute several cycles before it transfers control to the evaluator routine.

After the GA executes several cycles, the parameter intervals may become very narrow, showing convergence. Occasionally, however, the value of a parameter (or more than one) will be close to one end of its interval. This could mean that the endpoint is the true parameter value. More likely, it implies that the true value lies outside the original interval. So, in this case, the program assigns new endpoints, shifting the interval up or down as indicated, and begins the entire process again. The program may execute several *passes* in this way.

Table 8.1 lists user inputs. Most of them are self-explanatory, or easy to understand after reading the foregoing discussion. The next few paragraphs explain a few that may not be obvious.

Convergence is often swifter with larger scale factors. Scaling may yield more accuracy, but there is a danger of premature convergence to a local minimum. The program allows for this in two ways. The scale factor, *scale_fac*, is not used initially while the genetic algorithm is in the exploratory phase. Scaling is initiated after *begin_scale* percent of *num_gen* generations. Further, the user can choose to apply a different scale factor, *nscale*, during the first 10% of generations. Following are some typical values (they may be problem-specific):

num_gen—100 to 200 with the standard model, 100 with the crowding model

nscale— No scaling (1.0) is generally used early, but "negative scaling" (0.10 to 0.50) encourages exploration

begin_scale—0.50 (begin scaling after 50% of num_gen generations)

scale_fac—A large scale factor (1000 to 100000) yields greater accuracy, but could promote convergence to a suboptimal solution. Lower scale factors (10 to 100) lead to less accuracy, but are "fooled" less often.

The procedure for shrinking parameter intervals attempts to find new upper and lower bounds based on the parameter values found in the GA runs made with the previous bounds. However, if the new range would be less than *shrinkage* times the old range, the program revises the proposed new bounds. A typical value for shrinkage is 0.50 with the standard model, and 0.75 to 0.90 with the crowding model.

The discussion of fitness functions describes the procedure the program uses to deal with the usual situation in model fitting, in which the objective function is minimized. The input variable *convert_to_max* allows the program to be used when the objective function is maximized. Inputting *convert_to_max* = 1 gives the usual case, in which the objective function needs to be "converted" to one that is maximized to use the GA "machinery," while *convert_to_max* = 0 tells the program that the objective function is already one for which maximizing its value gives the optimum solution.

Table 8.1. User Inputs to GA Model-Fitting Program

Variable Name	Description
pop_size	Size of population
num_gen	Number of generations
scale_fac	Scale factor
nscale	Scale factor for first 10% of generations
beginScale	This percent of generations before scaling
genGap	Generation gap; 1.0 turns off
cF	Crowding factor
prob_xover	Probability of crossover
xover_type	Type of crossover
pr_u	Probability of uniform mutation
pr_b	Probability of boundary mutation
pr_n	Probability of non-uniform mutation
timesRepeat	Number of runs per cycle

maxCycles	Maximum number of cycles
shrinkage	Amount of shrinkage (of intervals) allowed each cycle
num_unknown	Number of unknowns (parameters) in fit function
convert_to_max	1 if obj func is minimized; 0 if maximized

for each parameter, i = 0 to num_unknown - 1:

lower_bound[i]	Lower endpoint of interval assumed to contain parameter i
upper_bound[i]	Upper endpoint of interval assumed to contain parameter i
freq_full	Frequency of full report
freq_brief	Frequency of brief report

The program prints various reports that summarize its progress. Among them are "brief" and "full" reports printed during the GA's execution. The input variable *freq_brief* tells how often the program prints a brief report, which gives statistics and information about the (current) best and average organisms. That is, every *freq_brief* generations, the program prints a brief report. In a similar way, *freq_full* controls the printing of full reports, which list all the organisms and their fitnesses along with the statistics. The program also prints a "complete" report after each pass through the parameter initial intervals loop. A complete report consists of all the statistical data and a list of the entire population.

Direct examination of the C code should clarify any details that are still unclear. While this is tedious, gaining familiarity with the program is the best way to understand its operation.

8.3 Application

This program was originally developed to aid in the solution of a problem in satellite altimetry. The satellite altimeter can provide useful oceanographic information (Lybanon, et al., 1990; Lybanon, 1994). The instrument is essentially a microwave radar that measures the distance between itself and the sea surface. The difference between that measurement and an independent measurement of satellite orbit height is the height of the sea surface relative to the same reference used for orbit height (typically the "reference ellipsoid"). Then, differencing this *sea level* with the marine geoid (a gravitational equipotential surface) yields *sea-surface-height* (SSH) *residuals*. The sea surface would coincide with the geoid if it was at rest, so the SSH signal can provide information about ocean dynamics.

A careful consideration of errors is critical, because SSH is the small difference of large numbers. This work considers a certain type of geoid error. The dynamic range of the geoid is several tens (perhaps more than a hundred) of meters, while that of the oceanographic SSH signal is a meter or less. So, a small geoid error can introduce a large error into the SSH calculation (a 1% geoid error is as big as the SSH signal, or bigger).

The true geoid is not precisely known. "Geoids" typically incorporate some information from altimetry, generally along with some *in situ* measurements. A detailed, ship-based geodetic survey might give the most accurate results, but it would take a very long time to complete and its cost would be prohibitive. An altimetric mean surface (the result of averaging altimeter measurements made over an extended period) is a good approximation to the geoid. But, as pointed out, small geoid errors can be large errors in the oceanographic information.

The portion of the geoid derived from altimeter data necessarily includes the time-independent (i.e., the long-term temporal mean) part of the oceanographic signal. So, in a region of the ocean that contains a permanent or semi-permanent feature, the temporal mean of that feature's signal contaminates the geoid. One such region is the Gulf Stream region of the North Atlantic; the Gulf Stream's meanders lie within a relatively small envelope of positions. The Mediterranean also possesses several long-lasting features.

Differencing the altimeter signal obtained from such a region with a geoid that contains this contamination produces "artifacts" such as apparent reverse circulation near the Gulf Stream, sometimes of a magnitude comparable to the Gulf Stream itself. This complicates interpretation of the SSH residuals, since the artifacts appear to represent features that have no oceanographic basis. Or, they may obscure real features. One remedy is to construct a "synthetic geoid" (Glenn et al., 1991).

Several synthetic geoid methods have been developed to deal with this problem (e.g., Kelly and Gille, 1990). Our procedure consists of modeling the SSH calculation and fitting the mathematical model to SSH profiles (Messa and Lybanon, 1992). The mathematical model is, schematically:

$$\text{SSH} = f_1(x) - f_2(x) + \varepsilon(x), \tag{9}$$

where x is an along-track coordinate, and

$f_1(x)$ = correct (i.e., without error) profile,

$f_2(x)$ = mean oceanography error introduced by geoid, and

$\varepsilon(x)$ = residual error (e.g., orbit error, other geophysical errors).

For the Gulf Stream, we used hyperbolic tangents for both $f_1(x)$ and $f_2(x)$, and we simply used a constant bias for $\varepsilon(x)$. Therefore, the explicit model fit to the altimeter profiles is:

$$\text{SSH} = a \tanh[b\,(x - d - e)] - f \tanh[c\,(x - e)] + g. \tag{10}$$

Interpretation of the parameters is straightforward. The parameter e defines an origin for x, d gives the displacement of the instantaneous Gulf Stream from the mean Gulf Stream, a and f give the amplitudes of the respective profiles, b and c are related to their slopes, and g is the residual bias. The same model should apply to other fronts, while a model that involves Gaussians should apply to eddies.

While this model is a simple idealization, it can give good results. For its use in this application, we refer the reader to the references. Because the model is a highly-nonlinear function of some of the parameters a-g, it is very difficult to fit using conventional least-squares algorithms. The GA approach, however, has proved successful.

8.4 Application of Program to Other Problems

Conventional least-squares programs work well for many problems, particularly those for which the model is a linear function of the fit parameters (e.g., polynomials; the coefficients are the fit parameters). In such cases there is no need to use the GA approach described here; the conventional techniques will give the correct answer, and probably much more quickly. But there are cases for which the standard techniques are very difficult (or impossible) to apply. The GA approach has the important advantage, in such cases, of working where other methods fail.

8.4.1 Different Mathematical Models

The program is presently set up to use a prefitness of the form shown in Eq. (4), with f($\mathbf{r}$; x) given by Eq. (10) (f($\mathbf{r}$; x) is SSH; the components of r are the parameters a, b, ..., g). The code for this is located in ga_obj.h. SSH is given by a *#define* preprocessor command. The function *double objfunc()* converts a chromosome into the numerical values of the parameters a-g, and calls the function *double pre_obj()* to calculate the prefitness and return its value. The function *double convert()*, which converts a chromosome to the corresponding parameter values, is located in ga_util.h. The functions *void revise_obj()* and *void scale()*, located in ga_util.h, calculate the raw fitness and fitness values, respectively. If the input quantity convert_to_max = 0, raw fitness = prefitness. The program automatically takes care of this.

The program also takes advantage of some problem-specific information: the fit parameters a, b, c, and f are never negative. The two functions *void adjustInterval()* and *int evaluateIntervals()*, located in ga_int.h, explicitly take advantage of this fact. They ensure that the intervals for these parameters are never redefined to allow any negative values. This is useful in this application, and possibly avoids some unnecessary computation, but it has meaning only for this specific problem.

Changing to fit a different mathematical model does not require changes to most of the program. The function *double objfunc()* needs to be changed to convert a chromosome into the numerical values of the appropriate number of fit parameters, and the function *double pre_obj()* must calculate the prefitness for the new mathematical model. The code in the two functions *void adjustInterval()* and *int evaluateIntervals()* that prevents the first, second, third, and sixth fit parameters from going negative should be removed. Optionally, code that takes advantage of any similar special characteristics of the new problem can be inserted. No other changes should be necessary to apply the program to a different fitting problem. In fact, the same fundamental technique was used by Messa (1991) to model fit functions of more than one independent variable.

8.4.2 Different Objective Functions and Fit Criteria

Situations exist for which one wants to minimize a weighted sum of squares of the residuals.

$$S = \sum_{i=1}^{N} W_i \left[f(a_1, a_2, \ldots, a_n; x_i) - y_i \right]^2 \tag{11}$$

One possible reason is that some observations may be "less reliable" than others. Then, it is plausible that one should allow the less reliable observations to influence the fit less than the more reliable observations. The statistical implications of this situation are discussed in textbooks on the subject (e.g., Draper and Smith, 1981); here we point out that choosing each W_i to be the measurement variance of observation i gives a so-called "maximum likelihood" fit.

The least-squares criterion is not always the best one to use. The theoretically correct criterion is determined by the measurement error distribution (Draper and Smith, 1981), but it is easy to see why, for practical reasons, one might want to use a different criterion. Each measurement differs from the "true" value of the variable because of measurement error. It is virtually impossible for a small set of measurements to accurately describe a distribution. In particular, one large error may be more than the theoretically expected number of errors in this range, and will significantly distort the (sample) distribution's shape. This will bias the fit toward large errors because deviations are squared in the least-squares criterion (Eq. 3). So, a single large error can significantly distort the fit. Several large errors can distort the fit in ways that are less easy to visualize, but are still distortions. To minimize this effect one may want to use an error measure that weights large errors less. One such measure is the sum of absolute deviations,

$$S = \sum_{i=1}^{N} | f[f(a_1, a_2, \ldots, a_n; x_i) - y_i]^2 \tag{12}$$

Changing to this error measure, or any other, merely requires the corresponding changes to the code in ga_obj.h. The comparable change to most conventional curve-fit programs would require actual changes to the "normal equations."

The program can also be readily changed to accommodate a more general problem. Again, we emphasize that the fit residuals are estimates of the measurement error. This interpretation means that f($\mathbf{a}$; x) ($\mathbf{a}$ is the vector ($a_1, a_2, \ldots, a_n$)), evaluated at x = x_i, is an "improved" estimate of the error-contaminated measured value y_i. Although the mathematical machinery used to solve fitting problems may seem to be a procedure for finding the values of the parameters, the parameters are secondary. The fitting process is a statistical procedure for adjusting the observations to produce values more suitable for some purpose (e.g., summarizing measurements, prediction, etc.). It "... is primarily a method of adjusting the observations, and the parameters enter only incidentally" (Deming, 1943).

Keeping this in mind, consider the simplest model-fitting case, fitting straight lines to measurements (commonly called "linear regression"). With the usual least-squares fit criterion, Eq. (3), one minimizes the sum of squares of y residuals and obtains a linear expression for y as a function of x.

$$y = a x + b. \tag{13}$$

A plot of the data and the fit shows a straight line with a number of points scattered about it. The residuals are shown as vertical lines from each point to the regression line. What can we do with this expression? Can we predict y, given a value of x? Can we predict x, given a value of y? To do the latter, one can algebraically solve Eq. (13) for x as a function of y.

$$x = c y + d, c = (1 / a), d = - (b / a), \tag{14}$$

as long as $a \neq 0$. We can also easily apply the least-squares "machinery," switching the roles of x and y, to obtain a straight-line fit.

$$x = e y + f. \tag{15}$$

If we do this, we find that (except in rare cases) e does **not** equal c and f does **not** equal d. And we should not be surprised. If we plot the new line on the same graph as before, we see that it is a straight line with different slope and intercept from the first, and the residuals are horizontal lines, not vertical lines. In regressing x on y we are doing something different from regressing y on x, so there is no reason to expect to obtain the same fitted line. The result of an ordinary least-squares fit is useful for some purposes, but it is not legitimate to treat it as an ordinary algebraic equation and manipulate it as if it were. The same is true for fits of other functions. The straight-line case just described is just a simple illustration of the principle.

Another point to consider is the interpretation of the residuals in the second straight-line fit, Eq. (15). Are they estimates of errors in x, as the y-residuals were estimates of errors in y in the first fit? And, what happened to those errors in y; shouldn't we somehow account for them in the second fit? (For that matter, if there are errors in x, why did we ignore them in the first fit?) These are not new questions. They were first considered over a hundred years ago by Adcock (1877, 1878), and almost that long ago by Pearson (1901). There is a more recent body of research on this topic, but many people who use least squares are still unaware of it.

There have been attempts to account for errors in both x and y by doing a weighted fit with a judicious choice of weights ("effective variance" methods). Sometimes this helps (but not completely), and sometimes it produces the same estimates as ordinary least squares (Lybanon, 1984b). A procedure that is less based on wishful thinking is outlined below.

In the problem of fitting mathematical models to observations, there are three kinds of quantities: the measured values, the (unknown) "true" idealized values of the physical observables, and estimates of those true values. Symbolically, for the case we have been considering (measurements of two quantities that we assume to be related by a mathematical formula)

$$x_i = u_i + error, \tag{16a}$$

$$y_i = v_i + error, \tag{16b}$$

and:

$$v_i = f(u_i). \tag{16c}$$

In these equations u and v are the true values and x and y are what we measure. If we could measure u and v exactly there would be no problem. For a linear relationship between u and v, two pairs of observations would provide enough information to specify the constants in f(u). Statisticians call an equation like (16c) a *functional relationship*. In practice, there are usually measurement errors in one or (often) both variables. If we were to substitute x and y into Eq. (16c) in place of u and v, we would need to add a term to account for the error. The result would be what statisticians call a *structural relationship*. The task nevertheless is to estimate the parameters in the mathematical model and, in doing so, to obtain estimates of the true values of the observables. In our discussion we will not distinguish notationally between the true values and their statistical estimates.

The distinction should be clear. Eq. (16c) represents a "law-like" relationship between u and v. In Eq. (16c) u and v are mathematical variables, there is no error, the relation is exact, and v is strictly a function of u (and sometimes also vice versa). Eq. (16c) may be a linear relation, such as Eq. (13), except with u and v replacing x and y. Then it would look tantalizingly like a regression equation, but the interpretation is different. Only where either x or y is known without error is a regression equation appropriate. The variable that is in error should be regressed on the one known exactly. If both are free of error and the model is correct, then there is no problem since the data fit the model exactly. But, in all other circumstances, regression is inappropriate.

The mathematical details are best left to the references (e.g., Jefferys, 1980, 1981; Lybanon, 1984a). A summary will put forth the basic ideas. Since all of the variables contain error, the objective function should involve all of the errors. So, for the case we have been considering (we measure pairs of variables x and y, whose true values are assumed to obey an explicit functional relationship of the form $v = f(u)$ that involves some unknown parameters $\mathbf{a}$), the objective function is

$$S = \sum_{i=1}^{N} \left[(x_i - u_i)^2 + (y_i - v_i)^2 \right], \tag{17}$$

where v_i satisfies Eq. (16c). We immediately note two things. First, the objective function used in "ordinary" least squares, Eq. (3), eliminates the first term in brackets. So, if Eq. (17) is right, we should use only Eq. (3) if x errors are negligible. Second, the graphical interpretation of Eq. (17) shows the residuals as lines from each point (x_i, y_i) perpendicular to the fit curve, not just the y-components of those lines as we had before. A fit that uses this objective function will generally be different from an ordinary regression.

Eq. (17) is symmetrical in x and y errors. Suppose $v = f(u)$ is a linear function ($v = a u + b$, as in Eq. (13), with u replacing x and v replacing y). If we now treat v as the independent variable and u as the dependent variable (with y and x their measured values, respectively), and fit a linear function u $= g(v)$ (i.e., $u = e v + f$) using Eq. (17) as the objective function, then the residuals are the same as before we reversed the roles of independent and dependent variables. Their geometrical interpretation is still the same: they are perpendiculars from the data points to the line. So, a fit of $u = g(v)$ *will result in the same* coefficients e and f as the coefficients obtained by mathematically inverting $v = a u + b$.

It is usually desirable to use a weighted fit, in which the objective function is

$$S = \sum_{i=1}^{N} \left[W_{xi} (x_i - u_i)^2 + W_{yi} (y_i - v_i)^2 \right], \tag{18}$$

instead of Eq. (17). Note that this form allows different weights for each component of each observation i. Choosing each weight to be the reciprocal of the associated measurement variance produces a "maximum likelihood" fit.

The "generalized" least-squares problem considered here has more unknowns than the "ordinary" least-squares problem (for which the fitness is based on Eq. (3) or Eq. (11)) considered previously. In the ordinary case, the improved values v_i of the dependent variable measurements y_i are given by $v_i = f(\mathbf{a}; x_i)$. That is, they are determined by the fit parameter values found by the fit process. In the generalized case, however, $v_i = f(\mathbf{a}; u_i)$ (i.e., $f(\mathbf{a}; u_i)$ is the best estimate of v_i); the function must be evaluated at the improved values u_i of the x_i. Note that u_i is required in the GA to evaluate the fitness function, which is based on Eq. (18). Thus the u_i must be found as well as the fit parameters. This can substantially increase the dimensionality of the problem, since $N > n$, perhaps much greater.

Fortunately, there is an alternative to adding the u_i to the chromosome. Lybanon (1984a) showed that the improved x values can be found by a simple geometrical construction. Specifically, suppose x_i' is an approximation to u_i. We can find a better approximation $x_i'' = x_i' + \Delta x_i$, where

$$\Delta x = \left\{ W_{yi}[y_i - f(x_i')]\frac{\partial f}{\partial x} + W_{xi}(x_i - x_i') \right\} * \left[W_{yi}\left(\frac{\partial f}{\partial x}\right)^2 + W_{xi} \right]^{-1} \quad (19)$$

and the partial derivative is evaluated at $x = x_i'$, using the "current" values of the parameters. Thus the x values can be updated in a "side" calculation and the parameter values found by the genetic algorithm. After several generations of sequential updates, the x_i'' approach the u_i and the parameters approach their correct values. The reader is referred to the references for more details and a more thorough discussion of the problem.

8.5 Conclusions

This chapter has described a C program, based on genetic algorithms, for fitting mathematical models to data. The software is tailored to a specific problem (the "synthetic geoid" problem described in the text), but can be easily adapted to other fitting problems. There is an extensive literature on fitting techniques, but there are cases for which conventional (i.e., derivative-based) approaches have great difficulty or fail entirely. The GA approach may be a useful alternative in those cases.

The three fundamental aspects of the fitting problem are the function to be fitted, the objective function (error criterion) to minimize, and the technique to find the minimum. A conventional genetic algorithm performs the third role. It remains the same even when the other two are changed. The fit function is supplied by the user. This software has been successfully applied to functions of one or several variables that can depend upon the adjustable parameters linearly or nonlinearly. The version of the software shown in the following listing uses the least-squares criterion, but it is simple to change to a different criterion. Only slight program changes are needed to use a different fit function or goodness-of-fit criterion.

The GA technique often finds solutions in cases for which other techniques fail. Like many other nonlinear curve fitting techniques, this one requires initial guesses of fit parameter values (strictly speaking, this one requires estimates of the ranges within which they lie). Poor estimates may preclude convergence with many techniques, but the GA technique can probably "discover" this fact and adjust its intervals to find the correct values. Thus, the GA technique is highly robust.

References

R. J. Adcock (1877), "Note on the method of least squares," *The Analyst (Des Moines)*, **4**, 183-184.

R. J. Adcock (1878), "A problem in least squares," *The Analyst (Des Moines)*, **5**, 53-54.

K. A. De Jong (1975), "An analysis of the behavior of a class of genetic adaptive systems," Ph.D. Dissertation, University of Michigan, *Dissertation Abstracts International*, **36**(10), 5104B.

K. A. De Jong (1992), "Genetic algorithms are NOT function optimizers," *Proc. of the Second Workshop on the Foundations of Genetic Algorithms (FOGA-2)*, Vail, Colorado, July 1992 (Morgan Kaufmann).

W. E. Deming (1943), *Statistical adjustment of data*, New York: Dover (1964 reprint of the book originally published by John Wiley and Sons, Inc., in 1943).

N. R. Draper and H. Smith (1981), *Applied regression analysis, second edition*, New York: Wiley.

K. Gallagher and M. Sambridge (1994), "Genetic algorithms: a powerful tool for large-scale nonlinear optimization problems," *Comput. & Geosci.*, **20**(7/8), 1229-1236.

S. M. Glenn, D. L. Porter, and A. R. Robinson (1991), "A synthetic geoid validation of Geosat mesoscale dynamic topography in the Gulf Stream region," *J. Geophys. Res.*, **96**(C4), 7145-7166.

D. Goldberg (1989), *Genetic algorithms in search, optimization, and machine learning*, Reading, MA: Addison-Wesley.

W. H. Jefferys (1980), "On the method of least squares," *Astron. J.*, **85**(?), 177-181.

W. H. Jefferys (1981), "On the method of least squares. II," *Astron. J.*, **86**(1), 149-155.

K. A. Kelly and S. T. Gille (1990), "Gulf Stream surface transport and statistics at 69°W from the Geosat altimeter," *J. Geophys. Res.*, **95**(C3), 3149-3161.

M. Lybanon (1984a), "A better least-squares method when both variables have uncertainties," *Am. J. Phys.*, **52**(1), 22-26.

M. Lybanon (1984b), "Comment on 'least squares when both variables have uncertainties'," *Am. J. Phys.*, **52**(3), 276-277.

M. Lybanon, R. L. Crout, C. H. Johnson, and P. Pistek (1990), "Operational altimeter-derived oceanographic information: the NORDA

GEOSAT ocean applications program," *J. Atmos. Ocean. Technol.*, **7**(3), 357-376.

M. Lybanon (1994), "Satellite altimetry of the Gulf Stream," *Am. J. Phys.*, **62**(7), 661-665.

K. Messa (1991), "Fitting multivariate functions to data using genetic algorithms," *Proceedings of the Workshop on Neural Networks*, Auburn, Alabama, February 1991, 677-686.

K. Messa and M. Lybanon (1992), "Improved interpretation of satellite altimeter data using genetic algorithms," *Telematics and Informatics,* **9**(3/4), 349-356.

Z. Michalewicz (1992), *Genetic algorithms + data structures = evolution programs*, Springer-Verlag, Berlin.

K. Pearson (1901), "On lines and planes of closest fit to systems of points in space," *Phil. Mag.* **2**, Sixth Series, 559-572.

8.6 Program Listings

```
/* ga_int.h */


/*********************************************

 * REVISE_INTERVALS          *

 *  Revises the interval in which the value lies, *

 *  based on the results of the GA runs. *

 *********************************************/

#define MIN_CUSH 1.0

void revise_intervals()


{
int run, coeff;

float min, max, newLo, newHi, diff, cushion, width;

float proposed_newLo, proposed_newHi;
```

```
INCR(coeff,0,num_unknown - 1){

    /* find the max and min of the runs */

    min = final_values[coeff][0];

    max = final_values[coeff][0];

    INCR(run,1,timesRepeat-1){

        if(final_values[coeff][run] < min)

            min = final_values[coeff][run];

        if(final_values[coeff][run] > max)

            max = final_values[coeff][run];}
```

```
    /* set new interval to [min-cushion,max+cushion] */ cushion =
MIN_CUSH;diff = max - min;width = upper_bound[coeff] -
lower_bound[coeff];

    proposed_newLo = min - cushion*diff;

    proposed_newHi = max + cushion*diff;if (proposed_newLo <
lower_bound[coeff])   proposed_newLo = lower_bound[coeff];if
(proposed_newHi > upper_bound[coeff])      proposed_newHi =
upper_bound[coeff];if ((proposed_newHi-proposed_newLo<width*shrinkage)
&& (diff!=0)){   /* interval is too small */    cushion =
(width/diff)*(shrinkage/2) - 0.5; newLo = min - cushion*diff;newHi = max
+ cushion*diff;}else {    newLo = proposed_newLo;  newHi =
proposed_newHi;}
```

```
    if(newLo > lower_bound[coeff]){ /* move up */   lower_bound[coeff] =
newLo;    if (trend[coeff] == UNKNOWN) trend[coeff] = UP; if (trend[coeff]
== DOWN) trend[coeff] = OK;}

    if(newHi < upper_bound[coeff]){ /* move down */upper_bound[coeff] =
newHi;    if (trend[coeff] == UNKNOWN) trend [coeff] = DOWN;    if
(trend[coeff] == UP) trend[coeff] = OK;}}
```

```
    }
```

```
/*******************************************

 * ADD_TO_TABLES              *

 * Adds the best results from the   *

 * current run to the table of final values *
```

```
******************************************/


double objfunc();


void add_to_table(pop,run)
  struct population *pop;
  int run;


  {
  double *best_chrom;
  int coeff;


  best_chrom = pop->individual[pop->raw_best].chrom;
  INCR(coeff,0,num_unknown - 1)
    final_values[coeff][run] =
      convert(best_chrom[coeff],lower_bound[coeff],upper_bound[coeff]);
  final_values[num_unknown][run] = objfunc(best_chrom);
  }


/******************************************
 *  Create Seed Table        *
 *  Copy final values for seeding   *
 ******************************************/


void createSeedTable()
  {
  int i, j;


  for (i=0;i<=MAX_UNK;i++)
    for (j=0;j<timesRepeat;j++)
```

```
        seedTable[i][j] = final_values[i][j];

  }


/*****************************************************
 * COMPUTE_THRESHOLD            *
 * Determines the maximum difference to terminate the cycling. *
 ******************************************************/

float compute_threshold()

  {
  float diff,t;
  int coeff;

  diff = upper_bound[0] - lower_bound[0];
  INCR(coeff,1,num_unknown - 1){
   t = upper_bound[coeff] - lower_bound[coeff];
    if(t>diff)diff = t;}
  return diff;
  }


/*********************************************
 * SET BOUNDS              *
 * Sets bound and trends with initial values.  *
 *********************************************/
void setBounds()
  {
  int i;

  INCR(i,0,num_unknown - 1){
    upper_bound[i] = beginning_upper_bound[i];
```

```
    lower_bound[i] = beginning_lower_bound[i];
    trend[i] = UNKNOWN; }
  }
/*********************************************
 *  ADJUST INTERVAL            *
 *  Shift interval upward or downward   *
 *********************************************/


void adjustInterval(coeff, direction)
 int coeff, direction;
 {
 float diff, bound;


  diff = beginning_upper_bound[coeff] - beginning_lower_bound[coeff];
  if (direction == DOWN){bound = beginning_lower_bound[coeff];
    beginning_lower_bound[coeff] = (((bound - diff) < 0.0) &&
    ((coeff == 0) || (coeff == 1) || (coeff == 2) || (coeff == 5)))
     ? 0.0 : bound - diff; /* Coeff A, B, C, F cannot be lt 0.0 */
    beginning_upper_bound[coeff] = bound + 0.1*diff;}
  else{ /* UP */bound = beginning_upper_bound[coeff];
    beginning_upper_bound[coeff] = bound + diff;
    beginning_lower_bound[coeff] = bound - 0.1*diff;}
  }


/****************************************************
 *  EVALUATE INTERVAL          *
 *  Determine if the coefficient is outside the original bounds. *
 ****************************************************/
 int evaluateIntervals()


  {
```

```
int bA = FALSE; /* whether to begin the cycles again */
int coeff;


INCR(coeff,0,num_unknown - 1){

   if (trend[coeff] == DOWN) {

      if ((coeff != 0 && coeff != 1 && coeff != 2 && coeff != 5) ||
   beginning_lower_bound[coeff] != 0.0){   adjustInterval(coeff, DOWN);/*
A, B, C, F are never < 0 */

      bA = TRUE;}}

   if (trend[coeff] == UP) {

      adjustInterval(coeff, UP);

      bA = TRUE;}


   if (trend[coeff] == UNKNOWN) bA = N
```

```
/* ga_gen.h */


/******************************************
 * UPDATE_ORGANISM              *
 * Module which updates a newly formed *
 *   organism with the chromosome, its  *
 *   fitness and its parents.        *
 ******************************************/


void update_organism(org,p1,p2,chr)
  struct organism *org;
  int p1, p2; /* parents */
  chromosome chr;


  {
  int i;
  double fit;


  (*org).parent1 = p1;
  (*org).parent2 = p2;
  fit = objfunc(chr);
  (*org).raw_fitness = fit;
  (*org).pre_fitness = fit;
  INCR(i,0,num_unknown - 1) (*org).chrom[i] = chr[i];
  }


/******************************************
 * GENERATE_FIRST              *
 * Computes the initial population.    *
 ******************************************/
```

```
void generate_first(pop,seed_val)
  struct population *pop;
  int seed_val; /* TRUE: seed; FALSE:no seed */


  {
  int i, j, pos, fit = num_unknown;
  chromosome chr;
  float min;


  if (seed_val == 0){ /* no seeding */
     INCR(i,0,pop_size){
        INCR(j,0,num_unknown - 1)
           chr[j] = random_float();
        update_organism(&(pop->individual[i]),0,0,chr);}
     compute_stats(pop);}
  else{  /* seeding */
     min = seedTable[fit][0];
     pos = 0;
     INCR(j,1,timesRepeat - 1) /* Choose best for seeding */
        if ( seedTable[fit][j] < min) {
           min = seedTable[fit][j];
           pos = j;}
     INCR(j,0,num_unknown - 1)
     if (upper_bound[j] == lower_bound[j]) chr[j] = 0.0; else
      chr[j] =
      (seedTable[j][pos]-lower_bound[j])/(upper_bound[j]-lower_bound[j]);
     update_organism(&(pop->individual[0]),0,0,chr);
     INCR(i,1,pop_size){
        INCR(j,0,num_unknown - 1)
           chr[j] = random_float();
```

```
            update_organism(&(pop->individual[i]),0,0,chr);}
        compute_stats(pop);}
    }


#define PopSize pop_size


/*********************************************
 * GENERATE              *
 * Forms a new generation from the      *
 * existing population.          *
 *********************************************/


void generate(new_pop,pop,gen)
    struct population *new_pop,*pop;
    int gen;


    {
    int i, j, best, mate1,mate2, remaining;
    double *chr1, *chr2;
    int chs[MAX_POP_SIZE];
    int *choices;
    chromosome new_chr1,new_chr2;


    choices = preselect(pop,chs);


    /* Elitist strategy:
       Find the best organism from the original population
       and place it into the new population.*/
    best = pop->raw_best;
    update_organism(&(new_pop->individual[pop_size]),best,best,
            pop->individual[best].chrom);
```

```
/* fill the positions using selection, crossover and mutation */
remaining = PopSize;
INCR(j,0,pop_size/2 - 1){
    mate1 = select(pop,choices,remaining--);
    chr1 = pop->individual[mate1].chrom;
    mate2 = select(pop,choices,remaining--);
    chr2 = pop->individual[mate2].chrom;
    if (flip(prob_xover) == TRUE) {/* then crossover */
        crossover(xover_type,chr1,chr2,new_chr1,gen);
        crossover(xover_type,chr2,chr1,new_chr2,gen);


        /* now mutate, if necessary */
        INCR(i,0,num_unknown - 1)
mutate_all(new_chr1,i,gen,pr_u,pr_b,pr_n);

        update_organism(&(new_pop-
>individual[2*j]),mate1,mate2,new_chr1);


        INCR(i,0,num_unknown - 1)
mutate_all(new_chr2,i,gen,pr_u,pr_b,pr_n);

        update_organism(&(new_pop-
>individual[2*j+1]),mate2,mate1,new_chr2);}
    else {  INCR(i,0,num_unknown - 1){ /* copy chroms unchanged */
    new_chr1[i] = chr1[i];    new_chr2[i] = chr2[i];}


        /* mutate, if necessary */
        INCR(i,0,num_unknown - 1)
mutate_all(new_chr1,i,gen,pr_u,pr_b,pr_n);

        update_organism(&(new_pop-
>individual[2*j]),mate1,mate1,new_chr1);


        INCR(i,0,num_unknown - 1)
mutate_all(new_chr2,i,gen,pr_u,pr_b,pr_n);
```

```
  update_organism(&(new_pop-
>individual[2*j+1]),mate2,mate2,new_chr2);}}
 compute_stats(new_pop);

 }
```

```
/*********************************************
 * CROWDING FACTOR            *
 * Compute the organism most like the  *
 * new one.                   *
 *********************************************/
```

```
int crowdingFactor(pop, ch, cf, slots, last)
 struct population *pop;
 chromosome ch;
 int cf, *slots, last;


 {
 int m1 = -1, m2, i, j, pos1 = last, pos2;
 double sum1 = 100000.0, sum2 = 0.0, diff, val;
 short remaining[MAX_POP_SIZE];


 INCR(i,0,last) remaining[i] = 1; /* available */
 INCR(j,1,cf){
   while ((remaining[pos2 = rnd(0,last)] == 0));
   /* choose new m2 */
   m2 = slots[pos2];
   remaining[pos2] = 0;
   INCR(i,0,num_unknown - 1){
   val = pop->individual[m2].chrom[i];
     diff = ch[i] - val;
```

```
    sum2 += diff*diff;}
    if (sum2 < sum1){ m1 = m2; pos1 = pos2; sum1 = sum2;}}
  slots[pos1] = slots[last];
  return (m1);
  }


/*******************************************
 * CROWDING GENERATE            *
 * Forms a new generation from the    *
 * existing population using generation  *
 * gap model.                *
 *******************************************/


void crowdingGenerate(new_pop,pop,gen,gap,cf)
  struct population *new_pop,*pop;
  int gen,cf;
  float gap;


  {
  int i, j, best, mate1,mate2, remaining;
  int slots[MAX_POP_SIZE]; /* places where new offspring are placed */
  double *chr, *chr1, *chr2, pf, rf, f;
  int chs[MAX_POP_SIZE];
  int *choices, mort1, mort2, last=PopSize;
  chromosome new_chr1,new_chr2;


  choices = preselect(pop,chs);
  INCR(i,0,PopSize) slots[i] = i; /* available */


  /* Fill the new pop with all the members of the old pop */
```

```
INCR(j,0,PopSize){chr = pop->individual[j].chrom;pf = pop-
>individual[j].pre_fitness;new_pop->individual[j].pre_fitness = pf;new_pop-
>individual[j].raw_fitness = pf;INCR(i,0,num_unknown - 1)      new_pop-
>individual[j].chrom[i] = chr[i];}


/*  Elitist strategy:
    Find the best organism from the original population
    and place it into the new population.*/
best = pop->raw_best;
update_organism(&(new_pop->individual[pop_size]),best,best,
          pop->individual[best].chrom);


/* fill gap % of the positions using selection, crossover and mutation */
remaining = PopSize;


INCR(j,0,(int)PopSize*gap/2){
    mate1 = select(pop,choices,remaining--);
    chr1 = pop->individual[mate1].chrom;
    mate2 = select(pop,choices,remaining--);
    chr2 = pop->individual[mate2].chrom;
    if (flip(prob_xover) == TRUE) {/* then crossover */
      crossover(xover_type,chr1,chr2,new_chr1,gen);
      crossover(xover_type,chr2,chr1,new_chr2,gen);


        /* now, mutate, if necessary */
    INCR(i,0,num_unknown - 1)
mutate_all(new_chr1,i,gen,pr_u,pr_b,pr_n);  mort1 =
crowdingFactor(pop,new_chr1,cf,slots,last--);

      update_organism(&(new_pop-
>individual[mort1]),mate1,mate2,new_chr1);


      INCR(i,0,num_unknown - 1)
mutate_all(new_chr2,i,gen,pr_u,pr_b,pr_n);
```

```
    mort2 = crowdingFactor(pop,new_chr2,cf,slots,last--);

    update_organism(&(new_pop-
>individual[mort2]),mate2,mate1,new_chr2);}

    else {  INCR(i,0,num_unknown - 1){ /* copy chroms unchanged */
    new_chr1[i] = chr1[i];    new_chr2[i] = chr2[i];}


        /* mutate, if necessary  */

    INCR(i,0,num_unknown - 1)
mutate_all(new_chr1,i,gen,pr_u,pr_b,pr_n);  mort1 =
crowdingFactor(pop,new_chr1,cf,slots,last--);

    update_organism(&(new_pop-
>individual[mort1]),mate1,mate1,new_chr1);


    INCR(i,0,num_unknown - 1)
mutate_all(new_chr2,i,gen,pr_u,pr_b,pr_n);  mort2 =
crowdingFactor(pop,new_chr2,cf,slots,last--);

    update_organism(&(new_pop-
>individual[mort2]),mate2,mate2,new_chr2);}}

  compute_stats(new_pop);

  }


#undef PopSize


/* sshga.c */


/**********************************************
 * List of global variables and      *
 *    some type declarations        *
 **********************************************/


#define EPS .001      /* Accuracy of convergence of x values */
#define BEGIN_SEED 10   /* Cycle in which seeding starts */
#define MAX_POP_SIZE 200+1  /* Maximum population size */
```

```c
#define MAX_UNK 10      /* Maximum number of unknowns */
#define MAX_PASS 3      /* Maximum number of passes */
#define MAX_TIMES 10   /* Maximum number of runs per cycle */
#define INCR(v,s,h)  for((v) = (s); (v) <= (h); (v)++)
#define TRUE 1
#define FALSE 0
#define MUT_LIMIT 9     /* Number of dec places for mutation */
#define UP 1
#define DOWN -1
#define OK 9
#define UNKNOWN 0
#define NOTSURE -1


#include <time.h>
#include <stdio.h>
#include <math.h>


typedef double chromosome[MAX_UNK];
typedef struct organism {
    chromosome chrom;
    int parent1,parent2; /* Parents of organism */
    double pre_fitness,
    raw_fitness,fitness; /* Fitnesses */
    };
typedef struct population {
    struct organism individual[MAX_POP_SIZE];
        /* List of organisms in population*/
    double max_fit,min_fit,avg_fit,sum_fit; /* Various stats */
    double max_raw,min_raw,avg_raw,sum_raw;
    double max_pre,min_pre,avg_pre,sum_pre;
    int raw_best,raw_worst;
```

```
        };


double random_number[55]; /* List of random numbers */

int next_rand = 0;  /* Position of next random number */

int pop_size;        /* Size of population */

int num_gen;         /* Number of generations */

float scale_fac;     /* Scale factor */

float nscale;        /* Scale factor for first 10% of generations */

float beginScale;    /* This percent of generations before scale */

float genGap;        /* Generation gap; 1.0 turns off*/

int cF;              /* Crowding factor */

float prob_xover;    /* Probability of crossover*/

float pr_u, pr_b, pr_n; /* Probabilities of uniform, boundary and non-
        uniform mutations */

int xover_type;      /* Type of crossover */

int num_xover;       /* Number of xover points */

int num_unknown;     /* Number of unknowns in function */

int num_items;       /* Number of data items */

int timesRepeat;     /* Number of runs per cycle */

int maxCycles;       /* Maximum number of cycles */

float shrinkage;     /* Amount of shrinkage allowed each cycle */

int convert_to_max;  /* 1, if obj func is a minimum; 0, if max */

float upper_bound[MAX_UNK],lower_bound[MAX_UNK]; /* Current
endpoints */

float
beginning_upper_bound[MAX_UNK],beginning_lower_bound[MAX_UNK]
;        /* Original endpoints */

double final_values[MAX_UNK+1][MAX_TIMES]; /* Final values of each
run*/

double seedTable[MAX_UNK+1][MAX_TIMES]; /*Best of each run for
seeding*/

int freq_full;     /* Frequency of full report */

int freq_brief;    /* Frequency of brief report */
```

```
double dataValue[500][2] ; /* Data of x,y pairs */
struct population oldpop,*old,newpop,*new,*temp;
        /* Previous and current populations */
FILE *outfile,*infile,*datafile;  /* IO files */
short trend[MAX_UNK]; /* Records trends in the successive intervals */


#include "ga_util.h"     /* Random number and statistics routines */
#include "ga_int.h"      /* Interval adjusting routines */
#include "ga_obj.h"       /* Computes the fitness function */
#include "ga_io.h"       /* Various reports and input modules*/
#include "ga_sel.h"      /* Selection routines */
#include "ga_xover.h"    /* Crossover and mutation */
#include "ga_gen.h"      /* Generates the populations */


main(argc,argv)
int argc;
char *argv[];
{
int coeff, gen, run, cycle, seed_val, beginAgain = TRUE,times =
MAX_PASS;
static char coeff_list[] = "ABCDEFGHIJ";
float orig_scale;
float threshold, elapsedTime;
long start, finish;

time(&start);
warm_up_random(); /* Initialize */
infile = fopen(argv[1],"r");
initpara(infile);
fclose(infile);
datafile = fopen(argv[2],"r");
```

```
initdata(datafile);

fclose(datafile);

outfile = fopen(argv[3],"w+");

header(argv[1],argv[2],argv[3]);

skip_line(2);

orig_scale = scale_fac;


while(beginAgain && (times-- > 0)){

cycle = 0;

setBounds();

threshold = compute_threshold();

while((cycle++ < maxCycles) && (threshold > EPS)){

INCR(run,0,timesRepeat-1) {

 old = &oldpop; new = &newpop;

 scale_fac = ((cycle == 1) ? 1.0 : 1.0);

 seed_val = (cycle >= BEGIN_SEED); /* begin to seed in appropriate cycle
*/

 generate_first(old,seed_val);

 INCR(gen,0,num_gen){

   interm_reports(gen,freq_full,freq_brief,old);

   scale_fac = ((gen < 0.10*num_gen) ? nscale : 1.0);

        /* scaling done early ? */

   if (gen > beginScale*num_gen) scale_fac = orig_scale; /* scaling after
beginScale % of generations */

   if (genGap == 1.0)    generate(new,old,gen);

   else    crowdingGenerate(new,old,gen,genGap,cF); /* Form next
generation using previous one */

   temp = new; new = old; old = temp;} /* change roles of new and old */

 add_to_table(old,run);}  /* add best of the run to the table */

print_table();

createSeedTable();

revise_intervals();/* revise the intervals based on the acquired values*/
```

```
threshold = compute_threshold();

skip_line(3);}

beginAgain = evaluateIntervals();

complete_report(threshold,beginAgain);

if (beginAgain) {fprintf(outfile,"Revision of intervals is needed. ");
fprintf(outfile,"Recalculating and starting over.\n\n");}}

time(&finish);

elapsedTime = (float)(finish - start) / 60.0;

fprintf(outfile,"\n\n\nApproximate runtime: %f min,"elapsedTime);

skip_line(4);


}
```

/* ga_xover.h */

```
/**************************************
 *  MUTATE              *
 *  These modules change the   *
 *  chromosome at the gene position. *
 **************************************/



/**************************************
 *  Version 0           *
 *  Uniform             *
 *  Replace gene with a random value. *
 **************************************/


void mutate0(chrom,gene,i)
  chromosome chrom;
  int gene,i;


  {
  chrom[gene] = random_float();
  }



/**************************************
 *  Version 1           *
 *  Boundary .          *
 *  Sets gene to an endpoint.   *
 **************************************/
```

```
void mutate1(chrom,gene,i)
  chromosome chrom;
  int gene,i;


  {
  chrom[gene] = (double) flip(0.5);
  }
```

```
/****************************
 * DECAY          *
 * Returns a value    *
 * between 0 and 1      *
 * that decreases over time *
 **************************/
double decay(i)
  int i;


  {
  double val;


  val = ((double) (num_gen - i))/num_gen;
  return val*val;
  }
```

```
/*************************************
 * Version 2              *
 * Same as version 2, but affect  *
 * decreases over time.     *
 * Non-uniform + decay.     *
 ************************************/
```

```
void mutate2(chrom,gene,i)
  chromosome chrom;
  int gene,i;

  {
  double epsilon, v = chrom[gene];

  if (flip(0.5) == TRUE) {epsilon = random_float()*(1 - v)*decay(i);
     chrom[gene] += epsilon;}
  else {epsilon = random_float()*v*decay(i);
     chrom[gene] -= epsilon;}
  }



/*************************************
 *  Version 3            *
 *  Non-uniform.         *
 *  Adds or Subtracts a small number  *
 *  to a gene.           *
 *************************************/

void mutate3(chrom,gene,i)
  chromosome chrom;
  int gene,i;

  {
  double epsilon, v = chrom[gene];

  if (flip(0.5) == TRUE) {epsilon = random_float()*(1 v);
     chrom[gene] += epsilon;}
```

```
  else {epsilon = random_float()*v;
     chrom[gene] -= epsilon;}
  }


/********************************
 *  MUTATE          *
 *  Changes chromosome.    *
 ********************************/


void mutate(kind, chrom, gene, i)
 chromosome chrom;
 int gene, i, kind;


 {
 switch (kind) {
   case 0:{ mutate0(chrom,gene,i);
        break;}
    case 1:{ mutate1(chrom,gene,i);
        break;}case 2:{ mutate2(chrom,gene,i);
        break;}case 3:{ mutate3(chrom,gene,i);
        break;}
  /* end switch */}
  }


/**********************************************
 *  MUTATE_ALL             *
 *  Mutates the organism at the gene   *
 *  position using each of the mutation   *
 *  versions.            *
 **********************************************/
void mutate_all(chrom,gene,i,pu,pb,pn)
```

```
chromosome chrom;

int gene, i;

float pu,pb,pn;


{
if (flip(pn) == TRUE)
   mutate2(chrom,gene,i);
else if (flip(pu) == TRUE)
   mutate0(chrom,gene,i);
else if (flip(pb) == TRUE)
   mutate1(chrom,gene,i);

}




/************************************************
 *  CROSSOVER0              *
 *  Uniform crossover.         *
 *  Forms the weighted mean between the   *
 *  chromosomes.            *
 ***********************************************/
void crossover0(chr1, chr2, new)
   chromosome chr1, chr2, new;


{
int i;
float p;

INCR(i,0,num_unknown - 1){p = random_float();
   new[i] = p*chr1[i] + (1-p)*chr2[i];}

}
```

```
/***********************************************
 *  CROSSOVER1                *
 *  Arithmetic crossover.     *
 *  Forms the weighted mean between the  chromosomes.  *
 ***********************************************/
void crossover1(chr1, chr2, new)
  chromosome chr1, chr2, new;


  {
  int i;
  float p;


  p = random_float();
  INCR(i,0,num_unknown - 1)
    new[i] = p*chr1[i] + (1-p)*chr2[i];
  }


/***********************************************
 *  CROSSOVER2                *
 *  Combination of arithmetic and one-point.  *
 *  (Note that this requires 3 or more    *
 *  unknowns.)              *
 ***********************************************/


void crossover2(chr1,chr2, new)
  chromosome chr1, chr2, new;


  {
  int i, xpt;
  float p;
```

```
  p = random_float();
  xpt = rnd(0,num_unknown - 1);
  if (xpt == 0){new[0] = p*chr1[0] + (1 - p)*chr2[0];
     INCR(i,1,num_unknown - 1)
        new[i] = chr2[i];}
  else if (xpt == num_unknown - 1){
     INCR(i,0,xpt - 1)
        new[i] = chr1[i];
     new[xpt] = p*chr1[xpt] + (1 - p)*chr2[xpt];}
  else {
     INCR(i,0,xpt - 1)
        new[i] = chr1[i];
     new[xpt] = p*chr1[xpt] + (1 - p)*chr2[xpt];
     INCR(i,xpt + 1, num_unknown - 1)
        new[i] = chr2[i];}
  }


/**************************************************
 *  CROSSOVER3                *
 *  One-point.               *
 *  Like traditional. Left segment of one    *
 *  organism is interchanged with the other.  *
 **************************************************/


void crossover3(chr1,chr2, new)
   chromosome chr1, chr2, new;

   {
   int i, xpt;
   float p;
```

```
xpt = rnd(1,num_unknown - 1);
INCR(i,0,xpt - 1)
   new[i] = chr1[i];
INCR(i,xpt, num_unknown - 1)
   new[i] = chr2[i];
}




/***********************************************
 *  CROSSOVER4              *
 *  Combination of arithmetic and two-point.  *
 *  (Note that this requires 3 or more    *
 *  unknowns.)             *
 ***********************************************/


void crossover4(chr1,chr2, new)
  chromosome chr1, chr2, new;


  {
  int i, j, xpt1, xpt2, first, second;
  float p,q;


  p = random_float();
  q = random_float();
  xpt1 = rnd(0,num_unknown - 1);
  while(xpt1 != (xpt2 = rnd(0,num_unknown - 1)));
  first = ((xpt1 < xpt2) ? xpt1 : xpt2);
  second = ((xpt2 < xpt1) ? xpt1 : xpt2);
  INCR(i,0,num_unknown - 1) new[i] = chr1[i];
  new[first] = p*chr1[first] + (1 - p)*chr2[first];
  new[second] = q*chr2[second] + (1 - q)*chr1[second];
```

```
   if (first < second - 1)
     for(i=first,j=second - 1; i<j;i++)
        new[i] = chr2[i];
   }


/***********************************************
 * CROSSOVER               *
 * Chooses the correct method of crossover. *
 *   Note that mutation is included in the  *
 *   crossover operator.           *
 ***********************************************/


void crossover(kind,chr1,chr2,new,gen)
  chromosome chr1,chr2,new;
  int kind; /* kind of crossover */
  int gen;

  {
  switch (kind) {
   case 0: crossover0(chr1, chr2, new);
     break;
   case 1: crossover1(chr1, chr2, new);
     break;
   case 2: crossover2(chr1, chr2, new);
     break;
   case 3: crossover3(chr1, chr2, new);
     break;
   case 4: crossover4(chr1, chr2, new);
     break;
   }
  }
```

```
/* ga_util.h */
```

```
/*******************************************************
 *  RANDOM NUMBER FUNCTIONS          *
 *                     *
 *  These functions generate various kinds of random  *
 *  numbers. They are based on the system function   *
 *  rand(). See Goldberg for a listing in Pascal.    *
 *******************************************************/
```

```
#define PRECISION 9
long seed;
```

```
/*******************************************************
 *  RANDOM_FLOAT          *
 *  Generates a random float (double) between 0 and 1.  *
 *******************************************************/
```

```
double random_float()
{
int i,x;
int digit_rand[PRECISION];
double y = 0;
double multiplier = 0.1;

  x = rand();
  INCR(i,1,PRECISION)
    {
    digit_rand[PRECISION-i] = x%10;
    x = (x - (x%10))/10;
```

```
  }

  INCR(i,1,PRECISION)
    {
    y = multiplier*digit_rand[i-1] + y;
    multiplier = multiplier*0.1;
    }
  return y;
}




/******************************************************
 * ADVANCE_RANDOM          *
 * Generates an array of random values between 0 and 1 *
 ******************************************************/


void advance_random()

  {
  int j;
  double new_random;

  INCR(j,0,23)
    {
    new_random = random_number[j] - random_number[j+31];
    if (new_random < 0.0) new_random = new_random + 1.0;
    random_number[j] = new_random;
    }
```

```
INCR(j,24,54)
 {
 new_random = random_number[j] - random_number[j-24];
 if (new_random < 0.0) new_random = new_random + 1.0;
 random_number[j] = new_random;
 }
}
```

```
/*******************************************************
 * WARM_UP_RANDOM            *
 * Fills in the random number list.        *
 * Note: This needs to be modified to match each site. *
 *******************************************************/
```

```
void warm_up_random()
 {

 int i,j;

 srand(time(&seed));
 INCR(i,0,54)
  random_number[i] = random_float();
 INCR(j,1,3)
  advance_random();
 }
```

```
/*******************************************************
 * RANDOM               *
 * Returns a random number from the random number list *
 *******************************************************/
```

```
double random()

 {
 next_rand++;
 if (next_rand > 54)
  {
  next_rand = 0;
  advance_random();
  }
 return random_number[next_rand];
 }


/******************************************************
 * FLIP            *
 * Returns 0 or 1 with a given probability.      *
 ******************************************************/


int flip(probability)
float probability;

 {
 if (probability == 1.0)
  return 1;
 else if (probability == 0.0)
  return 0;
 else
  return ceil(probability - random());
 }


/******************************************************
 * RND             *
```

```
*  Returns an integer between the given integer values. *
*********************************************************/


int rnd(low,high)
int low,high;


  {
  int flr;
  float a;


  if (low > high)
   return low;
  else
   {
   a = random();
   flr = floor(a*(high - low + 1) + low);
   if (flr > high)
     return high;
   else
     return flr;
   }
  }




#define PopSize pop_size


/***********************************************
 *  COMPUTE_RAW_STATS              *
 *                  *
 *  This function computes statistics for a    *
```

```
 *   population's raw fitnesses.        *
 ***********************************************/
#define org (*pop).individual
void compute_raw_stats(pop)
  struct population *pop;


  {
  int i, best=0, worst=0;
  double raw_max, raw_min, raw_avg, raw_sum;


  raw_max = org[0].raw_fitness;
  raw_min = org[0].raw_fitness;
  raw_sum = 0.0;
  INCR(i,0,PopSize)
   {
   if (org[i].raw_fitness > raw_max)
     {
     best = i;
     raw_max = org[i].raw_fitness;
     }
   if (org[i].raw_fitness < raw_min)
     {
     worst = i;
     raw_min = org[i].raw_fitness;
     }
   raw_sum += org[i].raw_fitness;
   }
  (*pop).max_raw = raw_max;
  (*pop).min_raw = raw_min;
  (*pop).avg_raw = raw_sum/(PopSize + 1);
  (*pop).sum_raw = raw_sum;
```

```
(*pop).raw_best = best;
(*pop).raw_worst = worst;
}
```

```
/**************************************************
 *    COMPUTE_PRE_STATS            *
 *                      *
 * This function computes statistics for a   *
 *    population's pre-fitnesses.        *
 **************************************************/
```

```
void compute_pre_stats(pop)
 struct population *pop;

 {
 int i,best=0,worst=0;
 double pre_max, pre_min, pre_avg, pre_sum;

 pre_max = org[0].pre_fitness;
 pre_min = org[0].pre_fitness;
 pre_sum = 0.0;
 INCR(i,0,PopSize)
  {
  if (org[i].pre_fitness > pre_max)
   {
   best = i;
   pre_max = org[i].pre_fitness;
   }
  if (org[i].pre_fitness < pre_min)
```

```
      {
      worst = i;
      pre_min = org[i].pre_fitness;
      }
    pre_sum += org[i].pre_fitness;
    }
  (*pop).max_pre = pre_max;
  (*pop).min_pre = pre_min;
  (*pop).avg_pre = pre_sum/(PopSize+1);
  }


/**********************************************
 *  REVISE_OBJ                *
 *  If the optimum objective value is a min *
 *  then this routine converts it to a max. *
 **********************************************/


void revise_obj(p)
  struct population *p;


  {
  int i;
  double max;
  max = p->max_raw;
  INCR(i,0,PopSize)
  p->individual[i].raw_fitness = max - p->individual[i].raw_fitness;
  }


/**********************************************
 *  SCALE                *
 *  Scales the fitnesses of the organisms.  *
```

```
 *  The scheme scales the max fitness to  *
 *  scale_fac times the average,        *
 *  when possible. If the minimum is     *
 *  negative, then a shift upward is done.  *
 ***********************************************/
void scale(p)
  struct population *p;


 {
 int i;
 double max,min,avg, scl,min_scl;
 double delta,m = 1.0,b = 0.0; /* scaling function is Y = mX + b */
 max = (*p).max_raw;
 min = (*p).min_raw;
 avg = (*p).avg_raw;


 if ((scale_fac != 1.0) && (max != avg)){
      /* default values, m = 1, b = 0 */ delta = max - avg;
   m = (scale_fac*avg - avg)/delta;
   b = avg/delta * (max - scale_fac*avg);
   }


 /* Use m and b to compute scaled values */


 INCR(i,0,PopSize)
  {
  scl = m * (*p).individual[i].raw_fitness + b;
  if (scl < avg)
   (*p).individual[i].fitness = (*p).individual[i].raw_fitness;
  else
   (*p).individual[i].fitness = scl;
```

```
 }
 }


/**************************************************
 *  COMPUTE_SCALE_STATS          *
 *                 *
 *  This function computes statistics for a   *
 *  population's scaled fitnesses.      *
 **************************************************/


void compute_scale_stats(pop)
  struct population *pop;

  {
  int i;
  double max, min, avg, sum;

  max = org[0].fitness;
  min = org[0].fitness;
  sum= 0.0;
  INCR(i,0,PopSize)
   {
   if (org[i].fitness > max) max = org[i].fitness;
   if (org[i].fitness < min) min = org[i].fitness;
   sum += org[i].fitness;
   }
  (*pop).max_fit = max;
  (*pop).min_fit = min;
  (*pop).avg_fit = sum/(PopSize + 1);
  (*pop).sum_fit = sum;
```

```
}



#undef org


/**************************************************
 *  COMPUTE_STATS                    *
 *  Computes the statistics needed in a population.        *
 **************************************************/



void compute_stats(p)
  struct population *p;


  {
  compute_pre_stats(p);
  compute_raw_stats(p);
  if (convert_to_max == 1)
   {
   revise_obj(p);
   compute_raw_stats(p);
   }
  scale(p);
  compute_scale_stats(p);
  }



/**************************************************
 * CONVERT                    *
 * Converts a chromosome into the appropriate *
```

```
 *  vector of reals.              *
 ************************************************/


double convert(seg,low,hi)
  double low,hi,seg;


  {
  double value;


  value = seg*(hi - low);
  value += low;
  return value;
  }




#define squared(x) (x)*(x)


/************************************************
 *  COMPUTE_AVG_EACH              *
 *  Computes the average value of each   *
 *  sub segment.             *
 ************************************************/


double compute_avg_each(pop,unk)
  struct population *pop;
  int unk; /* number of the unknown */


  {
  int i;
  double sum = 0.0;
  double val;
```

```
double *chr;

INCR(i,0,PopSize)
 {
 chr = pop->individual[i].chrom;
 val = convert(chr[unk],lower_bound[unk],upper_bound[unk]);
 sum += val;
 }
 return sum/(PopSize + 1);
 }


#undef squared
#undef PopSize



/* ga_sel.h */

#define PopSize pop_size

/****************************************
 * PRESELECT              *
 * Used in expected value selection *
 * to determine the frequency of the*
 * selection of each organism.   *
 ****************************************/


int *preselect(pop,choices)
 struct population *pop;
 int choices[];

 {
```

```
int j,assign,k = 0;
float expected, fraction[MAX_POP_SIZE];

if (pop->avg_fit == 0.0) /* then all organisms are the same */
INCR(j,0,PopSize)
   choices[j] = j;
else
   {
INCR(j,0,PopSize)
    {
   expected = (pop->individual[j].fitness / pop->avg_fit);
   assign = (int) expected;
   fraction[j] = expected - assign;
   while (assign > 0)
      {
      choices[k] = j;
      k++; assign--;
      }
    }
   j = 0;  /* now fill in the remaining slots using fractional parts */
   while (k <= PopSize)
      {
      if (j > PopSize) j = 0;
      if (fraction[j] > 0.0)
      {
      if (flip(fraction[j]) == TRUE)
         {
         choices[k] = j;
         k++;
         fraction[j] -= 1.0;
         }
```

```
        }
        j++;
        }
    }
    return &choices[0];
    }
```

```
/****************************************
 *  SELECT                      *
 *  Chooses  the index of an organism  *
 *  based on its fitness and a random  *
 *  number.                     *
 ****************************************/
```

```
int select(pop,chs,rem)
    struct population *pop;
    int rem, chs[]; /* last remaining of the choices */
```

```
/****************************************
 *   Version 1                  *
 *   Stochastic sampling without    *
 *   replacement. "Expected values"  *
 ****************************************/
```

```
    {
    int pick,chosen;

    pick = rnd(0,rem);
    chosen = chs[pick];
    chs[pick] = chs[rem];
```

```
    return chosen;

    }
```

```
/****************************************
 *   Version 2                    *
 *   Stochastic sampling with        *
 *   replacement.  "Roulette"        *
 ****************************************/
```

```
/*

    {
    float partsum = 0.0;
    float rand;
    int i = 0;

    rand = random() * (*pop).sum_fit;
    do   {
     partsum += (*pop).individual[i].fitness;
      i++;
        }
    while ((partsum < rand) && (i <= pop_size));
    return (i - 1);
    }

*/
```

```
#undef PopSize
/*  ga_obj.h  */
```

```
#define SSH(u) a*tanh(b*(300-(u)-d-e)) - f*tanh(c*(300-(u)-e)) + g


double pre_obj(a,b,c,d,e,f,g)
   double a,b,c,d,e,f,g;


   {
   register double sumSq,diff;
   register int i;


   for(i=0,sumSq=0;i<num_items;i++){diff = SSH(dataValue[i][0]) -
dataValue[i][1];sumSq += (diff*diff);}
   return sumSq;

   }




/**************************************************
 *  OBJFUNC                          *
 * Calculates the objective or fitness function*
 *  for the GA.                      *
 ***************************************************/


double objfunc(chr)
   chromosome chr;


   {
   double a,b,c,d,e,f,g,obj;


   a = convert(chr[0],lower_bound[0],upper_bound[0]);
   b = convert(chr[1],lower_bound[1],upper_bound[1]);
   c = convert(chr[2],lower_bound[2],upper_bound[2]);
```

```
d = convert(chr[3],lower_bound[3],upper_bound[3]);

e = convert(chr[4],lower_bound[4],upper_bound[4]);

f = convert(chr[5],lower_bound[5],upper_bound[5]);

g = convert(chr[6],lower_bound[6],upper_bound[6]);


obj = pre_obj(a,b,c,d,e,f,g);

return obj;

   }


#undef SSH
```

```
/*  ga_io.h  */


#define READLINE(in,var,typ) \fscanf(in,"%typ ,"&var); \fgets(temp,70,in)


/************************************************
 *  INITDATA                        *
 *  Reads data                      *
 ***********************************************/


void initdata(datafile)
    FILE *datafile;


    {
    int i=0;


    while(!feof(datafile)){
      fscanf(datafile,"%lf %lf ,"&dataValue[i][0],&dataValue[i][1]);
        i++; }
    num_items = i;
    }


/************************************************
 * INITPARA                         *
 *  Reads parameter information (population size,*
 *  chromosome length, etc.) from the data      *
 *    file.                         *
 ***********************************************/


void initpara(infile)
    FILE *infile;
```

```
{
int i;
float a,b;
char temp[70];

READLINE(infile,pop_size,d);
READLINE(infile,num_gen,d);
fgets(temp,70,infile);    /* consume (empty) line */
READLINE(infile,scale_fac,f);
READLINE(infile,nscale,f);
READLINE(infile,beginScale,f);
READLINE(infile,genGap,f);
READLINE(infile,cF,d);
fgets(temp,70,infile);    /* consume (empty) line */
READLINE(infile,prob_xover,f);
READLINE(infile,xover_type,d);
fgets(temp,70,infile);    /* consume empty line */
READLINE(infile,pr_u,f);
READLINE(infile,pr_b,f);
READLINE(infile,pr_n,f);
fgets(temp,70,infile);    /* empty line */
READLINE(infile,timesRepeat,d);
READLINE(infile,maxCycles,d);
READLINE(infile,shrinkage,f);
fgets(temp,70,infile);    /* consume (empty) line */
READLINE(infile,num_unknown,d);
READLINE(infile,convert_to_max,d);
fgets(temp,70,infile);    /* consume */
INCR(i,0,num_unknown 1){
  READLINE(infile,a,f);
```

```
      beginning_lower_bound[i] = a;
      READLINE(infile,b,f);
      beginning_upper_bound[i] = b;}
    fgets(temp,70,infile);    /* consume */
    READLINE(infile,freq_full,d);
    READLINE(infile,freq_brief,d);
    }


/***********************************
 *  SKIP_LINE                    *
 *  Skips a line in output.      *
 ***********************************/


void skip_line(k)
    int k;


    {
    int i;


    INCR(i,1,k) fprintf(outfile,"\n");
    }


/*************************************************
 *  HEADER  Prints the header for the main report *
 *************************************************/


void header(para,data,out)char *para,*data,*out;
    {
    int i;


    fprintf(outfile,"          GENETIC ALGORITHM          ");
```

```
    skip_line(2);
    fprintf(outfile,"    population size:          %d \n,"pop_size);
    fprintf(outfile,"    number of generations:    %d \n,"num_gen);
    fprintf(outfile,"    probability of crossover: %4.2f \n,"prob_xover);
    fprintf(outfile,"    crossover type:           %d \n,"xover_type);
    fprintf(outfile,"    prob of uniform mutation: %5.3f \n,"pr_u);
    fprintf(outfile,"    prob of boundary mut.  :  %5.3f \n,"pr_b);
    fprintf(outfile,"    prob of non-uniform mut.: %5.3f \n,"pr_n);
    fprintf(outfile,"    scale factor:             %4.2f \n,"scale_fac);
    fprintf(outfile,"    early scale factor:       %4.2f \n,"nscale);
    fprintf(outfile,"    begin scaling after:      %4.2f \n,"beginScale);
    fprintf(outfile,"    generation gap:           %4.2f \n,"genGap);
    fprintf(outfile,"    crowding factor:          %d \n,"cF);
    fprintf(outfile,"    shrinkage:                %4.2f \n,"shrinkage);
    skip_line(1);
    fprintf(outfile,"    parameter file:           ");
    fprintf(outfile,"%s,"para);
    skip_line(1);
    fprintf(outfile,"    data file:                ");
    fprintf(outfile,"%s,"data);
    skip_line(1);
    fprintf(outfile,"    output file (this file):  ");
    fprintf(outfile,"%s,"out);
    skip_line(2);
    }


/***********************************************
 *  PRINT_STATS  Prints the statistics for a     *
 *  population.                          *
 ***********************************************/
```

```
void print_stats(p)
      struct population *p;


   {
   fprintf(outfile,"        pre-fitness   scaled fitness\n");
   fprintf(outfile,"maximum      %10.6f
%10.6f\n,"(*p).max_pre,(*p).max_fit);
   fprintf(outfile,"average      %10.6f
%10.6f\n,"(*p).avg_pre,(*p).avg_fit);
   fprintf(outfile,"minimum      %10.6f
%10.6f\n,"(*p).min_pre,(*p).min_fit);

   }


/***********************************************
 *  PRINT_INDIVIDUALS  Prints the organisms'     *
 *  chromosomes and fitnesses.            *
 ***********************************************/


void print_individuals(p)
      struct population *p;


   {
   double seg;
   int i,j;
   struct organism *org;

   fprintf(outfile,
      "-------organism------->      pre-fitness      scaled fitness \n");
   INCR(i,0,pop_size)
   {
   org = &((*p).individual[i]);
   INCR(j,0,num_unknown -1)
```

```
      {
      skip_line(1);
      seg = (*org).chrom[j];
      fprintf(outfile,"%10.6f  ,"seg);
      }
   fprintf(outfile," %10.6f,"(*org).pre_fitness);
   fprintf(outfile," %10.6f,"(*org).fitness);
   fprintf(outfile," %3d,"(*org).parent1);
   fprintf(outfile," %3d,"(*org).parent2);
   skip_line(2);
   }
   }


/************************************************
 * BRIEF_REPORT                      *
 * Prints the statistics.            *
 ***********************************************/

void brief_report(pop,gen)
   struct population *pop;


   {
   static char coeff_list[] = "ABCDEFGHIJ";
   double *best_chrom;
   float outval=0;
   int i,j,k,all;

   fprintf(outfile,"Generation: %d\n,"gen);skip_line(2);
   print_stats(pop);
   skip_line(2);
```

```
/* Output value of best and average organism */
fprintf(outfile,"The offline results are ...\n");
fprintf(outfile,"                best        average\n");
best_chrom = pop->individual[pop->raw_best].chrom;
INCR(i,0,num_unknown - 1)
 {
 outval = convert(best_chrom[i],lower_bound[i],upper_bound[i]);
 fprintf(outfile,"     %1c = %10.6f    %10.6f\n,"coeff_list[i],
     outval,compute_avg_each(pop,i));
 }
skip_line(1);
fprintf(outfile,"Pre-fitness:%10.6f    %10.6f\n,"
pop->individual[pop->raw_best].pre_fitness,pop->avg_pre);
skip_line(3);
fprintf(outfile,"_____\n\n");
 }


/***********************************************
 *  FULL_REPORT                       *
 *  Prints each organism and their fitnesses *
 *  along with the statistics.             *
 ***********************************************/


void full_report(pop,gen)
   struct population *pop;
   int gen;


   {
   static char coeff_list[10] =
{'A','B','C','D','E','F','G','H','I','J'};
   float outval=0;
```

```
    double seg,*best_chrom;
     int i,j;

    printf("Generation = %d :,"gen);skip_line(1);
    print_individuals(pop);
    skip_line(2);
    print_stats(pop);
    skip_line(2);

    /* Output best organism */
    fprintf(outfile,"\nThe best organism so far is ...\n");
    INCR(j,0,num_unknown - 1) {
       seg = pop->individual[pop->raw_best].chrom[j];
      fprintf(outfile,"%10.6f ,"seg);}
    skip_line(2);

    /* Output value of best organism */
    fprintf(outfile,"It has values ...\n");
    best_chrom = pop->individual[pop->raw_best].chrom;
    INCR(i,0,num_unknown - 1) {
       outval = convert(best_chrom[i],lower_bound[i],upper_bound[i]);
       fprintf(outfile," ..... %1c = %10.6f\n,"coeff_list[i],outval);}
    fprintf(outfile,"It has pre-fitness: %10.6f\n,"
              pop->individual[pop->raw_best].pre_fitness);
    skip_line(2);
    }


/***********************************************
*   INTERM_REPORTS                      *
*   Prints brief and full reports with the   *
*   desired frequency.                  *
```

```
**********************************************/


void interm_reports(gen,full,brief,pop)
   int gen,full,brief;
   struct population *pop;


   {
   if (full != 0) { /* print full report at correct generations */
     if (gen % full == 0)   /* output results */
       full_report(pop,gen);
     else
       if (brief != 0)
         if (gen % brief == 0)
           brief_report(pop,gen);}
   else   /* print only brief reports */
     if (brief != 0)
       if(gen % brief == 0)
         brief_report(pop,gen);
   }


/***********************************************
 *   FINAL_REPORT                       *
 *   Lists the best organism and the alleles  *
 *   that were present in a certain percentage*
 *   of the final population (goal_thresh).   *
 **********************************************/


void final_report(p)
   struct population *p;


   {
```

```
    int i,best,j,k;
    int count,count1,in,stop;
    double seg;
    char best_char;
    double solution_value,best_val;
    static char coeff_list[10] =
{'A','B','C','D','E','F','G','H','I','J'};



/*

    Output best online value (threshhold method)
    solution = compute_online_thresh(p);
    fprintf(outfile,"The online solution (using threshholds) is......\n");
    INCR(j,0,num_unknown -1)
    {
    stop = sub_length;
    INCR(k,0,sub_length - 1)
        if ((*solution)[k + j*sub_length] == -1)
            {
            stop = k;
            break;
            }
    solution_value = convert(solution,j*sub_length,stop,
            lower_bound[j],upper_bound[j]);
    fprintf(outfile," %1c = %10.6f\n,"
            coeff_list[j],solution_value);
    }
    skip_line(2);
*/


/*
```

Output the online solution using averages
fprintf(outfile,"The online solution (using averages) is ......\n");
INCR(j,0,num_unknown - 1)

{

solution_value = compute_avg_each(p,j);
fprintf(outfile," %1c = %10.6f\n,"
        coeff_list[j],solution_value);

}

fprintf(outfile,

```
"_____\n\
n");
```

        */

/* Output best organism */
fprintf(outfile,"\nThe best organism is ...\n");
INCR(j,0,num_unknown -1)

    {

    seg = p->individual[p->raw_best].chrom[j];
    fprintf(outfile,"%10.6f ,"seg);

    }
skip_line(2);

/* Output value of best organism */
fprintf(outfile,"It has values ...\n");
INCR(i,0,num_unknown -1)
fprintf(outfile," ..... %1c = %14.10f\n,"coeff_list[i],
    convert(p->individual[p->raw_best].chrom[i],
    lower_bound[i],upper_bound[i]));
best_val = p->individual[p->raw_best].pre_fitness;

```
    fprintf(outfile,"It has pre-fitness:  %12.9f\n,"best_val);
    best_val = sqrt(best_val);
    fprintf(outfile,"        and error:  %12.9f\n,"best_val);
    skip_line(2);


fprintf(outfile,

"_____\n\
n");
      }



/*************************************************
 *  PRINT_TABLE                       *
 *  Prints the table of final values for each *
 *  run for each cycle.                *
 ************************************************/
void print_table()

  {
  static char coeff_list[]="ABCDEFGHIJ";
  int run,coeff;
  float error;

    fprintf(outfile,"            Run          ");
    fprintf(outfile,"                   Endpoints\n");
    fprintf(outfile," Coefficient   ");
    fprintf(outfile,"   1 ");
    INCR(run,2,timesRepeat) fprintf(outfile,"        %d,"run);
    fprintf(outfile,"     left        right");
    skip_line(1);
```

```
     INCR(coeff,0,num_unknown-1){

     fprintf(outfile," %c      ,"coeff_list[coeff]);

     INCR(run,0,timesRepeat-1)

          fprintf(outfile," %10.6f,"final_values[coeff][run]);fprintf(outfile,"
%9.4f        %9.4f,"        lower_bound[coeff],upper_bound[coeff]);

     skip_line(1);}

     fprintf(outfile,"\n  Error    ");

     INCR(run,0,timesRepeat-1){error =
pre_obj(final_values[0][run],final_values[1][run],

               final_values[2][run],final_values[3][run],
final_values[4][run],final_values[5][run],

               final_values[6][run]);

     fprintf(outfile," %10.6f,"sqrt(error));}


     skip_line(2);

     }



/*********************************************
 *  COMPLETE_REPORT                    *
 *  Prints the final report         *
 *********************************************/


void complete_report(x,y)
    int x,y;


    {
    static char coeff_list[]="ABCDEFGHIJ";
    int coeff,run,best;
    float error,lowestError;
    double bestRun[MAX_UNK];
```

```
if (y == 0){  /* Results good */fprintf(outfile,"          Summary of
Results\n");

    skip_line(2);

    fprintf(outfile," Coefficient   Best Value\n");

    INCR(coeff,0,num_unknown - 1)

        bestRun[coeff] = final_values[coeff][0];

    lowestError = pre_obj(bestRun[0],bestRun[1],bestRun[2],bestRun[3],
        bestRun[4],bestRun[5],bestRun[6]);

    best = 0;

    INCR(run,1,timesRepeat-1){

        INCR(coeff,0,num_unknown - 1)

            bestRun[coeff] = final_values[coeff][run];

        error = pre_obj(bestRun[0],bestRun[1],bestRun[2],bestRun[3],
        bestRun[4],bestRun[5],bestRun[6]);

        if (error < lowestError) {    best = run;        lowestError = error;}}


    INCR(coeff,0,num_unknown - 1)  fprintf(outfile,"   %c
%10.6f\n,"        coeff_list[coeff],final_values[coeff][best]);

    skip_line(1);

    fprintf(outfile," Error      %10.6f\n,"sqrt(lowestError));}

    else  if(y == 1) /* Poor results, so adjust original intervals
*/{fprintf(outfile,"Results indicate that the original intervals");

        fprintf(outfile," are not accurate. \n These have been revised and ");

        fprintf(outfile,"the cycle is being re-run.\n\n\n");}

    else {fprintf(outfile,"Convergence not achieved. ");

        fprintf(outfile,"Please check all parameters.\n\n");}

}
```

Chapter 9 A Hybrid Genetic Algorithm, Simulated Annealing and Tabu Search Heuristic for Vehicle Routing Problems with Time Windows

Sam R. Thangiah
Artificial Intelligence and Robotics Laboratory
Computer Science Department
Slippery Rock University
Slippery Rock, PA 16057
U.S.A.

9.1 Introduction

The vehicle routing problem with time windows (VRPTW) is an extension of the vehicle routing problem (VRP) with earliest, latest, and service times for customers. The VRPTW routes a set of vehicles to service customers having earliest, latest service times. The objective of the problem is to minimize the number of vehicles and the distance travelled to service the customers. The constraints of the problem are to service all the customers after the earliest release time and before the latest service time of each customer without exceeding the route time of the vehicle and overloading the vehicle. The route time of the vehicle is the sum total of the waiting time, the service time and distance travelled by the vehicle. A vehicle that reaches a customer before the earliest release time incurs waiting time. If a vehicle services a customer after the latest delivery time, the vehicle is considered to be tardy. The service time is the time taken by a vehicle to service a customer. A vehicle is said to be overloaded if the sum total of the customer demands exceed the total capacity of the vehicle. The quality of the solution is measured in terms of the minimization of the number of vehicles followed by the minimization of the total distance travelled, respectively, in that order. That is, a solution for the VRPTW with a lower total number of vehicles and greater total distance travelled is preferred over a solution that requires, greater number of vehicles and smaller total distance travelled.

Applications of routing and scheduling models arise in a wide range of practical decision-making problems. Efficient routing and scheduling of vehicles can save the public and private sectors millions of dollars per year. The VRPTW arises in retail distribution, school bus routing, mail and newspaper delivery, municipal waste collection, fuel oil delivery, dial-a-ride service and airline and railway fleet routing and scheduling. Surveys on classifications and applications of VRP can be found in [Bodin et al. 1983], [Laporte, 1992] and [Fisher, 1993].

Savelsbergh [1985] has shown that finding a feasible solution to the traveling salesman problem with time windows (TSPTW) is a NP-complete problem. Therefore, VRPTW is more complex as it involves servicing customers with time windows using multiple vehicles that vary with respect to the problem. VRPTW has been the focus of intensive research and special purpose surveys can be found in Desrosiers et al. [1993], Desrochers, Desrosiers and Solomon [1992], Desrochers et al. [1988], Golden and Assad [1988], Solomon and Desrosiers [1988], Solomon [1987], and Golden and Assad [1986]. Though optimal solutions to VRPTW can be obtained using exact methods, the computational time required to solve a VRPTW to optimality is prohibitive [Desrochers, Desrosiers and Solomon, 1992; Fisher, Jornsten and Madsen, 1992]. Heuristic methods often produce optimal or near optimal solutions in a reasonable amount of computer time. Thus, there is still a considerable interest in the design of new heuristics for solving large-sized practical VRPTW.

Heuristic approaches for the VRPTW use route construction, route improvement or methods that integrate both route construction and route improvement. Solomon [1987] designed and analyzed a number of route construction heuristics; namely, the extended savings, time-oriented nearest neighbor, time-oriented insertion, time-oriented sweep heuristic and giant tour heuristics for solving the VRPTW. In his study, the time-oriented nearest-neighbor insertion heuristic was found to be very successful. Other route construction procedures that have been employed to solve VRPTW are the parallel insertion method [Potvin and Rosseau, 1993], the greedy randomized adaptive procedure [Kontoravdis and Bard, 1992] and the Generalized Assignment heuristic [Koskosidis, Powell and Solomon, 1992]. A number of route improvement heuristics have been implemented for the VRPTW, namely, branch exchange procedures [Russell, 93; Solomon, Baker and Schaffer, 88; Baker and Schaffer, 1986] and cyclic-transfer algorithms [Thompson and Psaraftis, 1989]. Heuristic search strategies based on Genetic Algorithms, Simulated Annealing and Tabu Search have also been explored for solving the VRPTW.

Potvin and Bengio [1993] implemented a genetic based algorithm to solve the VRPTW, in which a population of solutions evolves from one generation to another by mating with parent solutions resulting in offspring that exhibit characteristics acquired from the parents. The implemented system, GENEROUS, exploits the general methodology used in genetic algorithms by substituting specific vehicle routing operators in place of the standard genetic operators.

A search strategy based on the Tabu Search [Potvin, Kervahut, Garcia and Rousseau, 1992] was implemented that uses a two-phase branch exchange procedure to improve VRPTW solutions. In the first phase of the search method, customers are moved out of routes to reduce the total number of vehicles for the solntion. The second phase does inter- and intra-customer exchanges to reduce the total distance travelled by the vehicles. Feasibility of the solution is maintained at each iteration of the solution improvement procedures.

A hybrid search strategy based on Simulated Annealing and Tabu Search [Chiang and Russell,1993]using neighborhood structures based upon the 7-interchange [Thangiah, 1993; Osman, 1993] and the k-node interchange method of Christofides and Beasley [1984] for generating the neighborhoods was used to improve initial VRPTW obtained using a parallel insertion method. The neighborhood search strategies interchange single customers between routes and accepted improvements if the solution was feasible.

A cluster-first route-second method using Genetic Algorithms and a local post-optimization process was introduced by Thangiah [Thangiah, Nygard and Juell, 1991; Thangiah 1993] to solve the VRPTW. The heuristic for clustering customers, termed Genetic Sectoring, required the number of vehicles for the problem to be known in advance and used a liner function to map the genetic structures to the customer clusters. The Genetic Sectoring method was used to find the set of clusters that reduced the travel time of the vehicles. The Genetic Sectoring method was not capable of reducing the number of vehicles from the required number of vehicles known in advance during the sectoring process. An extension of the Genetic Sectoring Heuristic using an exponential function for mapping the genetic structures to the customer clusters was shown to obtain better solutions for time deadline problems [Thangiah, Osman, Vinayagamoorthy and Sun, 1993]. This method has the capability to reduce the number of vehicles available during the clustering process. The λ-interchange method was used to significantly improve initial VRP solutions. The λ-interchange method allowed for interchange of single customers between routes in combination with non-monotonic Simulated Annealing and Tabu Search.

In this chapter we develop a hybrid search strategy that combines Genetic Algorithms, Simulated Annealing and Tabu Search methods for solving the VRPTW. The hybrid search strategy uses Genetic Algorithms as a global search method to find an initial solution to the VRPTW. The initial solution is improved using a customer interchange method guided by Tabu Search combined with non-monotonic Simulated Annealing with an evaluation function that allows for acceptance of infeasible solutions with a penalty.

The hybrid strategy was structured in this manner to exploit the strengths of the meta-heuristic search strategies. The Genetic Algorithm is a global search strategy that is capable of finding good solutions if a good representation can be found between the genetic structure and the problem. The Genetic Algorithm is known to obtain good global solutions or solutions that lie close to regions with good solutions. Tabu Search, in combination with a non-monotonic Simulated Annealing produces efficient local neighborhood search strategies. The proposed hybrid heuristics differs from the Chiang and Russel approach [1993] mainly in the method used for obtaining the initial solution, the number of customers used in the λ-interchange method, the cost function that allows for acceptance of infeasible solutions and the structure of the Simulated Annealing and Tabu search methods.

The Genetic Sectoring heuristic used for the VRPTW was adapted from the sectoring method used for clustering customers for the VRPTD [Thangiah, Osman, Vinayagamoorthy and Sun, 1993]. An exponential function was used to map the genetic structures to the customer clusters as it started from an initially specified number of clusters and obtained customer clusters that were equal or lower than the initially required number of vehicles.

The initial number of vehicles required for the Genetic Sectoring heuristic was obtained using a Push-Forward Insertion heuristic (PFIH) similar to the insertion heuristic of Solomon [1987]. The initial solution obtained by the Genetic Sectoring method was improved using a customer interchange method that allowed exchange of up to two or more customers. The customer interchange method used an evaluation function that allowed infeasible solutions to be accepted with a penalty. The neighborhood of the customer interchange method was guided by the Tabu Search method combined with non-monotonic Simulated Annealing. The combination of the Tabu search and non-monotonic Simulated Annealing with a cost function that allows infeasible solutions with penalty allows for search by strategic oscillation. Traditional algorithms improve solutions using search strategies that allow acceptance of only feasible moves. Strategic oscillation allows a search strategy to go beyond the feasible region into infeasible regions within a defined boundary in search of better solutions. The boundaries of the infeasible regions should not be too large as probing too deep into infeasible regions might result in the search not returning back to the feasible region. For highly constrained problems, such as the VRPTW, using strategic oscillation allows one to exit the feasible region from one point and reenter the feasible region from a different point. The bounds within which the oscillation will occur are defined by the cost function and the temperature of the Simulated Annealing process.

The proposed hybrid method based on Genetic Algorithms, Tabu Search, non-monotonic Simulated Annealing, and customer interchange methods obtains 30 new best known solutions and 9 previously known best solutions for a set of 60 VRPTW problems obtained from the literature with customer sizes varying from 100 to 417 and the number of vehicles varying from 2 to 55.

The chapter is organized as follows. Section 9.3 describes the PFIH used to obtain the initial number of vehicles required for the Genetic Sectoring heuristic. Section 9.2 describes the structure of the local search methods. The λ-interchange generation mechanism, which is the core strategy for searching a neighborhood to improve the initial solution is also described. Section 9.3 describes the non-monotonic Simulated Annealing and Tabu

Search implementations that guide the descent algorithm using the λ-interchange generation mechanism. The Genetic Sectoring heuristic that is used to obtain initial solutions for the VRPTW is also described. Section 9.4 reports the computational experience on 60 VRPTW problems obtained from the literature and the analysis of the results with other competing methods and known optimal solutions. Section 9.5 gives the summary and concluding remarks.

9.2 Push-Forward Insertion Heuristic (PFIH)

The following notations will help in the description of the methods used for solving VRPTW.

K = total number of vehicles.

N = total number of customers.

C_i = customer i, where $i = 1,...,N$.

C_0 = central depot.

d_{ij} = Euclidean distance (proportional to the travel time) from customer i to j, where $i,j = 0,...,N$. 0 is the central depot.

e_i = earliest arrival time at customer i, where $i = 1,...,N$.

l_i = latest arrival time at customer i, where $i = 1,...,N$.

t_i = total travel time to reach customer i, where $i = 1,...,N$.

u_{ij} = urgency of the customer j, i.e., $u_{ij} = l_j - (t_i + d_{ij})$, where $i,j = 1,...,N$.

a_i = service time for customer i, where $i = 1,...,N$.

b_{ij} = waiting time for customer j, i.e., $b_{ij} = \text{Max } [e_j - (t_i + d_{ij}), 0]$ where $i,j = 1,...,N$.

p_i = polar coordinate angle of customer i, where $i = 1,...,N$.

R_k = vehicle route k, where $k=1,...,K$.

O_k = total overload for vehicle route k, where $k = 1,...,K$.

T_k = total tardiness for vehicle route k, where $k = 1,...,K$.

D_k = total distance for a vehicle route k, where $k = 1,...,K$.

W_k = total travel time (total distance+total waiting time+total service time) for a vehicle route k, where $k = 1,...,K$.

$C(R_k)$ = cost of the route R_k based on a cost function.

$C(S)$ = sum total cost of individual routes $C(R_k)$.

α = weight factor for the total distance travelled by a vehicle.

β = weight factor for the urgency of a customer.

γ = weight factor for the polar coordinate angle of a customer.

ϕ = weight factor for the total travel time of a vehicle.

η = penalty weight factor for an overloaded vehicle.

κ = penalty weight factor for the total tardy time in a vehicle route.

The Push-Forward Insertion method for inserting customers into a route for the VRPTW was introduced by Solomon [1987]. It is an efficient method for computing the cost of inserting a new customer into the current route. Let us assume a route $R_p = \{C_1,...,C_m\}$ where C_1 is the first customer and C_m is the last customer with their earliest arrival and latest arrival time defined as e_1, l_1 and e_m, l_m, respectively. The feasibility of inserting a customer into route R_p is checked by inserting the customer between all the edges in the current route and selecting the edge that has the lowest travel cost. For a customer C_i to be inserted between C_0 and customer C_1, the insertion feasibility is checked by computing the amount of time that the arrival time of t_1 is pushed forward. A change in the arrival time of t_1 could affect the arrival time of all the successor customers of C_1 in the current route. Therefore, the insertion feasibility for C_i needs to be computed by sequentially checking the Push-Forward values of all the successor customers C_j of C_i. The Push-Forward value for a customer C_j is 0 if the time propagated by the predecessor customer of C_j, by the insertion of C_i into the route, does not affect the arrival time t_j. The sequential checking for feasibility is continued until the Push-Forward value of a customer is 0 or a customer is pushed into being tardy. In the worst case, all customers are checked for feasibility.

The Push-Forward Insertion Heuristic (PFIH) starts a new route by selecting an initial customer and then inserting customers into the current route until either the capacity of the vehicle is exceeded or it is not time feasible to insert another customer into the current route. The cost function for selecting the first customer C_i is calculated using the following formula:

$$\text{Cost of } C_i = -\alpha \, d_{0i} + \beta l_i + \gamma((p_i/360)d_{0i}) \tag{1}$$

The unrouted customer with the lowest cost is selected as the first customer to be visited. The weights for the three criteria were derived empirically and were set to $\alpha = 0.7$, $\beta = 0.1$ and $\gamma = 0.2$. The priority rule in (1) for the selection of the customer depends upon the distance, polar coordinate angle and latest time. The polar coordinate angle of the customer with respect to the depot in (1) is normalized in terms of the distance. This normalization allows comparison of the distance, latest deadline and angular value of the customer in terms of a common unit.

Once the first customer is selected for the current route, the heuristic selects from the set of unrouted customers the customer $j*$ which minimizes the total insertion cost between every edge $\{k, l\}$ in the current route without violating the time and capacity constraints. The customer $j*$ is inserted into the least cost position between $\{k*, l*\}$ in the current route and the selection process is repeated until no further customers can be inserted. At this stage, a new route is created and the above is repeated until all customers are routed. It is assumed that there is an unlimited number of vehicles, K, which is large and determined by the heuristic to route all the customers. The flow of the PFIH is described below.

Step PFIH-1: Begin with an empty route starting from the depot. Set $r = 1$.

Step PFIH-2: If{all customers have been routed} then go to step PFIH-8. For all unrouted customers j: Compute the cost according to (1), and sort them in ascending order of their costs.

Step PFIH-3: Select the first customer, $j*$ from the ordered list with the least cost and feasible in terms of time and capacity constraints.

Step PFIH-4: Append $j*$ to the current route r and update the capacity of the route.

Step PFIH-5: For all unrouted customers j: For all edges $\{k, l\}$ in the current route, compute the cost of inserting each of the unrouted customers between k and l.

Step PFIH-6: Select an unrouted customer $j*$ at edge $\{k*, l*\}$ that has the least cost
If{insertion of customer $j*$ between $k*$ and $l*$ is feasible in terms of time and capacity constraints} then insert customer $j*$ between $k*$ and $l*$, update the capacity of the current route r, and go to Step PFIH-5,

    else

        go to Step PFIH-7.

Step PFIH-7: Begin a new route from the depot. Set $r = r + 1$. Go to Step PFIH-2.

Step PFIH-8: All Customers have been routed. Stop with a PFIH solution.

9.3 Structure of Local Search Methods

This section describes the various notations and features that are common to each of the implemented search strategies. An λ-interchange local search descent method is used to improve initial solutions from a route construction method. This section describes the route construction heuristic for obtaining an initial solution to the VRPTW, the λ-interchange mechanism to generate neighboring solutions for improving the initial solution, the evaluation method for computing cost of changes, and the two different strategies for selecting neighbors.

9.3.1 λ-Interchange Generation Mechanism

The effectiveness of any iterative local search method is determined by the efficiency of the generation mechanism and the way the neighborhood is searched. The λ-interchange generation mechanism is based on customer interchange between sets of vehicle routes and has been successfully implemented with a special data structure to other problems in [Thangiah, Osman, Vinayagamoorthy and Sun 1994]. The λ-interchange generation mechanism for the VRPTW is described as follows.

Given a solution for the VRPTW represented by S = $\{R_1,...,R_p,...,R_q,...,R_k\}$ where R_p is a set of customers serviced by a vehicle route p, a λ interchange between pair of routes R_p and R_q is a replacement of subsets $S_1 \subseteq R_p$ of size $|S_1| \leq \lambda$ by another subset $S_2 \subseteq R_q$ of size $|S_2| \leq \lambda$, to get two new route sets $R'_p = (R_p - S_1) \cup S_2$, $R'_q = (R_q - S_2) \cup S_1$ and a new neighboring solution S' = $\{R_1,...,R'_p,...,R'_q,...,R'_k\}$. The neighborhood $N_\lambda(S)$ of a given solution S is the set of all neighbors S' generated by the λ-interchange method for a given integer λ.

The order in which the neighbors are searched is specified as follows. Let the permutation σ be the order of vehicle indices in a given solution S = $\{R_1,...,R_p,...,R_q,...,R_k\}$ (say, $\sigma(p) = p$, $\forall p \in K$). An ordered search selects all possible combination of pairs (R_p, R_q) according to (2) and σ without repetition. A total number of (K x (K - 1))/2 different pair of routes (R_p, R_q) are examined to define a cycle of search in the following order:

$$(R_{\sigma(1)}, R_{\sigma(2)}),...,(R_{\sigma(1)}, R_{\sigma(k)}),(R_{\sigma(2)}, R_{\sigma(k)}),... (R_{\sigma(k-1)}, R_{\sigma(k)}) \quad (2)$$

For heuristics based on a descent algorithm and Tabu Search the same permutation ~ is used after each cycle of search is completed. Furthermore, for a given pair (R_p, R_q) we must also define the search order for the customers to be exchanged. We consider the case of $\lambda = 1$ and $\lambda = 2$ for the neighboring search. The λ-interchange method between two routes results in customers either being shifted from one route to another, or customers being exchanged with other customers. The operator (0,1) on routes (R_p, R_q) indicates a shift of one customer from route q to route p. The operators (1,0), (2,0) and (0,2) indicate shifting customers between two routes. The operator $(1,1)$ on routes (R_p, R_q) indicates an exchange of one customer between route p and q. The operators $(1,2)$, $(2,1)$ and $(2,2)$ indicate exchange of customers between vehicle routes.

The customers in a given pair of routes are searched sequentially and systematically for improved solutions by the shift and exchange process. The order of search we implemented uses the following order of operators $(0,1)$, $(1,0)$, $(1,1)$, $(0,2)$, $(2,0)$, $(2,1)$, $(1,2)$ and $(2,2)$ on any given pairs to generate neighbors. After a solution is generated, a criterion is required for accepting or rejecting a move. An acceptance criterion may consider many solutions as potential candidates. Two selection strategies are proposed to select between candidate solutions.

(i) The First-Best (FB) strategy will select the first solution in S' in $N\lambda(S)$ in the neighborhood of S that results in a decrease in cost with respect to a cost function.

(I) The Global-Best (BG) strategy will search all solutions S' in $N\lambda(S)$ in the neighborhood of S and select the one which will result in the maximum decrease in cost with respect to a given cost function.

9.3.2 Evaluation of a Cost Move

A move which is a transition from one solution to another in its neighborhood may cause a change in the objective function values measured by $\Delta = C(S') - C(S)$. As the λ-interchange move involves insertion of customers into routes, the following cost function is used to compute the cost of inserting customer C_i into route R_k:

$$\text{insertion cost of } C_i = D_k + \phi W_k + \eta O_k + \kappa T_k \tag{3}$$

The insertion cost function (3) will accept infeasible solutions if the reduction in total distance is high enough to allow either a vehicle to be overloaded or tardy. Overloading and tardiness in a vehicle route are penalized in the insertion cost function (3). The weight factor for total travel time ~b was set to one percent of the total distance D_k. When calculating the penalty weight factors r_1 and ~ in (3), rl was set to ten percent of D_k and ~ to one percent of D_k. The penalty values were chosen in this manner to allow penalization relative to the total distance travelled by the vehicle. This cost function (3) can be similarly generalized for other cases and values of L.

9.3.3 A λ-Interchange Local Search Descent Method (LSD)

In this section, we describe the λ-interchange local search descent method (LSD) that starts from an initial feasible solution obtained by the Push-Forward Insertion Heuristic. The PFIH is further improved using the λ-interchange mechanism. The steps of the λ interchange LSD are as follows:

Step LSD-1: Obtain a feasible solution S for the VRPTW using the PFIH.

Step LSD-2: Select a solution S' ∈ Nλ(S) in the order indicated by (2).

Step LSD-3: If {C(S') < C(S)}, then

   accept S' and go to Step LSD-2,

  else go to Step LSD-4.

Step LSD-4: If{neighborhood of Nλ(S) has been completely searched (there are no moves that

      will result in a lower cost} then

    go to Step LSD-5

  else go to Step LSD-2.

Step LSD-5: Stop with the LSD solution.

The LSD solution is dependent upon the initial feasible solution. The LSD uses two different selection strategies for selection of neighbors. The two strategies are First-Best (FB) and GlobalBest (GB). The LSD with the GB search strategy (LSD-GB) is computationally more expensive than the LSD with FB strategy (LSD-FB), as LSD-GB has to keep track of all the improving moves while the LSD-FB is a blind search that accepts the first improving move.

As the LSD-FB and LSD-GB strategies accept only improving moves, the disadvantage of using such methods is that they could get stuck in a local optima and never have the means to get out of it. Simulated Annealing is a search strategy that allows non-improving moves to be accepted with a probability in order to escape the local optima while improving moves are always accepted as in the local search descent method.

9.4 Meta-Heuristic Methods

The past decade saw the rise of several general heuristic search schemes that can be adopted with a variable degree of effort to a wide range of combinatorial optimization problems. These search schemes are often referred to as "meta-heuristics," "meta-strategies," or "modern heuristics". We refer for further details to the recent bibliography on these techniques by Reeves [1993]. The following sections describe the meta-heuristie search strategies that were implemented to solve the VRPTW. The structure of each search strategy and its parametric values are described individually.

9.4.1 Simulated Annealing

Simulated Annealing (SA) is a stochastic relaxation technique which has its origin in statistical mechanics [Metropolis et al., 1953; Kirkpatrick, Gelart and Vecchi, 1983]. The Simulated Annealing methodology draws its analogy from the annealing process of solids. In the annealing process, a solid is heated to a high temperature and gradually cooled in order for it to crystallize. As the heating process allows the atoms to move randomly, if the cooling is done too rapidly it prevents the atoms from reaching thermal equilibrium. If the solid is cooled slowly, it gives the atoms enough time to align themselves in order to reach a minimum energy state. This analogy can be used in combinatorial optimizations with the states of the solid corresponding to the feasible solution, the energy at each state corresponding to the improvement in objective function and the minimum energy being the optimal solution.

The interest in SA to solve combinatorial optimization problems began with the work of Kirkpatrick et al. [1983]. Simulated Annealing (SA) uses a stochastic approach to direct the search. It allows the search to proceed to a neighboring state even if the move causes the value of the objective function to become worse. Simulated annealing guides the original local search method in the following way. If a move to a neighbor S' in the neighborhood $N_\lambda(S)$ decreases the objective function value, or leaves it unchanged, then the move is always accepted. More precisely, the solution S' is accepted as the new current solution if $\Delta \leq 0$, where $\Delta = C(S') - C(S)$. To allow the search to escape a local optimum, moves that increases the objective function value are accepted with a probability $e^{(-\Delta/T)}$ if $\Delta > 0$, where T is a parameter called the "temperature." The value of T varies from a relatively large value to a small value close to zero. These values are controlled by a cooling schedule which specifies the initial and temperature values at each stage of the algorithm.

We use a non-monotonic cooling schedule similar to that outlined in [Reeves, 1993], [Connolly, 1990] and [Osman and Christofides, 1994]. The non-monotonic reduction scheme reduces the temperature after each generated move (one iteration) with occasional temperature increases (or higher temperature resets) after the occurrence of a special neighborhood without accepting any moves. The design of the non-monotonic cooling schedule to induce an oscillation behavior in the temperature and, consequently, in the objective function is a kind of strategic oscillation concept borrowed from Tabu search [Glover 1986; Glover 1993]. This combination has been shown to yield better and improved performance over other standard SA approaches on a number of problems [Osman and Christofides, 1994].

The SA implementation starts from an initial solution which is generated by the Push-Forward Insertion Heuristic. It systematically searches the 2-interchange generation mechanism and allows infeasible moves to be to be accepted/rejected according to a SA criterion using the cost function (3) for move evaluations. The following notation will help in the explanation of the SA algorithm.

$T_S =$ 　　　 *Starting temperature of the SA method.*

$T_f =$ 　　　 *Final temperature of the SA method.*

$T_b =$ *Temperature at which the best current solution was found.*

$T_r =$ *Reset temperature of the SA method.*

$S =$ *Current solution.*

$S_b =$ *Best current route found so far in the search.*

$R =$ *Number of resets to be done.*

$\tau =$ *A decrement constant in the range of $0 < \tau < 1$.*

The flow of the Simulated Annealing method can be described as follows:

Step SA-1: Obtain a feasible solutions for the VRPTW using the PFIH heuristic.

Step SA-2: Improve S using the 2-interchange local search descent method with the First-Best selection strategy.

Step SA-3: Set the cooling parameters.

Step SA-4: Generate systematically a $S' \in N_\lambda(S)$ and compute $\Delta = C(S') - C(S)$.

Step SA-5: If{($\Delta \leq 0$) or ($\Delta > 0$ and $e^{(-\Delta/T_k)} \geq 0$), where 0 is a random number between [0, 1]} then

     set $S = S'$.

     if{$C(S') < C(S_b)$} then

     improve S' using the local search procedure in Step-SA2,

     update $S_b = S'$ and $T_b = T_k$.

Step SA-6:   Set $k = k+1$.

    Update the temperature using: $T_{k+1} = T_k/(1 + \tau T_k)$

     If{$N_2(S)$ is searched without any accepted move} then

    set $T_r =$ maximum {$T_r/2$, T_b, and set $T_k = T_r$.

Step SA-7:   If{R resets were made since the last S_b was found} then

     go to Step SA-8,

   else go to Step SA4.

Step SA-8:   Terminate the SA algorithm and print the S_b routes.

The above hybrid procedure combines the SA acceptance criteria with the local search descent algorithm and the strategic oscillation approach embedded in the cooling schedule. In our implementation, and after experimentation, the initial parameters for the cooling schedule are set at $T_s = 50$, $Tf = 0$, $T_r = T_s$, $R = 3$, $S_b = S$ and $k = 1$. After each iteration k, the temperature is decreased according to a parameter τ which was set to 0.5. Initial experiments using only feasible moves resulted in solutions that were not competitive with solutions obtained from competing heuristics. The LSD-FB selection strategy was used by the SA method to select candidate moves. The Simulated Annealing method, at times, could get caught in a succession of moves that could result in a move being made in state S that is reversed in state S'. In order to avoid moves that result in cycles and also force the search to explore other regions, a hybrid combination of the Tabu Search and Simulated Annealing method was implemented.

9.4.2 Tabu Search

Tabu search (TS) is a memory based search strategy to guide the local search descent method to continue its search beyond local optimality [Glover, 1989; Glover, 1990]. When a local optimum is encountered, a move to the best neighbor is made to explore the solution space, even though this may cause a deterioration in the objective function value. The TS seeks the best available move that can be determined in a reasonable amount of time. If the neighborhood is large or its elements are expensive to evaluate, candidate list strategies are used to help restrict the number of solutions examined on a given iteration.

Tabu search uses memory structures to record attributes of the recent moves when going from one solution to another in a Tabu list. Attributes of recently visited neighborhood moves are designated as Tabu and such moves are not permitted by the search strategy for the duration that is considered to be Tabu. The duration that an attribute remains on a Tabu list is determined by the Tabu List Size (TLS). A special degree of freedom is introduced by means of an aspiration concept and the Tabu status of a move can be overruled if certain aspiration conditions are met. Since TS is a heuristic method, the rule for execution is generally expressed as a pre-specified limit on the number of iterations or on the number of iterations since the last improvement was found. More details on recent developments and applications can be found in Glover, Taillard and De Werra (1993).

A new modification to the TS was implemented to solve the VRPTW. The TS algorithm was combined with the SA acceptance criterion to decide which moves to be accepted from the candidate list. The combined algorithm uses a special data structure which identifies the exact candidate list of moves rather than using a sampling approach of the neighborhood. The hybrid algorithm using the TS elements was implemented as follows. Given a solution $S = \{R_1,...,R_k\}$ with K vehicles, there are $(K \times (K - 1))/2$ pairs of $\{Rp, R_q\}$ routes considered by the 2-interchange mechanism to generate the whole neighborhood $N_2(S)$. The best solution in terms of the objective function value generated from each $\{Rp, R_q\}$ is recorded. The set of all these best solutions form the candidate list of solutions. Two matrices are used to record the $(K \times (K - 1))/2$ best solutions. *BestCost* and *BClist* are matrices with dimensions $K \times K$, and $(K \times (K - 1))/2 \times 8$ The top triangular part of the matrix *BestCost* (p, q), $1 \le p < q \le K$, stores the objective value Δ_{pq} associated the best move between (Rp, R_q).

The lower triangular part of *BestCost(q,p)* stores an index l indicating the row in *BClist* where the information (customers exchanged, route indices, etc.) associated with *BestCost(q,p)* are recorded for faster retrieval. The minimum attributes associated with a move are the indices of the maximum number of customers to be interchanged and of the two routes leading to six column entries in *BClist*. However, more information can be stored in *BClist* by adding more columns if necessary. The importance of *BestCost and BClist* is that they can be updated by evaluating only moves in the 2 x K pairs of routes rather than looking at $(K \times (K - 1))/2$ pairs in $N_2(S)$. After a move involving $\{Rp, Rq\}$ is done, the pairs of routes $N_2(S)$ which do not involve either Rp or Rq remain intact. The pairs of routes which need evaluations are $\{Rp, R_t\}, \{R_t, Rq\}$ $\forall t$ in K.

The Tabu list structure is represented by a matrix *TabuList* with dimension $K \times N$. Each *TabuList* (p, i) records the iteration number at which a customer i is removed from ronte R_p plus the value of the tabu list size TSL. As the neighborhood size generated by the 2-interchange mechanism is large, a value for TSL was set to 10 and found to be sufficient. When a move involves four customers exchanged between routes, the appropriate elements are added to the *TabuList*. The above structure can be easily checked to identify the Tabu status of the current move. For example, at iteration m, if *TabuList* (p, i) is greater than m, then a move which returns customer i to route R_p is considered Tabu. A Tabu status of a move S' is overruled if its objective function value $C(S')$ is smaller than the objective value of the best solution $C(S_b)$ found so far.

The hybrid Tabu Search and Simulated Annealing (TSSA) algorithm steps can be described as follows:

Step TSSA-1: Obtain an initial PFIH solution S.

Step TSSA-2: Improve S using the 2-interchange local search descent algorithm with the First-Best selection strategy.

Step TSSA-3: Initialize TabuList to zeros, set $S_b = S$ and set the iteration counter m = 0.

Step TSSA-4: Set the Simulated Annealing cooling schedule parameters.

Step TSSA-5: If (m = 0) then

> update the BestCost and BClist matrices from information in Step TSSA

else

> update the matrices of the candidate list as necessary to reflect the changes due to the performed move.

Step TSSA-6: Select $S' \in N_2(S)$ with the largest improvement or least non-improvement from the candidate list of moves in BestCost.

Step TSSA-7: If{S' is tabu} then

> if {$C(S') < C(S_b)$} then

>> go to Step TSSA-9,

> else

>> go to Step TSSA-6 to select the next best solution $S' \in$ BestCost.

> else if{S' is not tabu} then

go to Step TSSA-8.

Step TSSA-8: Accept or Reject S' according to the Simulated Annealing criterion.

> If{S' is accepted} then

>> go to Step TSSA-9.

> else

>> go to Step TSSA-6 to select the next best solution $S' \in$ BestCost.

Step TSSA-9: Update S = S', and TabuList and other SA parameters.

> If(c(s) < c(sp] then

>> set $S_b = S$,

>> set m = m + 1, and

>> go to Step TSSA-10.

Step TSSA-10: If[m is greater than a given number of iterations K x N] then

>>> go to Step TSSA-11.

>> else go to Step TSSA-5.

Step TSSA-11: Terminate the hybrid TSSA algorithm and print the S_b
   routes.

9.4.3 Genetic Algorithms

The Genetic Algorithm (GA) is an adaptive heuristic search method based on
population genetics.

The basic concepts of a GA were primarily developed by Holland [1975].
Holland's study produced the beginning of the theory of genetic adaptive
search [DeJong, 1980; Grefenstette, 1986; Goldberg, 1989].

The GA is an iterative procedure that maintains a population of P candidate
members over many simulated generations. The population members are
string entities of artificial chromosomes. The chromosomes are fixed length
strings with binary values (or alleles) at each position (or locus). Allele is
the 0 or 1 value in the bit string, and the locus is the position at which the
0 or 1 value is present in each location of the chromosome. Each
chromosome has a fitness value associated with it. The chromosomes from
one generation are selected for the next generation based upon their fitness
value. The fitness value of a chromosome is the payoff value that is
associated with a chromosome. For searching other points in the search
space, variation is introduced into the population chromosomes by using
crossover and mutation genetic operators. Crossover is the most important
genetic recombination operator. After the selection process, a randomly
selected proportion of the chromosomes undergo a two-point crossover
operation and produce offsprings for the next generation.

Selection and crossover effectively search the problem space exploring and
exploiting information present in the chromosome population by selecting
and recombining primarily the offsprings that have high fitness values.
These two genetic operations generally produce a population of
chromosomes with high performance characteristics. Mutation is a
secondary operator that prevents premature loss of important information by
randomly mutating alleles within a chromosome. The adaptations in a GA
are achieved by exploiting similarities present in the coding of the
chromosomes. The termination criteria of a GA are convergence within a
given tolerance or realization of the maximum number of generations to be
simulated.

A clustering method using the GA has been highly successful in solving
vehicle routing problems with time constraints, multiple depots and
multiple commodities [Thangiah, 1993; Thangiah, Vinayagamoorthy and
Gubbi, 1993; Thangiah and Nygard, 1993; Thangiah and Nygard, 1992a,
1992b; Thangiah, Nygard and Juell, 1991]. We investigate the use of the
genetic clustering method and show how it can be combined with other
meta-heuristics for solving VRPTW for a large number of customers.

The GA clustering method is based on the cluster-first route-second approach. That is, given a set of customers and a central depot, the heuristic clusters the customers using the GA, and the customers within each sector are routed using the cheapest insertion method [Golden and Stewart, 1985]. The GA solution can be improved using the LSD-FB, LSD-GB, SA, or TSSA heuristics. The clustering of customers using a GA is referred to as Genetic Sectoring. The Genetic Sectoring Heuristic (GSH) allows exploration and exploitation of the search space to find good feasible solutions with the exploration being done by the GA and the exploitation by local search meta-heuristics. The following notations will help in the description of the Genetic Sectoring heuristic.

S_k = pseudo polar coordinate angle of customer i, where $i = 1,...,N$.

F = fixed angle for Genetic Sectoring, Max$[s_i,...,s_n]/2K$, where $n = 1,...,N$.

M = maximum offset of a sector in Genetic Sectoring, $M = 3F$.

B = length of the bit string in a chromosome representing an offset, $B = 5$.

P = population size of the Genetic Algorithm, $P = 50$.

G = number of generations the Genetic Algorithm is simulated, $G = 1000$.

E_k = offset of the kth sector, i.e., decimal value of the kth bit string of size B, where $k = 1,...,K-1$.

S_k = seed angle for sector k, where $k = 1,...,K-1$.

S_0 = initial seed angle for Genetic Sectoring, $S_0 = 0$.

The GENESIS [Grefenstette, 1987] genetic algorithm software was used in the implementation of the GSH. The chromosomes in GENESIS are represented as bit strings. The sectors (clusters) for the VRPTW are obtained from a chromosome by subdividing it into K divisions of size B bits. Each subdivision is used to compute the size of a sector. The fitness value for the chromosome is the total cost of serving all the customers computed with respect to the sector divisions derived from it. The GSH is an extension of the clustering method of Fisher and Jaikumar [1981] and Gillett and Miller [1974].

In a N customers problem with the origin at the depot, the GSH replaces the customer angles $P_1,...,P_N$ with pseudo polar coordinate angles. The pseudo polar coordinate angles are obtained by normalizing the angles between the customers so that the angular difference between any two adjacent customers is equal. This allows sector boundaries to fall freely between any pair of customers that have adjacent angles, whether the separation is small or large. The customers are divided into K sectors, where K is the number of vehicles, by planting a set of "seed" angles, $S_0...,S_K$, in the search space and drawing a ray from the origin to each seed angle. The initial number of vehicles, K, required to service the customers is obtained using the PFIH. The initial seed angle S_0 is assumed to be 0^0. The first sector will lie between seed angles S_0 and S_1. the second sector will lie between seed angles S_1 and S_2, and so on. The Genetic Sectoring process assigns a customer, C_i, to a sector or vehicle route, R_k, based on the following equation:

C_i is assigned to R_k if $S_k < s_i \leq S_{k+1}$, where $k = 0,...,K-1$.

Customer C_i is assigned to vehicle R_k if the pseudo polar coordinate angle s_i is greater than seed angle S_k but is less than or equal to seed angle S_{k+1}. Each seed angle is computed using a fixed angle and an offset from the fixed angle. The fixed angle, F, is the minimum angular value for a sector and assures that each sector gets represented in the Genetic Sectoring process. The fixed angle is computed by taking the maximum polar coordinate angle within the set of customers and dividing it by $2K$. The offset is the extra region from the fixed angle that allows the sector to encompass a larger or a smaller sector area.

The GA is used to search for the set of offsets that will result in the minimization of the total cost of routing the vehicles. The maximum offset, M, was set to three times the fixed angle to allow for large variations in the size of the sectors during the genetic search. If a fixed angle and its offset exceeds 360^0, then that seed angle is set to 360^0 thereby allowing the Genetic Sectoring process to consider vehicles less than K to service all its customers. Therefore, K, the initial number of vehicles with which the GSH is invoked, is the upper bound on the number of vehicles that can be used for servicing all the customers.

The bit size representation of an offset in a chromosome, B, was derived empirically and was set at 5 bits. The decimal conversion of 5 bits results in a range of integer values between 0 and 31. The offsets are proportionately derived from the decimal conversion of the bit values using the decimal value 0 as a 0^0 offset and the bit value 31 as the maximum offset. Figure 9.1 describes the chromosome mapping used to obtain the offsets.

The seed angles are derived from the chromosome using the following equation:

$$S_i = S_{i-1} + F + \left(3^{\left(e_i \left(\frac{Log_M}{\log 3} \right) \right)} \right) \left(\frac{M}{2^B} \right) \tag{5}$$

The fitness value of a chromosome is the total cost of routing K vehicles for servicing N customers using the sectors formed from the set of seed angles derived from the chromosome. The seed angles are derived using the fixed angle and the offsets from the chromosomes. The cost function (Equation 5) for calculating the seed angles uses an exponential function.

The exponential function allows for large fluctuations in the seed angles with respect to the offsets derived from the chromosomes during the Genetic Sectoring process. The large fluctuations during the search for a feasible solution allow the Genetic Sectoring method to reduce the number of vehicles with respect to the initially defined value. The customers within the sectors, obtained from the chromosomes, are routed using the cheapest insertion method described in Section 9.2.4.

In the GSH, a chromosome represents a set of offsets for the VRPTW. Therefore, a population of P chromosomes usually has P different solutions for a VRPTW. That is, there may be some chromosomes in the population that are not unique. At each generation P chromosomes are evaluated for fitness. The chromosomes that have the least cost will have a high probability of surviving into the next generation through the selection process. As the crossover operator exchanges a randomly selected portion of the bit string between the chromosomes, partial information about sector divisions for the VRPTW is exchanged between the chromosomes. New information is generated within the chromosomes by the mutation operator. The GSH uses selection, crossover and mutation to adaptively explore the search space for the set of sectors that will minimize the total cost of the routes over the simulated generations for the VRPTD. The GSH would utilize more computer time than the PFIH for obtaining a solution because it has to evaluate $P \times G$ vehicle routes, where P is the population size and G is the number of generations to be simulated.

The parameter values for the number of generations, population size, crossover and mutation rates for the Genetic Sectoring process were derived empirically and were set at 1000, 50, 0.6 and 0.001. During the simulation of the generations, the GSH keeps track of the set of sectors obtained from the genetic search that has the lowest total route cost. The genetic search terminates either when it reaches the number of generations to be simulated or if all the chromosomes have the same fitness value. The best set of sectors obtained after the termination of the genetic search does not always result in a feasible solution. The infeasibility in a solution arises because of overloading or tardiness in a vehicle route. The solution obtained from the GA is improved using the LSD-FG, LSD-GB, SA and TSSA methods. The GSH method can be described as follows.

Step GSH-1: Set the bit string size for the offset: Bsize = 5. Set the variable, NumVeh, to the number of vehicles required by the PFIH to obtain a feasible solution.

Step GSH-2: Sort the customers in order of their polar coordinate angles and assign pseudo polar coordinate angles to the customers. Set the lowest global route cost to infinity: $g = \infty$.

Set the lowest local route cost to infinity. $l = \infty$

Step GSH-3: For each chromosome in the population:

For each bit string of size BSize,

  calculate the seed angle,

  sector the customers, and

  route the customers within the sectors using the cheapest

  insertion method.

If{cost of the current set of sectors is lower than 1} then

  set l to the current route cost, and

  save the set of sectors in lr.

If{cost of the current set of seclors is lower than g} then

  set g to the current route cost, and

  save the set of sectors in gr.

If{all the chromosomes have not been processed} then

  go to Step GSH-3.

else go to Step GSH-4.

Step GSH-4: Do Selection, Crossover and Mutation on the chromosomes. Go to Step GSH-3.

Step GSH-5: Improve the routes using LSD-FB, LSD-GB, SA and TSSA methods.

Step GSH-6: Terminate the GSH and print the best solution found.

9.5 Computational Results

The LSD-FB, LSD-GB, SA and TSSA heuristic methods were applied to the six data VRPTW sets R1, C1, RC1, R2, C2, and RC2 generated by Solomon [1987] consisting of 100 customers with Euclidean distance. In these problems, the travel time between the customers are equal to the corresponding Euclidean distances. The data consist of geographical and temporal differences in addition to differences in demands for the customers. Each of the problems in these data sets has 100 customers. The fleet size to service them varied between 2 and 21 vehicles. The VRPTW problems generated by Solomon incorporate many distinguishing features of vehicle routing with two-sided time windows. The problems vary in fleet size, vehicle capacity, travel time of vehicles, spatial and temporal distribution of customers, time window density (the number of demands with time windows), time window width, percentage of time constrained customers and customer service times. Problem sets R1, C1 and RC1 have a narrow scheduling horizon. Hence, only a few customers can be served by the same vehicle. Conversely, problem sets R2, C2 and RC2 have large scheduling horizon, and more customers can be served by the same vehicle.

In addition to the problem sets generated by Solomon, the heuristics were applied to four real world problems consisting of 249 and 417 customers. The first problem consisting of 249 customers is reported by Baker and Schaffer [1986]. Two problems, D249 and E249, are generated from this problem by setting the capacity of the vehicles to 50 and 100. The second problem consisting of 417 customers is based on a route for a fast food industry located in southeastern United States. The two problems, D417 and E417, consisting of 417 customers differ in that E417 has a larger number of tight time windows than D417. The problem with 217 customers has a demand of one for all the customers with the service time varying between 4 and 27 units with each customer having a large time window. The 417 customer problem has a service time of 60 units for each customer with demands varying between 100 and 500 units with tight and loose time windows. All of the problems were solved using the same parametric values for the heuristics. The CPU time reported is the total time for solving each of the problems. The distances for the problem were calculated in real value format and are reported with the final distance value rounded to two decimal places.

Table 9.1: Average number of vehicles and distance obtained for the six data sets using Genetic Sectoring for obtaining an initial solutions and four heuristics LSD-FB, LSD-GB, SA and TSSA for improving the solution.

Prob.	LSD-FB	LSD-GB	SA	TSSA
R1	12.4	12.5	12.4	12.4
	1289	1289	1287	1242
C1	10.0	10.0	10.0	10.0
	969	993	937	874
RC1	12.3.	12.2	12.1	12.0
	1511	1499	1471	1447
R2	3.0	3.1	3.2	3.2
	1023	1056	1052	1031
C2	3.0	3.0	3.0	3.0
	673	709	684	676
RC2	3.4	3.4	3.4	3.4
	1254	1255	1307	1261
Total Vehicles	420	423	422	419
Average CPU Time	1:2	0:9	3.9	7:3

(see legend, p.368)

Legend:
Prob.:   VRPTW data set.
LSD-FB: Local descent with First-Best strategy.
LSD-GB: Local descent with Global-Best search strategy.
SA: Simulated Annealing+ local descent with First Best Strategy.
TSSA: Tabu Search/Simulated Annealing with Global Best.

Comparison of Local Search and Meta-Strategy Heuristics

Table 9.1 lists the average number of vehicles, distance travelled and CPU time for the VRPTW problems using the PFIH and GSH to obtain initial solutions and the LSD-FB, LSD-GB, SA and TSSA methods to improve the solutions. The best average solutions in terms of the number of vehicles and total distances are obtained by the heuristics with respect to the data sets are highlighted in bold. All of the computations for the problems were computed in real value and the final values are reported after rounding off to two decimal places. The quality of the solution is measured in terms of the lowest number of vehicles required to solve the problem followed by the total distance travelled. The comparison is done in this manner as it is possible to reduce the total distance travelled by increasing the number of vehicles. The TSSA heuristic obtains the best average solution for problems that have uniformly distributed customers, tight time windows and require a large number of vehicles. The LSF-FB heuristic obtains the best average solution for problems that have larger time windows, customers that are clustered and use a smaller number of vehicles.

The average CPU time for the heuristic methods is high for the TSSA in comparison to the LSD methods. The TSSA method takes almost five times more CPU time than the LSD methods to obtain a solution to the problem sets. In comparison to the total number of vehicles required to solve all the problem sets, the TSSA requires one less vehicle than the LSD-FB solution.

Table 9.2 Average number of vehicles and distance obtained for the six data sets by the GenSAT system and six other competing heuristics.

Prob.	GRASP	CTA	PARIS	Solomon	GIDEON	PTABU	G-ROUS	Chiang/Russell	TSSA
R1	13.1	13.0	13.3	13.6	12.8	12.6	12.6	12.5	12.4
	1427	1357	1696	1437	1300	1378	1297	1309	1242
C1	10.6	10.0	10.7	10.0	10.0	10.0	10.0	10.0	10.0
	1401	917	1610	952	892	861	838	910	874
RC1	12.8	13.0	13.4	13.5	12.5	12.6	12.1	12.4	12.0
	1603	1514	1877	1722	1474	1473	1446	1474	1447
R2	3.2	3.2	3.1	3.3	3.2	3.1	3.0	2.9	3.2
	1539	1276	1513	1402	1125	1171	1118	1166	1031

C2	3.4	3.0	3.4	3.1	3.0	3.0	3.0	3.0	3.0
	858	645	1478	693	749	604	590	684	676
RC2	3.6	3.7	3.6	3.9	3.4	3.4	3.4	3.4	3.4
	1782	1634	1807	1682	1411	1470	1368	1401	1261
Total Vehicles	447	444	439	431	430	427	422	422	419

Legend:
Prob.:   VRPTW data set.
GRASP: Kontoravdis and Bard, 1992.
CTA: Thompson and Psaraftis, 1993.
PARIS: Potvin and Rousseau, 1993.
Solomon: Solomon, 1987.
GIDEON: Thangiah, 1993.
PTABU: Potrin et al, 1993.
G-ROUS: Potvin and Bengio, 1994.
Chiang/Russell: Chiang and Russell, 1994.
TSSA: Tabu Search/Simulated Annealing with Global Best.
Total Vehicles: Sum Total of Vehicles from all the problems.

Comparison of the TSSA Heuristic with Competing Heuristics

Table 9.2 compares the average number of vehicle and distance obtained by eight different heuristics and the TSSA hybrid heuristic. The eight different heuristics are diverse in their approach to solving VRPTW. The TSSA hybrid heuristic obtains the best average solutions for data sets R1, C1, RC1 and RC2 in comparison to the eight competing heuristics. The Chiang-Russell heuristic obtains the best average solution for data set R2 and the PTABU heuristic for data set C2. The GenSAT system obtains the best average number of vehicles and total distance travelled for data sets R1, C1, RC1 and R2. For data set R2, the Chiang-Russell algorithm, in comparison to the GenSAT system, obtains better average number of vehicles but has a larger average total distance. For data set C2, the GENEROUS algorithm obtains the best average number of vehicles and total distance in comparison to all the competing algorithms.

Table 9.3 Comparison of the best known solutions with the TSSA solutions for data sets R1, C1, RC1, R2, C2, RC2 and problems D249, E249, D417 and E417.

Prob.	Best Known	TSSA	Prob.	Best Known	TSSA

	NV	TD	NV	TD		NV	TD	NV	TD
R101	18	1608[o]	18	1677	RC101	15	1676[d]	**14**	**1669**
R102	17	1434[o]	17	1505	RC102	14	1569[b]	**13**	**1557**
R103	13	1319[b]	**13**	**1207**	RC103	11	1138[b]	**11**	**1310**
R104	10	1065[d]	**10**	**1056**	RC104	10	1204[d]	10	1226
R105	14	1421[d]	14	1477	RC105	14	1612[b]	14	1670
R106	12	1353[c]	**12**	**1350**	RC106	12	1486[e]	13	1420
R107	11	1185[c]	**11**	**1146**	RC107	11	1275[d]	**11**	**1410**
R108	10	1033[e]	**10**	**1013**	RC108	11	1187[d]	**10**	**1310**
R109	12	1205[d]	12	1228					
R110	11	1136[c]	**11**	**1115**					
R111	11	1184[c]	10	1223					
R112	10	1003[f]	**10**	**992**					

	NV	TD	NV	TD		NV	TD	NV	TD
C101	10	827[o]	10	829	R201	4	1478[b]	**4**	**1358**
C102	10	827[o]	10	829	R202	4	1279[b]	**4**	**1176**
C103	10	873[b]	10	968	R203	3	1167[b]	3	1195
C104	10	865[d]	10	926	R204	2	**904[d]**	3	803
C105	10	829[a]	10	862	R205	3	1159[d]	3	**1176**
C106	10	827[o]	10	829	R206	3	1066[d]	3	**1018**
C107	10	827[o]	10	965	R207	3	954[d]	3	**904**
C108	10	829[d]	10	829	R208	2	759[d]	2	826
C109	10	829[d]	10	829	R209	3	1108[d]	2	**993**
					R210	3	1146[d]	3	1071
					R211	3	898[b]	3	**816**

	NV	TD	NV	TD
D249	4	492[e]	**4**	**457**
E249	5	546[e]	6	574
D417	55	5711[e]	**54**	**4783**
E417	55	5749[e]	**55**	**4730**
C201	3	590[a]	3	592

Prob.	NV	TD	NV	Best Known
C202	3	591^c	3	781
C203	3	592^d	3	836
C204	3	591^d	3	817
C205	3	589^c	3	589
C206	3	588^c	3	618
C207	3	588^c	3	588
C208	3	588^c	3	588
RC201	4	1734^e	**4**	**1638**
RC202	4	1459^e	**4**	**1368**
RC203	3	1253^d	**3**	**1211**
RC204	3	1002^d	**3**	**897**
RC205	4	1594^d	**4**	**1418**
RC206	3	1298^e	3	1348
RC207	3	1194^e	**3**	**1181**
RC208	3	1038^b	**3**	**1027**

Legend:
Prob.:   VRPTW data set.
NV: Number of vehicles.
TD: Total Distance.
a: CTA [Thompson and Psaraftis, 1993]
c: TABU [Potvin et al, 1993].
e: Chiang/Russell [Chiang and Russell, 1994].
Best Known: Best known solution.
TSSA: Best solution obtained by the TSSA heuristic.
o: Optimal solution [Desrochers et al., 1992].
b: GIDEON [Thangiah, 1993].
d: GENEROUS [Potvin and Bengio, 1994].

Table 9.3 compares each of the solutions obtained by the TSSA heuristic against six competing heuristics that have reported individual solutions obtained for the 60 problems. In Table 9.3 "NV" indicates the total number of vehicles and "TD" the total distance required to obtain a feasible solution. The column "Best Known" lists the previously best known solution from six different heuristics that have reported individual solutions for each of the problems. The "GenSAT" column contains the best solution obtained by using two different heuristics to obtain an initial solution, Push-Forward Insertion and Genetic Sectoring, and four different local heuristics LSDFB,

LSD-GB, SA, TAAS for improving the solution. The solutions obtained by GenSAT, better than previously known best solutions in terms of the number of vehicles and total distance, are highlighted in bold.

The TSSA system obtained 30 new best known solutions for the VRPTW problems. In addition, it attained previously known best solutions for 9 problems and failed in the other 21 problems. The average performance of GenSAT system is better than other known competing methods.

Comparison of GenSAT with Optimal Solutions

In Table 9.4 the optimal solutions for some of the VRPTW problems reported in [Desrochers, Desrosiers and Solomon, 1992] are compared with the best known solutions and those obtained by the GenSAT system.

For problems C1, C2, C6 and C8, the best known solutions and the TSSA solutions are near optimal. It is worth noting that the optimal solutions were obtained by truncating the total distance to the first decimal place as reported by [Desrochers, Desrosiers and Solomon, 1992]. For the R101 problem, the best known solution is 6% over the optimal solution in terms of the number of vehicles and the total distance travelled, and GenSAT obtains the optimal number of vehicles and is 4% away from the optimal distance travelled. For R102 problem, both of the best known solutions and the GenSAT system obtain the optimal number of vehicles required for the customers, but the best known solution is 8% over the optimal distance travelled while the GenSAT solution is only 5% away from the optimal distance.

Table 9.4 Comparison of the optimal, best known, and TSSA solution for problems R101, R102, C101, C102, C106 and C108.

Prob.	Optimal	Best Known	TSSA	% over optimal for best known solution	% over optirnal for TSSA solution
R101	18	19[d]	18	6%	0%
	1608	1704	1677	6%	4%
R102	17	17[b]	17	0%	0%
	1434	1549	1505	8%	5%
C101	10	10[a]	10	0%	0%
	827	829	829	0.2%	0.2%
C102	10	10[c]	10	0%	0%
	827	829	829	0.2%	0.2%
C106	10	10[c]	10	0%	0%
	827	829	829	0.2%	0.2%

C108	10	10c	10	0%	0%
	827	829	829	0.2%	0.2%

Legend:
Prob.:   VRPTW data set.
NV: Number of vehicles
TD: Total Distance
b: GIDEON [Thangiah, 1993.]
d: Chiang/Russell [Chiang and Russell, 1994].
Best Known: Best known among competing heuristics.
TSSA: Best solution obtained using the TSSA heuristic.
a: CTA [Thompson and Psaraftis, 1993.]
c: GENEROUS [Potvin and Bengio, 1994].

Computational Time for the GenSAT System

The TSSA heuristic was written in the C language and implemented on a NeXT 68040 (25Mhz) computer system. The average CPU time required to solve the VRPTW varied with the problems. For problems that have a large number of customers, such as D249, E249, D417 and D417, the TSSA took an average CPU time of 26 minutes to solve the problems. The TSSA took an average of 7 minutes of CPU time to solve problems that had 100 customers.

The Chiang/Russell algorithm was implemented on a 486DX/66 machine and required an average of 2 minutes of CPU time for the 100 customer problems. The GENEROUS system was implemented on a Silicon Graphics workstation and required an average of 15 minutes of CPU time. The GIDEON system was implemented on a SOLBOURNE 5/802 and required an average of 2 minutes and 47 seconds, PTABU, on a SPARC 10 workstation, required an average of 13 minutes and 5 seconds, PARIS required 45 seconds on an IBM-PC, GRASP took 0.02 seconds on a RISC 6000 and SOLOMON used 24.7 seconds on a DEC system. It is very difficult to compare the CPU times for the TSSA heuristic with those of the competing methods due to differences in the language of implementation, the architecture of the system and the amount of memory available in the system. One can approximate the computing power of the different system to get an approximate efficiency of the algorithms. In comparison to the NeXT system, the 486/66 system is about seven times as fast, the SUN/SPARC-10 is about ten times as fast, the Silicon Graphics Systems is about twelve times as fast and the SOLBOURNE 5/802 about four times as fast.

Table 9.5 Comparison of the best known solutions using PFIH and GSH to obtain initial solutions and improving the solutions using LSD-FB, LSD-GB, SA and TSSA for data sets R1, C1, RC1, R2, C2, RC2 and problems D249, E249, D417 and E417.

Prob.	Best Known		GenSAT	
	NV	TD	NV	TD
R101	18	1608^o	**18**	**1644**
R102	17	1434^o	**17**	**1493**
R103	13	1319^b	**13**	**1207**
R104	10	1065^d	**10**	**1048**
R105	14	1421^d	14	1442
R106	12	1353^c	**12**	**1350**
R107	11	1185^c	**11**	**1146**
R108	10	1033^e	**10**	**898**
R109	12	1205^d	12	1226
R110	11	1136^c	**11**	**1105**
R111	11	1184^c	**10**	**1151**
R112	10	1003^f	**10**	**992**
C101	10	827^o	10	829
C102	10	827^o	10	829
C103	10	873^b	**10**	**835**
C104	10	865^d	**10**	**835**
C105	10	829^a	10	829
C106	10	827^o	10	829
C107	10	827^o	10	829
C108	10	829^d	10	829
C109	10	829^d	10	829
D249	4	477^e	**4**	**457**
E249	5	506^e	**5**	**495**
D417	55	4235^e	**54**	**4866**
E417	55	4397^e	**55**	**4149**
C201	3	590^a	3	591
C202	3	591^c	3	707
C203	3	592^d	3	791
C204	3	591^d	3	685

C205	3	589^c	3	589
C206	3	588^c	3	588
C207	3	588^c	3	588
C208	3	588^c	3	588
RC201	4	1734^e	**4**	**1294**
RC202	4	1459^e	**4**	**1291**
RC203	3	1253^d	**3**	**1203**
RC204	3	1002^d	**3**	**897**
RC205	4	1594^d	**4**	**1389**
RC206	3	1298^e	**3**	**1213**
RC207	3	1194^e	**3**	**1181**
RC208	3	1038^b	**3**	**919**
RC101	15	1676^d	**14**	**1669**
RC102	14	1569^b	**13**	**1557**
RC103	11	1138^b	**11**	**1110**
RC104	10	1204^d	10	1226
RC105	14	1612^b	**14**	**1602**
RC106	12	1486^e	13	1420
RC107	11	1275^d	**11**	**1264**
RC108	11	1187^d	**10**	**1281**
R201	4	1478^b	**4**	**1354**
R202	4	1279^b	**4**	**1176**
R203	3	1167^b	**3**	**1126**
R204	2	904^d	**3**	**803**
R205	3	1159^d	**3**	**1128**
R206	3	1066^d	**3**	**833**
R207	3	954^d	**3**	**904**
R208	2	759^d	2	823
R209	3	1108^d	**2**	**855**
R210	3	1146^d	**3**	**1052**
R211	3	898^b	3	816

Legend:
Prob.:   VRPTW data set.
NV: Number of vehicles.
TD: Total Distance.
a: CTA [Thompson and Psaraftis, 1993]
b: TABU [Potvin et al., 1993].
c: Chiang/Russell [Chiang and Russell, 1994].
Best Known: Best known solution.
TSSA: Best solution obtained by the TSSA heuristic.
o: Optimal solution [Desrochers et al., 1992].
b: GIDEON [Thangiah, 1993].
d: GENEROUS [Potvin and Bengio, 1994].

The heuristic algorithms that are close competitors to the TSSA heuristic are the GENEROUS [Potvin and Bengio, 1994] and the Chiang/Russel heuristic [Chiang and Russel, 1994]. In comparison to the TSSA heuristic, the GENEROUS and Chiang/Russel heuristic require considerably more CPU time for solving the VPRPTW.

Table 9.5 compares the best solutions obtained with heuristics that used initial solutions obtained from PFIH and GSH and improved the initial solutions using the LSD-FB, LSD-GB, SA and TSSA heuristics.

9.6 Summary and Conclusion

In this chapter, we developed a hybrid heuristic that uses meta-strategies for obtaining good solutions to Vehicle Routing Problems with Time Windows (VRPTW). The hybrid heuristic search strategy uses Genetic Algorithms (GA) to find an initial solution using a clustering method. The initial solution from the GA is improved using a customer-interchange method guided by a Tabu Search heuristic in combination with a non-monotonic simulated annealing method. Strategic oscillation is induced into the search process by using a cost function that allows infeasible solutions to be accepted with a penalty.

A computational study on the performance of the hybrid heuristic was conducted using 60 VRPTW problems obtained from the literature. The VRPTW problems had customer sizes varying from 100 to 417 and vehicle sizes varying from 2 to 55. The hybrid heuristic obtained 30 new best known solutions and 9 solutions that equaled previously best known solutions. The heuristic also obtained the lowest total number of vehicles required to solve all of the problems in comparison to known alternate heuristics. The average CPU time required by the heuristic to obtain solutions to the VRPTW was considerably lower in comparison to similar heuristics based on meta-strategies.

Finally, our results indicate that a hybrid combination of Genetic Algorithms as a global search strategy in combination with Tabu

Search/Simulated Annealing as local solution improvement methods provide good solutions to VRPTW.

Acknowledgments

The author wishes to thank R. Russell for providing the 249 and 417 customer problem sets. This material is based upon work partly supported by the Slippery Rock University Faculty Development Grant No. FD2E-030, State System of Higher Education Faculty Development Grant No. FD3I-051-1810 and the National Science Foundation Grant No. USE-9250435. Any opinions, findings, and conclusions or recommendations expressed in this material are those of the authors and do not necessarily reflect the views of Slippery Rock University, State System of Higher Education or the National Science Foundation.

References

Baker, E.K. and J.R. Schaffer(1986). Solution Improvement Heuristics for the Vehicle Routing Problem with Time Window Constraints. *American Journal of Mathematical and Management Sciences* (Special Issue) 6, 261-300.

Bodin, L., B. Golden, A. Assad and M. Ball (1983). The State of the Art in the Routing and Scheduling of Vehicles and Crews. *Computers and Operations Research* 10 (2), 63 -211.

Chiang, W. and R. Russell (1993). Simulated Annealing Metaheuristics for the Vehicle Routing Problem with Time Windows, Working Paper, Department of Quantitative Methods, University of Tulsa, Tulsa, OK 74104.

Christofides, N. and J. Beasley (1984). The Period Vehicle Routing Problem. *Networks 4,* 237-256.

Connolly, D.T. (1990). An Improved Annealing Schedule for The QAP, *European Journal of Operations Research* 46, 93-100.

Davis, L. (Ed.) (1991). Handbook of Genetic Algorithms, Van Nostrand Reinhold, New York.

DeJong, K. (1980). Adaptive System Design: A Genetic Approach. *IEEE Transactions on Systems, Man and Cybernetics* 10 (9), 566-574.

Desrochers, M., J. Desrosiers and M. Solomon (1992). A New Optimization Algorithm for the Vehicle Routing Problem with Time Windows, *Operations Research* 40(2), 342-354.

Desrochers, M., J.K. Lenstra, M.W.P. Savelsbergh and F. Soumis (1988). *Vehicle Routing with Time Windows. Optimization and Approximation.* Vehicle Routing: Methods and Studies, B. Golden and A. Assad (Eds.), North Holland.

Desrosier, J., Y. Dumas, M. Solomon, F. Soumis (199?). Time Constrained Routing and Scheduling. Forthcoming in Handbooks on Operations Research and Management Science. Volume on Networks, North-Holland, Amsterdam.

Fisher M. (199?). Vehicle Routing. Forthcoming in Handbooks on Operations Research and Management Science. Volume on Networks, North-Holland, Amsterdam.

Fisher, M., K.O. Jornsten and O.B.G. Madsen (1992). Vehicle Routing with Time Windows, Research Report, The Institute of Mathematical Statistics and Operations Research, The Technical University of Denmark, DK-2800, Lyngby.

Fisher, M and R. Jaikumar (1981). A Generalized Assignment Heuristic for the Vehicle Routing Problem. *Networks,* 11, 109-124.

Gillett, B. and L. Miller (1974). A Heuristic Algorithm for the Vehicle Dispatching Problem. *Operations Research* 22, 340-349.

Glover, F. and M. Laguna (1993). *Tabu Search.* In Modern Heuristic Techniques for Combinatorial Problems, Reeves, C. (Ed.), John Wiley, New York.

Glover, F., E. Taillard and D. de Werra (1993). A User's Guide to Tabu Search, *Annals of Operations Research,* 41, 3-28.

Glover, F. (1993). Tabu Thresholding: Improved Search by Non-monotonic Trajectories, Working Paper, Graduate School of Business, University of Colorado, Boulder.

Glover, F. (1989). Tabu Search - Part I, *ORSA Journal on Computing,* 1 (3), 190-206.

Glover, F. (1990). Tabu Search - Part II, *ORSA Journal on Computing,* 2(1), 4-32.

Glover, F and H.J. Greenberg. (1989). New Approaches for Heuristic Search: A Bilateral Link with Artificial Intelligence, *European Journal of Operations Research* 39, 119-130.

Glover, F. (1986). Future Paths for Integer Programming and Link to Artificial Intelligence. *Computers and Operations Research,* 13,533-554.

Goldberg D.E. (1989). Genetic Algorithms in Search, Optimization, and Machine Learning. Addison-Wesley Publishing Company, Inc.

Golden B. and A. Assad (Eds.) (1988). Vehicle Routing: Methods and Studies. North Holland, Amsterdam.

Golden, B and A. Assad (Eds.) (1986). Vehicle Routing with Time Window Constraints: Algorithmic Solutions. *American Journal of Mathematical and Management Sciences* 15, American Sciences Press, Columbus, Ohio.

Golden B. and W. Stewart (1985). *Empirical Analysis of Heuristics.* In The Traveling Salesman Problem, E. Lawler, J. Lenstra, A. Rinnooy Kan and D. Shmoys (Eds.), Wiley-Interscience, New York.

Grefenstette, J.J. (1987). A Users Guide to GENESIS. Navy Center for Applied Research in Artificial Intelligence, Naval Research Laboratory, Washington D.C. 20375-5000.

Grefenstette, J.J. (1986). Optimization of Control Parameters for Genetic Algorithms. *IEEE Transactions on Systems, Man and Cybernetics,* 16(1), 122-128.

Holland, J.H. (1975). Adaptation in Natural and Artificial Systems. University of Michigan Press, Ann Arbor.

Kirkpatrick, S., Gelart, C.D. and Vecchi, P.M.(1983), Optimization by Simulated Annealing. *Science* 220, 671-680.

Koskosidis, Y., W.B. Powell and M.M. Solomon (1992). An Optimization Based Heuristic for Vehicle Routing and Scheduling with Time Window Constraints. *Transportation Science* 26 (2), 69-85.

Kontoravdis, G. and J.F. Bard (1992). Improved Heuristics for the Vehicle Routing Problem with Time Windows, Working Paper, Operations Research Group, Department of Mechanical Engineering, The University of Texas, Austin, TX 78712-1063.

Laporte, G. (1992). The Vehicle Routing Problem: An Overview of Exact and Approximate Algorithms. *European Journal of Operational Research,* 59, 345-358.

Lin, S and B.W. Kernighan (1973). An Effective Heuristic Algorithm for the Travelling Salesman Problem. *Operations Research,* 21, 2245-2269.

Lundy, M and A. Mees (1986). Convergence of an Annealing Algorithm, *Mathematical Programming,* 34, 111-124.

Metropolis, N., A.W. Rosenbluth, M.N. Rosenbluth, A.H. Teller and E. Teller (1953). Equation of State Calculation by Fast Computing Machines. *Journal of Physical Chemistry* 21, 1087-1092.

Osman, I.H. and N. Christofides (1994). Capacitated Clustering Problems by Hybrid Simulated Annealing and Tabu Search. *International Transactions in Operational Research,* 1 (3).

Osman, I.H. (1993b). Metastrategy Simulated Annealing and Tabu Search Algorithms for the Vehicle Routing Problems. *Annals of Operations Research,* 41, 421-451.

Potvin, J. and S. Bengio (1993). A Genetic Approach to the Vehicle Routing Problem with Time Windows, Technical Report, CRT-953,

Centre de Recherche sur les Transports, Universite de Montreal, Canada.

Potvin, J, T. Kervahut, B. Garcia and J. Rousseau (1992). A Tabu Search Heuristic for the Vehicle Routing Problem with Time Windows. Technical Report, CRT-855, Centre de Recherche sur les Transports, Universite de Montreal, Canada.

Potvin, J. and J. Rousseau (1993). A Parallel Route Building Algorithm for the Vehicle Routing and Scheduling Problem with Time Windows. *European Journal of Operational Research,* 66, 331-340.

Reeves, C. (Ed.) (1993). Modern Heuristic Techniques for Combinatorial Problems, John Wiley, New York.

Russell, R.A. (in press). Hybrid Heuristics for the Vehicle Routing Problem with Time Windows. To appear in *Transportation Science.*

Savelsbergh M.W.P. (1985). Local Search for Routing Problems with Time Windows. *Annals of Operations Research,* 4, 285-305.

Semet, F. and E. Taillard (1993). Solving Real-Life Vehicle Routing Problems Efficiently using Tabu Search. *Annals of Operations Research* 41,469-488.

Solomon, M. M, Edward K. Baker, and Joanne R. Schaffer (1988). *Vehicle Routing and Scheduling Problems with Time Window Constraints. Efficient Implementations of Solution Improvement Procedures.* In Vehicle Routing: Methods and Studies, B.L. Golden and A. Assad (Eds.), Elsiver Science Publishers B.V. (North-Holland), 85-90.

Solomon, M.M. and Jacques Desrosiers (1986). Time Window Constrained Routing and Scheduling Problems: A Survey. *Transportation Science* 22 (1), 1-11.

Solomon, M.M. (1987). Algorithms for the Vehicle Routing and Scheduling Problems with Time Window Constraints. *Operations Research* 35 (2), 254-265.

Thangiah, S.R., I.H. Osman, R. Vinayagamoorthy and T. Sun (1993). Algorithms for Vehicle Routing Problems with Time Deadlines. *American Journal of Mathematical and Management Sciences,* 13 (3&4), 323-355.

Thangiah, S.R. (1993). *Vehicle Routing with Time Windows using Genetic Algorithms.* Application Handbook of Genetic Algorithms: New Frontiers, Volume II, L. Chambers (Ed.), CRC Press, Florida.

Thangiah, S.R., R. Vinayagamoorthy and A. Gubbi (1993). Vehicle Routing with Time Deadlines using Genetic and Local Algorithms. *Proceedings of the Fifth International Conference on Genetic Algorithms,* 506-513, Morgan Kaufman, New York.

Thompson, P.M. and H. Psaraftis (1989). Cyclic Transfer Algorithms for Multi-Vehicle Routing and Scheduling Problems. *Operations Research,* 41,935-946.

Appendix A

/* Evaluation function for the Genetic Sectoring Algorithm used in the Genesis Package*/

```
double evaluate (chromosome, length)
{
    float const ADD CONST = 1.5;
    int chromosome[];          //chromosome is an array consisting of
                               length bits
    int length;                //length of the chromosome
    int *Seed_Points;          //location for holding the seed points
    double cost;               //cost of routing using the seed points

    //dynamically declare the number of locations required to store the seed
    points Seed_Points = (int *) calloc ((unsigned) No_Vehicles,
    sizeof(int));

    //Maximum angle that each sector can have
    Maximum_Angle = Max_Theta / (No_Vehicles);
    for (i=0; i < (No_Vehicles -1); i++)
    {
        //convert the first three bits from i to i+3 to an integer value
        Seed_Points[i] = Ctoi(outbuf,(i*3));
        //range the integer value between 0 and 10, using a constant of 1.5,
        Seed_Points[i] = (int) (Const * (i+1)) + (Seed_Points[i] *
        ADD_CONST));
        //if the break point exceeds Max_Theta, set it to Max_Theta
        if (Seed_Points[i] > Max_Theta)
        Seed_Points[i] = Max_Theta + 1;
    }
    //the last break point always has the maximum possible theta to capture
    all the customers Seed_Points[i] = Max_Theta+ 1;

    //get the cost of routing with the seed points
```

```
    cost = Get_Cost(Seed_Points);

    //Return the cost of the routes as the fitness value of the chromosome
    return(cost);

}
/******************************** end of file ****/
```

Chapter 10 Doing GAs with GAGS

J.J. Merelo and G. Romero
Grupo GeNeura
Department of Electronics and Computer Technology
Facultad de Ciencias
Campus Fuentenueva, s/n
18071 Granada (Spain)

e-mail: jmerelo@kal-el.ugr.es
http://kal-el.ugr.es/jj/jj.html

10.1 Introduction

GAGS is a C++ class library designed to make programming a Genetic Algorithm (GA) [Heitköeter and Beasley, 1996] easy, and at the same time, flexible enough to add new objects, which are treated in the same way as *native* ones. As many other class libraries, GAGS includes the following features:

- *Chromosomes,* which are the basic building blocks of a genetic algorithms. Chromosomes are bit strings, and have a variable length.

- *Genetic operators,* which are not part of the chromosome class, but are outside it (actually, they are halfway outside: being `friends`, they conceptually belong to the same class); this way, operators are not reduced to mutation and crossover: there are many predefined genetic operators, like bitflip mutation, crossover (uniform, 1- and *n*-point, and gene-boundary respecting), creep (change gene value by 1), transpose (permute genes), and kill, remove, and randomAdd, which alter the length of the chromosome. New operators can be added and used in the same way as the predefined. Operators can be applied blindly, or to a certain part of a chromosome.

- *Views,* which represent the chromome and are used to evaluate and print it; views are objects used to watch the chromosome as a float or int array or whatever is needed to evaluate it.

- *Population,* which includes a list of chromosomes to be evaluated and sorts them in another list when they are; operators for generational replacement, and a list of genetic operators together with their rates; the number of chromosomes, the list of operators and their rates can be changed in runtime.

- The evaluation or fitness function can be any C or C++ function, and it can return any class or type; thus, fitness is not reduced to a floating point number: it can be an array or any other user-defined type; the only requirement is that it has ordering operators defined.

The second and last feature set it apart from other GA class libraries, like Matthew's GAlib [Wall, 1995] and TOLKIEN [Tang, 1994]. At the same time, it lacks some features like a user interface and provision for more than one kind of chromosome.

In this chapter we will see the way a genetic algorithm is programmed using these building blocks. We will take a bottom-up approach, starting with a chromosome, and building up from there.

10.2 Creating Chromosomes

Chromosomes are the basic building blocks of a genetic algorithm and are used to represent a problem solution. You will not usually need to create them directly (that is the role of the Population class), but, just in case, this is the way it would be done.

```
#include <gagschrm. hpp>
main ( ) {
    const unsigned numBits = 12;
    chrom aChromosome( numBits );
    cout << aChromosome << endl;
}
```

which would output somethig like

```
(kal-el) ~/txt/tex/gags/examples> ex1
001011111001
```

that is, a 12-bit string. Chromosomes are always bit strings and they are not much more than that. They have no structure and no operators to work with them, other than getLength(), operator [] and getValue. Since chromosomes can be changed only at birth time, there are no *murators*, only *inspectors*.[1]

Actually, there are other ways of creating chromosomes: through copy constructors (from other chrosmosomes) or using a constructor from a bitString, which is the internal representation used for the chromosomes' bit strings. A bitString is actually a set of ordered bits, whose value can be true or false, but they can be defined using char strings.

```
bitString aBS: "101001111111";
```

---

[1] Actually they can be changed. We will later see the way it is done.

```
chrom bChromosome ( aBS );
```

Actually, this can be done in a single sentence, since automatic conversion is applied to the character string; but the actual way it is done is as shown.

There is not much you can do with only a chromosome, so we will go ahead to the next section, which deals with how to operate on a chromosome.

10.3 Genetic Operators

Genetic operators are used in genetic algorithms to generate diversity (*mutation-like* operators) and to combine existing solutions into others, decreasing diversity (*crossover-like* operators). The main difference among them is that the former operate on one chromosome, that is, they are *unary*, while the latter are binary operators. In GAGS, genetic operators need not be coerced into the mutation/crossover paradigm; they are just functions that map chromosomes onto chromosomes.

In GAGS, operators constitute a class, that is, they are not methods of the usual chromosome class. That means that they can be created, destroyed and changed in runtime and they can be subclassed to create new operators. This class is designed to act as a *functor,* i.e., function-syntax objects, which have `operator` () overloaded; this is only syntactic sugar, but it allows the look of the C++ implementation to approach the actual algorithm.

They act in two different ways: *non-directed,* which means that they act on a random part of the chromosome; and *directed,* that is, they act on a preselected part of the chromosome. Besides, many operators consider the chromosome is divided into chunks, usually called *genes;* the size of those chunks must be passed to their constructors as a parameter.

Some operators are variable-length operators: they alter the length of the chromosome. Chromosome length is always computed in every method that needs it, which means that GAGS is prepared for variable length chromosomes. Length is always changed by a discrete amount of genes. Note also that binary operators acting on variable length chromosomes can also change the length of the resulting chromosome.

An example that includes binary and unary operators is as follows:

```
#include <genop. hpp> // Chromosomes already
    included

main () {
const unsigned NUMGENES = 4;
const bitLength_t SIZEGENES - 3;
chrom aChromosome( NUMGENES, SIZEGENES ); // Create
```

```
chromosome

cout << "aChromosome\t\t\t" << aChromosome << endl;

genOp creeper( SIZEGENES, genOp::CREEP );   //
Apply unary genOp

creeper( &aChromosome );

cout << "aChromosome creeped\t\t" << aChromosome <<
endl;

genOp mutator( (mutRate_t) 0.1 );        // Another
unary genOp

mutator( &aChromosome );

cout << "aChromosome mutated 0.1\t" << aChromosome
<< endl;

genOp SGA2pt( (unsigned char) 2 );       // Binary
genOp, 2-point crossover

bitString aBS ="111111111111";

chrom anotherChrom( aBS );          // Define other
Chromosome from string

cout << "Crossovering with anotherChrom\n\t\t\t"

    << aChromosome << "\n\t\t\t" << anotherChrom <<
endl;

SGA2pt( &aChromosome, &anotherChrom );   // Result
on the first

cout << "Result\t\t" << aChromosome << endl;
```

This program defines two unary operators (creeper, which changes a gene by plus or minus one and mutate, which does the usual thing) a binary operator, SGA2pt, and a simple two-point crossover, and applies them to the defined chromosomes. A gene has been defined as a 3-bit segment, and each chromosome has 4 genes. Output would look like this.

```
aChromosome                         100011011110

aChromosome creeped                 100010011110

aChromosome mutated 0.1             100010011010

Crossovering with anotherChrom      100010011010
                                    111111111111

Result                              100010011110
```

The genetic operator type is computed in two different ways: if its constructor has unique parameters, the genetic operator type is deduced; if not, typed enums are used.

Using the modes shown in Figure 10.1, genetic operators use three different constructors.

```
genOp( mutRate_t _rate,

      genOpMode_mode = MUT);     // Mutation and
          uniform crossover

genOp( bitLength_t _lenSits,

      genOpMode_mode = TRANSP);   // Most operators
          acting on genes

genOp( unsigned char _numPts );  // number of
   xOver points > 1
```

Genetic operators can also be applied in a directed way, by using the applyAt method

```
   const unsigned toKill = 1;

   chrom thisChrom( "111000111000" );

   genOp killer( SIZEGENES, genOp::KILL );

   killer.applyAt( toKill*SIZEGENES, &thisChrom );

   cout << "Result\t\t\t" << thisChrom << endl;
```

which would output something like

```
Result 111111000
```

genOp::DUP	duplicates a gene with mutation
genOp::KILL	eliminates a gene
genOp::RANDINC	adds a random gene
genOp::TRANSP	transposes two genes, i.e., permutes its contents
genOp::CLONE	copies the chromosome without doing anything
genOp::MUT	usual bit-flip mutation
gcnOp::UXOVER	uniform crossover. interchanges bits between the two chromosomes

genOp::XOVER	usual *n*-point crossover: interchanges the part of the chromosome between two or more randomly generated points
genOp::GXOVER	same as before, but respects gene boundaries, which means that only whole genes are interchanged
genOp::ZAP	changes the value of a gene or part of a chromosome to another value
genOp::CREEP	changes the value of a gene by plus or minus one.

Figure 10.1 Genetic operators and modes.

Once again, you will not be able to solve many problems with chromosomes if the only thing you can do is watch its binary face and change it at will. Usually, running a genetic algorithm involves decoding a chromosome, gene by gene, and applying some function to it. We will show how to do this in the next section.

10.4 Looking at Chromosomes

The most straightforward way of doing it is using chrom:: getValue ( ) function, which turns the chromosome raw bits into unsigned longs; this obviously means that the gene size can never be bigger than the size of an unsigned in the particular machine, but don't worry, GAGS will warn you if you attempt to define huge genes. Thus, chromosomes can be fully decoded as follows:

```
cout << "Gene values: ";        //Printing chromosome
as unsigned

for ( unsigned i = 0; i < NUMGENES; i ++ ) {
  cout << aChromosome.getValue( i*SIZEGENES,
SIZEGENES ) <<" ";

}
```

Most usually, a chromosome will be watched as an array of floats, with fixed or variable range; each gene in the chromosome must be decoded to this range before evaluating it. views are used with this aim: decoding chromosomes to evaluate them; and they are not part of the chromosome, but separate objects, mainly so that new views can be defined, without delving into the chrom code.

For instance, if you want to print your aChromosome as if it were an array of floats in the range -1,1 (that is: start -1, range 2), a view could be defined and used in this way.

```
view<float> vista( SIZEGENES, -1, 2 );

for ( unsigned i = 0; i <
vista.size(&aChromosome); i ++ ) {

        cout << vista( &aChromosome, i) <<" ";

} cout << endl;
```

views are also functors, that is, they have function syntax, and several are defined by default: fixed-range array, variable-range array, and travelling salesman (which prints letters, instead of numbers). size computes the size in genes of the chromosome that is being decoded; remember that GAGS always uses variable-length chromosomes. They are also defined in such a way that different types are created depending upon the parameters you give it: fixed and variable-range array are defined using the same constructor.

The classes with which you have to instantiate views depend upon the kind you have defined, but, in any case, they must have all arithmetic operations defined.

With views, which define a standard decoding, chromosomes can be easily evaluated and fitness computed, as explained in the next section.

10.5 Computing Fitness

Once chromosomes have been created, massaged, and watched in many different ways, it's about time we evaluate them. In genetic algorithms, success is usually defined by the evaluation of a function, which tells us how good the chromosome is at solving the problem it has been instructed to solve.

Following the spirit of the classes described in past sections, fitness will also be a class in most GA applications. Usually fitness is only a floating point number, but, in some cases, there are several things to take into account when evaluating a chromosome: success rate and size, for instance. If fitness is a scalar number, nothing needs to be defined anew, but if we are going to use a structured fitness, some functions need to be defined: mainly comparison functions, since fitness is used to rank chromosomes. This would be an example of how to define and use fitness.

```
#include <gagsview.hpp>    //Chromosomes already
   included

struct fitness_t {

  float evaluation;// Scalar fitness value
```

```
unsigned len;      // Vector length

fitness_t( float _d = 0, unsigned _l = 0 ):
evaluation(_d), len(_l) {};

int biggerThan( fitness_t_f ) const {

if (evaluation > _f.evaluation) {

  return 1;

} else if ( evaluation == _f.evaluation )

return ( len < _f.len )71:0;

return 0;

};

int operator > ( fitness_t _f ) const { return
biggerThan( _f ); };

int operator < ( fitness_t _f ) const { return
!biggerThan( _f ); };

int operator == ( fitness_t _f ) const {

return (( evaluation == _f.evaluation) && (len ==
_f.len ));

}
};


inline ostream& operator <<(class ostream& s,
   fitness_t _f) {

  return s << _f.evaluation << " & " << _f.len <<
  endl;

}


// Fitness function
fitness_t fitness( const chrom& _chrom, const
view<float>& _vista ) {

  // Evaluates fitness by adding gene values

  for ( unsigned i = 0, float eval = 0; i <
_vista.size(&_chrom); i ++ ) {
  eval += _vista( &_chrom, i);

  }
```

```
    return fitness_t ( eval, _vista.size (&_chrom));

}


main () {

   const unsigned NUMGENES = 4;        // Usual
prologue

   const bitLength_t SIZEGENES = 3;

   seed_random( time( (time_t) 0 ) );// From randcl

   chrom aChromosome( NUMGENES*SIZEGENES ); //
Create chromosome


   view<float> vista( SIZEGENES, -1, 2 );


   cout << "Fitness is " << fitness( aChromosome,
vista );

}
```

Obviously, you don't always need to go to these lengths to define fitness; most of the times, a simple scalar fitness is used. This fitness will usually be returned by an evaluation function or fitness function and will usually be a method of an object that will be constructed from a chromosome.

The previous example describes how to define a fitness `class`, which in this case is defined as a `struct` which has all members public by default. Three operators: <, > and == are defined, a default constructor and a function to print it. A function to evaluate fitness is also defined; in this case, it adds the total value of the genes and sets it as fitness and then determines the length of the chromosome, which will be the second component in the fitness.

The main program then creates a chromosome and evaluates it. A `view` must be created so that the fitness function can evaluate the chromosome, or else write the fitness function using the `getValue` method of `chrom`; both chromosome and instantiated view are passed to the fitness function.

Most of the GA components are already in place, so we can now create a population, which is the class that really deals with the genetic algorithm in the next section.

10.6 Population

A population has a list of chromosomes to evaluate, usually a ranked list of evaluated chromosomes and a list of operators; since they are all lists, they can be changed at runtime. Creating a population is just a matter, then, of giving parameters to the creation of those lists. The easiest way of creating a population is as follows:

```
const unsigned MAXGENES = 10, MINGENES = 10;

const bitLength_t SIZEGENES = 4;

const unsigned NUMCHROMS = 20;

mutRate_t mutProb = 1. 0/(SIZEGENES*NUMGENES);

Population<float>

    people( NUMCHROMS, SIZEGENES, MINGENES,
    MAXGENES, mutProb );
```

For example, default parameters are the number of chromosomes in the population, the size of each gene, and the max and min number of genes in each chromosome; remember that GAGS works by default with variable-size chromosomes. The next parameter is the mutation probability, that is, the number of bits that will be flipped each time a chromosome is selected for mutation. Population is a class template that must be instantiated with the fitness type; in this case, fitness is simply a floating-point number, but it could be something more complex. In any case, Population will be instantiated with fitness_t, which must have been defined in advance as the type returned by the fitness evaluation function.

At the same time, since Population is a *container* class, it must have *iterators* to run on it. Instead of defining an *iterator* class by default for each Population object, as is done in the standard template library, it must be created by hand (but this might change in the future). In any case, there are three iterator classes defined for the population, corresponding to the three lists defined before.

All in all, here is an example of how to use the Population class.

```
#include <gagspop.hpp>        // For Population
class and iterators

#include <gagsview.hpp>       // For views
  main () {
  const unsigned NUMGENES = 10;
  const bitLength_t SIZEGENES = 4;
  const unsigned GENERATIONS = 100;
  const unsigned numChroms= 20;

  seed_random( time( (time_t) 0 ) ); // From randcl

  // Set up population

  Population<float>

    people( numChroms, SIZEGENES, NUMGENES,
```

```
   NUMGENES, 1.O/(SIZEGENES.NUMGENES) );

 people.setElite( 0.6 ); // Sets Elite rate

 genOp adder( SIZEGENES, genOp::DUP); // Creates
 gene duplicationn operator...

 people.addOp( &adder, 0.1 ); // ...and adds it to
 the population

 // Starts GA

 popIter<float> censo( people );//Creates iterator

 view<float> vista( SIZEGENES, -1.0, 2.0 ); //
 View with range -1.0, 1.0

 for ( unsigned i = O; i < GENERATIONS; i ++ ) {

   censo.reset( Iter::FIRST);      // Resets
   iterator

   while( censo ) {  // Evaluates fitness

      float val = 0-0;

      chrom. trap = censo.current(); //
      Chromosome to evaluate

      for ( unsigned j = O; j < tmp-
      >getLength()/SIZEGENES; j ++ ) {

        val += vista( tmp, j);

      }

      censo.setFitness( val ); // This erases it
      from censo

      censo++; // Ahead to the next chromosome

   }

   evaliter<float> eval( people ); // Iterator on
   evaluated chromosomes

   // mapItem contains a pointer to the
   chromosome, key the

   // fitness value

   for ( unsigned j = O;

       j < eval.current().mapItem-
       >getLength()/SIZEGENES; j ++ ) {

     cout << vista(eval.current().mapItem, j) <<
```

```
        "";
    }
    cout << " = " << eval.current().key << endl;
    if ( i < GENERATIONS - i)
        people.newGeneration(); // Performs
        selection and genetic op application
    }
}
```

It is also an example of how to run a complete genetic algorithm. The Population class, which could actually be a Genetic Algorithm class, is defined in three steps.

- Creation of a population object, giving it information about the number of chromosomes, its size, and parameters for the mutation operator. The population has a mutation and a 2-point crossover operator in its list by default, with equal *priority* (see below).

- Setting of the *elite* parameter – the number of chromosomes that will remain in each generation. Only steady-state selection is supported: (l-elite)*population will be eliminated, and substituted by the off-spring of the (l-elite)*population best. Operators are applied according to *priority,* which means that, depending upon the operator rate, a roulette wheel is created with slices corresponding to operator probability, and each time a chromosome is selected for reproduction, a random number is drawn and a genetic operator selected.

- Creating new genetic operators, which, at the same time, are added to the population list. In this step, operator probabilities can be changed (in fact, they can be changed at any time during the genetic algorithm) using the setRate function. Rates are actually more like *priorities,* since all rates are added and normalized to one before applying them in each generation. That is, if you have rates of 0.5, 1 and 1.5, each operator will be applied to approximately 18, 33 and 72% of each generation of the population that is being created .

The genetic algorithm part of the program is just a for loop over the number of generations that has been selected in advance, with an inner loop to evaluate each member of the population in turn. Note that the loop uses the nonevaluated population iterator, which means that each chromosome is only evaluated once; using the setFitness function moves the chromosome from the nonevaluated list to the evaluated list. Finally, the best chromosome, that is, the one which is biggerThan any other, is printed using the evaluated-population iterator.

This is more or less how far you can go by using GAGS in your own programs, that is, you need to write a fitness function, select the population parameters, and then the genetic operators and their rates. However, as any good-behaviored class library, you should be able to extend GAGS to adapt to your own purposes. The next section explains how to do it.

10.7 Extending GAGS

Both genetic operators and view use a letter/envelope structure (see [Coplien, 1994]); different "letter" and, thus, different implementations are used for different views and genetic operators, but the client-programmer need not worry about it, since all *letters* can be accessed using *envelopes*, that is, genOp and view. This is done mainly because views and genOps are friends of chromosomes and, since friendship is not inherited, only view base objects can access tim chromosome internal representation (i.e., tile bitstring). This causes a problem with templates since in, gnu's g++, there is no such a thing as friendship-to-template. Currently, only templates instantiated with float and char can be used; to use any other, it must be added to the friendship list at the beginning of chrom.

This letter/envelope structure, at the same time, makes it easier to program new objects, since the interface is simplified, and most, if not all, needed protected functions are already defined.

10.7.1 Extending Views

If you want to properly use views, the same interface should be kept, which means that the envelope class will stay more or less the same, except for new constructors, and only a new "letter" will be created. To do so, several steps have to be taken. We will build a Travelling Salesman view, in which there are $log_2 n$ size genes for n cities and they are represented as letters, a being the first city, b the second and so on.

1. Add a new constructor to the View class. This constructor should have a different calling structure than the others; if it does not, an enum will have to be created to differentiate it from the constructors for other objects with the same calling convention. The declaration will be included in gagsview.hpp and the definition in gagsview.icc, which includes all the template definitions. Once the name of the view is decided, the constructor includes a forward to the constructor of the derived class, and a check.

```
template<class Type>

view<Type>::view( unsigned _numCities ) {

    inview = new TSPView( _numCities );
```

```
    check( inview );

}
```

In this case, `TSPView` is not itself a template, since it will always return characters; usually, `Type` will be used for the construction of the letter object.

2. Declaration and definition of `TSPView` itself. The public virtual functions `operator ()` and `size` have to be redefined; the first one will return the *i*th gene, and the other the total number of genes. Take into account that only methods in the base class can access private parts of the chromosome, so that usually these functions will forward request to the base class. In this case, it accesses the `getValue` protected method, that, in turn, accesses the internal representation of the chromosome. Since the base and envelope class has only a pointer to the letter class, its default constructor is called. Check GAGS sources, to which `TSPview` has already been incorporated, to see the way it is implemented: it basically involves converting strings to a suitable representation.

3. Testing the new view. Insert an object of the new class into the program `pgview.cc`, which tests view objects, compiles and runs it.

```cpp
//pgview.cc

#include "gagsview.hpp"

main () {
    const unsigned NUMGENES = 5;

    const bitLength_t SIZEGENES: 8;

    chrom aChrom( NUMGENES.SIZEGENES );

    view<unsigned long> vista( SIZEGENES, -1, 2 );

    for ( unsigned i = 0; i < vista.size(&aChrom); i
    ++ ) {

        cout << vista( &aChrom, i) <<" ";

    }

    cout << endl;


    const unsigned numCities = 8;
    const unsigned numBits = 3;

    chrom anotherChrom( numCities • numBits );


    view<char> TSPvista( numCities );
```

```
for ( unsigned i = 0; i <
    TSPvista.size(&anotherChrom); i ++ ) { cout
    << TSPvista( &anotherChrom, i) << "_";
```

```
cout << endl;
```

10.7.2 Extending Genetic Operators

Extending genetic operators involves the same operations views: it is a matter of creating new *letter* classes, and changing some stuff in the envelope class. We will create a new operator, shift, which adds or substracts one from one gene, substracting or adding one to the next at the same time.

- The operator name, for instance, SHIFT, should be added to the enum in the genOp class header, that is, line 30 of the genop.hpp file.

- A new genOp constructor for the new class will be needed. If so, add it to the genOp header and declare it; if parameters are more or less the same as already defined operators, use one of the existing constructors, but in this case, since shift needs only the gene size, which uses the second constructor. This means that this constructor must be modified anyway. Go to the genop.cc file, around line 77, and add something like this.

      ```
      case SHIFT:
          op = new shiftOp( _lenBits );
          arity = FALSE;
          break;
      ```

  which creates the *letter* object for the new operator class, and assigns it FALSE arity, that is, arity one. Binary operators would have TRUE arity.

- Go to the genopimp.hpp header to create the class interface and copy whatever other class interface suits you best. This means that you already have operator () with arity one and two defined, constructor, but *not* copy constructor. It does not make much sense to copy genetic operators. In this case, you can directly copy creepOp declaration and change names.

- Write an implementation of the genetic operator and put in genopimp.cc or your own file. In this implementation, try to use the functions in genop.cc; remember that the *letters* cannot directly access chroms, only the envelope. Take a look at the implementation of the other genetic operators so that you have an idea of how things are done.

10.8 Solving Real Problems with GAGS

Some applications can be solved in a straighforward way using GAGS, but
others need some programming, although many of the problems come from
the fitness function, not the genetic algorithm itself. Here is a showcase of
some problems to which GAGS has been successfully applied.

10.8.1 Optimizing Neural Networks

Designing a good neural network (NN) for patter recognition has always
been a problem in the neural net community [Ripley, 1994]. Most of the
times, NN parameters and policies (like initialization or selection of the
training samples) are chosen resorting to past experience. A good algorithm
that automatically designs at least a kind of neural network would be
desirable. That is why we applied genetic algorithms to Kohonen's LVQ,
Learning Vector Quantization, a supervised codebook training algorithm,
yielding G-LVQ, an optimized version of the same algorithm [Merelo and
Prieto, 1995, Merelo et al., 1997a], which creates a population of variable-
size neural networks, trains them, and selects them according to several
quantities: first accuracy, then neural net size.

This application is the kind of problem GAGS is designed for: complex-
fitness, variable-length chromosome problems. However, the main program
is quite straightforward. Some sections have been suppressed and substituted
by a short description.

```
//File includes ...

main( int argc, char **argv ) {

    //Variable declaration...

    //Command line processing...

    //NN Training and test file loading...

    //Population setup

    Population<fitness_t>

      myPop( popSize, SIZEGENES, numGenes / 2,
      rangeMax, mutRate );

    myPop.setElite( elitePerc );       //Never forget
    to do this

    popIter<fitness_t> censo( myPop );


    //Genetic operator setup
```

```
genOp adder( SIZEGENES, genOp::DUP);

genOp killer( SIZEGENES, genOp::KILL);

if ( ! directed ) {            // Non-directed
genop application

  myPop.addOp( &adder, addRate*100 ); // Command
  line rates have 0-1 range myPop.addOp(
  &killer, killRate*100 );


//Genetic algorithm with directed operator
application

for ( unsigned g = 0; g < GENERATIONS; g++ ) {

  censo.reset( Iter::FIRST);      // Evaluate
  now,

  while( censo ) {               // Compute fitness
  and correct

    unsigned minWinner, maxWinner;

    chrom* tmp = censo.current();

            censo.setFitness(fitness( tmp,
            trSample, tstSample, dim,
            numClasses, startR, endR, epochs,
            minWinner, maxWinner, NULL));

    if ( ( g < GENERATIONS - 1 ) && directed) {
      // Before last generation
      if ((myrand(1000)/1000.0) < killRate ) {

        if ( (SIZEGENES < tmp->getLength())

          && ( minWinner*SIZEGENES < tmp-
          >getLength()) ) {// If not too small

        killer.applyAt( minWinner*SIZEGENES,
        tmp );

        }

      }

      if ((myrand(1000)/1000.0) < addRate ) {

        if ( maxWinner*SIZEGENES < tmp-
        >getLength() )
```

```
                    adder.applyAt( maxWinner*SIZEGENES,
                    tmp );
        }

    censo++;


    }
    //Print out best


    if ( g < GENERATIONS - 1)
    myPop.newGeneration();

}
```

As can be seen, population is set up in the usual way, except that the chromosome size range is effectively used, giving the max and min number of genes for each chromosome. Then, besides the two default operators, crossover and mutation, the operators that eliminate and randomly or not add genes to the chromosome are added to the list. The GA loop is also the usual, except that for each chromosome, the gene that gets most and least hits (maxWinner, minWinner) is stored, and used with some probability to be duplicated (using adder) or eliminated (using killer). Fitness takes into account three things: first, accuracy achieved by the neural network in classification of the test file; then neural net length, and then distortion, which represents the average distance from the test file to the codebook.

This application, at the same time, combines genetic algorithms and neural networks, that is, global search procedures and local search procedures, in a meaningful way; instead of making the GA set the NN weights, it sets only the NN *initial* weights, which makes search, to a certain point, faster and more precise.

10.8.2 Optimizing Ad Placement

The problem can be defined in this way: given M media, which can be printed, broadcasted, or other, place N ads in such a way that the audience is maximized, taking into account several constraints, like a maximum or approximate bound for money spent, and a maximum audience reached. Different media have different *ratings,* or audience, and obviously, different prices for an advertising unit, or *module.* This makes advertising placement a combinatorial optimization problem. Besides, the objective is not only to reach the possible consumer, but to reach him or her a certain number of times (called *impacts),* so that he or she will remember the ad afterwards, and modify his or her behavior accordingly.

This application [Merelo et al., 1997b] is even more straightforward than before. In this case, all elements intervening in the fitness were combined in a formula, so no vectorial fitness was used, default operator rates had to be adjusted, and creepOp was added to the mix, so that the number of ad placements changed smoothly.

```
// include files ...

main( int argc, char** argv ) {

   unsigned

      popSize = 400,       // population size

      generations = 100,  // number of generations

      sizeGenes = 4;    // size in bits of each
      locus/gene


   mutRate_t mutRate = 0.1;   // mutation rate;
   xOver is uniform with prob 0.01
   // Command line checking ...


   // Creation of media objects ...


   // Population declaration and setup

   Population<float>

      myPop( popSize, sizeGenes, chromSize,
      chromSize, mutRate );

   myPop.setElite( 0.6 ); // Never forget to do


   // Add new operators

   genOp creeper( sizeGenes, genOp::CREEP);

   myPop.addOp( &creeper, 0.1 );   // A new


   // Change rates using the genOp iterator

   popIter<float> censo( myPop );

   opIter<float> oi( myPop );

   while (oi) {

      switch ( oi.current().mapItem->getMode() ) {

      case genOp::MUT:

         oi.setRate( 0.12 );
```

```
      break;
   case genOp::XOVER:
      oi.setRate( 0.1 );
      break;
   default:
      break;
   }
   oi++;

}


// Genetic algorithm loop .........
for ( unsigned g= 0; g < generations ; g++ ) {
   censo.reset( Iter::FIRST);    // Evaluate
   fitness
   while( censo ) {              // Compute fitness and
   correct
      chrom* tmp = censo.current();
      censo.setFitness( unMedio->fitness( tmp,
      sizeGenes ));
      censo++;
   }
   // Print best ...
   if ( g < generations - 1)
      myPop.newGeneration();
   }
}
```

10.8.3 Playing Mastermind

Solving the game of MasterMind using GAs is quite a difficult problem [Bernier et al., 1996], since there is only one correct solution. Along the game, there are several partial solutions, and the GA will strive to find them. This problem required a lot of tweaking of the GA, mainly to keep diversity, but also to overcome the problem of having discrete fitness and to keep the number of generated solutions to a minimum, since the success of a MasterMind solving program lies not only in the number of guesses made before the final solution, but also on the number of combinations generated to find it. Using GAGS, the following design decisions were taken:

- Population was huge, in order to keep diversity, around 500 individuals, and it increased with the length of the combination, that is, the size of the space to search.

- Besides usual operators, *transposition* was also used; since it permutes the values of two gene positions, it was quite adequate for combinatorial optimization problems.

- Some operators were not adequate for some phases of the search: for instance, it did not make much sense to use mutation when the combination was correct except for pin position. That is why the application rate of all the operators changed with the number of correct positions and colors, to become zero except for transposition when all the colors were correct.

This program has been working online for a long time at the URL http:// kal-el .ugr.es/mastermind.

10.9 Conclusion and Future Work

GAGS is a C++ class library which can be easily used for solving many problems using Genetic Algorithms, and, at the same time, can easily add new operators or new interpretations of the chromosome. So far, it has been used in several applications, allowing the rapid development of new applications, usually in less than one week for an expert C++ programmer.

This does not mean that it lacks some things. Some of these features orders of importance might be added in the future.

- STL *compliance* STL has been recently adopted as the standard C++ library. GAGS could use many of its data structures, like lists, maps, vectors and so on. STL involves some changes in mentality and, obviously, in interface.

- Adding new selection strategies as functors and taking selection strategies out of the Population class. Conceptually, selection and reproduction operators should be outside the Population class and, besides, this would allow changing population operators in the same way that chromosome-level operators can be changed now.

- Adding a user interface.

References

[Bernier et al., 1996] Bernier, J.L., Herraiz, C. 1., Merelo, J.J., Olmeda, S., and Prieto, A. (1996). Solving *mastermind* using GAs and simulated annealing: a case of dynamic constraint optimization. In *Proceedings PPSN, Parallel Problem Solving from Nature IV,* number 1141 in

Lecture Notes in Computer Science, pages 554 563. Springer-Verlag.

[Coplien, 1994] Coplien, J.O. (1994). *Advanced C++: programming styles and idioms.* Addison Wesley.

[Heitköeter and Beasley, 1996] Heitköeter, J. and Beasley, D. (1996). The Hitchhiker's Guide to Evolutionary Computation, v. 3.4. Technical report, Available at the ENCORE sites.

[Merelo and Prieto, 1995] Merelo, J.J. and Prieto, A. (1995). G-LVQ, a combination of genetic algorithms and LVQ. In D.W.Pearson, N. and R.F. Albrecht, editors, *Artificial Neural Nets and Genetic Algorithms,* pages 92-95. Springer-Verlag.

[Merelo et al., 1997a] Merelo, J.J., Prieto, A., and Morán, F. (1997a). A GA-optimized neural network for classification of biological particles from electron-microscopy images. In Prieto, Mira, C., editor, *Proceedings IWANN 97,* number 1240 in LNCS. Springer-Verlag.

[Merelo et al., 1997b] Merelo, J.J., Prieto, A., Rivas, V., and Valderrábano, J.L. (1997b). Designing advertising strategies using a genetic algorithm. In *6th AISB Workshop on Evolutionary Computing,* Lecture Notes in Computer Science. Springer.

[Ripley, 1994] Ripley, B.D. (1994). Neural networks and related methods for classification. *J.R. Statist. Soc. B,* 56(3):409-456.

[Tang, 1994] Tang, A. (1994). Constructing GA applications using TOLKIEN. Technical report, Dept. Computer Science, Chinese University of Hong Kong.

[Wall, 1995] Wall, M. (1995). Overview of Matthew's genetic algorithm library. found at http://lancet.mir.edu/ga.

Chapter 11 Memory-Efficient Code for GAs

Stuart J Clement
Transport Systems Centre
University of South Australia
Stuart.Clement@unisa.edu.au

Jonathan Halliday
Department of Computing Science
University of Newcastle upon Tyne
Jonathan.Halliday@ncl.ac.uk

Robert E Smith
Intelligent Computing Systems Centre
Computer Studies and Mathematics Faculty
The University of West England
robert.smith@uwe.ac.uk

11.1 Introduction

Despite ongoing increases in the speed and power of computers, the memory and processor requirements of large GA-based programs remain a concern for programmers. Presented here is a C language implementation of bit-level chromosome storage and manipulation routines that offer substantial memory savings over the more commonly used array-based representations. The code is that of SGA-C developed by Robert Smith of the University of Alabama and subsequently improved by Jeff Earickson of the Boeing Company (Smith, Goldberg and Earickson 1994a and 1994b). In distributed and parallel processing GAs, this code offers the additional benefit of reducing message size when exchanging chromosome information between nodes or processes.

With reference to the SGA-C source code (Smith, Goldberg and Earickson 1994a and 1994b), a detailed commentary and step-by-step analysis of the operation of a number of basic GA functions is presented. The explanations for chromosome initialisation, mutation, crossover and value extraction (for the case where a chromosome contains several concatenated integer values) are given. From this base the reader may derive more complex operations suited to their particular application and two such routines are given: one to initialise a chromosome to specified values and the other to write out the chromosome string in a form conducive to western number recognition. The SGA-C code contains older C style function headers – here the function headers are presented in C++ style for greater compiler compatibility.

0-8493-2539-0/98/$0.00+$.50
© 1999 by CRC Press LLC

Following the explanations of the code (accompanied by portions of the code), detailed results are presented to show the benefits of its use. Theoretical and benchmarked examples are supported by our experience with Cabal (Clement 1997), a GA-based system intended for realtime traffic signal optimisation at road junctions. The use of bit-level chromosome representation using modified SGA-C code was an essential element of Cabal, being theoretically practical for realtime operation. Initially, Cabal was implemented using an array-based representation written in Pascal on a machine with 4M of memory. This limited the possible population size and curbed the effectiveness of the GA search to the extent that realtime operation was not possible. Cabal was then implemented using the SGA-C bit-based representation (Smith, Goldberg and Earickson 1994a and 1994b) which alleviated the memory problem and provided the bonus of being much faster than the Pascal array-based code. The price of computer memory has since fallen to the extent that optimisation of traffic signals for a single intersection (Cabal's function) can easily be performed on one machine using array-based code without seriously running into memory limits due to population size. But the memory efficiency of the SGA-C code makes the extension of Cabal's function to optimising the signals of a network of intersections theoretically possible. Details of this method can be found in Clement (1997) and need not be repeated here suffice to say that the scheme uses many GAs. It should be noted that an array-based C language implementation of Cabal was consistently two seconds faster than the bit-based code. This was independent of the population size or chromosome configuration. Further development of Cabal towards realising a road traffic network optimisation capability was to implement Cabal in parallel computing configurations (Halliday 1997). This work also benefited from the reduced chromosome sizes by achieving speed gains.

11.2 Background to the code

SGA-C is a freely available (see Smith, Goldberg and Earickson 1994a for Internet location) suite of C language functions for building GA-based programs. SGA-C was developed using as a basis Goldberg's SGA (simple genetic algorithm) code written in Pascal (Goldberg 1989). What then are the differences between Goldberg's original SGA code and the SGA-C code of Smith, Goldberg and Earickson? The SGA code implements a chromosome as an array of ones and zeros using a boolean data type for each chromosome bit. Hence a population is an array of arrays of booleans.

The SGA-C code represents each chromosome as the actual bits of one or more unsigned integer data types. The number of unsigned integers needed to store each chromosome depends on the length of the chromosome, what range of values the integral bit-groups of the chromosome are to handle and the memory architecture of the machine on which the GA is implemented.

11.3 Computer Memory

To aid in understanding the code and its operations and functions, a simplified model of a part of a computer's memory is employed. This is an abstract representation and should not be taken as corresponding to any particular hardware. The C language is independent of memory architecture with respect to the number of bytes used for some data types. For example, on a 16-bit architecture the *unsigned int* type has a size of 2 bytes but in 32-bit memory 4 bytes is used. Therefore, to keep the examples to manageable proportions *sizeof(unsigned int)* will be set to 2 bytes. The code will work equally well on 32-bit architectures with the appropriate compiler. In addition the code will work if a data type of *unsigned long* is used.

	Low Memory			High Memory	
Memory	\<one byte\>	\<one byte\>	\<one byte\>	\<one byte\>	...
Address	0x0000	0x0001	0x0002	0x0003	...

Figure 11.1   Linear model of computer memory

Although memory can be thought of as a linear model (as in Figure 11.1) of blocks of one byte memory, a vertical representation will be used in the examples (as in Figure 11.2).

High Memory

\<one byte\>	\<one byte\>
0x000C	
\<one byte\>	\<one byte\>
0x000A	
\<one byte\>	\<one byte\>
0x0008	
\<one byte\>	\<one byte\>
0x0006	
\<one byte\>	\<one byte\>
0x0004	
\<one byte\>	\<one byte\>
0x0002	
\<one byte\>	\<one byte\>
0x0000	

Low Memory

Figure 11.2   Vertical model of computer memory

This illustrates clearly the use of consecutive unsigned integers (each of 2 bytes) to hold the chromosomes of a population. Dynamic memory allocation (i.e., that performed with the C function *malloc()*) is assumed to take place on a heap which grows from low memory towards high memory. For simplicity our memory allocation begins at virtual address 0x0000 and all required memory will be regarded as occupying contiguous physical memory though, of course, the code compiler decides where exactly in memory the chromosomes are held. The C language uses the 0x prefix to denote a hexadecimal constant value.

11.3.1 Memory Allocation

We begin by reserving the memory we will need to hold an example population of 100 chromosomes, each of which is 22 binary digits long. As each integer (and unsigned integer) in our system can hold 16 bits we have the option of either close packing the chromosomes or beginning each new chromosome on an integer boundary. To achieve a compromise between memory space and simplicity of code, the latter method was implemented.

Each of our 22-bit chromosomes requires 2 unsigned integers worth (32 bits) of memory. Figure 11.3 shows a model of the required memory structure. The block of memory holding the total population will be manipulated using a pointer to *unsigned int*. This pointer can be made to return the address of any *unsigned int* in the block of memory holding the population. For our example, incrementing the pointer to *unsigned int* by 2 (i.e., by 4 bytes) would cause it to point to the first byte of the next chromosome in the population.

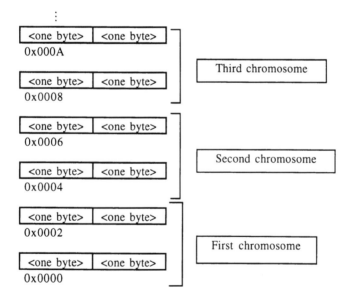

Figure 11.3   Memory model of example chromosome population

There is no need to delve further into the memory allocation and addressing intricacies of computers but a memory model diagram of how we will represent the chromosomes and the position of the significant bits is necessary. This is shown in Figure 11.4.

	Unsigned Integer 0		Unsigned Integer 1	
chromosome	01011001	01101001	msb	00110110
		lsb	10001011	

Figure 11.4   Chromosome representation

The unsigned integers are labelled from zero to be consistent with the C language addressing scheme. The number represented by the chromosome has its least significant bit in the rightmost position of Unsigned Integer 0 and its most significant bit in the leftmost position of Unsigned Integer 1. Thus the number x, represented by the above bit-string would be

$$x = 2^0 + 2^3 + 2^5 + 2^6 + 2^8 + 2^{11} + 2^{12} + 2^{14} + 2^{17} + 2^{18} + 2^{20} + 2^{21} + 2^{24} + 2^{25} + 2^{27} + 2^{31}$$

In reality, the chromosome would often be divided into bit-groups, each bit-group representing an integer number which reflects some parameter of the GA objective function.

With the preliminary descriptions of memory models dispensed with, the code needed for memory allocation can be presented. The following variable declaration code is contained in the original sga.h SGA-C header file and referenced in external.h.

⋮

int popsize;   /* population size */

int lchrom;   /* length of the chromosome per individual */

int chromsize;   /* number of bytes needed to store lchrom string */

⋮

Also in external.h and sga.h is the code for setting up the size of the unsigned integer

#define BITS_PER_BYTE 8   /* number of bits per byte on this machine */

#define UINTSIZE (BITS_PER_BYTE*sizeof(unsigned))   /* # of bits in unsigned */

To gel with the memory explanation above and to make the examples which follow clearer:

int UINTSIZE = 8 * sizeof(unsigned int);   /* System dependent. Here UINTSIZE = 16 */

In SGA-C the length of the chromosome and the size of the population are set in the function *initdata( )* of the initial.c file by reading in values from the keyboard or a file. For the example they are set in code as lchrom = 22 /* The length of the chromosome, in bits. */

popsize = 100 /* The number of chromosomes per population */

unsigned int * population; /* A pointer to our set of chromosomes. */

Note also that we have dispensed with the pointers to the *individual* data structures of the populations as given in SGA-C and simply implemented a pointer to *unsigned int* to reference the population of chromosomes. This is done purely to simplify the description of the bit-level manipulations to follow.

Also in initial.c of SGA-C the chromosome size in *unsigned ints* is calculated.

```
{
    ⋮

    /* define chromosome size in terms of machine bytes, i.e.,  */

    /* length of chromosome in bits (lchrom)/(bits-per-byte) */

    /* chromsize must be known for malloc() of chrom pointer */

    chromsize = (lchrom/UINTSIZE);

    if(lchrom%UINTSIZE) chromsize++;

    ⋮

}
```

The comments in the original code above are relevant to a machine where the number of bytes used for an *unsigned int* is one. Therefore **chromsize** in this case is the number of bytes needed to store the chromosome bit-string as commented in the SGA-C header files. But for machines of different *unsigned int* representation (e.g., 16-bit and 32-bit systems), the variable **chromsize** is a misnomer as it actually contains the number of unsigned integers needed to store the chromosome bit-string. The code of initial.c is again presented but with comments that are applicable to our example.

```
/* Calculate the number of unsigned ints needed to store the chromosome */

    chromsize = ( lchrom / UINTSIZE );    /* chromsize = 22 / 16 = 1 */

if( lchrom % UINTSIZE ) /* 22 % 16 = 6 = true because it is non-zero */

    chromsize++;                /* now chromsize = 2 */
```

The memory for the chromosomes is allocated in the memory.c file but since we have changed the representation of the population, the memory allocation must differ, too. Note that here we use **numBytesPerPop** where the original code used **nbytes**.

/* Allocate the memory for the chromosomes. */

int numBytesPerPop = popsize * chromsize * sizeof(unsigned int);

/* 100* 2 * 2 = 400 bytes */

population = (unsigned int *) malloc(numBytesPerPop);

11.3.2 Shift Operators in C

The SGA-C code contains several functions that use the left-shift and right-shift bit operators of the C language. These operators alter the bit patterns of data structures (*ints, chars,* etc.) in C and are integral to the SGA-C functions. The left-shift (language symbol <<) and right-shift (>>) operators are detailed in Kernighan and Ritchie (1988). Relevant portions of the full descriptions that appear in the Microsoft Visual C++ Reference Volume II manual (Microsoft Corporation 1993) are given below as an understanding of the operators is fundamental to understanding the operation of many SGA-C functions including *ithruj2int( ), mutation( )* and *crossover( ).*

'The left-shift operator causes the bit pattern in the first operand to be shifted left the number of bits specified by the second operand. Bits vacated by the shift operation are zero-filled. The shift is a logical shift...'

'The right-shift operator causes the bit pattern in the first operand to be shifted right the number of bits specified by the second operand. Bits vacated by the shift operation are zero-filled for unsigned quantities... The shift is a logical shift if the left operand is an unsigned quantity...'

11.4 Initialisation of Chromosomes

Having allocated the necessary memory for our population, initialisation of the chromosomes to random values is performed. For this step, the provision of a simple random number generator function, *int flip(float* **prob**), is assumed. This function randomly returns a value of 1 with a probability of **prob** and a value of 0 with probability 1 - **prob**. It is, in effect, a biased coin flip (or a fair coin flip if a **prob** value of 0.5 is used). Although care should be taken with boundary conditions, as these may be system dependent, a sample implementation would be:

```
int flip(float prob)
{
    if( (randnumgen() / RAND_MAX) <= prob )
        return 1;
    else
        return 0;
}
```

It is well known that on some machines the 'standard' *rand()* and its seed-setting conjunct function *srand()* lack performance and hence we use the function name *randnumgen()* as a generic name to express what the function is supposed to do. It is assumed that a good function to return the required values is available to the reader.

The initialisation routine *initpop()* for the population is found in initial.c. The code is characterised by three nested *for* loops which iterate through

1. every chromosome in the population;

2. each *unsigned int* of the chromosome; and

3. each relevant bit of the current integer.

Note that all of the bits of the *unsigned int* are set to 0 in the middle loop but not all of the bits in the chromosome's last unsigned integer may be used. In this example all 16 bits of the first integer are used along with the first 6 bits of the second integer, leaving 10 bits spare. These do not need initialising and in fact they are required to be 0 for the other genetic operations. The position of the end of the chromosome is found in this loop and this is used as the stop point for the inner loop (loop 3). Note also in the inner loop the OR operation with the mask value of 1 to set the least significant bit.

Now we can step through this code for a single chromosome while referring to the states of the unsigned integers of the chromosome as shown in the diagrams of Figure 11.5. As before, the LSB is the rightmost bit of Integer 0 and the MSB is the leftmost bit of Integer 1.

Integer 0		Integer 1		State
????????	????????	????????	????????	One
00000000	00000000	????????	????????	Two
00000000	00000001	????????	????????	Three
00000000	00000010	????????	????????	Four
10110101	10001011	????????	????????	Five
10110101	10001011	00000000	00000000	Six
10110101	10001011	00000000	00100101	Seven

Figure 11.5   Results of chromosome initialisation procedure

Initially, i.e., at the start of the *initpop()* function, the value of the bits is random and unknown (State One). On the first pass through the middle of the three loops (for each *unsigned int* of the chromosome), attention is focused on the first of the integers making up the chromosome. This is initialised to 0 (State Two).

On first entering the innermost loop, the left-shift operation has no visible effect since all the bit values are already 0 and the left-shift operator automatically fills the least significant bit of the integer with a zero. If the random coin flip returns a value of 1, then the LSB of the integer is set to 1 by using the OR operator on the integer and an integer mask set to 1 (State Three). If 0 is returned from the random coin flip, then there is no change as the LSB would already be 0 (State Four).

The inner loop is repeated another 15 times, i.e., up to the value of the variable **stop**. This leaves all the bits of the first integer set to 1 or 0 with equal probability (e.g., State Five). On the second pass through the middle loop, **stop** will not be 16 but instead is determined by the number of unset bits in the chromosome. We have

stop = lchrom - (**k** * UINTSIZE)   /* stop = 22 – (1 * 16) = 6 */

Hence the innermost loop will execute only six times. As the integer is initially set to zero (State Six), the repeated left-shift operations leave zero values in the 10 most significant bit positions, which are not used by the chromosome. The six positions on the right are occupied by randomly initialised values when the function completes (State Seven).

11.4.1 Mutation Operator

As we now have a population of randomly initialised chromosomes, it is time to perform the necessary genetic operations on them. The simpler mutation operator will be examined first, followed by the single point crossover operator. Both of these are found in the operator.c file of SGA-C.

The following function header is C++ compiler-compatible

void mutation(unsigned int * child)

The structure of the mutation operator has many of the same elements as the initialisation routine: indeed the algorithm it uses is very similar. Part of the same loop structure is employed but mutation – the changing of a single bit of the chromosome from a 1 to a 0 or vice versa with some probability, pmutation - occurs only if the *flip()* function returns a 1 in which case the mask value, initially all zeros, has its jth bit set to a one. Mutation is accomplished by using the XOR operator ^ between the chromosome and the mask. The point to note about the XOR operator is that whatever the value of the chromosome bit, if the corresponding mask bit value is 1, then the chromosome bit is inverted.

Table 11.1 contains the truth table for the XOR operator.

The point to note about the XOR operator is that whatever the value of the chromosome bit, if the corresponding mask bit value is 1, then the chromosome bit is inverted.

Table 11.1   XOR operator truth table

Chromosome bit	Mask bit	Result
0	0	0
0	1	1
1	0	1
1	1	0

This code can, in theory, mutate every bit of each chromosome since it references each bit (counter j varying from zero to either one less than the number of bits in the unsigned integer structure or one less than the number of bits used in the last unsigned integer of the chromosome) of every chromosome and the *flip()* function is called at each bit reference. A typical low value of **pmutation** in practice means few bits are inverted.

If *flip()* returns a 1, then the bits of **temp** are left-shifted **j** times. The OR operator is then used with **mask** and **temp**. The **mask** variable is incrementally built during the for(j=0;...) loop and could well contain more than one 1 in its bit string. An example of how the mask operator is built is shown in Figure 11.6 for the first eight iterations using an unusually high number of mutations. The illustration uses only one unsigned integer of each of **temp** and **mask**.

j	Mutate?	temp<<j		mask=mask|(temp<<j)	
0	No	00000000	00000001	00000000	00000000
1	No	00000000	00000001	00000000	00000000
2	Yes	00000000	00000100	00000000	00000100
3	No	00000000	00000001	00000000	00000100
4	Yes	00000000	00010000	00000000	00010100
5	No	00000000	00000001	00000000	00010100
6	No	00000000	00000001	00000000	00010100
7	Yes	00000000	10000000	00000000	10010100
⋮	⋮	⋮	⋮	⋮	⋮

Figure 11.6   Building the mask for the mutation operation

Figure 11.7 illustrates the action of the XOR operator using the mask (built in Figure 11.6) on the first unsigned integer (Integer 0) of the example chromosome.

	Integer 0		Integer 1	
chromosome	10110101	10001011	00000000	00100101
mask	00000000	10010100	00000000	00000000
Result of XOR	10110101	00011111	00000000	00100101

Figure 11.7   Result of using the mutation function

Mutation is performed one unsigned integer at a time until the number of unsigned integers (**chromsize**) comprising the chromosome is reached. The mask is built one unsigned integer at a time and hence the XOR operator applies the mask to one integer of the chromosome at a time.

11.4.2 Single-point Crossover Operator

The single-point crossover operator takes two parent chromosomes and produces two child chromosomes. The children will be identical to their parents with probability 1 - **pcross**. A utility function, *int rnd(int* **lowerbound**, *int* **upperbound**) is assumed to be available and to return a uniform random value between and including its two arguments **lowerbound** and **upperbound**.

The function header is C++ compiler-compatible

int crossover (unsigned int *parent1, unsigned int *parent2, unsigned int *child1, unsigned int *child2)

The code worthy of explanation is taken directly from SGA-C. Added explanatory comments begin with '/*//'.

⋮

```
if(jcross >= (k*UINTSIZE))
```

/* Crossover point is not in this unsigned integer (i.e., it is in another unsigned integer to be referenced in a subsequent iteration) therefore do straight parent to child copy. */

```
    {
        child1[k-1] = parent1[k-1];
        child2[k-1] = parent2[k-1];
    }
    else if((jcross < (k*UINTSIZE)) && (jcross > ((k-1)*UINTSIZE)))
```

```
/*// Crossover point is somewhere in this unsigned integer. */
{
    mask = 1;
    for(j = 1; j <= (jcross-1-((k-1)*UINTSIZE)); j++)
    {
        /*// Set the mask value. */
        temp = 1;
        mask = mask<<1;
        mask = mask|temp;
    }
    /* Construct two offspring chromosomes by using the mask to
    copy the bits from the corresponding parent up to the crossover
    point, and then invert the mask to copy the bits after the
    crossover point from the other parent. */
    child1[k-1] =(parent1[k-1]&mask)|(parent2[k-1]&(~mask));
    child2[k-1] =(parent1[k-1]&(~mask))|(parent2[k-1]&mask);
}
else
    /* Crossover point was in an unsigned integer referenced and
    processed during a previous iteration therefore do a swapped
    parent to child copy. */
{
    child1[k-1] = parent2[k-1];
    child2[k-1] = parent1[k-1];
}
  ⋮
```

The example that follows illustrates how the crossover operator works. The example chromosome is 22 bits long with **chromsize** = 2. The example shows how crossover is effected when the crossover point is 10, i.e., in the first unsigned integer of the chromosome. Crossover occurs in exactly the same way when the crossover point is in the second unsigned integer and the code will work with chromosomes utilising any number of integers.

For the example **jcross** is set to 10 and the interesting function action occurs when the counter **k** is 1. Inside the else if... statement, **mask** is set to 1 and **j** will iterate from 1 to 9. Figure 11.8 shows some of the process of building the required mask.

j	$Mask=mask<<1$		**mask=maskltemp**	
1	00000000	00000010	00000000	00000011
2	00000000	00000110	00000000	00000111
⋮	⋮	⋮	⋮	⋮
9	00000011	11111110	00000011	11111111

Figure 11.8   Building the mask for the crossover operation

Figure 11.9 shows how the two offspring are constructed using the mask, the logical AND operator (&), the bit inversion operator (~) and the OR operator. The result of the AND operation on parent2 is transferred to the left chromosome in Figure 11.9 to illustrate clearly the OR operation.

child1	parent1	01111001	01100001
	mask	00000011	11111111
	&	00000001	01100001
	(transfer)	10000000	00000000
	OR	10000001	01100001
	parent2	10000011	11101000
	(~mask)	11111100	00000000
	&	10000000	00000000
child2	parent1	01111001	01100001
	(~mask)	11111100	00000000
	&	01111000	00000000
		00000011	11101000
	OR	01111011	11101000
	parent2	10000011	11101000
	mask	00000011	11111111

&	00000011	11101000

Figure 11.9   Construction of two offspring

The example shows crossover applied to the *unsigned int* containing the crossover point. For all other *unsigned int*s of the chromosome, straight parent to child copying occurs and requires no description.

11.4.3 The Value Extraction Function *ithruj2int()*

This function converts a portion of a chromosome bit string into its integer value. The function works irrespective of the number of unsigned integers (memory structures) comprising that chromosome and the positions of the bit string portions within that chromosome. The function will correctly convert a group of bit strings that fall across a memory structure boundary.

This function header is C++ compiler-compatible:

int ithruj2int(int i, int j, unsigned int *from)

The portion of code is copied from the SGA-C file utility.c and this is followed by detailed descriptive examples of its operation.

⋮

```
/* check if bits fall across a word boundary */

if(iisin == jisin)

    bound_flag = 0;

else

    bound_flag = 1;

if(bound_flag == 0)

{

    mask = 1;

    mask = (mask<<(j1-i1+1))-1;

    mask = mask<<(i1-1);

    out = (from[iisin]&mask)>>(i1-1);

    return(out);

}

else
```

```
    {
        mask = 1;
        mask = (mask<<j1)-1;
        temp = from[jisin]&mask;
        mask = 1;
        mask = (mask<<(UINTSIZE-i1+1))-1;
        mask = mask<<(i1-1);
        out = ((from[iisin]&mask)>>(i1-1)) |
(temp<<(UINTSIZE-i1+1));
        return(out);
    }
    ⋮
```

The function has error trapping code to check that i and j have the correct interrelationship and are within the desired range.

The 22-bit chromosome used in the examples is shown in Figure 11.10.

	Integer 0	Integer 1
chromosome	01111001 01101001	00000000 00110110

Figure 11.10 Chromosome used in examples of function *ithruj2int()*

As before the unsigned integers of the chromosome memory structure are base-referenced from the LSB (right-hand end). The integer numbers in this example are contained in the bit groups of 5, 5, 5 and 7 bits grouped from the LSB. Therefore we expect values of 9, 11, 30 and 108 to be returned from the *ithruj2int()* function.

Table 11.3   Variable value calculations for processing of the first group of 5 bits

Variable	Calculation	Value
i		1
j		5
iisin	1/16	0
jisin	5/16	0
i1	1-(0*16)	1

j1	5-(0*16)	5
bound_flag		0

The first example shows how the first group of 5 bits is converted to the value 9. Table 11.3 shows calculations for setting some of the variables used in the code execution.

The variables **iisin** and **jisin** are used to reference the unsigned integers of the chromosome. The variables **i1** and **j1** are used to govern the left- and right-shift operations. The results of the processing steps are shown in Figure 11.11. Note that the code is broken down into its component steps.

Step #	Operation	Result	
1	mask	00000000	00000001
2	mask = mask<<5	00000000	00100000
3	mask = mask − 1	00000000	00011111
4	a = mask = mask<<0	00000000	00011111
5	b = from[iisin]	01111001	01101001
6	out = a AND b	00000000	00001001
7	out = out>>0	00000000	00001001

Figure 11.11 Processing steps for the first 5-bit group

The integer value of **out** is therefore 9.

Next let's look at the case where the integer group extends across the unsigned integer boundary: the 7-bit group from bit 16 through to bit 22. Calculations for variables are shown in Table 11.4.

Table 11.4   Variable value calculations for processing of 7-bit group

Variable	Calculation	Value
i		16
j		22
iisin	16/16	1
jisin	22/16	1
i1	16-(1*16)	0
j1	22-(1*16)	6

Further processing occurs for the values of **iisin** and **i1** and **bound_flag** is set as shown in Table 11.5.

Table 11.5   Further calculations for iisin and i1 variables and setting of bound_flag

Variable	Calculation	Value
iisin	1-1	0
i1	16-(0*16)	0
bound_flag		1

The results of the processing steps are shown in Figure 11.12. Processing is over two unsigned integers and hence **iisin** had to be recalculated to one less than **jisin** for correct referencing.

Step #	Operation	Result	
1	mask	00000000	00000001
2	mask = mask<<6	00000000	01000000
3	a = mask = mask – 1	00000000	00111111
4	b = from[jisin]	00000000	00110110
5	temp = a AND b	00000000	00110110
6	mask	00000000	00000001
7	mask = mask<<1	00000000	00000010
8	mask = mask – 1	00000000	00000001
9	a = mask = mask<<15	10000000	00000000
10	b = from [iisin]	01111001	01101001
11	c = a AND b	00000000	00000000
12	d = c>>15	00000000	00000000
13	e = temp<<1	00000000	01101100
14	out = d OR e	00000000	01101100

Figure 11.12 Processing steps for the 7-bit group

The integer value of **out** is therefore 108.

11.4.4 Loading Integer Values Into Chromosomes

There are instances, such as when testing an objective function, where integer values need to be loaded into chromosomes. The function *initLoad()* was written during the building of Cabal to initialise the chromosome population to known values. This was one way to test the coding of the objective function which evaluated chromosomes – whose integer groups contained values for green time lengths of traffic signals – for their effectiveness in handling the given vehicular traffic flows at isolated junctions. The *initLoad()* function is a good example of how the user can use the SGA-C code as a basis and modify it to suit a specific application.

Parameters passed to the *initLoad()* function are the same as that for *ithruj2int()* with the addition of the number to load into the bit group denoted by the parameters **i** and **j**.

```
void initLoad(unsigned int *chrom, int numbertoload, int i, int
j)
{
    unsigned int number;

    int bound_flag;

    int iisin, jisin;

    int i1, j1;

    if(j < i)
    {
        fprintf(stderr,"Error in initLoad: j< i\n");

        exit(-1);
    }
    if(j-i+1 > UINTSIZE)
    {
        fprintf(stderr,"Error in initLoad: j-i+1 > UINTSIZE\n");

        exit(-1);
    }
    iisin = i/UINTSIZE;

    jisin = j/UINTSIZE;

    i1 = i - (iisin*UINTSIZE);

    j1 = j - (jisin*UINTSIZE);
```

```
if(i1 == 0)
{
    iisin = iisin-1;
    i1 = i - (iisin*UINTSIZE);
};

if(j1 == 0)
{
    jisin = jisin-1;
    j1 = j - (jisin*UINTSIZE);
};

/* Check if bits fall across a word boundary. */
if(iisin == jisin)
    bound_flag = 0;
else
    bound_flag = 1;

Number = (unsigned int) numbertoload;
if(bound_flag == 0)
{
    number = number << (i1-1);
    chrom[iisin] = chrom[iisin] | number;
}
else
{
```

```
        number = number << (i1-1);

        chrom[iisin] = chrom[iisin] | number;

        number = (unsigned int) numbertoload;

        number = number >> (UINTSIZE -i1 + 1);

        chrom[jisin] = chrom[jisin] | number;

    }
```

`}` // End void initLoad(unsigned int *chrom, int numbertoload, int i, int j)

The code is very similar to *ithruj2int()* and therefore a description of its operation is unnecessary here. In Cabal the function was used to load the population with a set of known values, thus the objective function operation could be checked at boundary conditions.

11.4.5 Reporting Chromosome Contents

The final piece of code is a method of reporting the bit-string composition of each chromosome. The original SGA-C code of the function *writechrom()* writes the string of ones and zeroes with the most significant bit on the right. The modified code used in Cabal (shown below) writes the bit-string with the most significant bit on the left, the convention for number representation in the western world.

```
void writechromwestern(unsigned int *chrom)

/* First the chromosome is converted into a chromosome where the
LEFTMOST bit is the most significant bit. Then it is reported bit by bit
and written to the output file report using the same code as in the original
SGA-C. It is assumed that the reporting file has been opened and is
available. */

{
    int Counter1, Counter2, Stop;

    unsigned int mask = 1, tmp, Tempchrom = 0;


    for(Counter2 = 0; Counter2 < chromsize; Counter2++)

    {
        tmp = chrom[Counter2];
```

```
if(Counter2 == (chromsize-1))
    Stop = lchrom - (Counter2*UINTSIZE);
else
    Stop = UINTSIZE;


for(Counter1 = 0; Counter1 < Stop; Counter1++)
{
    Tempchrom = Tempchrom << 1;
    if(tmp&mask)
        Tempchrom = Tempchrom + 1;
    tmp = tmp>>1;
}


tmp = Tempchrom;
if(Counter2 == (chromsize-1))
    Stop = lchrom - (Counter2*UINTSIZE);
else
    Stop = UINTSIZE;


for(Counter1 = 0; Counter1< Stop; Counter1++)
{
    if(tmp&mask)
        fprintf(report,"1");
    else
        fprintf(report,"0");
    tmp = tmp>>1;
```

```
    }

 }

 } /* End void writechromwestern(unsigned int
 *chrom) */
```

11.4.6 Memory Size and Efficiency Savings

This section contains some calculations for the memory savings realised by using the bit-level code compared to an array-based chromosome representation. Examples are given for 16-bit, 32-bit and 64-bit machine architectures (these are becoming increasingly common) and memory requirements are calculated as follows:

For the array-based representation the calculation is straightforward.

$$Totalmemory = popsize * chromlength * integerlength$$

The total memory is calculated in bits and the integer length depends on the architecture. A 16-bit system has an integer length of 16 bits, a 32-bit system 32 bits...etc.

For the bit-based representation, the minimum number of integers to house each chromosome must be calculated. This depends on the byte-size of integer storage in the particular architecture and the length of the chromosome. The calculation was described earlier in the discussion of the SGA-C variable **chromsize.** Therefore

$$Totalmemory = popsize * numberofintegers$$

Table 11.6 contains memory efficiency values for various system architectures and population sizes for GA chromosomes whose bits are grouped evenly into integers. The Bits per Int column contains parameters that give the range of the integer representation for each group. For example, a Bits per Int value of 7 means that one integer in the chromosome can be any number in the range $[0, 2^7\text{-}1]$. The Bits per Int and the Num of Ints columns are shown to give the reader a sense of what saving can be made for different chromosome-integer representations. The minimum possible bit requirement values are shown in kilobytes as are the memory requirements for array-based and bit-based representations. The final column shows the bit-based representation as a fraction of the array-based requirement given in percentage form. Thus for a 16-bit system where a chromosome length of 60 represents 5 integers ranging from 0 - 4096, the bit-based representation consumes 6.7% of the memory that an array-based representation would.

The values given for populations of 460 are included to show the memory savings that were made for one configuration of a working GA optimisation system, Cabal. In addition, many GAs hold two populations concurrently.

In these cases, the number of populations should be included in the calculations.

Table 11.6 Memory efficiency values

Syst	Pop'n Size	Bits per Int	Num of Ints	Chrom Length
16	100	7	3	21
16	100	12	5	60
16	100	22	25	396
16	460	7	5	35
32	100	7	3	21
32	100	12	5	60
32	100	22	13	396
32	460	7	5	35
64	100	7	3	21
64	100	12	5	60
64	100	22	7	396
64	460	7	5	35

Syst	Min Mem Req'd (kB)	Array-based (kB)	Bit-Based (kB)	Bit/Array (%)
16	2.1	33.6	3.2	9.5
16	6.0	96.0	8.0	6.7
16	39.6	633.6	40.0	6.3
16	16.1	257.6	22.1	8.6
32	2.1	67.2	3.2	4.8
32	6.0	192.0	6.4	3.3
32	39.6	1,267.2	41.6	3.3
32	16.1	515.2	29.4	5.7
64	2.1	134.4	6.4	4.8
64	6.0	384.0	6.4	1.7
64	39.6	2,534.4	44.8	1.8
64	16.1	1,030.4	29.4	2.9

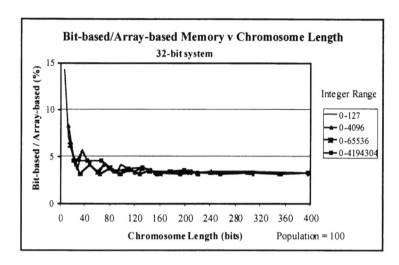

Figure 11.13 Memory reduction v chromosome length curves for a 32-bit system and for varying GA integer range representations

Figure 11.13 shows the values calculated for Table 11.6 for the 32-bit system. A full range of values is plotted for each of four integer ranges. For small chromosome lengths the ratio of bit-based to array-based memory size is highest and varies much more than it does for longer lengths. The memory taken by the bit-based representation is typically around 3.7% of that required for the array-based code.

Graphs for other integer representations and system architectures (16- and 64-bit systems) have similar shapes to that of Figure 11.13.

11.5 Conclusion

Bit-based SGA-C code gives the GA practitioner significant memory savings over an array-based representation. As an example, the Cabal software for optimisation of traffic signal timing for a single intersection was capable of operating in realtime once it was implemented in the bit-based configuration. This was due to the size of the population being made large enough to enable the GA search to be effective in realtime. Even though memory for a single GA implementation is now not a significant problem on PCs, the memory saving theoretically enabled extensions of Cabal's function to optimising the signals of a road network.

The memory ratio value for long chromosomes differs for different machine architectures. Typically the ratio of bit-based to array-based memory requirements on a 16-bit system is about 6.4%; for the 32-bit system, it is about 3.3%; and for the 64-bit system, it is often under 2.0%. As computers become larger architecturally, the bit/array ratio decreases.

The SGA-C code is a good example of bit-based GA implementation and it makes extensive use of the left- and right-shift operations available in the C language. The explanations given here begin with a general description of the memory space allocation and reference mechanisms of computers before discussion of the code workings of the relevant portions of SGA-C. The descriptions are sufficiently detailed to enable the GA practitioner to build useful routines for specific applications. Two examples of such utilities are given.

References

Clement, SJ (1997) Genetic algorithms in dynamic traffic signal timing: the Cabal model. Master of Engineering thesis, University of South Australia, Australia.

Goldberg, DE (1989) Genetic Algorithms in Search, Optimization and Machine Learning. Addison-Wesley Publishing Company Inc., Reading, Massachusetts, U.S.A.

Halliday, JJ (1997) Genetic algorithms on parallel processors. Master of Computing Science thesis, University of Newcastle upon Tyne, England.

Kernighan, BW and Ritchie, DM (1988) The C Programming Language, Second Edition. Prentice Hall, New Jersey, U.S.A.

Microsoft Corporation (1993) Microsoft Visual C++ Reference Volume II manual v1.0.

Smith, RE; Goldberg, DE and Earickson, JA (1994a) SGA-C source code http://www.cs.cmu.edu/afs/cs/project/ai-repository/ai/areas/genetic/ga/systems/sga/sga_c/0.html. Site last updated 13 Feb 1995 and last visited 13 October 1998.

Smith, RE; Goldberg, DE and Earickson, JA (1994b) SGA-C: A C-language Implementation of a Simple Genetic Algorithm. TCGA Technical Report Number 91002, 2 March. The Clearing House for Genetic Algorithms, University of Alabama, U.S.A.

Chapter 12 Adaptive Portfolio Trading Strategies

Arthur Rabatin

Rabatin Investment Technology Ltd.

11 Grosvenor Place, 5th Floor

Spectra Capital Ltd.

London SW1X 7HH

England, United Kingdom

http://www.rabatin.com

ar@rabatin.com

Abstract

This chapter describes a Genetic Algorithm-based implementation of adaptive trading models, specifically designed for trading of multi-currency trading portfolios. The chapter describes the aspects of the decision-making process that must be incorporated into the trading model in order to accurately simulate the decisions a human portfolio trader is required to make in this position. The chapter describes the different types of learning processes for market timing and risk management and how these can be incorporated into the same GA-based learning process.

The basic concept of a distributed, object-oriented learning process is demonstrated, as well as specific fitness value calculations to increase consistency and predictability of portfolio trading performance, which the reader can implement in their own testing and development procedure. Two different concepts of designing the adaptive process are shown, with the effect described in the model portfolio described below.

The performance of such an adaptive system is demonstrated using a diversified foreign exchange trading portfolio, which yields acceptable levels of risk-adjusted return, under realistic assumptions of portfolio constraints and transaction costs.

12.1 Introduction

Portfolio Trading deals with trading decisions made within the context of a diversified portfolio. The aim of any such trading operation is to achieve an above-average return on the available trading capital. More precisely, the aim is to achieve an above-average *risk-adjusted* return, that takes into account the accepted risk parameters of the trading desk's management, shareholders, or even constraints imposed by financial regulators.

Because the requirements for portfolio trading strategies focus on portfolio risk and portfolio return, employing market forecasting systems alone (through traditional or AI methods, such as Neural Networks) is *not sufficient* to develop an autonomous, intelligent trading model.

Within a portfolio, every trading decision requires a decision to be made on

- market selection
- market timing (buy/sell decision)
- accepted price risk
- portfolio allocation
- portfolio risk exposure

All these aspects are highly relevant for the performance of the portfolio. It is important to note that every trading decision always includes *assumptions* on price risk, portfolio allocation and the resulting portfolio risk exposure, even if the *decision* has not been explicitly made.

Risk and allocation decisions have a very significant impact on the performance of the portfolio. Although clearly no risk management strategy can turn a losing trade into a profit (or v.v.), the actual performance is the result of a stream of profits and losses. These profits or losses are a function of both the gain/loss in terms of price *and* the quantity traded.

The trading quantity (position size) for each individual portfolio component is a function of the allocation and risk decisions made for that portfolio and instrument. Because each individual portfolio component affects the value of the entire portfolio, a trading decision in one individual market is always a function of the performance in all other instruments traded within the portfolio.

As a consequence, the amount gained or lost on an individual trading decision is influenced *simultaneously* by risk and allocation decisions, and is *not independent* of other components of the portfolio. If a trading strategy does not incorporate the risk and allocation decision in addition to the market timing model, it would not be possible to replicate the success of the market forecasting model in real-time performance. Whatever the actual performance would be, the decision on the position size would be unrelated to the market timing decision and would turn the forecasting strategy into an unpredictable stream of profits or losses.

We have designed a development framework that allows the development of intelligent trading models, which incorporate all aspects of the trading decision into one complete decision-making process. This Adaptive Portfolio Trading (APT) framework is an object-oriented framework providing the underlying Genetic Algorithm framework, as well as the accounting and reporting functions necessary for any trading model.

We are employing a parallel, distributed learning process based on a master/slave design. Genetic Algorithms lend themselves very well to an object-oriented distributed learning process. Since, within a generation, each member of the population is evaluated independent of the other individuals, a network of processes (i.e., slave processes) can evaluate a generation in a parallel process. The master process has the responsibility of preparing the individual trading model objects for the slave workstations to process and to perform the genetic operations after the generation is completed.

12.2 Portfolio Trading Learning Process – Overview

An automatic learning process evolves around a basic function in the form of if <condition> then <action>.

Applied to a real-world environment, such as portfolio trading, both components of the function usually develop a very complex shape. To derive the <condition>, a large number of environment data have to be analysed and interpreted, plus data the system generates as a result of its own behaviour, i.e., its own trading and performance history. The <action> also represents a large set of possible decisions that the system takes simultaneously. Most important, this is a decision to buy, sell, or to adjust an existing position, as well as a decision on positions size, portfolio allocation and position risk.

For an adaptive model to learn appropriate behaviour, a payoff value must be available that will allow the system to interpret its actions as success or failure. In many robotics applications, such a payoff value may be immediately available after an action is taken. For portfolio trading systems, no immediate feedback is available. Even though, clearly, we would wish each decision to be as profitable as possible (or at least avoiding losses), we must expect a real-world trading strategy to result in a stream of both profits and losses. In other words, a certain decision strategy might have resulted in a loss today, but it still was the best decision to take because it had the highest long-term expectation. What we should expect is stable, profitable performance over a number of trades, where we can measure the distribution of profits and losses. This essentially implies evaluating performance over a longer time frame.

Because each action taken by the trading model includes a number of different decisions and because the payoff is measured after a stream trading decisions, the performance payoff cannot be directly attributed to a single type of action or decision. This is not necessarily an aspect of machine learning based trading models; as every trading decision – systematic or discretionary – is simultaneously influenced by multiple factors.

The challenge of real-time portfolio trading is the nature of the constraints the trader (system) is subject to.

While the individual trading decision (including a decision on price risk, portfolio allocation, portfolio risk) is always made for one single market instrument, the defined constraints – such as overall risk thresholds or exposure limits – are defined for the entire portfolio. It cannot be predicted, if such global thresholds are exceeded, how each individual portfolio component is affected. More precisely, we do not know how the performance of each individual portfolio component is affected by the constraints placed on the portfolio as a whole. The only way to simulate the effect of such constraints is to implement them in the original learning process. In other words, the trading model object used during the learning process must include the risk management parameters placed upon the system during the real-time execution of this strategy.

12.3 Performance Measurement/Fitness Measurement

Choosing the appropriate tool for portfolio performance measurement is also an important tool for selecting a fund, trader or trading system for investment purposes. With hindsight, every investor would like see the highest possible return on the account that could have been achieved under given market circumstances. In reality, for the purpose of selecting a trader or system for investing, the investor must define a level of risk he/she is prepared to take. This risk expectation defines the parameters within which the investor would accept the trader/system performance. It also serves as a threshold to define when the trader/system does not perform as expected.

The appropriate notion in this context is that of "risk adjusted return." It means any rational investor would expect the highest return given his/her level of accepted risk. Risk is typically defined as variance of returns. This concept has become subject to controversy, because it relies on a defined distribution of portfolio value changes as risk measurement. In reality, the distribution of portfolio value changes does not resemble a normal distribution. As a result, a system relying on estimating risk through variance will always grossly under-estimate the real risk to which the system is exposed.

Within a GA-based learning process, performance measurement is especially relevant because it also yields the fitness value through which the "survival of the fittest" process is implemented. The choice of fitness value also determines the success of the trading model during cross-validation. Because the GA process has shown to be a very effective optimisation tool, the success of the learning process must always be interpreted relative to the evaluation on out-of-sample data periods.

In terms of a portfolio trading model, we are, therefore, interested in the future performance of the portfolio when applying the parameters and rules the system has learned during the training process. Because such performance can never be perfectly predicted, the investor into a trading strategy therefore seeks consistency of performance.

The performance benchmark must therefore measure this consistency. To create a trading model that adapts without human interference, the performance benchmark must also measure the absolute level of performance relative to the expected return and the accepted risk. Because the accepted risk is largely a user-defined value (because it is a result of each investor's preference), the trading model must balance return with this risk level.

For evaluating a trading system's performance consistency in an automatic self-learning process, the fitness must take into account the time structure of performance (consistency of performance) as well as the absolute level of performance.

The target function of the learning process applied in our trading models is based on a user-defined Return Path (RP). This return path is a monthly or quarterly range of expected returns within which the system ideally performs. The fitness of the trading model is measured by the error of tracking this target range, the Return Path Error (RPE). Formally, the RPE is defined as

$$RPE = \sqrt{\sum_{n=0}^{N-1} e_n^2} \Big/ N,$$

where

N ... number of periods (either calendar quarters or calendar months),

n is the nth calendar period (indexed between 0 and N - 1) and

e_n is the actual tracking error for the nth period.

e_n is defined as follows: if $(r_n > RP^+)$: $e_n = (r_n - RP^+)W$, if $(r_n < RP^-)$: $e_n = (r_n - RP^-)$, where r_n is the calculated actual percentage return of the portfolio for the nth period, RP^+ is the upper limit of the return path target range, RP^- is the lower limit of the return path target range and W is a weighting applied to smooth the effect of *upside* errors of the portfolio (typically $0.3 \leq W \leq 1.0$; this model uses an error weighting of 0.4).

The GA process seeks to minimise the RPE, which ideally equals zero, when the system performs completely within the desired return path.

The advantage of using RPE as performance benchmark is that it emphasises and measures the consistency of performance in that it matches the user's expectation on return with any risk thresholds attached to the portfolio. If the return expected from the model is not compatible with the risk constraints placed upon the portfolio, this discrepancy can then be already detected during the learning process. Either the portfolio constraints or the performance expectations will then have to be adjusted.

12.4   APT   Object-Oriented   Distributed   Parallel Processing

The underlying GA library, Evolving Programming Library (EPL), is a domain independent object oriented (OO) GA framework implemented in standard C++. The specification for a problem domain is achieved through deriving classes from the base class collection in the GA framework, implementing the desired functionality by overriding the appropriate virtual functions.

The basic class of the EPL framework is CUserData, which – in the application – represents the problem domain. CUserData is implemented as an abstract base class, requiring the derived class to override exactly those member functions which are specific to the problem domain. Because of this clear OO design, the EPL is designed to handle all functions relating to the GA process without any domain specific adjustment to the GA code itself.

The main virtual public functions of CUserData are

```
void CUserData::Copy ( const CUserData& _That )
```

> // a virtual copy function. Avoids assignment operator overloading

```
CUserData* CUserData::Clone (  void )
```

> // returns a new object of correct run time type as copy of this object

> // derived function contains return (new CUserData_derived (*this))

```
void CUserData::Write (...)
```

> // writes the contents of the object into a binary stream

```
void CUserData::Read (...)
```

> // reads the contents of this object from a binary stream

```
TFitnessvalue CUserData::FitnessFunction ( void )
```

> // performs the actual fitness calculation and returns fitness measurement

> // the return type TFitnessvalue is currently implemented as a typedef of double

```
void CUserData::RegisterVar ( ... )
```

> // registers the variables for optimisation with the underlying GA engine.

The implementation of such copy, cloning and stream functions does not implement a significant overhead to any application because these functions are normally required in some form for a standard OO architecture. The RegisterVar(...) function contains the registration of each variable prepared for optimisation with the underlying GA engine.

Because the EPL framework defines the fitness function as pure virtual function, the GA process is completely shielded from the implementation of the problem domain. In OO terms, the implementation of the problem domain is entirely encapsulated.

The ability to write to and read from binary streams is the basic requirement for the distributed learning process. By using streams, CUserData objects (i.e., trading model objects) can be exchanged between applications and workstations within a network. Because the fitness function is already defined within the base class, a client process can read the object from a stream (e.g., a file) and execute the fitness function independent of the application containing the GA engine. This represents the basic design of a distributed parallel process based on a master/slave architecture.

The APT framework extends the EPL library by implementing a base class trading model, CStrategyTmpl (short for class Strategy Template, although not implemented as template in C++ sense).

The CStrategyTmpl class provides a definition for the CUserData:: FitnessFunction() member function, as well as for other member functions, to implement the basic framework for the trading model to execute a strategy. This mainly includes the accounting functions as well as the database function.

The trading model data access is implemented using a datafeed object that simulates a real-time datafeed by accessing the defined database and returning data to the trading model based on the defined time frames.

The CStrategyTmpl base class provides virtual functions for any specific trading model implementation, which are called by the framework during execution of a trading model. Specifically, these are functions called by the framework at specific points in time, such as the beginning of trading during a specific period, end of trading per period, end-of-day procedures.

Currently implemented public virtual event functions are

```
virtual void
CStrategyTmpl::Before_Portfolio_AllMarkets ( void
)// called as trading model initialisation

virtual void
CStrategyTmpl::Daily_BeginOfDay_AllMarkets ( void
)// called at begin of each trading period (
trading day )

virtual void
CStrategyTmpl::Daily_BeginOfDay_EachMarket ( int
_MarketIndex )// called only for markets which are
trading in this period ( today )
```

```
virtual void CStrategyTmpl::Daily_AllMarkets ( void
)// called for all markets in a particular trading
period ( day )
```

```
virtual void CStrategyTmpl::Daily_EachMarket (int
_MarketIndex )// called only for markets which are
trading in this period ( day )// this function
contains the main routines for executing the
trading// model during training and application
process
```

```
virtual void
CStrategyTmpl::Daily_EndOfDay_EachMarket (int
_MarketIndex )
```

```
        // called for active markets at the end of
        the trading period ( day ) for each
        // market
```

```
virtual void
CStrategyTmpl::Daily_EndOfDay_AllMarkets ( void )
```

```
        // called at the end of each trading period (
        day )
        // the base class contains mainly accounting
        end-of-day functions, such as
        // portfolio evaluation and reporting
        functions
```

```
virtual void
CStrategyTmpl::After_Portfolio_AllMarkets ( void
)// called after the trading model is completely
evaluated
```

These functions are called by the framework in connection with the datafeed object. The APT framework controls these functions through a pointer to the trading model object of type of the base class CStrategyTmpl. Any specific trading model implementation will override these event functions to implement the specific functionality required. This design makes the actual trading model implementation largely independent of the functionality provided by the framework.

In addition to the GA-based reporting functions of top and average fitness, and learning process performance, the APT framework also includes a standardised portfolio reporting object using standard portfolio performance measurements.

Trading Models derived from CStrategyTmpl can be designed independently of the GA engine and can also contain nonoptimised parameters or rule sets.

This article concentrates on the implementation of a GA-based learning process.

The distributed architecture is currently implemented based on file streams, which are locked/unlocked by the current process, which is either a slave or the master process, performing the genetic operators, although the file-based process creates a larger overhead.

12.5 Learning Process Implementation

The decision-making process of the adaptive trading models simulates the human decision-making process by learning behaviour patterns that are matched against patterns the system detects in the environment data.

The behaviour of the trading model is therefore a result of events that take place in the environment – both being the market data and its own accounting database.

It is the purpose of the system's training process to learn to detect what constitutes an *event* and what is the appropriate behaviour in response to it.

Because *behaviour* in the context of a trading model is always a decision to sell, to buy, or to do nothing (keep the current position unchanged) in a particular financial instrument, it also means that risk for the portfolio may either be increased (by establishing a new net position in a market) or decreased (by reducing an open net position in any market).[1] The trading model is designed to retrieve information from both external data (market prices and other data) and from its own trading performance to decide on its trading decisions. These trading decisions (and subsequent decisions on any open positions) are also subject to the risk threshold placed upon the system by the trading manager or system supervisor.

Every trading decision has four components that simultaneously define the portfolio performance: market timing, price risk, portfolio allocation and portfolio risk. Below, the learning process for each of the components, and the integration into a final trading decision are described.

12.5.1 Market Timing Decision

The process of mapping environment data patterns to a trading behaviour pattern is implemented through two layers of objects that perform these functions: a layer of objects performing calculations on data (*Calculation Node Objects*) and a layer of objects interpreting the results of calculations as events and mapping these events to possible decisions.

Calculation Node Objects (CNOs) are an array of objects that retrieve data and perform mathematical, statistical and logical operations on

---

[1] Portfolio Risk here is referred to as the aggregated risk of all positions, not the variance of the portfolio returns itself.

these data, returning a set of numerical values that will be interpreted as events by another layer. It is through the GA process that it is decided which data are retrieved and which operator is applied for each CNO. The calculation that is performed within each CNO can be either very simple (such as a logical comparison { e.g., ">" or "="} or a simple arithmetic operation {e.g., "+" "/" ...}), or it can involve complete statistical calculations, such as the historical volatility of a retrieved time series. Boolean values (true/false) are represented as numerical values (0/1).

CNOs are not limited to only retrieving input from an external database. The output of each CNO can also be selected in the learning process as input for calculations. This allows the trading model to detect complex patterns by connecting a large number of CNOs that each perform very simple calculations. In an array of (n > 1) number of CNOs, the nth CNO can use the output of (CNO[0] ... CNO[n-1]) as input.

The actual number of Calculation Node Objects is a matter of trading model design and is largely dependent on available machine time, as increased complexity also dramatically increased the time required to train these trading models. The advantage of using an array of CNOs is the flexibility that otherwise would be restricted by a fixed-length Genetic Algorithm.

Event Node Objects (ENOs) are an array of objects that interpret output values of CNOs by learning to map the CNO output to a logical value of true/false in terms of an *event* occurring yes/no.

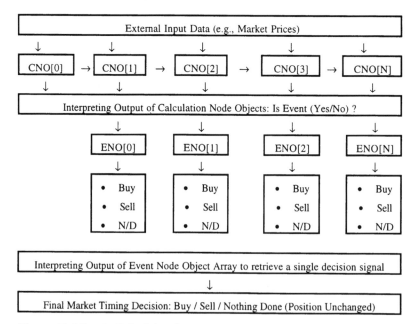

Figure 12.1 Buy/sell decision flowchart.

Because the output of each CNO is a fixed structure of values that is always assigned valid data (independent of the calculation performed with a CNO), ENOs can always interpret the result of a calculation in terms of an event occurring or not. Event Node Objects also learn to map a particular CNO output set to a particular type of decision – to buy, sell or do nothing in a market.

After both layers of objects have been constructed, the trading model possesses a pattern to interpret data it retrieves from the external database in terms of possible trading decisions. Each Event Node Object that, for the actual data, returns *true* for an *event* to exist, also returns a signal for a decision to buy, to sell or to remain unchanged in the current position (if any). The trading model then chooses the one decision that occurs most frequently as ENO result, as a final decision to be made in the moment the calculation is performed. Figure 12.1 illustrates the process described.

N/D indicates a decision result of "nothing done," i.e., current position unchanged.

12.5.2 Risk and Portfolio Allocation Decisions

Risk and allocation decisions are learned differently by the trading model, as these decisions are not "event-driven," but depend only on a position in a market to exist.

Learning market timing behaviour differs from the other components in the way that we have little knowledge of how a buy and sell decision should be made. The system can and should therefore have a large degree of freedom to develop these decision-making rules by learning to develop appropriate adaptive behaviour patterns. For risk and allocation decisions, we do, however, have some knowledge about possible rules (e.g., restricting risk exposure, diversifying portfolios), and we especially have certain constraints that we want to enforce on the trader/the system. Consequently, the trading system can have freedom to decide on allocation strategies and risk exposure within a portfolio, but it must do so by observing the global risk and allocation thresholds.

The learning process for risk and allocation decision is a rule-based learning process that uses a "bottom-up" approach by combining simple rules ("sub-rules") using mathematical and logical operators, resulting in a more complex decision rule, onto which certain risk thresholds are applied by the trading model.

For the GA process, each sub-rule and each operator (mathematical or logical) is recognised as a simple integer value indexing the rule within the risk management object of the trading model.

The GA learning process of the trading model creates a rule which is a formula consisting of data parameters (i.e., calculation results of the sub-rules) and the operators. Although the structure of the formula is limited in its flexibility compared to the market timing decision, the trading model can still create a relatively complex structure of rules very different from the simple components on which it is based.

Although we will focus here on the risk and allocation decision to be made as part of a new trading decision (resulting from a market timing signal), the same calculations are repeated at the end of each trading period (usually end of the trading day), in order to adjust existing open positions. These adjustments are particularly important because they make sure that the portfolio will always be maintained within the desired risk thresholds. Our testing has further found that the constant application of risk thresholds is one of the most significant contributing factors to creating a consistent and predictable performance pattern, largely independent of the actual market timing strategy used.

The trading model uses three levels of risk calculation.

- expected risk (calculating an estimation of the risk a certain trading position carries)

- accepted risk (a risk threshold which the trading model defines for each trading position)

- portfolio risk threshold (the overall constraint that the system supervisor or trading manager puts on the entire portfolio which must be observed by the trading model).

The *expected risk* is used by the trading model to make an initial decision on the size of a trading position, similar to a human trader. Through the learning process described below, the model is trained to improve its risk-forecasting ability. The *accepted risk* is a threshold that is associated with a given market position, and is calculated by the trading model as part of the learning process. This is not necessarily a "worst-case-scenario" threshold. More likely, the trading model uses risk thresholds to actively manage the size of open positions in a market to create a less volatile equity curve.

Portfolio risk thresholds are defined by the supervisor of the system, the trading manager. The trading manager normally has a clear view of the amount of risk he/she is prepared to accept for a given portfolio. This, however, is a threshold for the aggregated risk of all positions and the trading model learns to translate this global constraint into position limits for each individual component of the portfolio.

The risk calculation for a portfolio is performed in three steps: price risk, portfolio allocation and portfolio risk.

12.5.3 Price Risk Calculation

Price risk is associated with the market in which a position is taken and is independent of the size of a trading position. It is, however, closely linked to the type of market timing strategy developed by the trading model, since the expected variance of prices is a function of the time horizon of an investment.[1]

The rules, which the trading model can learn to use for calculating risk use as input are

- price volatility (in absolute terms over previous n number of days)
- price volatility (in terms of standard deviation)
- previous trading range (price high/low over previous n number of days)
- maximum/minimum price changes (over previous n number of days)
- average/total/minimum/maximum price risk calculations of other components of the portfolio
- price and volatility correlation to other components of the portfolio

The "sub-rules" are combined in a rule structure developed by the GA process that result in a risk estimate for a position in that market. Initially this risk estimate is also the maximum accepted price risk for the position. In trading terms, that price risk is the "Stop-Loss" level at which a position is closed out, because the market forecast, implied by the decision taken by the market timing strategy, is accepted as wrong.

After the initial position is established, trading models recalculate price risk estimates every time the trading model is updated with new prices. However, the trading model does enforce to cut the position size, if the estimated price risk exceeds the accepted risk. The accepted risk is allowed to increase only by a limited margin when applied to an existing open position.

12.5.4 Portfolio Allocation Decision

Portfolio allocation decides the percentage share of the overall portfolio value that is allocated to one portfolio component. Portfolio allocation within asset investment strategies is a defined process which could easily be implemented through any GA (or other) learning algorithm. The decision for a trading portfolio is more complex, because the system is not invested into all markets at the same time. The trading model must learn how to allocate a share of the portfolio for a new position to be taken, but also take into account existing positions and any portfolio limits placed onto the system.

---

[1] In absolute terms, variance of prices is dependent upon the time horizon, not the variance itself. Assuming a normal distribution of price changes, both a short term trader and a long term trader face the same probability of a 1-σ event in the market. In absolute terms (and in percentage of the portfolio), these will be different values.

For any given portfolio component, the allocated share of the portfolio is calculated by the trading model as

$$A_i = F_i \bigg/ \sum_{n=1}^{N} F_n$$

where

A_i ..... allocation for the ith market

N ..... number of markets available to the portfolio

F_i ..... the function expressing the rule used to calculate the allocation

Each rule developed by the GA learning process is based on a combination of sub-rules which are based on the input of

- price data and time series statistics of the given market
- time series statistics in relation to the other market's in the portfolio (cross-correlation and relative strength)
- relative value of each market's performance within the portfolio

The result of each calculation is always the recalculation of allocation for the entire portfolio. This makes sure the trading model not only strictly observes any portfolio constraints, but it also ensures that the trading model can re-balance the entire portfolio after the addition of a new position to the portfolio. Since performance consistency is typically the most important criteria for evaluating trading models, the automatic reallocation ensures the highest possible level of diversification within the portfolio.

Market Selection is a special case of the portfolio allocation decision, as it is also possible for the system to allocate a share of zero to a market, thus effectively excluding a market from the portfolio. When setting up the adaptive trading model, a number of markets are made available to the system. However, the trading model can, in every stage of the calculations, decide to reject a certain market either by not creating a buy or sell decision or by allocating a very small share of the portfolio to that market that cannot be traded in the market.

12.5.5 Portfolio Risk Decision

After the trading model estimates the price risk associated with a position and calculates the share of the portfolio allocated to a particular trade, it can calculate the actual quantity to trade, i.e., the size of the trade and of the open position.

In learning to perform this calculation, the trading model uses the calculated price risk, the allocated equity and all input data used by previous calculations to create the decision-making rule for the trading quantity.

The decision on the size of each position has been shown to be among the most relevant decisions determining the consistency of portfolio performance.

In general terms, the position size is a function of the available (allocated) equity for a given market and the price risk associated with the market (which the trading model both learns to calculate/estimate), i.e., $U_i = f(A_i, P_i, R)$ where U_i is the number of trading units (position size) for the ith market within the portfolio, P_i is the calculated Price Risk for the ith market, A_i is the calculated allocation for the ith market and R is the portfolio risk accepted by the system expressed as a percentage of total capital accepted to be at risk.

$R_{actual} = f(A_i, P_i, U_i)$ is a way of rewriting the above function. In other words, the actual percentage of the portfolio at risk (R_{actual}) is a function of a market's price risk, the allocated share of the portfolio and the number of units bought or sold in the market. For the purpose of the learning process, the R_{actual} value should be smaller or equal to the accepted portfolio risk value, R. Therefore the learning process of the trading model first finds a percentage portfolio risk value R that is compatible with the global parameters of the portfolio, and then, using both other parameters to the function, A_i, P_i, calculates the actual number of trading units (U_i) that would create the desired exposure.

The calculation of the trading size is the link between the market timing behaviour developed by the system and the required risk management strategy.

For every market timing decision made by the system, it will calculate the associated market risk (per unit price risk) and the associated portfolio allocation.

It is through the variation of the trading size (position size) that the system is performing a trade off between higher (lower) per unit price risk and smaller (larger) size of the trading position.

Such an approach to managing the portfolio risk is, in our view, an essential component of a consistently developed trading strategy. It removes the uncertainty that is normally associated with profitable open positions; if profits should be kept "running" or positions liquidated in order to ensure that open profits are protected from any more risk.

In order to develop consistent performance behaviour, the trading model must have the ability to manage constant risk exposure during changing portfolio composition and market events. This process enables the trading model to consistently balance the market price risk and the portfolio risk, by changing the actual position size it will have in any market. Increased risk per unit traded (price risk) can be matched by a decrease in number of units traded and vice versa. As with the calculation of portfolio allocation, portfolio risk management is always performed over the entire portfolio.

Therefore, a change in one portfolio component can be matched by shifting (i.e., reducing) exposure in other markets. This allows the trading model to rebalance the portfolio either when a new market position is to be taken or when the daily mark-to-market process is performed.

12.6 Adaptive Process Design

12.6.1 Dynamic Parameter Optimisation

Since the trading system is designed to learn behaviour patterns, the optimisation of fixed numerical parameters (such as the number of days over which historical volatility is calculated) should be avoided. Not only is such a parameter unlikely to be usable across different markets, it is also very dependent on the actual distribution of prices within the training period. It must therefore be expected that the optimum parameter changes dramatically when market conditions change. Using statically optimised parameters leads to inflexible curve fitting which, in our previous research, has shown not to hold up in subsequent out-of-sample tests.

The concept used by these trading models is "dynamic parameters;" i.e., parameters for which the value is not directly optimised, but for which the model learns to develop rules for calculation. This is done within the Calculation Node Objects. For any calculations that cannot be developed as rules within the trading model, a hard coded time frame of 200 data periods is used as the starting point for calculations.

Example: The system may learn to use a moving average of market prices as part of an estimate of near-term market direction. For this moving average, a parameter is needed for the number of days over which the calculation is performed. Rather than optimising this parameter directly (within a given range of possible values), the system learns to calculate the value based on another calculation, e.g., market volatility (calculated by a different Calculation Node Object). The system retrieves current market volatility reading, and the min/max value for that calculation (over the time span calculated by the other CNO). The last reading is then expressed as a ratio within the historical min/max values. If min = 8 and max = 20, a current volatility value of 15 would result in a ratio value of 0.58. The general formula is: Ratio = (Current - Min) / (Max - Min).

This ratio is the current value expressed as a percentage of the range. This percentage value is then applied to the min/max range of possible moving average values used by the system. Assuming we define possible moving average parameters between (10;100) periods, the actual moving average parameter used in this calculation would be ((100 - 10)*0.58) ≈ 52 period moving average. Note that this resulting value changes every time the underlying calculation data (here, volatility reading) changes.

The benefit of dynamic parameter optimisation is that the learning process focuses on developing calculation *rules* rather than optimising parameters. Such rules can be applied to all markets, since they will automatically adjust to new data by recalculating the actual parameters used for calculation of events.

12.6.2 Cross-Validation

Cross-validation is an important tool in the evaluation of machine learning processes, dividing the available time series into periods for training and periods for testing, during which the performance of the trained parameters is validated.

An adaptive process uses a continuous training and application process to learn behaviour and apply this behaviour on new data. In the Foreign Exchange portfolio demonstrated in Table 12.1, the following setup for training/evaluation periods is being used.

The choice of 3 adaptation periods is largely arbitrary and frequently depends upon the available machine time for performing the testing. A higher frequency of re-training and adaptation would create a more continuous adaptive behaviour. We have opted for 3 periods to keep the training of the portfolio within the parameters of our available computer resources.

	Training (Learning)		Testing (Application)	
	From	*To*	*From*	*To*
Period A	09-05-1990	04-06-1993	07-06-1993	08-11-1994
Period B	09-05-1990	08-11-1994	09-11-1994	12-04-1996
Period C	05-09-1991	12-04-1996	15-04-1996	17-09-1997

Table 12.1 Setup and Training Evaluation Periods for the Foreign Exchange portfolio.

Each test period uses a defined training period for learning. The learning algorithm starts with a random initialisation (i.e., a state of no knowledge and random behaviour). An initial training period of 800 trading days (9 May 1990 to 4 June 1993) is assigned to the first testing period. During this period the system learns to develop basic rules and already eliminates a large number of consistently unsuccessful behaviour patterns. 800 trading days as an initial period represent about 1/3 of the database available. After that, a maximum amount of 1200 trading days is allowed for the training period to create similar training environments for each testing phase.

12.6.3 Top Fitness/Closest Fitness Behaviour Patterns

After the training process, the trading model selects one single rule system to be applied to new data. Typically, this is the rule set which had resulted in the optimum theoretical performance during the training phase. At the beginning of the application phase of each new period, the trading model will adapt its behaviour according to the rules it has learned during the training process. Since every learning process is an optimisation process, the system always carries the risk of over optimisation during the learning process.

Using over-optimised behaviour patterns on new data is very likely to result in undesirable, negative performance. We have therefore developed a concept of not using the optimised rule set for the actual trading period ("top-fitness" rule set), but to select one rule set, which is likely to be more robust in its real-time performance than a highly optimised behaviour, the "closest-fitness" rule set.

To calculate the "closest-fitness rule set," the trading model uses an internal threshold to find a range of behaviour rules, which results in acceptable performance during the training period, including the best strategy, i.e., the top-fitness rule set. Within this group of rule sets, the system then tries to find a smaller group with similar performance results. If such a group is found, the system selects the best rule set of this group to adapt to, for the new period. This rule set is referred to as the closest-fitness rule set. It may be the case that the selected closest-fitness rules set and the top-fitness rule set are identical, but more often this is not the case.

Because this closest-fitness rule sets is not as highly optimised as the top-fitness behaviour, we found a very significant increase in consistency of performance, when comparing the training results with the application periods.

12.6.4 Continuous Strategy Evaluation

During a real-time application of the trading model, at the end of each actual period (when the system prepares to adapt new behaviour patterns), the trading model already has generated a stream of trading decisions, which have resulted in a profit or loss, and the system may also have open positions in any of the markets of the portfolio. Although the division of the database into several training/application periods is necessary to create an adaptive learning process, it does not correctly reflect how the system would be applied in a real-time environment.

To replicate real-time behaviour, the trading model has the ability to dynamically adapt new behaviour while keeping all existing open positions and existing accounting values. This creates a continuous performance measurement and allows measurement of the effect switches in the behaviour patterns would have on existing market positions.

12.6.5 GA Parameters

These main parameters for the Genetic Algorithm are currently uniformly used across all learning processes.

Crossover Rate	0.6
Mutation Rate	0.001
Equal Parents Probability	0.5
Population Size	100
Elitist Selection	TRUE
Selection Method	Tournament Selection

Table 12.2 Main parameter set.

The termination condition of each process is defined by a time stop. Because a higher number of generations is essential for an improved learning process, we have set a time stop to let the system calculate only as many generations as possible within acceptable computer resource requirements. A configuration as shown here requires one week of training on a distributed process using 4 workstations in order to achieve any level of acceptable results.

Although the results by the trading models are profitable and relatively stable, the available hardware configuration has not yet generated a learning process of what we believe is sufficient depth. We have seen that a more demanding fitness target (such as the return path error) does indeed require far greater computational resources be made available to the learning process.

12.7 Foreign Exchange Trading Portfolio Simulation

We have concentrated on foreign exchange markets to create a portfolio of liquid (i.e., easily tradable), but only low correlated currency pairs. Foreign Exchange (FX) represents the world's largest and continuous financial market with no restriction on buying or selling of currencies in most OECD economies. FX markets take a special role in portfolio trading and investing. FX rates are not asset prices (such as stocks, bonds or commodities), but represent, as a ratio, the relative purchasing power between the monetary base of two economies. As a consequence, FX markets do not lend themselves very well to traditional asset allocation techniques. FX markets are a very good example of defining a trading strategy in terms of risk taking and risk aversion, which allow the system to fully exploit its risk management abilities during the development of the trading model.

12.7.1 Portfolio Specification

GBP-USD	GBP-DEM
USD-CHF	DEM-JPY
USD-CAD	DEM-CHF

AUD-USD	

Table 12.3 Currency Pairs Available for Trading.

Database History: Daily Open/High/Low/Last Data 09-05-1990 through 17-09-1997.

Data Source: Bride Information Systems.

Portfolio Base Currency	U.S. Dollar, Profits/Losses are converted to U.S.$ at prevailing exchange rates as they are realised.
Individual Market Constraints	No internal restrictions on individual position size. Each market could be allocated any position size between zero and 100% of the available trading capital.
Global Portfolio Constraints	Maximum portfolio exposure must not exceed 3 times current portfolio value (including open positions evaluated at current market prices).
Transaction Costs	Each transaction is assumed to carry 0.1% of the price as transaction costs.[1] Swap costs/gains have not been included.
Return Path Specification	Quarterly Return Path of 3%-15%.
Drawdown Limit	A drawdown of 30% is considered a total loss on the portfolio. In other words, if at any time the trading model would lose 30% from the last equity peak, trading for this system is to be stopped.

Table 12.4 Trading Parameter/Risk Parameter Inputs.

Fitness Value	Reciprocal value of the Quarterly Return Path Error (see Table 12.4).
Yield	Annualised compound yield of return.
YieldDD	Ratio of Yield/Maximum Drawdown that ever occurred in the trading model (measured on a daily basis).

---

[1] Typical Forex transaction costs may be around this value; however, this assumption also allows for *slippage*, that is, an actual execution price worse than the desired trading price (e.g., due to volatile market conditions).

P/M	Percentage of Months Profitable.
P/Q	Percentage of Calendar Quarters Profitable (more important here because fitness value is based on quarterly return path optimisation).
No. Trades	Number of Trades during the relevant period (includes transactions which resulted in partial close-out of existing position due to risk management adjustment).

Table 12.5 Portfolio Performance Measurements.

12.7.2 Performance Result – Overview

	Period	Begin	End	Fitness	Yield (%)	Yield DD	P/M (%)	P/Q (%)	No. Trades
Train	A	09-05-1990	04-06-1993	245.74	14.19	2.95	72.97	100.00	81
Apply	A	07-06-1993	08-11-1994	39.83	3.26	0.59	70.59	66.67	29
Train	B	09-05-1990	08-11-1994	112.08	12.62	1.73	72.22	88.89	56
Apply	B	09-11-1994	12-04-1996	37.17	5.01	0.92	52.94	50.00	25
Train	C	05-09-1991	12-04-1996	60.89	7.96	1.00	67.27	84.21	106
Apply	C	15-04-1996	17-09-1997	22.17	-4.26	-	41.18	40.00	36
Apply	All	07-06-1993	17-09-1997	31.28	0.51	0.04	52.94	47.06	94

Table 12.6 Results of Selection of *Top-Fitness Rule* Set for Trading during Application Period

	Period	Begin	End	Fitness	Yield (%)	Yield DD	P/M (%)	P/Q (%)	No. Trades
Train	A	09-05-1990	04-06-1993	210.97	14.52	2.95	70.27	100.00	91
Apply	A	07-06-1993	08-11-1994	40.25	3.64	0.64	70.59	66.67	31
Train	B	09-05-1990	08-11-1994	112.07	12.64	1.74	72.22	88.89	56
Apply	B	09-11-1994	12-04-1996	38.66	5.69	1.20	52.94	50.00	22
Train	C	05-09-1991	12-04-1996	58.25	7.54	0.99	63.64	73.68	86
Apply	C	15-04-1996	17-09-1997	43.16	6.86	1.00	52.94	60.00	18

| Apply | All | 07-06-1993 | 17-09-1997 | 42.12 | 5.11 | 0.76 | 58.82 | 64.71 | 86 |

Table 12.7 Results of Selection of *Closest-Fitness* Rule Set

It can be seen from Tables 12.6 and 12.7 that although both methods of selecting the trading model during training have yielded positive results, choosing the closest-fitness rule set has resulted in a more consistent and, also in absolute terms, more profitable strategy during all hold-out periods.

It is interesting to note that during the training periods, both top-fitness and closest-fitness rule sets have shown similar *Yields* but clearly different fitness values. This confirms, in our view, the importance of an appropriate type of fitness target for the learning process, as we believe the return path target is.

Choosing the closest-fitness target has also yielded another desired result: increasing the predictability of portfolio performance itself, which is shown in more detail in the following tables.

12.7.3 Performance Result – Consistency across Training/Application Periods

We measure the consistency across training/application sets by calculating the ratio between the performance measurement of the hold-out period over the measurement of the training period. A ratio of 1.00 would mean exactly the same performance; a ratio of > 1.00 means better performance. Although a higher real-time return, compared to the training set, would be desirable for practical reasons, it is not desirable for measuring the predictability of performance. Generally, however, it must be assumed that real-time performance (or hold-out tests) perform considerably less profitably than the training results suggest.

Fitness		PeriodA	PeriodB	PeriodC	Average
	Training	245.74	112.08	60.89	139.57
	Application	39.83	37.17	22.17	33.06
	Ratio	0.16	0.33	0.36	0.29

Yield		PeriodA	PeriodB	PeriodC	Average
	Training	14.19%	12.62%	7.96%	11.59%
	Application	3.26%	5.01%	-4.26%	1.34%
	Ratio	0.23	0.40	#N/A	0.31

Profitable		PeriodA	PeriodB	PeriodC	Average
Months	Training	72.97%	72.22%	67.27%	70.82%

	Application	70.59%	52.94%	41.18%	54.90%
	Ratio	0.97	0.73	0.61	0.77

Profitable		PeriodA	PeriodB	PeriodC	Average
Quarters	Training	100.00%	88.89%	84.21%	91.03%
	Application	66.67%	50.00%	40.00%	52.22%
	Ratio	0.67	0.56	0.48	0.57

Table 12.8 Top Fitness Rule Set – Training/Application Result Ratio.

Fitness		PeriodA	PeriodB	PeriodC	Average
	Training	210.97	112.07	58.25	127.10
	Application	40.25	38.66	43.16	40.69
	Ratio	0.19	0.34	0.74	0.43

Yield		PeriodA	PeriodB	PeriodC	Average
	Training	14.52%	12.64%	7.54%	11.57%
	Application	3.64%	5.69%	6.66%	5.33%
	Ratio	0.25	0.45	0.88	0.53

Profitable		PeriodA	PeriodB	PeriodC	Average
Months	Training	70.27%	72.22%	63.64%	68.71%
	Application	70.59%	52.94%	52.94%	58.82%
	Ratio	1.00	0.73	0.83	0.86

Profitable		PeriodA	PeriodB	PeriodC	Average
Quarters	Training	100.00%	88.89%	73.68%	87.52%
	Application	66.67%	50.00%	60.00%	58.89%
	Ratio	0.67	0.56	0.81	0.68

Table 12.9 Closest Fitness Rule Set – Training/Application Result Ratio

Comparing the performance ratios training set/hold-out set shows that the closest-fitness parameter set significantly increases the consistency and predictability of performance.

Calculating an average of all training/hold-out ratios, the average ratio for the top-fitness parameter set is 0.48, whereas the average ratio for the closest fitness set is 0.62.

12.8 Results of Other Portfolio Types

We have tested the same trading model configuration on other types of portfolios, including various individual currency pairs (IAW a portfolio with just one component) and equity index portfolios.

It has emerged that the diversity of the portfolio is an important contributing factor for the consistency of performance. Although a pure equity index portfolio yielded better returns (in absolute terms) over the same period, the consistency of a number of performance measurements has been lower than in the FX model described here. Portfolios containing fewer markets (or just one single market) have generally performed less well than diversified portfolios, even when correlation among individual portfolio components is relatively high (as it is with major western stock markets, such as New York, London, Paris or Frankfurt).

Although the positive effect of diversification on portfolio performance is known in portfolio theory, this would only partially explain the observed results: although the system may have a number of markets available for allocation of trades, the system does not keep positions in all low-correlated markets at the same time. The actual effect of diversification is therefore much less than the theoretical.

It seems that the GA-based model can learn better, i.e., more flexible, behaviour rules, if a larger amount of different data is available within the same learning period. Providing the system with more data of different types of distribution reduces the risk of over optimisation of a specific type of price distribution. After having faced a more complex environment during the learning process, the trading model seems to be more capable of dealing with the new environment during hold-out periods.

12.9 Results of Different Fitness Targets and Variation of GA Parameters

12.9.1 Different Fitness Target Calculations

Performance comparisons of fund managers or trading advisors typically use a number of different benchmarks to analyse structure of risk and return. Most commonly used are the Sharpe Ratio, which measures risk as variance of returns relative to above risk-free returns, and various types of yield/ drawdown calculations.

We have not been able to develop any acceptable performance during hold-out periods when using these portfolio measurements as fitness targets. Although the training process itself typically yields very high values for these benchmarks (proving that the GA process is an effective search process), these results have generally not translated into appropriate hold-out period behaviour.

It seems that the type of behaviour the trading model develops is very different for various types of fitness targets, although, with hindsight any trading model delivering performance within a defined return path will always have high ratings compared with other performance measurements.

Consequently, we see the concept of optimising a trading model towards a minimum Return Path Error (RPE) as the most appropriate fitness target definition for the learning process.

12.9.2 Variation of GA Parameters

Over a larger number of generations, we have not observed significant changes in the result of the learning process when adjusting GA parameters, except for the use of elitist selection, which seems to contribute significantly to the speed of the learning process. A slight advantage in the learning process has emerged using the tournament selection method, with the roulette-selection performing comparatively inefficiently.

The requirement of computer resources increases substantially when the number of portfolio components is increased. This is most likely due to the increased complexity of the solution space. It has however (as described before) a positive effect on the result of the portfolio simulation.

The fully scalable, distributed GA library on which the APT framework is based provides the development environment to increase the performance of the learning process by adding hardware resources to the learning process.

Conclusion

It has been shown Genetic Algorithms lend themselves very well to complex decision-making simulations, due to the parallel nature of the algorithm. GAs also allow the integration of very different types of learning processes into one optimisation process.

The foreign exchange trading portfolio described here has yielded an acceptable level of risk/return and, compared to other portfolio tests, shows a clear direction for further development.

By implementing all important requirements the human trader faces in the process of making trading decisions, the GA-based learning process proves to be more stable and more efficient, thus leading to lower and more predictable risk for the portfolio management industry.

Chapter 13 Population Size, Building Blocks, Fitness Landscape and Genetic Algorithm Search Efficiency in Combinatorial Optimization: An Empirical Study

Jarmo T. Alander
University of Vaasa
Department of Information Technology and Industrial Economics
P.O. Box 700
FIN-65101 Vaasa, Finland

phone +358-6-324 8444
fax +358-6-324 8467
E-mail: Jarmo.Alander@uwasa.fi

Abstract

In this chapter we analyse empirically genetic algorithm search efficiency on several combinatorial optimisation problems in relation to building blocks and fitness landscape. The test set includes five problems of different types and difficulty levels, all with an equal chromosome length of 34 bits. Four problems were quite easy for genetic algorithm search while one, a folding problem, turned out to be a very hard one due to the uncorrelated fitness landscape.

The results show that genetic algorithms are efficient in combining building blocks if the fitness landscape is well correlated and if the population size is large enough. An empirical formula for the average number of generations needed for optimization and the corresponding risk level for the test set and population sizes are also given.

13.1 Introduction

Genetic algorithms have gained nearly exponentially growing popularity among researchers and engineers as a robust and general optimisation method [4,5]. They have been applied in a wide range of difficult problems in numerous areas of science and engineering.

There does not, however, exist much theoretical or experimental analysis of the properties of the optimisation problems that would, in practice, guide those trying to apply genetic algorithms to practical problem solving. The purpose of this study is to shed some more light on the basic fitness function properties with respect to the functioning of genetic algorithms and, thereafter, give some advice for future applications. The emphasis is put on population size, building blocks and fitness landscape, which further influence search efficiency and success rates.

The optimisation efficiency will be empirically studied using population sizes ranging from 25 to 1600 and five different type and complexity combinatorial problems. The problem set consists of 1) Onemax, 2) maximum sum, 3) boolean satisfiability, 4) polyomino tiling, and 5) a folding problem. The fitness landscape has been analysed by evaluating autocorrelation along random one-bit mutation paths.

The results show that GAs are efficient in combining building blocks, if the fitness landscape is correlated especially around the solutions and the population size is large enough to provide shelter to all "hibernating" but scattered building blocks of the solution.

13.1.1 Related Work

In our previous study we optimized some parameters of a genetic algorithm including population size and mutation rate. The optimization was done by a genetic algorithm using a tiny travelling salesman problem (TSP) as a test function [1]. In a later study, we analysed the effect of population size on genetic algorithm search [3]. The object problem was an Onemax type with the chromosome length n_g varying in the range [4,28]; n_g was used as a measure of problem complexity. The empirical results suggested that the optimal population size n_p seems to be included in the interval [log(N), 2 log(N)], where N is the size of the search space. For the Onemax problem N $= 2^{n_g}$ and thus the optimal population size interval is approximately [n_g, $2n_g$]. Perhaps the most important conclusion from these empirical studies was that the search efficiency does not seem to be too sensitive to either population size or other parameters, but more to the properties of the object problem fitness function.

In [35] Bryant Julstrom derived a formula for the population size when solving TSP. Julstrom's estimate gives a somewhat smaller upper bound for the population size than our estimate, especially for larger search spaces.

De Jong and Spears analysed the interacting roles of population size and crossover [34]. Goldberg et al. have statistically analysed the selection of building blocks [23].

Reeves analysed a small population using an experimental design approach [48]. Hahner and Ralston noticed that small population size may be the most efficient in rule discovery [28].

Also, infinite populations have been theoretically analysed [52].

Weinberger has analysed autocorrelations both in Gaussian and in the NK-landscapes [54].

Other studies with fitness distributions include [13,20,26,44,45]. Work on fitness landscapes includes [17,19,32,36,37,40,41,42,43,47].

References to other studies on parameters of genetic algorithms and search efficiency can be found in the bibliography [8].

13.2 Genetic Algorithm

The genetic algorithm used in this study was coded in C++ specifically for this study keeping in mind possible further studies, e.g., on other fitness functions. In practise this means that the results of this study should be as much as possible independent of the results of the previous studies [1,3] done by another program, also coded in C++.

A somewhat simplified body of the genetic algorithm used is shown at the end of this chapter, while its parameters are shown in Table 13.1.

For a near complete list of references on basics of genetic algorithm see [7] and for implementations see [6] (part of which is included in this volume).

parameter	typical value range
population size n_p	[25,3200]
elitism	[1 , $n_p/2$]
maximum generations	[0,1000]
crossover rate	[0,1]
mutation rate	[0,1]
swap rate	[0,1]
crossover type	binary/genewise

Table 13.1 The parameters of the genetic algorithm used in the experiments.

13.3 Theory

Here we will deduce a stochastic search dynamics model for both selection efficiency and genetic algorithm search risk in relation to population size. In spite of the simplicity of the models, we will see that they fit quite nicely to the empirical results.

13.3.1 Risk Estimation

In order to analyse proper population size with respect to the risk that the solution is not found, let us assume that the initial population contains all the building blocks necessary for the solution. These hibernating but scattered building blocks should also survive until the solution is rejoined by the selection and crossover process.

The probability P (risk level) of having all necessary parameter values or alleles present in the population is most easily evaluated by the complement: what is the probability of a missing solution building block?

$$1 - P = \sum_{i=1}^{n_g} (1 - p_i)^{n_p}$$

where n_p is the size of the population, n_g is the number of parameters, i.e., genes and p_i is the probability of the solution parameter(s) at gene i. In case of homogenous genes, for which $\forall i : p_i = p$, this equation reduces to

$$1 - P = n_g (1 - p)^{n_p}$$

n_a\P	$n_g = 1$			$n_g = 7$			$n_g = 34$		
	0,99	0,999	0,9999	0,99	0,999	0,9999	0,99	0,999	0,9999
2	7	10	13	9	13	16	12	15	18
4	16	24	32	23	31	39	28	36	44
8	34	52	69	49	66	84	61	78	95
16	71	107	143	102	137	173	126	162	197
32	145	218	290	206	279	351	256	329	401
64	292	439	585	416	562	708	516	663	809
128	587	881	1174	835	1129	1422	1037	1330	1624
1024	4713	7070	9427	6705	9062	11418	8323	10679	13036
10240	47155	70732	94309	67080	90657	114234	83263	106840	130417

Table 13.2 Population size estimates $n_g = (\log(1 - P) - \log(n_g))/\log(1 - 1/n_a)$ at several risk levels P, allele probabilities $p = 1/n_a$ and number of genes n_g.

In the further special case, when only one parameter value is valid for a solution, holds $p = 1/n_a$, where n_a is the number of all possible values of a parameter i.e. the number of alleles. Usually $n_a = 2^{n_{bb}} - 1$, where n_{bb} is the length of the building block in bits. By taking a logarithm of the above equation we get an equation for n_p as function of the risk level P, number of genes n_g and solution allele probability p:

$$n_p(P, n_g, P) = \frac{\log(1 - P) - \log(n_g)}{\log(1 - p)}$$

The values of this population size estimate are shown for several risk levels P and number of alleles $n_a = 1/p$ in Table 13.2. From the above equation we can easily solve the risk level P as function of population size n_p and allele probability p.

$$P(n_P, P) = 1 - n_g(1 - p)^{n_P}$$
$$= 1 - n_g e^{n_P \ln(1-p)}$$
$$\approx 1 - n_g e^{-p n_P}$$

This exponential behaviour can be seen more or less clearly in Figure 13.4, where the histogram of the number of fitness function evaluation n_f at different population sizes are quite similar in shape so that they nicely scale with the above risk level function. Another way of using the above relation is to estimate the length of the building blocks n_{bb} via p. Solving the above equation with respect to p, we get

$$p = \frac{\ln(1 - P) - \ln(n_g)}{-n_P}$$

Now, the average building block size $n_{\hat{b}b}$ is approximately given by

$$n_{\hat{b}b} = {}^2\log\left(\frac{1}{p}\right)$$
$$= {}^2\log\left(\frac{-n_P}{\ln(1 - P) - \ln(n_g)}\right)$$

13.3.2 Number of Generations

In order to estimate the speed of genetic search, let us assume that the search is, at first, primarily done only by crossover and selection. The motivation behind this assumption is the empirical fact that the role of mutation in both natural and artificial evolution seems to be much smaller than the role of crossover, when the solution should be found relatively fast. For the long run, the roles interchange and mutation becomes the driving force of evolution (re)creating extinct and eventually totally new alleles.

Let us thus assume that, in each generation i after the initial one the number of building blocks of the solution is increased by a factor s_i. This increase is continually driven by crossover and selection until either the hibernating but scattered solution is happily rejoined or, otherwise, the search transits to the next phase, where the missing building blocks are primarily searched by mutation and selection. This two-phase functioning can be clearly seen in Figure 13.4. It also comfortably resolves the "crossover or mutate" dilemma: which one is more important in genetic algorithm search. For efficient processing, both are vital: crossover for the first few starting generations and mutations thereafter.

Let the size of the initial population be n_p and the selection efficiency (ratio) acting on the hibernating building blocks during each generation i be s_i.

Search by crossover ceases naturally with the diversity of the population, i.e., when the population is filled with more or less identical specimens by the selection process. Assuming a constant selection efficiency s, this happens at generation n_G for which

$$s^{n_G} = n_P$$

i.e., when

$$n_G = \frac{\log(n_P)}{\log(s)}$$

In the somewhat idealistic case, when $s = 2$, then

$$n_G = {}^2\log(n_P)$$

which is the tile number of steps in binary search among 'rip items in cord with the assumption that genetic algorithm search can be seen as a stochastic parallel binary search process. Using the above equation for the expected number of fitness function evaluations, we get

$$n_f = n_P n_G = n_P{}^2 \log(n_P)$$

which seems to be more or less empirically valid for reasonable population sizes np.

13.4 Problem Set

In order to analyse genetic algorithm search efficiency we have selected the following five different type and complexity combinatorial optimisation problems to be solved exactly in our experiments:

- Onemax,
- 7 fields maximum sum,
- seven polyomino block tiling on a square,
- a 233 clause 3-SAT problem, and
- a 16 degrees of freedom toy brick problem we call snake folding.

All problems share a 34-bit long chromosome vector structure consisting of seven genes (ints of C, see Table 13.3 for one example problem encoding) of lengths 4, 4, 5, 5, 6, 6, and 4 bits correspondingly. 34 bits give approximately 16 x 10^9 possible combinations, which is quite a large number but not too much to allow massive "search until success" repetition of experiments to get some significance level for statistical analysis.

13.4.1 All are Ones

Our first test problem is the well-known Onemax problem: for a binary string x of length l, this is the problem of maximising

$$\sum_{i=1}^{l} x_i, \ x_i \in \{0,1\}^l$$

The fitness is the number of one bits in the chromosome vector (maximum = 34 = solution). This problem should be ideal in terms of building blocks and fitness landscape correlation; each bit position contributes to the fitness independent of any other position. But, as we will see, this simple problem is not very simple for a genetic approach, perhaps because the fitness landscape is actually highly multimodal [14].

Onemax has been used by many GA researchers as a test problem [14, 30, 32], because, in many respects, it is a simple and well-suited problem for genetic algorithms.

13.4.2 Maximum Sum

The next problem is to find the maximum sum of the seven genes, the lengths of which vary from 4 to 6 bits (see Table 13.3 for chromosome structure). The fitness of the solution is 233 = (15 + 15 + 31 + 31 + 63 + 63 + 15), when all bits = 1. So, while the Onemax and the maximum sum problem share exactly the same solution vector, their fitness function values and distributions (see Figure 13.2) are clearly different.

Maximum sum is actually a linear problem, which can be solved best by linear programming.

13.4.3 Tiling Polyominoes

The fitness of our polyomino tiling or packing problem is defined as the number of unit squares covered by the seven polyomino blocks. The blocks have integer coordinate values (0, 1, 2, 3). In addition, polyominoes 3 and 4 (dominoes) can be in vertical or horizontal orientation ($\phi = 0$, π) and polyominoes 5 and 6 (corner trominoes) may be in any of the four orientations $\phi = 0$, $\pm\pi/2$, π. The tiling of all seven polyominoes is thus encoded in 34 bits as shown in Table 13.3.

Polyominoes provide many interesting combinatorial problems. To GA research they seem to be quite new, however. The author knows only one other study of solving polyomino tiling problems using genetic algorithms [27]. The problem is to fill a 4 x 4 square using the following blocks called polyominoes [24]:

i = 1 2 3 4 5 6 7

i	int	bin	ϕ	x	y
1	7	01 11	-	1	3
2	12	11 00	-	3	0
3	1	0 00 01	0	0	1

4	4	0 01 00	0	1	0
s	26	01 10 10	$-\pi/2$	2	2
6	44	10 11 00	$-\pi$	3	0
7	11	10 11	-	2	3
bits	34	34	6	14	14

Table 13.3 An example of a polyomino tiling (fitness $f = 12$) and the parameters of the polyominoes: i = block index, int = parameter as an integer, bin = parameters as a binary string, ϕ = angle of rotation, x and y the x- and y-coordinates of the bottom left corner of the blocks. This chromosome structure is shared by all test problems.

13.4.4 3-SAT Problem

The next problem is a 3-variables-per-clause boolean satisfiability problem (3-SAT) consisting of 233 clauses. The 3-SAT problem means the following: we have a set of three variable boolean expressions of the form

$$v_i + v_j + v_k$$

where v_i is a boolean variable or its negation and '+' stands for conjunction, i.e., boolean or-operation. The problem now is to find such a truth assignment to the variables that maximum number of clauses evaluates to true, i.e., our problem is actually of MAX 3-SAT type.

In our test case we have chosen 34 variables to meet the chromosome compatibility requirement and 233 clauses to be comparable with the maximum sum problem. The C-language routine SATinitialize that generates our 3-SAT problem instance is shown below

```
void SATinitialize(int t)
// Initialize SAT problem cases: t gives the number of clauses (=233).
// Variables and their complements are represented by 2D arrays
// VARS and NVARS, which are assumed to be reset.
// setBit() sets the bits in these arrays so as to meet the chromosome
// structure. Nbits is the number of bits or variables (=34).
{
    inti,j
    for (i=0; i<3,t/2; i++) {
        j = (i+i/Nbits)%Nbits;
        setBit (len,VARS [i%t], j );
        setBit (len, NVARS [( i+1 )%t], j );
    }
}
```

The fitness of our 3-SAT problem is an integer function, the maximum value of which is 233, when all the boolean clauses are true, i.e., satisfied.

3-SAT is proven to be an NP-complete problem [22], but only a small fraction of all possible cases are actually that hard to solve [2]. It seems that when the ratio of the number of clauses to the number of variables is approximately 4.5, then the most difficult cases appear [50]. In our 3-SAT problem case, the ratio is $233/34 \approx 6.85$, which should not imply an extremely difficult case. It turned out that our problem was actually quite easy.

Being a basic NP-hard problem boolean satisfiability problems have been also quite popular in genetic algorithm research [12, 157 18, 21, 25, 29, 33, 39, 46, 51].

13.4.5 Folding Snake

The last and most difficult problem for a genetic algorithm is what we call the folding snake. The "toy snake" consists of 27 small cubic bricks arranged like a string of pearls, each brick able to rotate with respect to its predecessor. The problem is to fold the snake into a 3 x 3 x 3 cube (see Figure 13.1). It turns out that there are sixteen degrees of freedom that are enough to solve the problem. Each degree of freedom may have rotation value 0, 1, 2, or 3 times $\pi/2$ giving total 2 x 16 = 32-bit long chromosome representation. This is 2 bits less than in the other problems, but that was not considered a primary factor for GA search efficiency. Due to symmetry reasons the solution is not unique, but there are 4 x 2 solutions: the first degree of freedom may have any of the 4 possible values while the fourth has 2 possibilities. In any case, the author feels that this type of problem may be NP-complete (the similar protein folding problem, namely, is [38]). At least it is the most difficult problem in our test set.

The snake can be thought of as a simple model of a multi-degree-of-freedom mechanism such as a robot or a macromolecule. For example, solving inverse kinematics can be studied by changing the fitness function to a function that measures the distance of the end brick from a given destination. Constraining the workspace would naturally lead to path-planning type problems.

Sad to say to both robot engineers and chemists, the snake folding seems to be quite a hard problem, at least for a purely genetic algorithm approach.

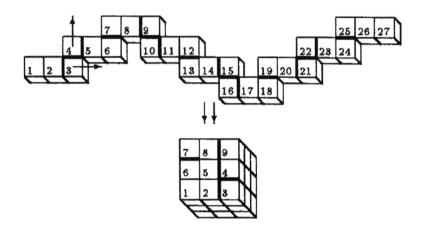

Figure 13.1 The toy snake folding problem of 27 bricks: initial unfolded (top) and final folded (bottom) states. The relevant rotational degrees of freedom are shown by emphasised lines between bricks. Axis of rotations are shown for brick pairs (2,3) and (3,4).

13.5 Experiments

The following properties of the test problems were analysed:

- distribution of fitness values in order to see how probable it is to find the solution by change (mutation alone) and

- autocorrelation of fitness values along a random one-bit mutation path in order to have a rough idea of the fitness landscape for search efficiency (crossover and selection).

In addition, the effect of population size on search efficiency and success was analysed by undertaking 1000 optimization runs using a different random number seed (and sequence) each time. This test was repeated on each problem of the test set.

13.5.1 Fitness Distributions

The first experiment was to reveal the fitness function distributions. This was done by generating 10×10^6 random chromosomes and making a histogram of the fitness values (Figure 13.2). In spite of the quite large number of trials no solution for any of the test set problems was actually found when evaluating the histograms suggesting that pure random search is not efficient to solve our test set problems or, at least, the random number sequence used does not contain the solution pattern.

Observe that the shapes of the histograms of the Onemax, maximum sum, and the polyomino problems are quite symmetrical and near second-order polynomials on the logarithmic scale. That means that their distributions are quite near e^{x^2}, i.e., Gaussian, as assumed in many statistical analysis texts, but this is true only near the peak.

Distributions of the 3-SAT and snake problems are markedly skewed towards the solution, but, as we will see, this seems to help the GA reveal the solution easily only in the 3-SAT case.

13.5.2 Fitness Landscape

In order to have a rough idea of the shape of the fitness landscapes, the following autocorrelation function `acorr`, as a function of the Hamming distance $h_{i,j}$ between chromosomes c_i and c_j, was evaluated:

$$acorr(h_{i,j}) = \frac{100\%}{n} \sum_{i \neq j} \left(2 \frac{\min(f_i, f_j)}{\max(f_i, f_j)} - 1 \right)$$

where n is the number of samples sharing the same Hamming distance and $f_i = f(c_i)$ is the fitness of chromosome c_i. Figure 13.3 shows this autocorrelation function for the test problems. Each histogram was evaluated by generating a 10^6 long single random mutation path and evaluating the above correlation index among 67 (= 2 x 34 - 1) consecutive chromosomes along this path.

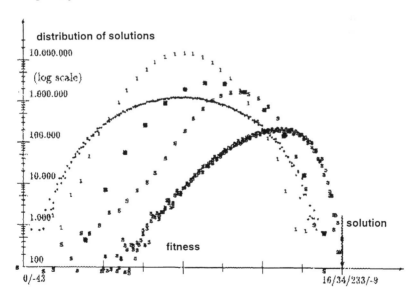

Figure 13.2 Distribution of problem fitness values (1 = Onemax, 3 = 3-SAT, ● = maximum sum, ■ = polyomino and s = snake folding) of 10 x 10^6 random samples. Observe that random sampling has not found any solution for any problem (the best trial for 3-SAT has the fitness 232). Solution fitness = 16/34/233/-9 is fixed and shown by a vertical arrow ($\downarrow$), while x-scale varies with the problem. All problems share $x = 0$, except the

snake problem, which has negative fitness values; otherwise, it has the same scaling as the Onemax problem.

Notice that the autocorrelations were evaluated only to $h = 29$ because of the way in which the evaluation was done: it is quite improbable that all the bits would be inverted during a relatively short (=67) single mutation path. For the same reason, the values for high values of h are much more noisy than the values for shorter h.

As can be seen, 3-SAT and Onemax problems have the highest correlation, while the polyomino tiling and snake problems have lower correlations for short correlation distances. Observe, however, that the definition of our autocorrelation differs from the traditional definition. For example, the skewness of the distribution implies a relatively higher overall correlation. This shows clearly in autocorrelation of the 3-SAT, snake and polyomino problems, while Onemax and, especially maximum sum, has quite a linear decreasing trend with increasing h.

The multimodality of the Onemax problem slowing GA search [14] can be seen as a small periodic component on the mainly decreasing trend of the autocorrelation function.

The snake-folding problem, which, as we will see, seems to be the most difficult problem in our test set has quite a high autocorrelation after all. It is nearly as high as that for the Onemax problem and even higher than for the polyomino problem, which will be much easier for GA.

In basic statistical analysis (e.g., in [54]), statistical isotrophy is usually assumed. That means that the autocorrelation, on the average, is independent of the point where it was evaluated. This isotropy is also the basic assumption in the NK-landscape model, where contributions of each gene position are assumed to be normally distributed and independent of each other. For an optimization problem it is not at all a priori certain that the statistical isotrophy holds, however.

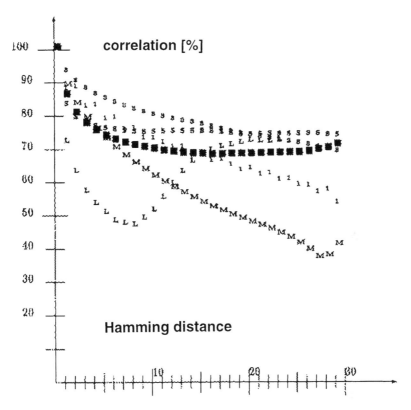

Figure 13.3 (a) Correlation of fitness values. (1 = Onemax, 3 = 3-SAT, M = maximum sum, ■ = polyomino, and s = snake folding, L = local autocorrelation of snake folding around a solution).

If the problem is not statistically isotrophic, then the interpretation of the autocorrelation is not straightforward. In order to test the isotrophy assumption, we have evaluated the autocorrelation in the vicinity of one of the solutions of the snake problem, which is potentially the most unisotrophic problem in our test set. The local autocorrelation, denoted by 'L,' is shown in Figure 13.3 among the "global" autocorrelations. As can be seen, the difference between the local and global autocorrelation is huge. Another point is that in the random path used in the evaluation of the autocorrelation, the most probable values around the mean determine the correlation thus greatly amplifying the possible nonisotrophism. This result immediately warns us not to rely too much on a priori assumptions when solving an optimization problem, the statistical properties of which are not known, which unfortunately is usually the case.

13.5.3 Search Speed

The effect of population size on search speed and success was analysed by 1000 optimization runs each time using a different random number seed (and sequence). The experiment was repeated on all test problems and on several population sizes, while the other parameters were kept fixed. No attempt was made to optimize the parameters of the genetic algorithm. For example, genewise crossover was used in all tests, even if it is not as good as binary crossover for problems like the Onemax.

The search was continued until either the solution was found or, approximately, the given number of fitness functions was evaluated. For all except the snake-folding problem, we had a limit of 50,000 fitness function evaluations. For snake folding, the limit was set at 160,000 evaluations. The results of the search speed experiment are shown in Tables 13.5 and 13.6 and Figure 13.4.

The parameters of the GA were: elitism = population size/2, while the other parameters were fixed for all experiments; crossover rate = 50%, single gene mutation rate = 10%, swap rate = 10%, and total chromosome mutation rate = 30%.

As can be seen, the solution is found using more or less fitness function evaluations depending upon the problem case and population size. Obviously, the speed of search gives a rough measure of the relative complexity of the test problems with respect to GA optimization.

One hypothesis is that the difference is due to the building blockability, i.e., separability of the problems. Unfortunately the estimate

$$n_{bb} = {}^2\log(n_P / \ln(1 - P))$$

cannot be used because for all test problems the selection efficiency is monotonic without any clear asymptoticity.

13.5.4 Population Size

Figure 13.4 shows the effect of population size on the number of function evaluations needed to find the solution for the polyomino problem. The histograms have been created by running the GA 1000 times using different random number seeds and population sizes ranging from 25 to 1600. As can be seen, the larger the population size, the more function evaluations are needed. From this we could conclude that the small population sizes are most effective, but that is true only if the solution is actually found within a reasonable time. Namely, the probability of finding the solution by using, at most, a given number of evaluations (50,000) also decreases with decreasing population size (see Table 13.6).

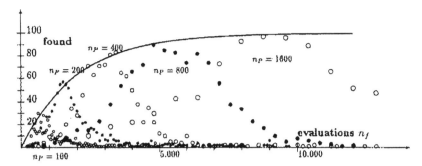

Figure 13.4 Distribution of evaluations needed for finding the solution to the polyomino layout problem. Parameters of the GA are the same as in the previous Figure except for the population size np, which was 50, 100, 200, 400, 800 and 1600, while elite = $np/2$. Solid line is $100\left(1 - e^{-n_f/3000}\right)$.

problem	$n_\infty \atop G$	$n_\infty \atop P$	n_0	ES	n_P
Onemax	7.25	520	9	1.11	50-1600
max sum	10.5	2000	5	1.46	100-1600
polyomino	10.5	2650	36	0.80	100-1600
3SAT/233	3.5	470	S	3.10	25-1600
snake	26	15000	72	0.58	200/800/3200

Table 13.4 Approximate parameter values $n_0, n_\infty \atop G$, $n_\infty \atop P$ and n_0 of the average generations model $\hat{n}_G = n_\infty \atop G + n_\infty \atop P / (n_P - n_0)$ for the test problem set.

13.5.5 Search Efficiency Model

Table 13.5 shows the average number of generations needed to solve the test problems. It seems that the average number of generations needed can be quite precisely approximated by the (empirical) formula

$$\hat{n}_G \atop G = n_\infty \atop G + \frac{n_\infty \atop P}{n_P - n_0}$$

where $n_\infty \atop G$ is the asymptotic number of generations needed for a very large population, n_0 is the minimum population size and $n_\infty \atop P$ is the size of a population that does not much reduce the number of generations, i.e., "a very large population size" typically ranging from a few hundred to over 2000 in our experiments.

Table 13.4 shows the model parameters for the average number of generations n_G needed. This estimation is also shown as a solid line in Figure 13.5 for the polyomino problem, where the effect is most clearly manifesting itself.

np	1max	maxs.	poly.	3-SAT	snake
the average number of generations					
25	47.61	-	-	31.18	-
50	19.97	50.81	71.31	16.07	-
100	12.66	31.50	51.88	7.49	-
* 100	10.09	14.89	24.32	5.48	-
200	10.07	22.10	26.63	5.84	142.76
400	8.91	15.53	18.06	4.87	-
800	8.05	13.03	14.05	4.23	46.63
1600	7.38	11.74	11.81	3.64	-
3200	-	-	-	-	30.92
standard deviation of n_G					
25	415.47	-	-	21.98	-
50	8.89	30.00	121.57	28.92	-
100	3.05	19.09	81.37	2.69	-
* 100	1.41	2.94	12.36	1.32	-
200	1.72	13.10	28.46	1.50	200.71
400	1.29	4.50	10.89	1.23	-
800	1.04	1.86	4.80	1.07	37.50
1600	0.90	1.43	4.08	0.97	-
3200	-	-	-	-	6.78

Table 13.5 The average number of generations n_G needed for finding the solution at different population sizes np and problems using unoptimized genetic algorithm and a more optimal (*100) one.

The data from Table 13.5 is also displayed in Figure 13.5(a). Figure 13.5(b) shows the corresponding average selection efficiency felt by specimens, i.e., s/np given by

$$s/n_P = [n_P]^{\left(1/\hat{n}_G - 1\right)}$$

As can be seen, the efficiency is at its maximum at relatively small population sizes actually not that far away from the so popular value $np = 50$.

Unfortunately, the search success rate also rapidly decreases with decreasing population size, making the definition of the exact location of the best search efficiency uncertain. It seems that the population size range $[n_c, 2n_c]$, where n_c is the length of the chromosome vector in bits, is quite reasonable as was already suggested in [1].

It must be reemphasised that no attempt to optimize the parameters of the genetic algorithm itself was done either before or during the experiments. Comparisons to other experiments seem to show that the genetic algorithm used was not that bad, however. Table 13.5 also shows the results of a more optimal genetic algorithm at population size 100 (denoted as 100). As can be seen, the relative difference to the corresponding nonoptimal results is of the order 2.

13.6 Discussion and Recommendations

13.6.1 Comparisons to Other Results

In our experiments we have partly used the same or similar problems to those in our previous studies and studies by other authors. Unfortunately, in our previous experiment, we measured the efficiency in CPU time units and not in the number of fitness function evaluations [1].

np	$1max$	$maxs.$	$poly.$	$3SAT$	$snake$
25	95.7	-	-	93.1	-
50	99.3	55.3	34.1	99.1	-
100	99.8	75.4	60.8	99.8	-
*100	99.8	86.3	43.1	99.8	-
200	99.8	96.2	86.7	99.8	5.1
400	98.8	99.4	98.2	99.8	-
800	98.8	99.8	99.7	99.2	37.6
1600	99.4	99.8	99.8	98.8	-
3200	-	-	-	-	87.2

Table 13.6 Problems and the success rate P at different population sizes np.

Höhn and Reeves used a standard GA to solve the Onemax problem of length $n_g = 9, 19, 29, 39$ and 49 bits. Their result showed that the number of function evaluations increased linearly with length [30] at population size 200. An interpolation for 34 bits after their data would give $n_f \approx 2447$ function evaluations, while in our experiment $n_f \approx 2014$, when population size is 200.

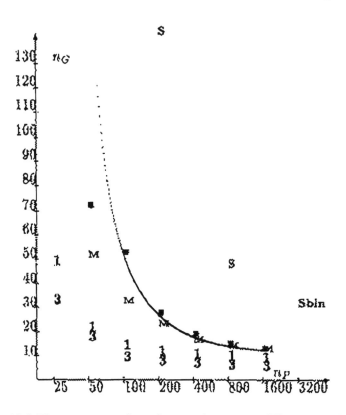

Figure 13.5 The average number of generations n_G at different population sizes np and test problems (1 = Onemax, 3 = 3-SAT, M = maximum sum, ■ = polyomino, and s = snake folding. Sbin = snake folding using binary crossover). The solid line shows the average number of generations needed for the polyomino tiling using the estimate $\hat{n}_G = 10.5 + 2650/(n_P - 36)$.

13.6.2 Autocorrelation Analysis

If the problem is not statistically isotrophic, then the interpretation of the autocorrelation is not straightforward. The local autocorrelation for the most difficult problem, snake folding, is shown in Figure 13.3 among the "global" autocorrelations. As can be seen, the difference between the local and global autocorrelation is huge. This result warns us not to rely too much on a priori assumptions when solving an optimization problem, the statistical properties of which are not known, which unfortunately usually is the case.

The other conclusion we can draw from the modest local autocorrelation is that there must be one major factor explaining the hardness of the snake problem.

The 34-variable MAX 3-SAT problem seems to be the easiest of the five test problems. This can be seen in the fast and reliable processing. This is somewhat surprising because 3-SAT is known to be an NP-complete problem. It must be kept in mind that the NP-completeness does not tell much about the given problem instance. A given MAX 3-SAT problem may be difficult, but not necessarily. The easiness may actually be a more general property of rule-based systems for which Hahnert and Ralston found that small population sizes are the most efficient ones [28]. The maximum sum and polyomino tiling problems are both quite difficult, but in a different sense. If we look at what happens during the active search phase, we see that the polyomino problem is faster, but, unfortunately, is also less reliable than the GA search on the maximum sum problem.

It is obvious that the polyomino problem better fulfills the building block hypothesis: once we have the building blocks available, the rest goes more reliably than in the search for the maximum sum problem solution, where the least significant bits are easily lost during the first few generations and where the most significant bits dominate search done by selection and crossover. When the population size is also large the least significant bits of the maximum sum problem survive and the behaviour of both problems is quite similar.

The snake-folding problem, interesting because of its relation to some important practical optimisation problems, turned out to be far more difficult than the rest of the test problems. It seems that the short correlation length around the solution explains this difficulty and also warns us not to rely too much on statistical isotrophy assumptions when using autocorrelation in estimation of problem difficulty.

13.6.3 Sensitivity

Perhaps the most positive property of genetic algorithms is their robustness, i.e., nonsensitivity to most of the parameters. It is actually difficult to find any parameter that would dramatically affect its functioning. That is a nice feature especially in pilot projects applying genetic algorithm optimisation to new problems. If the genetic approach is suited at all, it usually gives quite good results. The processing time might be quite long, but that is not usually a great disadvantage in engineering design work where part of the engineers' work could be replaced in this way by an automatic optimisation system as in [10].

13.6.4 Recommendations

In this chapter we will give a few recommendations for those who plan to use genetic algorithms in combinatorial optimization.

Population Size

In the case of a random initial population, it is wise to use a population size that gives a high enough probability for each solution building block to be included. In the case of integer or real number encoded problems, remember to also take into account the effect of encoding.

A combinatorial problem may not be isotrophic. This has a profound affect on population size. The shorter the correlation length, the larger the population size needed.

In combinatorial problems we should also consider local hill-climbing, which effectively makes fitness landscape correlation much longer.

The acceptable risk level of not solving the problem within a given number of function evaluations is also an essential factor having influence on proper population size. The lower the risk level the larger the population needed, which, unfortunately, also leads to slower processing on the average.

In many real valued problems, the fitness landscape is much more strongly correlated than in discrete combinatorial problems and this has a profound affect on the search efficiency and optimal population size, but that is beyond the scope of this study. It must be noted, however, that in many cases the real encoded problem can be effectively solved by using a surprisingly small population size and local hill-climbing might even make convergence considerably faster.

Number of Generations

Easy problems should be solvable usually within a few tens of generations. Longer processing time may be due to insufficient population size or might be caused by problems with fitness function and/or short correlation length and/or nonisotrophism. The less correlated or local fitness landscape, the larger the population size needed.

Fitness Function

Be aware that in constrained problems the constraints themselves affect fitness values and further landscape correlation. By careless penalty values we can make landscape correlation length really short and thus prevent selection from catching those trials that are "just round the corner".

Genetic Operators

In binary encoded problems, use binary uniform crossover. Be sure that during the later generations the mutation operator is capable of producing the missing building blocks while not too much disturbing the already found blocks. Use elitism but do not prevent creation of new building blocks too much either.

13.7 Conclusions

In this study we analyse empirically genetic algorithm search efficiency on five combinatorial optimisation problems with respect to building blocks and fitness landscape. Four of the problems were quite easy for genetic algorithm search while the fifth, the folding problem, turned out to be very hard due to the uncorrelated fitness landscape around the solution.

The results show that genetic algorithm is efficient in combining building blocks if the fitness landscape is also well correlated around the solution and if the population size is large enough to shelter all necessary solution building blocks. An empirical formula for the average number of generations needed for optimization and the corresponding risk level for the test set and population sizes was further given.

Acknowledgments

The author wants to acknowledge Miss Lilian Grahn for proofreading a version of this chapter and the Atk-keskus of the University of Vaasa for providing tools and facilities to test runs.

Bibliography

[1] Jarmo T. Alander. On finding the optimal genetic algorithms for robot control problems. In *Proceedings IROS '91 IEEE/RSJ International Workshop on Intelligent Robots and Systems '91,* volume 3, pages 1313-1318, Osaka (Japan), 3-5 November 1991. IEEE Cat. No. 91TH0375-6.

[2] Jarmo T. Alander. On Boolean SAT and GA. In Jarmo T. Alander, (Ed.), *Geneettiset algoritmit - Genetic Algorithms,* number TKO-C53, pages 190-207. Helsinki University of Technology (HUT), Department of Computer Science, 1992.

[3] Jarmo T. Alander. On optimal population size of genetic algorithms. In Patrick Dewilde and Joos Vandewalle, editors, *CompEuro 1992 Proceedings, Computer Systems and Software Engineering, 6th Annual European Computer Conference,* pages 65-70, The Hague, 4.-8. May 1992. IEEE Computer Society, IEEE Computer Society Press.

[4] Jarmo T. Alander. *An indexed bibliography of genetic algorithms: Years 1957-1993.* Art of CAD Ltd., Vaasa (Finland), 1994. (over 3000 GA references).

[5] Jarmo T. Alander. In Lance Chambers, editor, *Practical Handbook of Genetic Algorithms: New Frontiers,* vol. II of *New Frontiers,* chapter Appendix 1. An indexed bibliography of genetic algorithms (Books, proceedings, journal articles, and PhD thesis), pages 333-427. CRC Press, Inc., Boca Raton, FL, 1995.

[6] Jarmo T. Alander. Indexed bibliography of genetic algorithm implementations. Report 94-1-IMPLE, University of Vaasa, Department of Information Technology and Production Economics, 1995. (ftp ftp.uwasa.fi: cs/report94-1/gaIMPLEbib.ps.Z).

[7] Jarmo T. Alander. Indexed bibliography of genetic algorithms basics, reviews, and tutorials. Report 94-1-BASICS, University of Vaasa, Department of Information Technology and Production Economics, 1995. (ftp ftp.uwasa.fi: cs/report94-I/gaBASICSbib.ps.Z).

[8] Jarmo T. Alander. Indexed bibliography of genetic algorithms theory and comparisons. Report 94-1-THEORY, University of Vaasa, Department of Information Technology and Production Economics, 1995. (ftp ftp.uwasa.fi: cs/report94-1/gaTHEORYbib.ps.Z).

[9] Jarmo T. Alander, editor. *Proceedings of the Second Nordic Workshop on Genetic Algorithms and their Applications (2NWGA),* Proceedings of the University of Vaasa, Nro. 11, Vaasa (Finland), 19.-23. August 1996. University of Vaasa. (ftp ftp.uwasa.fi: cs/2NWGA/*.ps. Z).

[10] Jarmo T. Alander and Jouni Lampinen. Cam shape optimization by genetic algorithm. In C. Poloni and D. Quagliarella, editors. *Genetic Algorithms and Evolution Strategies in Engineering and Computer Science,* pages 153-174, Trieste (Italy), 28 November - 5 December 1997. John Wiley & Sons, New York.

[11] Richard K. Belew and Lashon B. Booker, editors. *Proceedings of the Fourth International Conference on Genetic Algorithms,* San Diego, 13-16 July 1991. Morgan Kaufmann Publishers.

[12] Thomas A. Bitterman. *Genetic algorithms and the satisfiability of large-scale Boolean expressions.* Ph.D. thesis, Louisiana State University of Agricultural and Mechanical College, 1993.

[13] Lashon B. Booker. Recombination distributions for genetic algorithms. In Whitley [55], pages 29-44.

[14] Joseph C. Culberson. Mutation-crossover isomorphisms and the construction of discriminating functions. *Evolutionary Computation,* 2(3), 1995.

[15] Raphaël Dorne and Jin-Kao Hao. Nouveaux operators genetiques appliques au SAT [New genetic operators based on stratified population for SAT]. In *Proceedings of the Evolution Artificielle '94,* Toulouse (France), September 1994.

[16] In *Proceedings of the Artificial Evolution 97 (EA '97) Conference,* Nimes (France), 22-24 October 1997.

[17] Ágoston E. Eiben and C.A. Schippers. Multi-parent's niche: n-ary crossovers on NK-landscapes. In Voigt et al. [53], pages 319-328.

[18] Ágoston E. Eiben and J.K. van der Hauw. Solving 3-SAT by GAs adapting constraint weights. In *Proceedings of 1997 IEEE International Conference on Evoutionary Computation,* pages 81-86, Indianapolis, IN, 13-16 April 1997. IEEE, New York, NY.

[19] Larry J. Eshelman, Richard A. Caruana, and J. David Schaffer. Biases in the crossover landscape. In Schaffer [49], pages 10-19.

[20] Patrik B.J. Floréen and Joost N. Kok. Tracing the moments of distributions in genetic algorithms. In Jarmo T. Alander, (Ed.), *Proceedings of the Second Finnish Workshop on Genetic Algorithms and their Applications,* Report Series 94-2, pages 51-60, Vassa (Finland), 16-18 March 1994. University of Vaasa, Department of Information Technology and Industrial Economics. (ftp ftp.uwasa.fi: cs/report94-2/Floreen.ps.Z).

[21] J. Frank. A study of genetic algorithms to find approximate solutions to hard 3-SAT problems. In E.A. Yfantis, editor, *Proceedings of the Intelligent Systems - Third Golden West International Conference, Edited and Selected Papers*, volume 1-2, Las Vegas, NV, 6-8 June 1994. Kluwer Academic Publishers, Dordrecht, Netherlands.

[22] Michael R. Garey and David D. Johnson. *Computers and Intractability, A Guide to the Theory of NP-Completeness.* W.H. Freeman and Co., San Francisco, 1979.

[23] David E. Goldberg, Kalyanmoy Deb, and James H. Clark. Accounting for noise in the sizing of populations. In Whitley [55], pages 127-140.

[24] Soloman W. Golomb. *Polyominoes, Puzzles, Patterns, Problems, and Packings.* Princeton University Press, Princeton, New Jersey, 2nd edition, 1994.

[25] J. Gottlieb and N. Voss. Representations, fitness functions and genetic operators for the satisfiability problem. In [16].

[26] John J. Grefenstette. Predictive models using fitness distributions of genetic operators. In *Proceedings of the Foundations of Genetic Algorithms 3 (FOGA 3)*, 1994. (ftp ftp.aic.nrl.navy.mil: /pub/papers/1994/AIC-94-001.ps(.Z), NCARAI Technical Report No. AIC-94-001).

[27] B.H. Gwee and Meng-Hiot Lim. Polyominoes tiling by a genetic algorithm. *Computational Optimization and Applications*, 6(3):273291, 1996.

[28] W.H. Hahnert, III and Patricia A.S. Ralston. Analysis of population size in the accuracy and performance of genetic training for rule-based control systems. *Computers & Operations Research*, 22(1):55-72, 1995.

[29] Jin-Kao Hao and Raphaël Dorne. An empirical comparison of two evolutionary methods for satisfiability problems. In ICEC'94 [31], pages 450-455.

[30] Christian Höhn and Colin R. Reeves. The crossover landscape for the Onemax problem. In Alander [9], pages 27-44. (ftp ftp.uwasa.fi: cs/2NWGA/Reeves.ps.Z).

[31] *Proceedings of the First IEEE Conference on Evolutionary Computation*, Orlando, FL, 27-29 June 1994. IEEE, New York, NY.

[32] Terry Jones. *Evolutionary Algorithms, Fitness Landscapes and Search.* Ph.D. thesis, University of New Mexico, 1995.

[33] Kenneth A. De Jong and William M. Spears. Using genetic algorithms to solve NP-complete problems. In Schaffer [49], pages 124-132.

[34] Kenneth A. De Jong and William M. Spears. An analysis of the interacting roles of population size and crossover in genetic algorithms. In Hans-Paul Schwefel and R. Männer, editors, *Parallel Problem Solving from Nature*, volume 496 of *Lecture Notes in Computer Science*, pages 38-47, Dortmund (Germany), 1-3 October 1991. Springer-Verlag, Berlin.

[35] Bryant A. Julstrom. A simple estimate of population size in genetic algorithms for the traveling salesman problem. In Alander [9], pages 3 14. (ftp ftp .uwasa.fi: cs/2NWGA/Julatrom1.ps.Z).

[36] Kenneth E. Kinnear, Jr. Fitness landscapes and difficulty in genetic programming. In ICEC'94 [31], pages 142-147.

[37] K. Kolarov. Landscape ruggedness in evolutionary algorithms. In *Proceedings of 1997 IEEE International Conference on Evolutionary Computation,* pages 19-24, Indianapolis, IN, 13-16 April 1997. IEEE, New York, NY.

[38] Richard H. Lathtop. The protein threading problem with sequence amino acid interaction preferences is NP-complete. *Protein Engineering,* 70:1059 1068, 1994.

[39] T.C.S. Lee and A.L.G. Koh. Genetic algorithm approaches to SAT problems. In *Proceedings of the 2nd Singapore International Conference on Intelligent Systems (SPICIS'94),* pages B171-B176, Singapore, 14-17 November 1994. Japan-Singapore AI Centre, Singapore.

[40] Bernard Manderick, Mark de Weger, and Piet Spiessens. The genetic algorithm and the structure of the fitness landscape. In Belew and Booker [11], pages 143-150.

[41] Bernard Manderick and Pier Spiessens. Computational intelligence imitating life. How to select genetic operators for combinatorial optimization problems by analyzing their fitness landscape, In Jacek M. Zurada, Robert J. Marks, II and Charles J. Robins, (Eds.), *Computational Intelligence Imitating Life,* pages 170-181. IEEE Press, New York, 1994.

[42] Keith E. Mathias and Darrell L. Whitley. Genetic operators, the fitness landscape and the traveling salesman problem. In R. Männer and B. Manderick, editors, *Parallel Problem Solving from Nature, 2,* pages 219-228, Brussels, 28-30 September 1992. Elsevier Science Publishers, Amsterdam.

[43] Melanie Mitchell, Stephanie Forrest, and John H. Holland. The royal road for genetic algorithms: Fitness landscapes and GA performance. In Francisco J. Varela and Paul Bourgine, editors, *Toward a Practice of Autonomous System: Proceedings of the First European Conference on Artificial Life,* pages 245-254, Paris, 11-13 December 1991. MIT Press, Cambridge, MA. (ftp ftp.santafe.edu: /pub/Users/mm/sfi-91-10-046.ps.Z)

[44] Heinz Mühlenbein, Jürgen Bendisch, and Hans-Michael Voigt. From recombination of genes to the estimation of distributions: II. In Voigt et al. [53], pages 188-197.

[45] Heinz Mühlenbein and Gerhard PaaB. From recombination of genes to the estimation of distributions: I. binary parameters. In Voigt et al. [53], pages 178-187.

[46] Kihong Park. A comparative study of genetic search. In Larry J. Eshelman, editor, *Proceedings of the Sixth International Conference on Genetic Algorithms*, Pittsburgh, PA, 15-19 July 1995.

[47] Lae-Jeong Park, Dong-Kyung Nam, Cheol Hoon Park, and Sang-Hoon Oh. An empirical comparison of simulated annealing and genetic algorithms on NK fitness landscapes. In *Proceedings of 1997 IEEE International Conference on Evolutionary Computation*, pages 147-151, Indianapolis, IN, 13-16 April 1997. IEEE, New York, NY.

[48] Colin R. Reeves. Using genetic algorithms with small populations. In Stephanie Forrest, editor, *Proceedings of the Fifth International Conference on Genetic Algorithms*, pages 92-99, Urbana-Champaign, IL, 17-21 July 1993. Morgan Kaufmann, San Mateo, CA.

[49] J. David Schaffer, editor. *Proceedings of the Third International Conference on Genetic Algorithms*, George Mason University, 4.-7. June 1989. Morgan Kaufmann Publishers, Inc.

[50] Bart Selman, David G. Mitchell, and Hector J. Levesque. Generating hard satisfiability problems. *Artificial Intelligence*, 81(1-2):17-29 March 1996. A special issue of phase transitions.

[51] Peter Smith. Finding hard satisfiability problems using bacterial conjugation. In *Evolutionary Computing, Proceedings of the AISB96 Workshop*, pages 236-244, Brighton, UK, 1-2 April 1996.

[52] Luminita State. Information theory analysis of the convergence and learning properties of a certain class of genetic algorithms in continuous space and infinite population assumption. In *Proceedings of the 1995 IEEE International Conference on Systems, Man and Cybernetics*, pages 229-234, Vancouver, BC (Canada), 22-25 October 1995. IEEE, Piscataway, NJ.

[53] Hans-Michael Voigt, Werner Ebeling, Ingo Rechenberg, and Hans-Paul Schwefel, editors. *Parallel Problem Solving from Nature - PPSN IV*, volume 1141 of *Lecture Notes in Computer Science*, Berlin (Germany), 22-26 September 1996. Springer-Verlag, Berlin.

[54] Ed Weinberger. Correlated and uncorrelated fitness landscapes and how to tell the difference. *Biological Cybernetics*, 63():325 336, 1990.

[55] Dartell Whitley, editor. *Foundations of Genetic Algorithms 2 (FOGA-92)*, Vail, CO, 24-29 July 1992 1993. Morgan Kaufmann: San Mateo, CA.

The GA used

```
int       GA()
//        the Genetic Algorithm
//        returns the best fitness found so far
//        parameters used here (as global variables):
//        Psize     population size (max MAXpopu)
//        elite     number of best selected in each generation [0,Psize]
//        maxGens   maximum number of generations
//        parameters used indirectly via subroutines called:
```

```
//              crossRate          crossover rate [0,1]
//              mutationRate       mutation rate [0,l]
//              swapRate           swap rate [0,1]
//              newRate            rate of new individuals [0,1]
//              FOUND              solution found (set in fitness())
(
    int i,j,c,best,s,f;

    FOUND = 0;

    // create initial population
    for (j=0; j<Psize; j++) {
        for (i=0; i<Npoly; i++){
            chromosome [j] [i] = newGene (i);
        }
    }
    // generations loop
    for (int g=0; ((g<=maxGens)&&(trials<=maxTrials)&&(!FOUND));
    g++) {
        int total = elite;
        while (total<Psize) {
            for (j=0; (j<elite)&&(total<Psize);j++) {
                // take the first (consecutively)
                // and the second (randomly) parent
                if (elitism) {
                    c = mini(elite-l,maxi(0,elite*random()));
                } else {
                    c = mini(Psize-1,maxi(0,Psize*random()));
                }
                cross(j,c,total);
                mutate(j,total);
                swap(j,total);
            }
        }
        //evaluate fitness
        for (j=0; j<Psize; j++) {
            F[I[j]] = f = fitness(I[j]);
            best = maxi(best,f);
        }
        //sort population by fitness value
        Sort(F,I ,0,Psize,Psize);
        //compact pop / create random new individuals:
        if (reInitialize) {
            for (i=elite; i<Psize; i++) {
                for (j =0; j >Npoly; j ++) {
                    chromo some [I [j] ] [i] = newGene (j);
```

```
                }
            }
        }

        if (diffuse) {
            for (j=0,i=0; i<elite; i++,j++) {
                if (i%2) {
                    swap(I[i], I[j] );
                    j++;
                }
            }
        }//diffuse end
    }//generation loop end
    return best;
}
```

Chapter 14 Experimental Results on the Effects of Multi-Parent Recombination: an Overview

A.E. Eiben
Leiden University and CWI Amsterdam
gusz@wi.leidenuniv.nl, gusz@cwi.nl

Abstract

This chapter is concerned with multi-parent recombination operators that can apply more than two parents to create offspring. After a brief overview of various operators from the literature, a summary of the results of experimental research into the advantages of using more parents is given.

14.1 Introduction

In natural evolution, reproduction mechanisms are either asexual or sexual. In the case of asexual reproduction one parent creates one or more offspring, whereas sexual reproduction requires two parents. The majority of the species on earth reproduces in an asexual manner, showing the power of asexual reproduction. However, it is the higher level species that use sexual reproduction, suggesting that sexual reproduction is more advanced. In simulated evolution, that is in evolutionary algorithms, many technical features are inspired by natural mechanisms. In particular, abstract variants of sexual and asexual reproduction are implemented as search operators. Some evolutionary techniques, e.g., evolutionary programming, work exclusively with mutation (i.e., they implement a simplification of asexual reproduction), while others, e.g., genetic algorithms and evolution strategies, use recombination (i.e., they implement a simplification of sexual reproduction) and mutation. There are several papers investigating the advantages and disadvantages of mutation with respect to crossover [18,19, 20,23,35,40]. At the moment, the question of whether mutation or crossover is preferable (or rather, which one is preferable under certain circumstances) is still an open research issue. Technically, the question concerns the arity of the reproduction operators. Mutation and crossover have arity one and two, respectively, and the question is whether unary or binary operators are preferable for typical optimization problems. From a purely technical point of view, there is no need to restrict the arity of reproduction operators to one or two. In general, a reproduction operator can have an arity from one up to the population size. Hereby the analogy with natural evolution breaks down; to our knowledge there are no species on earth that would apply multi-parent reproduction mechanisms where genetic material of more than two parents is mixed in *one* reproductive action. Simulating such reproduction operators, however, is no problem. The main goal of this chapter is to present experimental results on the performance of multi-parent reproduction mechanisms. To this end, let us start with setting some conventions on terminology. The term *population* is used for a multiset of individuals that undergo selection and reproduction. This terminology is maintained in genetic algorithms, evolutionary programming and genetic programming, but in evolution strategies, all m individuals in a (m,l) or (m+l) strategy are called parents. Here, however, the term *parents* is used only for those individuals that are selected to undergo recombination. That is, parents are those individuals that are actually used as inputs for a recombination operator; the *arity of a recombination operator* is the number of parents it uses. An individual is called a *donor* if it is a parent that actually contributes to (at least one of) the alleles of the child(ren) created by the recombination operator. This contribution can be, for instance, the delivery of an allele, as in uniform crossover in bitstring GAs, of participating in an averaging operation, as in intermediate recombination in ES. As an illustration, consider a steady-state GA where 100 individuals form the population and two of them are chosen as parents to undergo uniform crossover to create one single child. If, by pure chance, the child only inherits alleles from parent 1, then parent 1 is a donor, and parent 2 is not.

14.2 Multi-Parent Reproduction Operators

An early paper mentioning multi-parent recombination is [8] on solving linear equations. It presents the definition of three different multi-parent recombination mechanisms, called *majority mating, mating by crossing over*, and *mating by averaging*. All of these oparators are defined in a general way, that is, they can be applied to any number of *m* parents. Unfortunately, only very little is reported on the performance of these operators. The recombination mechanism of [27] for evolving models for a given process utilizes four models to create one new model. The mechanism called stochastic iterated genetic hill-climbing (SIGH) applies a sophisticated probabilistic voting mechanism, where *m* "voters" (*m* being the size of the population) determine the values of a new bit-string, [1]. SIGH is shown to be better than a GA with 1-point and uniform crossover on four out of six test functions and the overall conclusion is that it is "competitive in speed with a variety of existing algorithms." *Global recombination* in evolution strategies is known from the seventies. It allows the use of more than two recombinants, [2,37] because the two donors are drawn randomly for each position (gene) of the new offspring. These drawings take the whole population of m individuals into consideration. The multi-parent character of global recombination is thus the consequence of redrawing the donors, therefore, probably more than two individuals contribute to the offspring, but their number is not defined. It is clear that investigations on the effects of different numbers of parents on algorithm performance could not be performed in the traditional ES framework. The option of using multiple parents can be turned on or off; that is, global recombination can be used or not, but the arity of the recombination operator is not tunable. Experimental studies on global versus two-parent recombination are possible, but so far there are almost no experimental results available on this subject. In [37] it is noted that "appreciable acceleration" is obtained by changing from the asexual to bisexual scheme (i.e., adding recombination using two parents to the mutation-only algorithm), but only "slight further increase" is obtained when changing from bisexual to multisexual recombination (i.e., using global recombination instead of the two-parent variant). Let us note that the terms bisexual and multisexual are not appropriate: individuals have no gender or sex, and recombination can be applied to any combination of individuals. The *p-sexual voting recombination* from [30] is applied for the quadratic assignment problem. The operator produces one child of p parents. Let us remark again that the name p-sexual is somewhat misleading, as there are no different genders and no restriction on having one representative of each gender for recombination. In the experiments it "worked surprisingly well," but comparison between this scheme and usual two-parent recombination is not performed. An interesting attempt to combine genetic algorithms with the Simplex Method in [5] resulted in the ternary *simplex crossover*. The simplex GA performed better than the standard GA on the DeJong functions. This idea has also been extended to an $n+1$ parents version, where n is the dimensionality of the space, [6, 33]. Uniform crossover with two as well as with three parents in a GA using integer representation is compared on the problem of placing actuators on space structures in [22]. Based on the experimental results, the authors conclude that the use of three parents did not improve the performance. *Scanning crossover* has been introduced in [13] as a generalization and extension of uniform crossover in GAs creating one child from r parents. The name is based on the following general procedure for scanning parents and thus building the child from left to right. Let $x^1, \ldots, x^r$ be the selected parents of length L and let x denote the child.

Procedure scanning
    begin
        INITIALIZE (put markers at the 1st position in each parent)
        for i = 1 to i = L
            CHOOSE j from 1, ...,r
            let i-th allele of x be the i-th allele of parent j
            UPDATE position markers
    end

The above procedure provides a general framework for a certain style of multi-parent recombination, where the precise execution, hence the exact definition of the operator, depends upon the mechanisms CHOOSE and UPDATE. In the simplest case, UPDATE can shift the markers one position to the right. This is appropriate for bit-strings, integer and floating-point representation. Scanning can also be easily adapted to order-based representation, where each individual is a permutation, if UPDATE shifts to the first allele which is not yet in the child. This guarantees that each offspring will be a permutation, if the parents are permutations themselves.

Depending upon the mechanism used to CHOOSE a parent (and thereby an allele), there are three different versions of scanning. The choice can be deterministic, choosing the allele with the highest number of occurrences and randomly breaking ties (*occurrence based scanning*). Alternatively, it can be random, either unbiased, following a uniform distribution thus giving each parent an equal chance to deliver its allele (*uniform scanning*), or be biased by the fitness of the parents, where the chance of being chosen is fitness proportional (*fitness based scanning*). Uniform scanning for $r = 2$ is the same as uniform crossover, although creating only one child, and the occurrence-based version is very much like the voting or majority mating mechanism discussed before. *Diagonal crossover* has been introduced in [13] as a generalization of 1-point crossover in GAs. In its original form, diagonal crossover creates r children from r parents by selecting $(r - 1)$ crossover points in the parents and composing the children by taking the resulting in r chromosome segments from the parents 'along the diagonals.' Later on, a one-child version was introduced and used, [42]. Figure 1 illustrates both variants. It is easy to see that for $r = 2$ diagonal crossover coincides with 1-point crossover and, in some sense, it also generalizes traditional 2-parent n-point crossover.

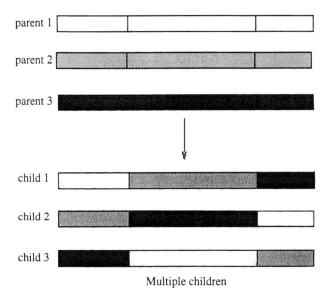

Multiple children

Figure 14.1 Diagonal crossover (top) and its one-child version (bottom) for 3 parents

A so-called *triadic crossover* is introduced and tested in [32] for a multimodal spin-lattice problem. The triadic crossover is defined in terms of two parents and one extra individual for creating a child, but technically the result is identical to the outcome of a voting crossover on these three individuals as parents. A comparison between triadic, 1-point and uniform crossover is done, where triadic crossover turned out to deliver the best results. *Gene-pool recombination* (GPR) and its variants were introduced in [31,43] as a multi-parent recombination mechanism for discrete domains. It is defined as a generalization of two-parent recombination (TPR). Applying GPR is preceded by selecting a gene-pool consisting of would-be parents. Applying GPR, two parent alleles are recombined to form the allele of an offspring, and the two parents are drawn from the gene-pool. Similar to global recombination in evolution strategies, the arity of the operator is not defined in advance. GPR is shown to converge about 25% faster than TPR for ONEMAX, and its extension to continuous domains outperforms the corresponding two-parent operator on the spherical function. A recombination mechanism with tunable arity in ES is proposed in [38]. The (m,k,l,r) - ES provides the possibility of freely adjusting the number of parents (called ancestors). The parameter r stands for the number of parents and global recombination is redefined for any given set of r parents. The discrete version randomly chooses one of the parent's alleles, while the intermediate version takes the average of all parent alleles as the allele of the child. Observe that r-ary discrete recombination coincides with uniform scanning crossover, while r-ary intermediate recombination is a special case of mating by averaging. At this time there are no experimental results available on the effect of r within this framework. Related work in Evolution Strategies also uses r as the number of parents as an independent parameter for recombination [7]. For purposes of a theoretical analysis, it is assumed that all parents are different, uniform, randomly chosen from the population of m individuals, and r-ary intermediate and discrete recombinations are defined similar to [38]. Investigations are limited to the special case of $r = m$ on the spherical function. By this assumption it is not possible to draw conclusions on the effect of r, but the analysis shows that the optimal progress rate is a factor m higher than that of the traditional (m, l) evolution strategy, for both recombination mechanisms. A very particular mechanism is the *Linkage Evolving Genetic Operator* (LEGO) as defined in [39]. The mechanism is designed to detect and propagate blocks of corresponding genes of potentially varying length during the evolution. Although the multi-parent feature is only a side effect, LEGO is a mechanism where more than two parents can contribute to an offspring. A recent generalization of global intermediate recombination in evolution strategies can be found in [11]. The new operator is applied after selecting r parent individuals from the population of m, and the resampling of the two donors for each i takes only these r individuals into consideration. In this way, the original mechanism is kept as intact as possible, while a gradual variation between the two extremes $r = 2$ and $r = m$ is facilitated. Note that this operator differs from the ones defined in [7,38].

14.3 The Effects of Higher Operator Arities

In the last few years a number of papers have studied the effect of operator arity on EA performance. Most papers consider numerical optimization problems as test functions. In the EC literature, there are a number of such functions that are repeatedly used for performance assesment of various agorithm variants. In the this survey we refer to these functions by their common names and omit the exact formulas that can be retreived from the referred articles. Extensive treatment of such test functions can be found in [3,37].

The performance of scanning crossover for different number of parents is studied in [13] in a generational GA with proportional selection. A canonical GA with bit representation is applied for function optimization (DeJong functions F1-F4 and a function from Michalewicz) and an order-based GA for graph coloring and the TSP. Different mechanisms to CHOOSE in the general scanning procedure are tested and compared. For the function optimization problems, the number of generations needed to reach a solution is used as a performance measure. For the graph coloring problem, the percentage of runs that found a solution forms the basis of comparison, while, for the TSP, the length of the best tour found is used. On the numerical optimization test suite, more parents perform better than two; for the TSP and graph coloring, 2 parents turn out to be advisable. Comparing different biases in choosing the child allele, on four out of the five numerical problems, fitness-based scanning outperforms the other two and occurrence-based scanning is the worst operator. In this chapter only the definition of diagonal crossover is given; there are no experiments with this operator.

Diagonal crossover is investigated in [12]. It is compared to the classical 2-parent n-point crossover and uniform scanning in a steady-state GA with linear ranked biased selection ($b = 1.2$) and worst-fitness deletion. The test suite consists of two 2-dimensional problems (DeJong's F2 and a function from Michalewicz) and four scalable functions (after Ackley, Griewangk, Rastrigin and Schwefel).

When monitoring the performance, two different measures were used; namely, efficiency (speed) and success rate (percentage of cases when an optimum was found). Speed was measured by the total number of function evaluations (averaged over all runs). The performance of diagonal crossover using r parents and n-point crossover (for two parents) showed a significant correspondence with r, respectively, n. The best performance was always obtained with high values, between 10 - 15, where 15 was the maximum tested. Besides, diagonal crossover is always better than n-point crossover using the same number of crossover points ($r = n - 1$), thus representing the same level of disruptiveness.

For illustration we present the optimal number of parents and the corresponding success rates in Table 14.1. An interesting observation in [12] is that for scanning the relation between r and performance is less clear than for diagonal crossover, although the best performance is achieved for more than two parents on five out of the six test functions. A concise overview of all experiments in this study can be found in [16].

The interaction between selection pressure and the parameters r for diagonal crossover, respectively, n for n-point crossover is investigated [42]. A steady-state GA with tournament selection (tournament size between 1 - 6) combined with random deletion and worst-fitness deletion is applied to the Griewangk and the Schwefel functions. The disruptiveness of both operators increases in parallel as the values for r and n are raised, but the experiments show that diagonal crossover consistently outperforms n-point crossover. The best option proves to be low selection pressure and high r in diagonal crossover combined with worst-fitness deletion.

Test	Scanning Xover		Diagonal Xover		N-point Xover	
function	#par	succ.	#par	succ.	#Xover pnts	succ.
F2	7	.91 (.73)	11	.88 (.38)	11	.84
Ackl	8	.90 (.84)	15	.89 (.00)	10	.24
Grie	10	.48 (.22)	14	.32 (.04)	10	.15
Mic	10	.72 (.57)	15	.76 (.34)	15	.60
Ras	5	.10 (.00)	13	.28 (.00)	15	.06
Schw	2	.02 (.02)	15	.24 (.00)	10	.10

Table 14.1 Optimal number of parents and corresponding success rates. Within brackets the results for 2 parents.

The aforementioned studies have given sufficient indication that using multi-parent operators with higher arities can increase GA performance. Such an increase, however, does not always occur, and the most recent studies shifted the focus of attention from showing that it occurs to investigating when it occurs. Such an analysis implicitly assumes a relationship between the characteristics of the test functions and the observed GA behavior. Unfortunately, it is very difficult to characterize the commonly used test functions.

Motivated by the difficulties to characterize the shapes of numerical objective functions, the effects of operator arity are studied on fitness landscapes with controllable ruggedness [14]. The NK-landscapes of Kauffman [26], where the level of epistasis, hence the ruggedness of the landscape, can be tuned by the parameter K, are used for this purpose. The multiple-children and the one-child version of diagonal crossover and uniform scanning are tested within a steady-state GA with linear ranked biased selection ($b = 1.2$) and worst-fitness deletion for $N = 100$ and different values of K. Two kinds of epistatic interactions, nearest neighbor interaction (NNI) and random neighbor interactions (RNI) are considered. As the NK-landscapes are randomly generated, the exact optimum is not known for any combination of N and K. This study uses the "practical global optimum" (being the highest value ever found during the tests) as the basis of performance comparison. The quality of a particular algorithm variant is evaluated by the distance to this practical global maximum which is computed by $D = (f_{maximal} - f_{obtained})/f_{maximal}$ where $f_{obtained}$ is the best value found by the given variant. Similar to earlier findings [12], the tests show that the performance of uniform scanning cannot be related to the number of parents. The two versions of diagonal crossover behave identically, and for both operators there is a consequent improvement when increasing r. However, as the epistasis (ruggedness of the landscape) grows from $K = 1$ to $K = 5$, the advantage of more parents becomes smaller. On landscapes with significantly high epistasis ($K = 25$), the relationship between operator arity and algorithm performance seems to diminish; see Figure 14.2 for illustration, where the error at termination D is shown for

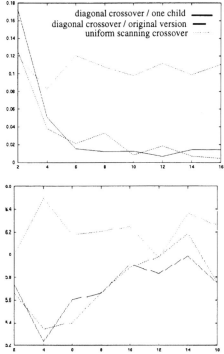

Figure 14.2 Effect of the number of parents (horizontal axis) on D (vertical axis) on NK-landscapes with nearest-neighbor interaction, $N = 100$, $K = 1$ (left), $K = 25$ (right).

nearest-neighbor interaction. The final conclusions of this investigation can be very well related to works of [18,35] and [23] on the usefulness of (2-parent) recombination. It seems that if and when crossover is useful, i.e., on mildly epistatic problems, then multi-parent crossover is more useful than the 2-parent variants.

A recent investigation analyses diagonal crossover in detail on numerical optimization problems pursuing two research objectives [17]. First, trying to find connections between the structure of the fitness landscape and the performance of diagonal crossover, i.e., establishing on what kind of landscapes it is advantageous to increase the number of parents. Second, trying to disclose the source of increased performance of the diagonal crossover with more parents if and when it is superior to 2-parent recombination. As for the first goal, the functions in the applied test suite are characterized by their modality, separability and the arrangement of local otpima. Regarding the second goal, a number of working hypotheses are formed based on two observations. It is observed that the increase in the number of parents in diagonal crossover automatically leads to an increased number of crossover points. It can be the case that higher performance is not the result of using more parents, but simply comes from being more disruptive by using more crossover points. Furthermore, it is noticed that when applying diagonal crossover, r parents create r children in one go. In a steady-state GA, the population is updated, i.e., offspring are inserted, after each application of crossover (followed by mutation), which means that a GA using 10 parents diagonal crossover has more information before performing the selection step than a GA using the 2 parent version. In other words, GAs with higher operator arity have a bigger generational gap which might cause a bias in their favor. Based on these observations the following working hypotheses are made.

Hypothesis 1: using more crossover points leads to better performance.
Hypothesis 2: bigger generational gap leads to better performance.
Hypothesis 3: using more parents leads to better performance.

These hypotheses are tested by using a steady-state GA with uniform random parent selection and worst fitness deletion on eight different numerical optimization test functions. Performance is measured by accuracy (error at termination), speed (median number of fitness evaluations before termination), and success rate, if the first two measures are inconclusive. Unfortunately, the outcomes do not provide sufficient basis for well-grounded conclusions on the relationship between the structure of the fitness landscape and the performance of diagonal crossover. A surprising result is that on Rosenbrock's saddle (DeJong's F2) increasing r decreases the performance. This function is low-dimensional ($n = 2$), unimodal, and nonseparable, but none of the other functions with these features led to such behavior. As for the working hypotheses, Hypothesis 2 is clearly rejected by the similar behavior of the one-child, respectively, original variant of diagonal crossover. Hypothesis 1 is supported by the increasing performance of n-point crossover if n is increased. This, however, does not imply rejection of Hypothesis 3, i.e., that better performance for higher ns would come *only* from having more crossover points. In fact, diagonal crossover was better than n-point crossover on all but two functions: on Rosenbrock's saddle (F2) and on the Fletcher-Powell function (multi-modal, nonseparable, with a random arrangement of local optima). On the Fletcher-Powell function, diagonal crossover was very similar to n-point crossover, indicating that increased performance for higher rs and ns seems to be the result of crossover's effect as macro mutation, which effect is intensified by more crossover points. An illustration is given in Figure 14.3.

The working of multi-parent recombination operators in continuous search spaces, in particular within evolution strategies, is investigated in [11]. This study compares *r*-ary global intermediate recombination as defined at the end of Section 14.2, *r*-ary discrete recombination, which is identical to uniform scanning crossover, and diagonal crossover with one child. The main working hypothesis is that increasing the number of parents leads to increased evolutionary algorithm performance in terms of achieved accuracy, i.e., distance from the global optimum at termination. This is divided into two sub-hypotheses

A)   Increasing the number of parents from one to two leads to increased evolutionary algorithm performance.

B)   Increasing the number of parents from two to larger numbers leads to increased evolutionary algorithm performance.

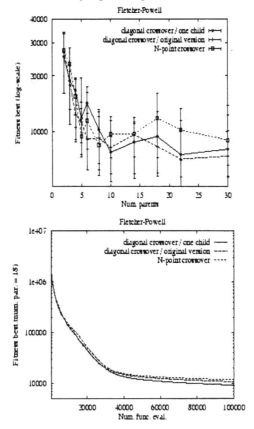

Figure 14.3 Fletcher-Powell function. Top: effect of the number of parents, respectively crossover points on accuracy. Bottom: populations best fitness during a run as a function of time (number of fitness evaluations), for r = 18, N = 17.

Note, that sub-hypothesis A amounts to hypothesising that recombination and mutation works better than mutation alone. Experiments are performed on unimodal landscapes (sphere model and Schwefel's double sum), multimodal functions with regularly arranged optima and a superimposed unimodal topology (Ackley, Griewangk and Rastrigin functions). Furthermore, two functions with an irregular, random arrangement of local optima are studied, the Fletcher-Powell function and the Langerman function. A classification of the investigated fitness landscapes is presented in order to find relationships between fitness landscape characteristics and performance of operators; see Table 14.2.

The results indicate that a diversity of possible outcomes can occur and whether the working hypotheses hold or not depends upon the particular combination of objective function and recombination operator. Sub-hypothesis A holds in more than 80% of the cases studied, but a further increase of the number of parents beyond two (sub-hypothesis B) does not necessarily have an advantageous affect on the accuracy achieved. In fact, there might be no significant impact, or even a negative impact; it can happen that A holds, but increasing the number of parents above two leads to worse results. All in all, out of the 21 cases (3 operators, 7 test functions), multi-parent recombination leads to a deterioration of solution quality in only 2 cases, it has no significant effect in 7 cases, and in the majority of the cases (12 out of 21 in total) it has a postive effect. These outcomes imply that, athough there is no guarantee for success, it is reasonable to try multi-parent recombination in an ES.

SEPARABILITY	MODALITY	
	UNIMODAL	MULTIMODAL
YES	Sphere	Rastrigin (regular)
NO	Schwefel	Ackley (regular) Griewangk (regular) Fletcher-Powell (irregular) Langerman (irregular)

Table 14.2 Characterization of the test functions.

It is very interesting to consider the results on the randomly arranged fitness landscapes. On the Fletcher-Powell function, there was no correlation between the number of parents and the ES performance [11], while in a genetic algorithm higher arities of diagonal crossover do lead to better performance [17]. This observation discloses that the same operator can behave differently on the same function, depending upon the evolutionary algorithm in which it is used. Furthermore, the Fletcher-Powell function and the Langerman function are of the same type, nonseparable and multimodal with an irregular, random arrangement of local optima. Still, the behavior of the evoution strategy is different: there is no clear effect of using more parents and the Fletcher-Powell function, while an advantage of higher arities, can be observed on the Langerman function. This prevents general conclusions on ES behavior on quasi-random landscapes.

Let us note that introducing operator arity as a new parameter implies an obligation of setting its value. Since, so far, there are no reliable heuristics on setting this parameter, finding good values may require numerous tests. A way to circumvent this problem can be based on previous work on adapting or self-adapting the frequency of applying different operators [10, 41], or using a number of competing subpopulations [36], each applying an operator with a different arity. A first assessment of this technique can be found in [15], where subpopulations with greater progress, i.e., with more powerful operators, become larger. As a "side-effect," this study also compares six-parent diagonal crossover and two parent one-point crossover within the traditional one population scheme on seven different fitness landscapes that have been specifically designed for studying the effect of crossover. On all of these landscapes (Onemax, Plateau, Plateau-d, Trap, Trap-d from [35], Twin Peaks from [18], and Royal Road from [28]), the multi-parent operator is superior. As an illustration we present the mean best fitness values found and the number of fitness evaluations needed to find an optimum for these two operators.

Problem	6-parent diagonal crossover		1-point crossover	
	mean best	nr. of steps	mean best	nr. of steps
Onemax	100.0	3095	100.0	5691
Twin Peaks	100.0	3839	100.0	6021
Plateau	100.0	8060	97.6	25021
Plateau-d	99.2	23479	93.9	35063
Trap	189.8	3701	168.3	-
Trap -d	138.8	-	136.7	-
Royal Road	595.8	10228	480.2	30050

Table 14.3 Diagonal crossover with 6 parents vs. 1-point crossover. Note: all problems are to be maximized; the number of steps is undefined (-) if no runs found the optimum.

14.4 Conclusions

Although the idea of using more than two parents for recombination in an evolutionary problem solver has occurred repeatedly in the history of the field, investigations explicitly devoted to the effect of operator arity on EA performance are still scarce. It is, of course, questionable whether multi-parent recombination can be considered as one single phenomenon showing one behavioral pattern. The present overview shows that there are (at least) three different types of multi-parent mechanisms with tunable arity.

1. Operators based upon allele frequencies among the parents.

2. Operators based upon segmenting and recombining the parents.

3. Operators based upon numerical operations of (real valued) alleles.

In general, it cannot be expected that these different schemes show the same response to raising operator arities. There are also experimental results supporting differentiation among various multi-parent mechanisms. For instance, there seems to be no clear relationship between the number of parents and the performance of uniform scanning crossover, while the opposite is true for diagonal crossover [14].

Studies on multi-parent operators also have to consider possibly different behavior on different types of fitness landscapes. So far, there is not enough experimental data to support general conclusions. Some studies indicate that on irregular landscapes, such as NK-landscapes with relatively high K values [14] or the Fletcher-Powell function [11], they do not work. Meanwhile there are also results indicating the opposite on the Fletcher-Powell function [17] and on the Langerman function [11]. This stresses the importance of more experimental and theoretical research, following the tradition of studying the (dis)advantages of two-parent crossovers under different circumstances, [18, 23, 35, 40].

Finally, let us conclude this chapter with the following note. Even though there are no biological analogies of recombination mechanisms where more than two parent genotypes are mixed in one single recombination act, computer simulations allow for studying such mechanisms. Although there is still much to be investigated, there is already substantial evidence that applying more than two parents can increase the performance of EAs. Considering multi-parent recombination mechanisms is thus a promising design heuristic for practitioners and a challenge for theoretical analysis.

Acknowledgments

I am grateful for the help of all those colleagues who contributed to this chapter, as co-authors of experimental research papers, or by exchanging ideas. Special thanks go to Thomas Bäck and Cees van Kemenade.

References

[1] D. H. Ackley. An empirical study of bit vector function optimization. In L. Davis, editor, *Genetic Algorithms and Simulated Annealing*, pages 170-215. Morgan Kaufmann, 1987.

[2] T. Bäck. *Evolutionary Algorithms in Theory and Practice*. Oxford University Press, New York, 1996.

[3] T. Bäck and Z. Michalewicz. Test landscapes. In T. Bäck, D. Fogel, and Z. Michalewicz, editors, *Handbook of Evolutionary Computation*, pages B2.7:14-B2.7:20. Institute of Physics Publishing Ltd; Bristol and Oxford University Press, New York, 1997.

[4] R. K. Belew and L. B. Booker, editors. *Proceedings of the 4th International Conference on Genetic Algorithms*. Morgan Kaufmann, 1991.

[5] H. Bersini and G. Seront. In search of a good evolution-optimization crossover. In R. Männer and B. Manderick, editors, *Proceedings of the 2nd Conference on Parallel Problem Solving from Nature*, pages 479-488. North-Holland, 1992.

[6] H. Bersini and F.J. Varela. The immune recruitment mechanism: A selective evolutionary strategy. In Belew and Booker [4], pages 520-526.

[7] H.-G. Beyer. Toward a theory of evolution strategies: On the benefits of sex- the $(m/m,l)$ theory. *Evolutionary Computation*, 3(1):81-111, 1995.

[8] H. J. Bremermann, M. Rogson, and S. Salaff. Global properties of evolution processes. In H. H. Pattee, E. A. Edlsack, L. Fein, and A. B. Callahan, editors, *Natural Automata and Useful Simulations*, pages 3-41. Spartan Books, Washington D.C., 1966.

[9] Y. Davidor, H.-P. Schwefel, and R. Männer, editors. *Proceedings of the 3rd Conference on Parallel Problem Solving from Nature*, number 866 in Lecture Notes in Computer Science. Springer-Verlag, Berlin, 1994.

[10] L. Davis. Adapting operator probabilities in genetic algorithms. In Schaffer [34], pages 61-69.

[11] A. E. Eiben and Th. Bäck. An empirical investigation of multi-parent recombination operators in evolution strategies. *Evolutionary Computation*, 5(3):347-365, 1997.

[12] A. E. Eiben, C. H. M. van Kemenade, and J. N. Kok. Orgy in the computer: Multi-parent reproduction in genetic algorithms. In Morán et al. [29], pages 934-945.

[13] A. E. Eiben, P-E. Raué, and Zs. Ruttkay. Genetic algorithms with multi-parent recombination. In Davidor et al. [9], pages 78-87.

[14] A. E. Eiben and C. A. Schippers. Multi-parent's niche: n-ary crossovers on NK-landscapes. In H.-M. Voigt, W. Ebeling, I. Rechenberg, and H.-P. Schwefel, editors, *Proceedings of the 4th Conference on Parallel Problem Solving from Nature*, number 1141 in Lecture Notes in Computer Science, pages 319-328. Springer, Berlin, 1996.

[15] A. E. Eiben, I. G. Sprinkhuizen-Kuyper, and B. A. Thijssen. Competing crossovers in an adaptive GA framework. In *Proceedings of the 5th IEEE Conference on Evolutionary Computation*. IEEE Press, 1998.

[16] A. E. Eiben and C. H. M. van Kemenade. Performance of multi-parent crossover operators on numerical function optimization problems. Technical Report TR-95-33, Leiden University, 1995. Also as ftp://ftp.wi.leidenuniv.nl/pub/CS/TechnicalReports/ 1995/tr95-33.ps.gz.

[17] A. E. Eiben and C. H. M. van Kemenade. Diagonal crossover in genetic algorithms for numerical optimization. *Journal of Control and Cybernetics*, 26(3), 1997. In press.

[18] L. J. Eshelman and J. D. Schaffer. Crossover's niche. In Forrest [21], pages 9-14.

[19] D. B. Fogel and J. W. Atmar. Comparing genetic operators with Gaussian mutations in simulated evolutionary processes using linear systems. *Biological Cybernetics*, 63:111-114, 1990.

[20] D. B. Fogel and L. C. Stayton. On the effectiveness of crossover in simulated evolutionary optimization. *Biosystems*, 32:171-182, 1994.

[21] S. Forrest, editor. *Proceedings of the 5th International Conference on Genetic Algorithms*. Morgan Kaufmann, 1993.

[22] H. Furuya and R. T. Haftka. Genetic algorithms for placing actuators on space structures. In Forrest [21], pages 536-542.

[23] W. Hordijk and B. Manderick. The usefulness of recombination. In Morán et al. [29], pages 908-919.

[24] *Proceedings of the 2nd IEEE Conference on Evolutionary Computation*. IEEE Press, 1995.

[25] *Proceedings of the 3rd IEEE Conference on Evolutionary Computation*. IEEE Press, 1996.

[26] S. A. Kauffman. *Origins of Order: Self-Organization and Selection in Evolution*. Oxford University Press, New York, NY, 1993.

[27] H. Kaufman. An experimental investigation of process identification by competitive evolution. *IEEE Transactions on Systems Science and Cybernetics*, SSC-3(1):11-16, 1967.

[28] M. Mitchell, S. Forrest, and J. H. Holland. The royal road for genetic algorithms: Fitness landscapes and GA performance. In F. J. Varela and P. Bourgine, editors, *Toward a Practice of Autonomous Systems: Proceedings of the 1st European Conference on Artificial Life*, pages 245-254. The MIT Press, 1994.

[29] F. Morán, A. Moreno, J. J. Merelo, and P. Chacón, editors. *Advances in Artificial Life. Third International Conference on Artificial Life*,

volume 929 of Lecture Notes in Artificial Intelligence. Springer, Berlin, 1995.

[30] H. Mühlenbein. Parallel genetic algorithms, population genetics and combinatorial optimization. In Schaffer [34], pages 416-421.

[31] H. Mühlenbein and H.-M. Voigt. Gene pool recombination in genetic algorithms. In I. H. Osman and J. P. Kelly, editors, *Meta-Heuristics: Theory and Applications*, pages 53-62, Boston, London, Dordrecht, 1996. Kluwer Academic Publishers.

[32] K. F. Pál. Selection schemes with spatial isolation for genetic optimization. In Davidor et al. [9], pages 170-179.

[33] J.-M. Renders and H. Bersini. Hybridizing genetic algorithms with hill-climbing methods for global optimization: two possible ways. In *Proceedings of the 1st IEEE Conference on Evolutionary Computation*, pages 312-317. IEEE Press, 1994.

[34] J. D. Schaffer, editor. *Proceedings of the 3rd International Conference on Genetic Algorithms*. Morgan Kaufmann, 1989.

[35] J. D. Schaffer and L. J. Eshelman. On crossover as an evolutionary viable strategy. In Belew and Booker [4], pages 61-68.

[36] D. Schlierkamp-Voosen and H. Mühlenbein. Adaptation of population sizes by competing subpopulations. In IEEE [25], pages 330-335.

[37] H.-P. Schwefel. *Evolution and Optimum Seeking*. Wiley & Sons, New York, 1995.

[38] H.-P. Schwefel and G. Rudolph. Contemporary evolution strategies. In Morán et al. [29], pages 893-907.

[39] J. Smith and T. C. Fogarty. Recombination strategy adaptation via evolution of gene linkage. In IEEE [25], pages 826-831.

[40] W. M. Spears. Crossover or mutation? In L. D. Whitley, editor, *Foundations of Genetic Algorithms - 2*, pages 221-238. Morgan Kaufmann, 1993.

[41] W. M. Spears. Adapting crossover in evolutionary algorithms. In J. R. McDonnell, R. G. Reynolds, and D. B. Fogel, editors, *Proceedings of the 4th Annual Conference on Evolutionary Programming*, pages 367-384. MIT Press, 1995.

[42] C. H. M. van Kemenade, J. N. Kok, and A. E. Eiben. Raising GA performance by simultaneous tuning of selective pressure and recombination disruptiveness. In IEEE [24], pages 346-351.

[43] H.-M. Voigt and H. Mühlenbein. Gene pool recombination and utilization of covariances for the Breeder Genetic Algorithm. In IEEE [24], pages 172-177.

Appendix 1 An Indexed Bibliography of Genetic Algorithms

Jarmo T. Alander
Department of Information Technology and Production Economics
University of Vaasa
P.O. Box 700, FIN-65101
Vaasa, Finland

e-mail: Jarmo.Alander@uwasa.fi
www: http://www.uwasa.fi/~jal
phone: +358-6-324 8444
fax: +358-6-324 8467

Trademarks

Product and company names listed are trademarks or trade names of their respective companies.

Warning

While this bibliography has been compiled with the utmost care, the editor takes no responsibility for any errors, missing information, the contents or quality of the references, nor for the usefulness and/or the consequences of their application. The fact that a reference is included in this publication does not imply a recommendation. The use of any of the methods in the references is entirely the user's own responsibility. Especially, the above warning applies to those references that are marked by trailing '†' (or '*'), which are the ones that the editor has unfortunately not had the opportunity to read. Only an abstract was available for references marked with '*'.

A1.1 Preface

The material in this bibliography has been extracted from the genetic algorithm bibliography [14], which, when this report was compiled contained 9832 items collected from several sources of genetic algorithm literature including Usenet newsgroup comp.ai.genetic and the bibliographies [46,50,200,503]. The following index periodicals have been systematically used:

- ACM: *ACM Guide to Computing Literature:* 1979 - 1993/4

- BA: *Biological Abstracts:* July 1996 - Aug. 1997

- CA: *Computer Abstracts:* Jan. 1993 - Feb. 1995

- CCA: *Computer & Control Abstracts:* Jan. 1992 - Nov. 1997 (except May   95)

- CTI: *Current Technology Index* Jan./Feb. 1993 - Jan./Fcb. 1994

- DAI: *Dissertation Abstracts International:* Vol. 53 No. 1 - Vol. 56 No. 10 (Apr. 1996)

- EEA: *Electrical & Electronics Abstracts:* Jan. 1991 - Nov. 1997

- P: *Index to Scientific & Technical Proceedings:* Jan. 1986 - Dec. 1997 (except Nov. 1994)

- A: *International Aerospace Abstracts:* Jan. 1995 - Nov. 1997

- N: *Scientific and Technical Aerospace Reports:* Jan. 1993 - Dec. 1995 (except Oct. 1995)

- El A: *The Engineering Index Annual:* 1987 - 1992

- EI M: *The Engineering Index Monthly:* Jan. 1993 - Nov. 1997 (except May 1997)

A1.1.1 Your Contributions Erroneous or Missing?

The bibliography database is updated on a regular basis and certainly contains many errors and inconsistences. The editor would be glad to hear from any reader who notices any errors, missing information, articles etc. In the future a more complete version of this bibliography will be prepared for the genetic algorithm implementations research community and others who are interested in this rapidly growing area of computer science.

When submitting updates to the database, paper copies of already published contributions are preferred. Paper copies (or ftp references) are needed mainly for indexing. We are also doing reviews of different aspects and applications of GAs where we need a complete as possible collection of GA papers. Please, do not forget to include complete bibliographical information: copy also proceedings volume title pages, journal table of contents pages, etc.

Complete bibliographical information is really helpful for those who want to find your contribution in their libraries. If your paper was worth writing and publishing, it is certainly worth referencing correctly in the major GA bibliographical database read daily by GA researchers, both new and established.

For further instructions and information see
ftp.uwasa.fi/cs/GAbib/README.

A1.1.2 How to Get this Report via Internet?

An updated version of this and some related bibliographies are available via anonymous ftp (and www) from the directory /cs/report94-1 at the ftp.uwasa.fi site:

ref	media	file	contents
[16]	ftp	gaIMPLEbib.ps.Z	this bibliography
[17]	ftp	gaENGbib.ps.Z	engineering applications
[19]	ftp	gaGPbib.ps.Z	genetic programming
[15]	ftp	gaPhRhbib.ps.Z	parallel and distributed GAs
[18]	ftp	gaTHEORYbib.ps.Z	GA theory, comparisons, etc.

The directory also contains a number of other indexed GA bibliographies arranged by application area, year of publishing, etc.

A1.1.3 Acknowledgment

The editor wants to acknowledge all who have kindly supplied references, papers and other information on genetic algorithm implementations literature.

The editor also wants to acknowledge Elizabeth Heap-Talvela for proofreading the manuscript of this bibliography.

A1.2 Indexes

A1.2.1 Authors

The following list contains all authors in alphabetical order.

Aarts, E.H.L.,	[603]	Anand, Vic,	[557]
Abe, Tamotsu,	[538]	Anderson, D.,	[27]
Abramson, David,	[5]	Anderson, P.G.,	[67]
Adeli, H.,	[258]	Andre, David,	[28, 29,
Adeli, Hojjat,	[6]	310]	
Aggarwal, Charu C.,	[7]	Angeline, Peter J.,	[30-32]
Aguado-Bayon, L.		Angus, J. E.,	[209]
Enrique,	[8]	Anheyer, Thomas,	[613]
Ait-Boudaoud, D.,	[86]	Ann, SouGuil,	[399]
Akamatsu, N.,	[178]	Anon.,	[33-36]
Alander, Jarmo T.,	[9, 11-13,	Ansari Nirwan,	[37]
23]		Antonisse, Hendrik	
Alfonseca, Manuel,	[25]	James,	[38]
Alippi, Cesare,	[471, 472]	Arabas, Jaroslaw,	[39, 40]
Almeida, F.,	[26]	Arslan, T.,	[596-598]

Yanagiya, Masayuki,	[647]	Yun, Wei-Min,	[656]
Yan-Da, Li,	[333]	Yurramendi, Yosu,	[324]
Yang, Hong-Tzer,	[648]	Zamparelli, M.,	[283]
Yang, Pai-Chuan,	[648]	Zanati, S.,	[328]
Yao, Xin,	[117-119,	Zeanah, Jeff,	[657]
336, 337]		Zell, A.,	[623]
Yasunaga, M.,	[650]	Zhai, W.,	[658]
Ye, Ju,	[651]	Zhang, Byoung-Tak,	[659]
Yeralan, Sencer,	[652]	Zhang, Liang-Jie,	[660]
Yoshikawa, Tomohiro,	[653]	Zhi-Hong, Mao,	[333]
Yoshizawa, Shuji,	[398]	Zhou, Yejin,	[661]
Yukiko, Y.,	[654]	Zhu, Zhaoda,	[646]
Yulu, Qi,	[655]	Zhuang, Wenjun,	[169]

Total of 598 articles by 831 different authors.

A1.2.2 Subject Index

All subject keywords given by the editor of this bibliography are in the references shown next in alphabetical order.

adaptation,	[248]	analysing GP	
adaptive coding,	[578.579]	mutation,	[346]
analysing GA,	[9, 44, 114,	population size,	[184]
143, 195, 252, 378, 392, 412,		schema theory,	[446]
489, 617, 619]		analyzing	
analysing GA		crossover,	[645]
coding,	[105, 246]	antennas	
continuous space,	[453, 454]	optimization,	[101]
convergence,	[488]	wire,	[338]
crossover,	[166, 228,	APLOGEN,	[566]
229, 237, 238, 273, 275, 332,		art,	[541]
336, 346, 362, 435, 441, 452,		artificial intelligence,	[324]
605, 609]		artificial life	
diploidy,	[431]	text book,	[323]
diversity,	[237, 238]	autonomous robot,	[611]
dominance,	[485]	Bayes networks,	[324]
fitness moments,	[563]	BEA,	[137]
infinite population size,	[612]	bibliography	
information theory,	[564]	implementation,	[16]
Markov chains,	[42, 43]	engineering	
mutation,	[246, 282,	applications,	[17]
283, 618, 660]		genetic	
mutation rate,	[531]	programming,	[19]
parameters,	[12]	theory,	[18]
population size,	[11, 13,	bin-packing,	[157, 316,
218, 224]		317]	
selection,	[332]	2D,	[524]

A1.3 Bibliography

[1] *Proceedings of the IEE Colloquium on Genetic Algorithms for Control and Systems Engineering,* vol. Digest No. 1993/130, London, 28. May 1993. IEE, London.

[2] *Proceedings of the Third Online Workshop on Soft Computing,* Nagoya (Japan), Aug. 1996 (to appear).†

[3] D. Abramson and J. Abela. A parallel genetic algorithm for solving the school timetabling problem. In *15th Australian Computer Science Conference,* Hobart (Australia), Feb. 1992. (Also as [4]; ftp:ftp.cit.gu. edu.au:pub/D.Abramson/SchoolGA.ps.Z)

[4] D. Abramson and J. Abela. A parallel genetic algorithm for solving the school timetabling problem. In *IJCAI Workshop on Parallel Processing in AI,* Sydney (Australia), Aug. 1992. (Also as [3]; ftp: ftp. cit.gu.edu.au: pub/D.Abramson/SchoolGA.ps.Z)

[5] D. Abramson, G. Mills, and S. Perkins. Parallelisation of a genetic algorithm for the computation of efficient train schedules. In D. Arnold, R. Christie, J. Day, and P. Roe, Editors, *Parallel Computing and Transputers,* Vol. 37 of *Transputer and Occam Engineering Series,* pages 139-149, Brisbane (Australia), 3-4. Nov. 1993. IOS Press. (ftp: ftp.cit.gu.edu.au: pub/D.Abramson/Trains.ps.Z)

[6] H. Adeli and S. Kumar. Concurrent structural optimization on massively parallel supercomputer. *Journal of Structural Engineering,* 121(11):1588-1597, Nov. 1995.

[7] C.C. Aggarwal, J.B. Orlin, and R.P. Tai. Optimized crossover for the independent set problem. *Operations Research,* 45(2):226-234, Mar.-Apr. 1997.

[8] L.E. Aguado-Bayon and P.G. Farrell. Reducing the complexity of trellises for block codes. In *Proceedings of the 1995 IEEE Symposium on Information Theory,* page 347, Whistler, BC (Canada), 17.-22. Sept. 1995. IEEE, Piscataway, NJ. *

[9] J.T. Alander. On finding the optimal genetic algorithms for robot control problems. In *Proceedings IROS '91 IEEE/RSJ International Workshop on Intelligent Robots and Systems '91,* vol. 3, pages 1313-1318, Osaka, 3.-5. Nov. 1991. IEEE Cat. No. 91TH0375-6.

[10] J.T. Alander, editor. *Geneettiset algoritmit, Genetic Algorithms,* number TKO-C53. Helsinki University of Technology (HUT), Department of Computer Science, 1992. (Proceedings of a GA Seminar held at HUT.)

[11] J.T. Alander. On optimal population size of genetic algorithms. In P. Dewilde and J. Vandewalle, editors, *CompEuro 1992 Proceedings, Computer Systems and Software Engineering, 6th Annual European Computer Conference,* pages 65-70, The Hague, 4.-8. May 1992. IEEE Computer Society, IEEE Computer Society Press.

[12] J.T. Alander. Optimal GA. In Alander [10], pages 164-179. Also as [9].

[13] J.T. Alander. Population size. In Alander [10], pages 180-189. Also as [11].

[14] J.T. Alander. *An indexed bibliography of genetic algorithms: Years 1957-1993.* Art of CAD Ltd., Vaasa (Finland), 1994, (over 3000 GA references).

[15] J.T. Alander. Indexed bibliography of distributed genetic algorithms. Report 94-1-PARA, University of Vaasa, Department of Information Technology and Production Economics, 1995. (ftp: ftp.uwasa.fi:cs/report94-1/gaPARAbib.ps.Z)

[16] J.T. Alander. Indexed bibliography of genetic algorithm implementations. Report 94-1-IMPLE, University of Vaasa, Department of Information Technology and Production Economics, 1995. (ftp: ftp.uwasa.fi: cs/report94-1/gaIMPLEbib.ps.Z)

[17] J.T. Alander. Indexed bibliography of genetic algorithms in engineering. Report 94-1-ENG, University of Vaasa, Department of Information Technology and Production Economics, 1995. (ftp: ftp.uwasa.fi: cs/report94-1/gaENGbib.ps.Z)

[18] J.T. Alander. Indexed bibliography of genetic algorithms theory and comparisons. Report 94-1-THEORY, University of Vaasa, Department of Information Technology and Production Economics, 1995. (ftp: ftp.uwasa.fi: cs/report94-1/gaTHEORYbib.ps.Z)

[19] J.T. Alander. Indexed bibliography of genetic programming. Report 94-1-GP, University of Vaasa, Department of Information Technology and Production Economics, 1995. (ftp: ftp.uwasa.fi: cs/report94-1/gaGPbib.ps.Z)

[20] J.T. Alander, editor. *Proceedings of the First Nordic Workshop on Genetic Algorithms and their Applications (1NWGA)* Proceedings of the University of Vaasa, Nro. 2, Vaasa (Finland), 9.-12. Jan. 1995. University of Vaasa. (ftp: ftp.uwasa.fi: cs/1NWGA/*.ps.Z)

[21] J.T. Alander, editor. *Proceedings of the Second Nordic Workshop on Genetic Algorithms and their Applications (2NWGA),* Proceedings of the University of Vaasa, Nro. 11, Vaasa (Finland), 19.-23. Aug. 1996. University of Vaasa. (ftp: ftp.uwasa.fi: cs/2NWGA/*.ps.Z)

[22] J.T. Alander, editor. *Proceedings of the Third Nordic Workshop on Genetic Algorithms and their Applications (3NWGA),* Helsinki (Finland), 18.-22. Aug. 1997. Finnish Artificial Intelligence Society (FAIS). (ftp: ftp.uwasa.fi: cs/3NWGA/*.ps.Z)

[23] J.T. Alander, M. Nordman, and H. Setälä. Register-level hardware design and simulation of a genetic algorithm using VHDL. In Osmera [428], pages 10-14.

[24] R.F. Albrecht, C.R. Reeves, and N.C. Steele, editors. *Artificial Neural Nets and Genetic Algorithms,* Innsbruck, Austria, 13.-16. Apr. 1993. Springer-Verlag, Wien.

[25] M. Alfonseca. Genetic algorithms. *APL Quote Quad,* 21(4):1-6, Aug. 1991.

[26] F. Almeida, F. Garcia, J. Roda, D. Morales, and C. Rodriguez. A comparative study of two distributed systems: PVM and transputers. In *Proceedings of the 1995 World Transputer Congress,* pages 244-258, Harrogate, U.K., 4.-6. Sept. 1995. lOS Press, Amsterdam. *

[27] D. Anderson. Systolic array IC for genetic computation. In *Proceedings of the 1991 3rd NASA Symposium on VLSI Design,* Idaho Univ., 1991. NASA. †

[28] D. Andre and J.R. Koza. A parallel implementation of genetic programming using the transputer architecture that achieves superlinear performance. In *Proceedings of the International Conference on Parallel and Distributed Processing Techniques and Applications (PDPTA '96),* Sunnyvale, CA, 9.-11. Aug. 1996. †

[29] D. Andre and A. Teller. A study in program response and the negative effects of introns in genetic programming. In Koza et al. [312]. †

[30] P.J. Angeline. An investigation into the sensitivity of genetic programming to the frequency of leaf selection during subtree crossover. In Koza et al. [312]. †

[31] P.J. Angeline. An alternative to indexed memory for evolving programs with explicit state representations. In Koza et al. [311]. †

[32] P.J. Angeline. Subtree crossover: Building block engine or macromutation In Koza et al. [311]. †

[33] Anon. *C Darwin H - Cenvägen till Kunskap,* 1991. †

[34] Anon. *MicroCA.* Palo Alto, CA, 1992. †

[35] Anon. Evolver™ 2.0 A genetic algorithm for spreadsheets. *Computers & Mathematics with Applications,* 26(12):94, 1993.

[36] Anon. *Cerius² Release 1.6, Drug Discovery Workbench QSAR+User's Reference, Chapter 16: Introduction to genetic function approximation,* 1994.

[37] N. Ansari and E.S.H. Hou. *Computational Intelligence for Optimization.* Kluwer Academic Publishers, London (U.K.), 1997. †

[38] H.J. Antonisse. A new interpretation of schema notation that overturns the binary encoding constraint. In Schaffer [506], pages 86-91.

[39] J. Arabas, Z. Michalewicz, and J.J. Mulawka. GAVaPS - a genetic algorithm with varying population size. In ICEC'94 [260], pages 73-78.

[40] J. Arabas, J.J. Mulawka, and J. Pokrasniewicz. A new class of crossover operators for numerical optimization. In Eshelman [150]. †

[41] T. Asveren and P. Molitor. New crossover methods for sequencing problems. In Voigt et al. [616], pages 290-299.

[42] H. Aytug, S. Bhattacharrya, and G.J. Koehler. A Markov chain analysis of genetic algorithms with power of 2 cardinality alphabets. *European Journal of Operational Research,* 96(1):195-201, 10. Jan. 1996.

[43] F.Q. Bac and V.L. Perov. New evolutionary genetic algorithms for NP-complete combinatorial optimization problems. *Biological Cybernetics,* 69(3):229-234, 1993.

[44] T. Bäck. The interaction of mutation-rate, selection, and self-adaptation within a genetic algorithm. In Männer and Manderick [360], pages 85-94.

[45] T. Bäck. A user's guide to GENESYS 1.0. Technical report, University of Dortmund, Department of Computer Science, 1992.

[46] T. Bäck. Genetic algorithms, evolutionary programming, and evolutionary strategies bibliographic database entries, (personal communication), 1993.

[47] T. Bäck. Optimal mutation rates in genetic search. In Forrest [170], pages 2-8.

[48] T. Bäck. *Evolutionary Algorithms in Theory and Practice.* Oxford University Press, New York, 1996. †

[49] T. Bäck and U. Hammel. Evolution strategies applied to perturbed objective functions. In ICEC'94 [260], pages 40-45.

[50] T. Bäck, F. Hoffmeister, and H.-P. Schwefel. Applications of evolutionary algorithms. Technical Report SYS-2/92, University of Dortmund, Department of Computer Science, 1992.

[51] T. Bäck and M. Schutz. Intelligent mutation rate control in canonical genetic algorithms. In *Proceedings of the 9th International Symposium, ISMIS 96,* vol., pages 158-167, Zakopane (Poland), 9.-13. June 1996. Springer-Verlag, Berlin (Germany). *

[52] T. Bäck and H.-P. Schwefel. Evolution strategies I: Variants and their computational implementation. In Winter et al. [639], pages 111-126.

[53] S. Baluja and R. Caruana. Removing the genetics from the standard genetic algorithm. Report CMU- CS-95-151, Carnegie-Mellon University, Department of Computer Science, 1995. (Also as [54], available via www URL: http://rose.mercury.acs.cmu.edu:80/)

[54] S. Baluja and R. Caruana. Removing the genetics from the standard genetic algorithm. In *Proceedings of the Twefth International Conference on Machine Learning,* vol., Lake Tahoe, CA, July 1995. (Also as [53], available via www URL: http://rose.mercury.acs.cmu.edu:80/)

[55] W. Banzhaf, F.D. Francone, and P. Nordin. The effect of extensive use of the mutation operator on generalization in genetic programming using sparse data sets. In Voigt et al. [616], pages 300-309.

[56] G. Bartlett. Practical handbook of genetic algorithms. In Chambers [88], chapter 1. Genie: A first GA, pages 31-56.

[57] S.E. Bayer and L. Wang. A genetic algorithm programming environment: SPLICER. In *Proceedings of the 1991 IEEE International Conference on Tools with Artificial Intelligence TAI'91,* pages 138-144, San Jose, CA, 10.-13. Nov. 1991. IEEE Computer Society Press, Los Alamitos, CA. *

[58] D. Beasley, D.R. Bull, and R.R. Martin. Complexity reduction using expansive coding. In T.C. Fogarty, editor, *Proceedings of the AISB Workshop on Evolutionary Computation; Selected Papers,* pages 304-319, Leeds (U.K.), 11.-13- Apr. 1994. Springer-Verlag, Berlin. *

[59] R.K. Belew and L.B. Booker, editors. *Proceedings of the Fourth International Conference on Genetic Algorithms,* San Diego, 13-16. July 1991. Morgan Kaufmann Publishers.

[60] P.J. Bentley and J.P. Wakefield. Hierarchical crossover in genetic algorithms. In *Proceedings of the First Online Workshop on Soft Computing (WSC1),* pages 37-42, WWW (World Wide Web), 19-30. Aug. 1996. Nagoya University.

[61] H. Bersini and G. Seront. In search of a good crossover between evolution and optimization. In Männer and Manderick [360], pages 479-488. †

[62] D. Bhandari and N.R. Pal. Directed mutation in genetic algorithms. *Information Sciences,* 79(3-4):251-270, July 1994.

[63] S. Bhattacharrya and G.J. Koehler. An analysis of non-binary genetic algorithms with cardinality υ. *Complex Systems,* 8(4):227-256, Aug. 1994. PUUTTUU(EEA 37285/95 CCA 36469/95)

[64] R. Bianchini and C.M. Brown. Parallel genetic algorithms on distributed-memory architectures. Technical Report 436, The University of Rochester, Computer Science Department, 1993.

[65] R. Bianchini and C.M. Brown. Parallel genetic algorithms on distributed-memory architectures. In *Transputer Research and Applications, NATUG-6, Proceedings of the Sixth Conference on the North American Transputer Users Group,* pages 67-82, Vancouver, BC (Canada), 10.-11. May 1993. IOS Press, Amsterdam.

[66] C. Bierwirth, D.C. Mattfeld, and H. Kopfer. On permutation representations for scheduling problems. In Voigt et al. [616], pages 310-318.

[67] J.A. Biles, P.G. Anderson, and L.W. Loggi. Neural network fitness functions for a musical IGA. In *Proceedings of the International ICSC Symposia on Intelligent Industrial Automation and Soft Computing,* pages B39-44, Reading, U.K., 26.-28. Mar. 1996. Int. Comput. Sci. Conventions, Millet, Alta. *

[68] T.A. Bitterman. *Genetic algorithms and the satisfiability of largescale Boolean expressions.* Ph.D. thesis, Louisiana State University of Agricultural and Mechanical College, 1993. PUUTTUU(DAI Vol. 54 No. 9).

[69] I.M. Bland and G.M. Megson. Implementing a generic systolic array for genetic algorithms. In *Proceedings of the First Online Workshop on Soft Computing (WSC1),* pages 268-267, WWW (World Wide Web), 19.-30. Aug. 1996. Nagoya University.

[70] I.M. Bland and G.M. Megson. Efficient operator pipelining in a bit serial genetic algorithm engine. *Electronics Letters,* 33(12):1026-1028, 1997. PUUTTUU(EEA72040/97).

[71] J. Born. ASTOP Ein interaktives Softwarepaket zur Optimierung mit adaptiven, stochastischen Verfahren. In *Vortragsauszüge Jahrestagung "Matematische Optimierung,"* pages 15-17, 1984. Humboldt Universität zu Berlin. †

[72] J.L. Breeden. Practical handbook of genetic algorithms. In Chambers [88], chapter 5. Optimal state space representations via evolutionary algorithms: Supporting expensive fitness functions, pages 115-141.

[73] J.L. Breeden. Uncertainty and multiple fitness function optimization. In McDonnell et al. [374]. †

[74] P. Brigger and M. Kunt. Morphological contour coding using structuring functions optimized by genetic algorithms. In *Proceedings of the 1995 IEEE International Conference on Image Processing,* vol. 1, pages 534-537, Washington, DC, 23.-26. Oct. 1995. IEEE, Los Alamitos, CA. *

[75] M. Bubak, W. Ciesla, and K. Sowa. Parallel object-oriented library of genetic algorithms. In *Proceedings of the Third International Workshop,* pages 135-146, Lyngby, Denmark, 18.-21. Aug. 1996. Springer-Verlag, Berlin (Germany). *

[76] M. Bubak, W. Ciesla, and K. Sowa. Object-oriented library of parallel genetic algorithms and its implementation on workstations and HP/Convex Exemplar. In *Proceedings of the International Conference and Exhibition,* pages 514-523, Vienna (Austria), 28.-30. Apr. 1997. Springer-Verlag, Berlin (Germany). *

[77] B.P. Buckles and F.E. Petry. Cloud identification using genetic algorithms and massively parallel computation. NASA Contract Report NASA-CR-201071, Tulane University, Department of Mechanical Engineering, 1996. †

[78] B.P. Buckles, F.E. Petry, and R.L. Kuester. Schema survival rates and heuristic search in genetic algorithms. In A. Dollas and N.G. Bourbakis, editors, *Proceedings of the 1990 IEEE International Conference on Tools for Artificial Intelligence TAI'90,* pages 322-327, Herndon, VA, 6.-9. Nov. 1990. IEEE Computer Society Press, Los Alamitos, CA.

[79] T.N. Bui and B.-R. Moon. On multi-dimensional encoding/crossover. In Eshelman [150]. †

[80] R. Calabretta, R. Galbiati, S. Nolfi, and D. Parisi. Two is better than one: a diploid genotype for neural networks. *Neural Process. Lett. (Netherlands),* 4(3):149-155, 1996. *

[81] R. Campanini, C. Di Caro, M. Villani, I.D. D'Antone, and G. Giusti. Parallel architectures and intrinsically parallel algorithms: genetic algorithms. *Int. J. Mod. Phys. C, Phys. Comput. (Singapore),* 5(1):95-112, Feb. 1994. *

[82] C.A.D. Carpio. A parallel genetic algorithm for polypeptide three dimensional structure prediction. A transputer implementation. *Journal of Chemical Information and Computer Sciences,* 36(2):258-269, Mar./Apr. 1996.

[83] R.A. Caruana and J.D. Schaffer. Representation and hidden bias: Gray vs. binary coding for genetic algorithms. In *Proceedings of the Fifth International Conference on Machine Learning,* pages 153-162, 1988. Morgan Kaufmann, Los Altos, CA. †

[84] R.A. Caruana, J.D. Schaffer, and L.J. Eshelman. Using multiple representations to improve inductive bias – Gray and binary coding for genetic algorithms. In A.M. Segre, editor, *Proceedings of the Sixth International Workshop on Machine Learning,* pages 375-378, Cornell University, Ithaca, NY, June 1989. Morgan Kauffman, San Mateo, CA. †

[85] W. Cedeno and V.R. Vemuri. On the use of niching for dynamic landscapes. In *Proceedings of 1997 IEEE International Conference on Evolutionary Computation,* pages 361-366, Indianapolis, IN, 13-16. Apr. 1997. IEEE, New York, NY. *

[86] R. Cemes and D. Ait-Boudaoud. A MATLAB based development tool for multiplier-less filter design using genetic algorithm. In *Proceedings of the 5th International Conference on Signal Processing Applications and Technology,* vol. 1, pages 408-413, Dallas, TX (U.S.A.), 18.-21. Oct. 1994. DSP Associates, Waltham, MA. PUUTTUU (EEA83243/96).

[87] L. Chambers. Practical handbook of genetic algorithms. [88], chapter 7. Strategic modelling using a genetic algorithm approach, pages 173-218.

[88] L. Chambers, editor. *Practical Handbook of Genetic Alqorithms,* vol. 1, Applications. CRC Press, Boca Raton, FL, 1995.

[89] L. Chambers, editor. *Practical Handbook of Genetic Algorithms,* vol. 2, Applications. CRC Press, Boca Raton, FL, 1995.

[90] H. Chan and P. Mazumder. A systolic architecture for high speed hypergraph partitioning using genetic algorithms. In Yao [649], pages 109-126. †

[91] K.C. Chan and H. Tansri. Study of genetic crossover operations on the facilities layout problem. *Computers & Industrial Engineering,* 26(3):537-550, July 1994. *

[92] M.-S. Chang and H.-K. Chen. A new encoding method of genetic algorithms towards parameter identification of fuzzy expert systems. In *Proceedings of the 1996 Asian Fuzzy Systems Symposium,* pages 406-411, Kenting, Taiwan, 11.-14. Dec. 1996. IEEE, New York, NY. *

[93] S. Chatterjee, C. Carrera, and L.A. Lynch. Genetic algorithms and traveling salesman problems. *European Journal of Operational Research,* 93(3):490-510, 20. Sept. 1996.

[94] K. Chellapilla. Evolutionary programming with tree mutations: Evolving computer programs without crossover. In Koza et al. [311].†

[95] R. Cheng and M. Gen. Crossover on intensive search and traveling salesman problem. *Computers & Industrial Engineering,* 27(14) :485-488, Sept. 1994. (Proceedings of the 16th Annual Conference on Computers and Industrial Engineering, Ashigaga (Japan) 7.-9. Mar.). *

[96] P.-C. Chi. Genetic search with proportion estimations. In Schaffer [506], pages 92-97.

[97] A.J. Chipperfield, P.J. Fleming, and T.P. Crummey. PARSIM: a parallel optimization tool. In *Proceedings of the IEEE/IFAC Joint*

Symposium on Computer-Aided Control System Design, pages 579-584, Tucson, AZ, 7.-9. Mar. 1994. IEEE Computer Society Press, Los Alamitos, CA. *

[98] A.J. Chipperfield, P.J. Fleming, and H.P. Pohlheim. A genetic algorithm toolbox for MATLAB. In *Proceedings of the International Conference on Systems Engineering,* pages 200-207, Coventry (U.K.) 6.-8. Sept. 1994.

[99] T. Clark and J.S. Mason. Adaptive uniform crossover in genetic algorithms. In *Proceedings of the IEE/IEEE Workshop on Natural Algorithms in Signal Processing,* Essex (U.K.), 14.-16. Nov. 1993. IEEE. †

[100] H.G. Cobb. An investigation into the use of hypermutation as an adaptive operator in genetic algorithms having continuous, time-dependent nonstationary environments. Memorandum Report 6760, Navy Research Laboratory, Washington, D.C., 1990. †

[101] N. Cohen. Fractal coding in genetic algorithm (GA) antenna optimization. In *Proceedings of the 1997 IEEE Antennas and Propagation Society International Symposium Digest,* vol. 1, pages 1692-1695, Que, Montreal. (Canada), 13.-18. July 1997. †

[102] J.P. Cohoon, S.U. Hegde, W.N. Martin, and D.S. Richards. Distributed genetic algorithms for the floorplan design problem. *IEEE Trans. on Computer-Aided Design of Integrated Circuits and Systems,* 10(4):483-492, Apr. 1991.

[103] M. Coli and P. Palazzari. Searching for the optimal coding in genetic algorithms. In ICEC'95 [261], pages 92-96. †

[104] A. Colin. Solving ratio optimization problems with a genetic algorithm. *Advanced Technology for Developers,* 2:18, May 1993.

[105] J.J. Collins and M. Eaton. Genocodes for genetic algorithms. In Osmera [430], pages 23-30.

[106] J.J. Collins and M. Eaton. A global representation scheme for genetic algorithms. In *Proceedings of the International Conference on Computational Intelligence,* Lecture Notes in Computer Science, Dordmund, 28.-30. Apr. 1997. Springer-Verlag, Berlin, (to appear).†

[107] D.G. Conway and M.A. Venkataramanan. Genetic search and the dynamic facility layout problem. *Computers & Operations Research,* 21(8):955-960, Oct. 1994.

[108] A.L. Corcoran, III and S. Sen. Using real-valued genetic algorithms to evolve rule sets for classification. In ICEC'94 [260], pages 120 124.

[109] A.L. Corcoran, III and R.L. Wainwright. LibGA: A user-friendly workbench for order-based genetic algorithm research. In K.M.G. Deaton and G.H.H. Berghel, editors, *Proceedings of the 1993 ACM/SIGAPP Symposium on Applied Computing,* pages 111-117, Indianapolis, IN, 14.-16. Feb. 1993. ACM, New York. *

[110] J. Cui and T.C. Fogarty. Optimization by using a parallel genetic algorithm on a transputer computing surface. In M. Valero, E.

Onate, M. Jane, J.L. Larriba, and B. Suarez, editors, *Transputer and Occam Engineering Series,* vol. 28, pages 246-254, Barcelona (Spain), 21-25. Sept. 1992. IOS Press, Amsterdam. *

[111] J. Cui, T.C. Fogarty, and J.G. Gammack. Searching databases using parallel genetic algorithms on a transputer computing surface. In *Proceedings of the Third Annual Conference of the Meiko User Society,* Manchester Business School, University of Manchester, 9.10. Apr. 1992. †

[112] J. Cui, T.C. Fogarty, and J.G. Gammack. Searching databases using parallel genetic algorithms on a transputer computing surface. *Future Generation Computer Systems,* 9(1):33 40, May 1993.

[113] J.C. Culberson. Crossover versus mutation: Fueling the debate: Tga versus giga. In Forrest [170], page 632.

[114] J.C. Culberson. Mutation-crossover isomorphisms and the construction of discriminating functions. *Evolutionary Computation,* 2(3):, 1995. †

[115] T. Dabs. Eine Entwicklungsumgebung zum Monitoring Genetischer Algorithmen [A development tool for monitoring genetic algorithms]. Diploma thesis, Universitat Wurzburg, Lehtstuhl fur Informatik II, 1994. †

[116] T. Dabs and J. School. A graphical user interface for genetic algorithms. Report 98, Universität Würzburg, Institut für Informatik, 1995.

[117] P.J. Darwen and X. Yao. A dilemma for fitness sharing with a scaling function. In ICEC'95 [261], pages 166-171. †

[118] P.J. Darwen and X. Yao. How good is fitness sharing with a scaling function. Technical Report CS8/95, Australian Defence Force Academy, Department of Computer Science, 1995, (presented at ICEC'95). †

[119] P.J. Darwen and X. Yao. Every niching method has its niche: Fitness sharing and implicit sharing compared. In Voigt et al. [616], pages 398-407.

[120] D. Dasgupta. Practical handbook of genetic algorithms. In Chambers [89], chapter 13. Incorporating redundancy and gene activation mechanisms in genetic search for adapting to non-stationary environments, pages 303-316.

[121] Y. Davidor. Analogous crossover. In Schaffer [506], pages 98-103.

[122] Y. Davidor, H.-P. Schwefel, and R. Manner, editors. *Parallel Problem Solving from Nature - PPSN 111,* vol. 866 of *Lecture Notes in Computer Science,* Jerusalem (Israel), 9.-14. Oct. 1994. Springer-Verlag, Berlin. †

[123] R. Davies. Parallel implementation of a genetic algorithm. *Control Engineering Practice,* 3(1):11 19, Jan. 1995. *

[124] L. Davis, editor. *Genetic Algorithms and Simulated Annealing,* London, 1987. Pitman Publishing.

[125] L. Davis. Genetic algorithm profiles: Matt Jensen and user-friendly evaluation functions. *Advanced Technology for Developers*, 1:7-10, Dec. 1992.

[126] L. Davis and J.J. Grefenstette. Concerning GENESIS and OOGA. Pages, 374-376, 1991.

[127] M. Davis. An empirical evaluation of the Gaussian mutation function in evolutionary programming. In Sebald and Fogel [528]. †

[128] K. Deb and D.E. Goldberg. mGA in C: A messy genetic algorithm in C. Report 91008, University of Illinois at Urbana-Champaign, 1991. (ftp:gal4.ge.uiuc.edu:/pub/papers/IlliGALs/91008.ps.Z).

[129] K. Deb and A. Kumar. Real-coded genetic algorithms with simulated binary crossover: studies on multimodal and multiobjective problems. *Complex Systems (U.S.A.)*, 9(6):431 454, 1995. *

[130] H. Delmaire, A. Langevin, and D. Riopel. Evolution systems and the quadratic assignment problem. Report G-95-24, Universite McGill, Ecole Polytechnique, GERAD, 1995. †

[131] P. D'haeseleer. Context preserving crossover in genetic programming. In ICEC'94 [260], pages 256-261.

[132] N. Dodd, D. Macfarlane, and C. Marland. Optimization of artificial neural network structure using genetic techniques on multiple transputers. In P. Welch, D. Stiles, T.L. Kunii, and A. Bakkers, editors, *Transputing '91. Proceedings of the World Transputer User Group (WOTUC)*, pages 687-700, Sunnyvale, CA, 22.-26. Apr. 1991. lOS Press, Amsterdam. †

[133] H. Doi, K.-N. Wada, and M. Furusawa. Asymmetric mutations due to semiconservative DNA replication: double-stranded DNA type genetic algorithms. In *Proceedings of the Fourth International Workshop on the Synthesis and Simulation of Living Systems*, pages 359-364, Cambridge, MA, U.S.A., 6.-8. July 1994. MIT Press, Cambridge, MA. *

[134] M. Dorigo. Using transputers to increase speed and flexibility of genetics-based machine learning systems. *Microprocessing and Microprogramming EURO-Micro Journal*, 34(1-5):147-152, 1992.

[135] M. Dorigo and E. Sirtori. ALECSYS: A parallel laboratory for learning classifier systems. Technical Report 91-004, Politecnico di Milano, Dipartimento di Elettronica, 1991.

[136] M. Dorigo and E. Sirtori. ALECSYS: A parallel laboratory for learning classifier systems. In Belew and Booker [59], pages 296-302.

[137] R. Drechsler, N. Göckel, E. Mackensen, and B. Becker. BEA: specialized hardware for implementation of evolutionary algorithms. In Koza et al. [311]. †

[138] V. Dvorák. Performance prediction of GA on transputer arrays. In Osmera [428], pages 35-40.

[139] V. Dvorák. Evaluating embedded parallel applications in TRANSIM. In Osmera [430], pages 44-49.

[140] A. Dymek. An examination of hypercube implementations of genetic algorithms. Master's thesis, Air Force Institute of Technology, Wright-Patterson Air Force Base, Ohio, 1992. (Report No. AFIT/ GCS/ENG/92M-02). †

[141] In *Proceedings of the Artificial Evolution 97 (EA '97) Conference,* Nimes (France), 22.-24. Oct. 1997. †

[142] I. East and D. Macfarlane. Implementation in Occam of parallel genetic algorithms on transputer networks. In Stender [568], chapter 3. Implementation, pages 43-64. †

[143] I.R. East and J. Rowe. Effects of isolation in a distributed population genetic algorithm. In Voigt et al. [616], pages 408-419.

[144] L.V. Edmondson. *Genetic algorithms with 3-parent crossover.* Ph.D. thesis, University of Missouri - Rolla, 1993. PUUTTUU (DAI Vol. 54 No. 9).

[145] Á.E. Eiben and C.A. Schippers. Multi-parent's niche: n-ary crossovers on NK-landscapes. In Voigt et al. [616], pages 319-328.

[146] Á.E. Eiben and C.H.M. van Kemenade. Performance of multi-parent crossover operators on numerical function optimization problems. Technical Report 95-33, Leiden University, Department of Computer Science, 1995. (ftp: ftp.wi.leidenuni.nl: /pub/CS/TechnicalReports/1995//tr95-33.ps.gz)

[147] S. Elo. A parallel genetic algorithm on the CM-2 for multi-modal optimization. In ICEC'94 [260], pages 818-822.

[148] T. Eloranta. Geneettisten algoritmien soveltaminen suuntaamattomien verkkojen piirtoon [Applying genetic algorithms to drawing undirected graphs]. Report Series B B-1996-1, University of Tampere, Department of Computer Science, 1996, (in Finnish, also as [149]).

[149] T. Eloranta. Geneettisten algoritmien soveltaminen suuntaamattomien verkkojen piirtoon [Applying genetic algorithms to drawing undirected graphs]. Master's thesis, University of Tampere, Department of Computer Science, 1996, (in Finnish, also as [148]).

[150] L.J. Eshelman, editor. *Proceedings of the Sixth International Conference on Genetic Algorithms,* Pittsburgh, PA, 15.-19. July 1995. †

[151] L.J. Eshelman, R.A. Caruana, and J.D. Schaffer. Biases in the crossover landscape. In Schaffer [506], pages 10-19.

[152] L.J. Eshelman and J.D. Schaffer. Real-coded genetic algorithms and interval-schemata. In D.L. Whitley [634], pages 187-202.

[153] L.J. Eshelman and J.D. Schaffer. Crossover's niche. In Forrest [170], pages 9-14.

[154] S.C. Esquivel, A. Leiva, and R.H. Gallard. Multiple crossover per couple in genetic algorithms. In *Proceedings of 1997 IEEE International Conference on Evolutionary Computation,* pages 103-106, Indianapolis, IN, 13.-16. Apr. 1997. IEEE, New York, NY. *

[155] A. Fairley. Comparison of methods of choosing the crossover points in the genetic crossover operation, 1991. †

[156] I.D. Falco, R.D. Balio, E. Tarantino, and R. Vaccaro. Simulation of genetic algorithms on MIMD multicomputers. *Parallel Processing Letters,* 2(4):381-389, Dec. 1992.

[157] E. Falkenauer and A. Delchambre. A genetic algorithm for bin packing and line balancing. In *Proceedings of the 1992 IEEE International Conference on Robotics and Automation,* vol. 2, pages 1186-1192, Nice, France, 12.-14. May 1992. IEEE Robotics and Automation Society, IEEE Computer Society Press, Los Alamitos, CA.

[158] R.J. Ficek. GENROUTE: *A hybrid approach for single and multiple depot routing.* Ph.D. thesis, North Dakota State University of Agriculture and Applied Sciences, 1993. PUUTTUU(DAI Vol. 55 No. 1).

[159] P. Field. Nonbinary transforms for genetic algorithm problems. In T.C. Fogarty, editor, *Evolutionary Computing. Selected Papers of the AISB Workshop,* pages 38-50, Leeds (U.K.), 11.-14. Apr. 1994. Springer-Verlag, Berlin. *

[160] J.R. Filho. GAME's library structure. In Stender [568], pages 111-116. †

[161] T.C. Fogarty. An incremental genetic algorithm for real-time optimization. In *Proceedings of the 1989 IEEE International Conference on Systems, Man, and Cybernetics,* vol. III, pages 321-326, Cambridge, MA, 14.-17. Nov. 1989. IEEE.

[162] T.C. Fogarty. Varying the probability of mutation in the genetic algorithm. In Schaffer [506], pages 104-109.

[163] T.C. Fogarty, editor. *Evolutionary Computing, Proceedings of the AISB96 Workshop,* Brighton, U.K., 1.-2. Apr. 1996, (to appear). †

[164] T.C. Fogarty and R. Huang. Implementing the genetic algorithm on transputer based parallel processing systems. In Schwefel and Manner [519], pages 145-149.

[165] D.B. Fogel. *Evolutionary Computation: Toward a New Philosophy of Machine Intelligence.* IEEE Press, Piscataway, N. J. 1995. †

[166] D.B. Fogel and L. Stayton. On the effectiveness of crossover in simulated evolutionary optimization. *BioSystems,* 32(3):171-182, 1994. †

[167] C. Fonlupt, D. Robilliard, and P. Preux. Fitness landscape and the behavior of heuristics. In [141].

[168] E. Fontain. Application of genetic algorithms in the field of constitutional similarity. *Journal of Chemical Information and Computer Sciences,* 32(6):748-752, 1992. (May 1992 Workshop on Similarity in Organic Chemistry.)

[169] H.Y. Foo, J. Song, W. Zhuang, H. Esbensen, and E.S. Kuh. Implementation of a parallel genetic algorithm for floorplan optimization on IBM SP2. In *Proceedings of the High Performance Computing on the Information Superhighway HPC Asia '97,* pages

456-459, Seoul (South Korea), 28. Apr - 2. May 1997. IEEE Computer Society Press, Los Alamitos, CA. *

[170] S. Forrest, editor. *Proceedings of the Fifth International Conference on Genetic Algorithms,* Urbana-Champaign, IL, 17.-21. July 1993. Morgan Kaufmann, San Mateo, CA.

[171] S. Forrest and M. Mitchell. Relative building-block fitness and the building-block hypothesis. In D. L. Whitley [634], pages 109-126. (ftp: ftp.santafe.edu: /pub/Users/mm/Forrest-Mitchell-FOGA.ps.Z).

[172] J. Freeman. Simulating a basic genetic algorithm. *The Mathematica Journal,* 3(2):52-56, 1993.

[173] B. Freisleben and M. Härtfelder. Optimization of genetic algorithms by genetic algorithms. In Albrecht et al. [24], pages 392-399.

[174] R.M. Friedberg. A learning machine: Part I. *IBM Journal,* 2:2 13, 1958. †

[175] R.M. Friedberg, B. Dunham, and T. North. A learning machine: Part II. *IBM Journal of Research and Development,* 3(3):282-287, 1959. †

[176] M. Friedman, U. Mahlab, and J. Shamir. Collective genetic algorithm for optimization and its electro-optic implementation. *Applied Optics,* 32(23):4423-4429, 1993.

[177] G. Fuat Üler and O.A. Mohammed. Ancillary techniques for the practical implementation of GAs to the optimal design of electromagnetic devices. *IEEE Trans. Magn.,* 32(3/1):1194-1197, 1996. *

[178] M. Fukumi and N. Akamatsu. A method to design a neural pattern recognition system by using a genetic algorithm with partial fitness and a deterministic mutation. In *Proceedings of the 1996 IEEE International Conference on Systems, Man and Cybernetics,* vol. 3, pages 1989-1993, Beijing, China, 14.-17. Oct. 1996. IEEE, New York, NY. *

[179] M. Fukumi, S. Omatsu, and Y. Nishikawa. A method to design neural network by the genetic algorithm with partial fitness. *Trans. Inst. Syst. Control Inf. Eng. (Japan),* 9(3):74-81, 1996. *

[180] A.S. Fukunaga and A.B. Kahng. Improving the performance of evolutionary optimization by dynamically scaling the evaluation function. In ICEC'95 [261], pages 182-187. †

[181] T. Furuhashi. Fuzzy evolutionary computation. Chapter 2.2 Development of *if-then* rules with the use of DNA coding, In Witold Pedrycz, Ed., *Fuzzy Evolutionary Computation,* pages 105-108. Kluwer Academic Publishers, New York, 1997.

[182] R. Galar. Simulation of local evolutionary dynamics of small populations. *Biological Cybernetics,* 65(1):37-45, 1991.

[183] C. Gathercole and P. Ross. An adverse interaction between crossover and restricted tree depth in genetic programming. In Koza et al. [312]. †

[184] C. Gathercole and P. Ross. Small populations over many generations can beat large populations over few generations in genetic programming. In Koza et al. [311]. †

[185] A. Geyer-Schulz. Holland classifier systems. *APL Quote Quad,* 25(4):43-55, June 1995. (Proceedings of the International Conference on APL, June 4.-8., 1995, San Antonio, TX.)

[186] A. Ceyer-Schulz. The next 700 programming languages for genetic programming. In Koza et al. [311]. †

[187] A. Geyer-Schulz and T. Kolarik. Distributed computing with APL. *APL Quote Quad,* 23(1):60-69, July 1992. (Proceedings of the International Conference on APL 6.-10. July 1992 St. Petersburg (Russia).)

[188] A. Ghozeil and D.B. Fogel. A preliminary investigation into directed mutations in evolutionary algorithms. In Voigt et al. [616], pages 329-335.

[189] J. Gillespie. A genetic algorithm solution to the project selection problem using static and dynamic fitness functions. In Koza [309]. †

[190] T. Gohtoh, K. Ohkura, and K. Ueda. An application of genetic algorithm with neutral mutations to job shop scheduling problems. In *Proceedings of the International Conference on Advances in Production Management Systems - APMS'96,* pages 563-568, Kyoto (Japan), 4.-6. Nov. 1996. Kyota Univ. (Kyota, Japan). *

[191] D.E. Goldberg. Optimal initial population size for binary-coded genetic algorithms. TCGA Report 85001, University of Alabama, 1985.

[192] D.E. Goldberg. A note on the disruption due to crossover in a binary-coded genetic algorithm. *TCGA* Report 87001, University of Alabama, 1987.

[193] D.E. Goldberg. Sizing populations for serial and parallel genetic algorithms. In Schaffer [506], pages 70-79.

[194] D.E. Goldberg. Real-coded genetic algorithms, virtual alphabets, and blocking. IlliGAL Report 90001, University of Illinois at Urbana-Champaign, 1990. Also as [196].

[195] D.E. Goldberg. A theory of virtual alphabets. In Schwefel and Männer [519], pages 13-22.

[196] D.E. Goldberg. Real-coded genetic algorithms, virtual alphabets and blocking. *Complex Systems,* 5(2):139-167, 1992. Also as [194].

[197] D.E. Goldberg, K. Deb, and J.H. Clark. Accounting for noise in the sizing of populations. In D.L. Whitley [634], pages 127-140.

[198] D.E. Goldberg, K. Deb, and J.H. Clark. Genetic algorithms, noise, and the sizing of populations. *Complex Systems,* 6(4):333-362, 1992. (Also TCGA Report No. 91010.)

[199] D.E. Goldberg and T. Kerzic. mGA1.0: A Common LISP implementation of a messy genetic algorithm. TCGA Report 90004, University of Alabama, 1990.

[200] D.E. Goldberg, K. Milman, and C. Tidd. Genetic algorithms: A bibliography. IlliGAL Report 92008, University of Illinois at Urbana-Champaign, 1992.

[201] D.E. Goldberg and W.M. Rudnick. Genetic algorithms and the variance of fitness. IlliGAL Report 91001, University of Illinois at Urbana-Champaign, 1991.

[202] D.E. Goldberg and W.M. Rudnick. Genetic algorithms and the variance of fitness. *Complex Systems,* 5(3):265-278, June 1991. Also as [201].

[203] M. Golub. An implementation of binary and floating point chromosome representation in genetic algorithm. In *Proceedings of the 18th International Conference on Information Technology Interfaces,* pages 417-422, Pula, Croatia, 18.-21. June 1996. Univ. Zagreb, Zagreb (Croatia). *

[204] M. Gorges-Schleuter. ASPARAGOS an asynchronous parallel genetic optimization strategy. In Schaffer [506], pages 422-427.

[205] M. Gotesman and T.M. English. Stacked generalization and fitness ranking in evolutionary algorithms. In McDonnell et al. [374]. †

[206] J. Gottlieb and N. Voss. Representations, fitness functions and genetic operators for the satisfiability problem. In [141].

[207] P. Graham and B. Nelson. A hardware genetic algorithm for the traveling salesman problem on Splash 2. In *Proceedings of the 5th International Workshop on Field-Programmable Logic and Applications,* pages 352-361, Oxford, U.K., 29. Aug.-l. Sept. 1995. Springer-Verlag, Berlin. *

[208] P. Graham and B. Nelson. Genetic algorithms in software and in hardware - a performance analysis of workstation and custom computing machine implementations. In *Proceedings of the 1996 IEEE Symposium on FPGAs for Custom Computing Machines,* pages 216-225, Napa Valley, CA, 17.-19. Apr. 1996. *

[209] R.N. Greenwell, J.E. Angus, and I. Finck. Optimal mutation probability for genetic algorithms. *Math. Cornput. Model. (U.K.),* 21(8):111, Apr. 1995. *

[210] J.J. Grefenstette. A user's guide to GENESIS. Technical Report CS-83-11, Vanderbilt University, Nashville, Department of Computer Science, 1983. †

[211] J.J. Grefenstette, editor. *Genetic Algorithms and their Applications: Proceedings of the Second International Conference on Genetic Algorithms and Their Applications,* MIT, Cambridge, MA, 28.- 31. July 1987. Lawrence Erlbaum Associates: Hillsdale, N J.

[212] J.J. Grefenstette. Incorporating problem specific knowledge into genetic algorithms. In Davis [124], pages 42-60.

[213] F.C. Gruau, D.L. Whitley, and L. Pyeatt. A comparison between cellular encoding and direct encoding for genetic neural networks. In Koza et al. [312]. †

[214] H.R. Gzickman and K.P. Sycara. Self-adaptation of mutation rates and dynamic fitness. In *Proceedings of the Thirteenth National*

Conference on Artificial Intelligence and the Eighth Innovative Applications of Artificial Intelligence Conference, vol. 2, page 1389, Portland, OR, 4.-8. Aug. 1996. MIT Press, Cambridge, MA. *

[215] J. Haataja. Geneettisten algoritmien simulointi Matlab 4.0IIa [On simulation of genetic algorithms by Matlab 4.0]. *SuperMenu,* (2):21 25, 1993, (in Finnish).

[216] J. Haataja. Evoluutiostrategiat Fortran 90:IIä [Evolution strategies in Fortran 90]. *@CSC,* (3):28-30, June 1997, (In Finnish).

[217] W.F. Hahnert and P.A.S. Ralston. Genetic algorithms for controller training - effects of population-size on training accuracy and efficiency. In B.J. Schneider and D.A. Stanley, editors, *Emerging computer techniques for the mining industry,* pages 21-30. Soc. Min. Engineers AIME, 1993. †

[218] W.H. Hahnert, III and P.A.S. Ralston. Analysis of population size in the accuracy and performance of genetic training for rule-based control systems. *Computers & Operations Research,* 22(1):65-72, 1995.

[219] T. Hämäläinen. *Implementation and Algorithms of a Tree Shape Parallel Computer.* Ph.D. thesis, Tampere University of Technology, 1996.

[220] T. Hämäläinen, H. Klapuri, J. Saarinen, P. Ojala, and K. Kaski. Accelerating genetic algorithm computation in tree shaped parallel computer. *Journal of Systems Architecture,* 42(1):19-36, Aug. 1996.

[221] T. Hämäläinen, J. Saarinen, P. Ojala, and K. Kaski. Implementing genetic algorithms in a tree shape computer architecture. In Alander [20], pages 259-284. (ftp: ftp.uwasa.fi:cs/1NWGA/Hamalainen.ps.Z)

[222] P.J.B. Hancock. *Coding strategies/or genetic algorithms and neural nets.* Ph.D. thesis, University of Stirling, Department of Computing Science and Mathematics, 1992. †

[223] P.J.B. Hancock. Genetic algorithms and permutation problems: A comparison of recombination operators for neural net structure specification. In J.D. Schaffer and D.L. Whitley, editors, *COGANN92, International Workshop on Combinations of Genetic Algorithms and Neural Networks,* pages 108-122, Baltimore, MD, 6. June 1992. IEEE Computer Society Press, Los Alamitos, CA. †

[224] N. Hansen, A. Gawelczyk, and A. Ostermeier. Sizing the population with respect to the local progress in $(1, \lambda)$-evolution strategies a theoretical analysis. In ICEC'95 [261], pages 80-85.

[225] N. Hansen, A. Ostermeier, and A. Gawelczyk. On the adaptation of arbitrary normal mutation distributions in evolution strategies: The generating set adaptation. In Eshelman [150]. †

[226] G. Harik, E. Cantu-Paz, D.E. Goldberg, and B.L. Miller. The gambler's ruin problem, genetic algorithms, and the sizing of populations. In *Proceedings of 1997 IEEE International Conference on Evolutionary Computation,* pages 7-12, Indianapolis, IN, 13.-16. Apr. 1997. IEEE, New York, NY. *

[227] C. Harris and B. Buxton. GP-COM: A distributed component-based genetic programming system in C++. In Koza et al. [312]. †

[228] R. Harris. An alternative description of the action of crossover. In *Proceedings of Adaptive Computing in Engineering Design and Control,* University of Plymouth (U.K.), 21.-22. Sept. 1994. †

[229] R. Harris and C. Ellis. An alternative description of the action of crossover. Internal Report PEDC-02-93, Plymouth Engineering Design Centre, 1993. †

[230] W.E. Hart. *Adaptive global optimization with local search.* Ph.D. thesis, University of California, San Diego, 1994. PUUTTUU(DAI Vol. 55 No. 7).

[231] R.L. Haupt and S.E. Haupt. Continuous parameter vs. binary genetic algorithms. In *Applied Computational Electromagnetics Symposium Digest,* vol. II, pages 1387-1392, Monterey, CA, 17.-21. Mar. 1997. †

[232] R. Hauser and R. Männer. Implementation of standard genetic algorithm on MIMD machines. In Davidor et al. [122], pages 504-513. *

[233] T. Haynes and S. Sen. Crossover operators for evolving a team. In Koza et al. [311]. †

[234] F. Herrera, E. Herrera-Viedma, M. Lozano, and J.L. Verdegay. Fuzzy tools to improve genetic algorithms. In *Proceedings of the Second European Congress on Intelligent Techniques and Soft Computing (EUFIT'94),* vol. 3, pages 1532-1539, Aachen (Germany), 20.-23. Sept. 1994. ELITE-Foundation. (ftp: decsai.ugr.es:pub/arai/tech_rep /ga-fl/eufit94.ps.Z).

[235] F. Herrera and M. Lozano. Heuristic crossovers for real-coded genetic algorithms based on fuzzy connectives. In Voigt et al. [616], pages 336-345. (ftp: decsai.ugr.es: pub/arai/tech_rep/ga-fl/paper_86. ps.Z).

[236] F. Herrera, M. Lozano, and J.L. Verdegay. Tackling real-coded genetic algorithms: operators and tools for behavioural analysis. Technical Report DECSAI 95107, Universidad de Granada, ETS de Ingeniería Informaática, 1994. (ftp: decsai.ugr.es:pub/arai/tech_rep/ga-fl/RCGh. ps.Z)

[237] F. Herrera, M. Lozano, and J.L. Verdegay. Dynamic and heuristic crossover operators for controlling the diversity and convergence of real-coded genetic algorithms. Technical Report DECSAI 95113, University of Granada, Department of Computer Science and Artificial Intelligence, 1995. (ftp: decsai.ugr.es: pub/arai/tech_rep/ga-fl/HD-crossovers.ps.Z) †

[238] F. Herrera, M. Lozano, and J.L. Verdegay. Fuzzy connectives based crossover operators to model genetic algorithms population diversity. Technical Report DECSAI-95110, University of Granada, Department of Computer Science and Artificial Intelligence, 1995. (To appear in *Mathware & Soft Computing.*) †

[239] F. Herrera, M. Lozano, and J.L. Verdegay. The use of fuzzy connectives to design real-coded genetic algorithms. *Mathware &*

Soft Computing, 1(3):239-251, 1995. (ftp: decsai.ugr.es:pub/arai/ tech-rep /ga-fl/Mathware95.ps.Z).

[240] F. Herrera, M. Lozano, and J.L. Verdegay. Dynamic and heuristic fuzzy connectives based crossover operators for controlling the diversity and convergence of real-coded genetic algorithms. *International Journal of Intelligent Systems,* 11:1013-1041, 1996. (ftp: decsai.ugr.es: pub/arai/tech_rep/ga-fl/IJIS.ps.Z)

[241] J. Hesser, J. Ludvig, and R. Männer. Real-time optimization by hardware supported genetic algorithms. In Osmera [429], pages 52-59.

[242] J. Hesser and R. Manner. An alternative genetic algorithm. In Schwefel and Männer [519], pages 33-37.

[243] J. Hesser and R. Männer. Towards an optimal mutation probability for genetic algorithms. In Schwefel and Männer [519], pages 23-32.

[244] J. Hesser and R. Männer. Investigation of the M-heuristic for optimal mutation. In Männer and Manderick [360], pages 115-126. †

[245] R. Hinterding. Gaussian mutation and self-adaption for numeric genetic algorithms. In ICEC'95 [261], pages 384-389. †

[246] R. Hinterding. Representation and self-adaptation in genetic algorithms. In Korea-Australia EC'95 [305], pages 77-90.

[247] R. Hinterding, H. Gielewski, and T.C. Peachey. The nature of mutation in genetic algorithms. In Eshelman [150]. †

[248] R. Hinterding, Z. Michalewicz, and T.C. Peachey. Self-adaptive genetic algorithm for numeric functions. In Voigt et al. [616], pages 420-429.

[249] M.F. Hobbs. Genetic algorithms, annealing, and dimension alleles. Master's thesis, Victoria University of Wellington, New Zealand, 1991.†

[250] F. Hoffmeister. KORR 2.1 – implementation of a (γ^+, λ) evolution strategy. Technical Report, University of Dortmund, Department of Computer Science, 1990. †

[251] F. Hoffmeister. *The User's Guide to* ESCAPADE *1.0 A Runtime Environment for Evolution Strategies,* Nov. 1990. †

[252] C. Höhn and C.R. Reeves. The crossover landscape for the Onemax problem. In Alander [21], pages 27-44. (ftp: ftp.uwasa.fi:cs/ 2NWGA/Reeves.ps.Z)

[253] I. Hong, A.B. Kahng, and B.R. Moon. Exploiting synergies of multiple crossovers: initial studies. In ICEC'95 [261], pages 245-250. †

[254] T.-P. Hong and H.-S. Wang. A dynamic mutation genetic algorithm. In *Proceedings of the 1996 IEEE International Conference on Systems and Cybernetics,* vol. 3, pages 2000-2005, Beijing, China, 14.-17. Oct. 1996. IEEE, New York, NY. PUUTTUU (EEA21093/ 97)

[255] H. Hörner. *A C++ Class Library for Genetic Programming: The Vienna University of Economics Genetic Programming Kernel,* 1996. (in German as [256]; available via URL: http://aif.wu-wien.ac. at/%/Egeyers/archive/gpk/vuegpk.html.)

[256] H. Hörner. Ein Kern für genetisches Programmieren in C++ [Genetic programming kernel in C++]. Master's thesis, Vienna University of Economics and Business Admimistration: Informationsverarbeitung und Informationswirtschaft, 1996. (In German; partly in English as [255].) †

[257] R. Huang and J. Ma. A distributed genetic algorithm over a transputer based parallel machine for survivable communication network design. In *Proceedings of the International Conference on Parallel and Distributed Processing Techniques and Applications (PDPTA '96),* Sunnyvale, CA, 9.-11. Aug. 1996. †

[258] S.-L. Hung and H. Adeli. A parallel genetic/neural network learning algorithm for MIMD shared memory machines. *IEEE Trans. on Neural Networks,* 5(6):900-909, Nov. 1994.

[259] S. Hurley. Taskgraph mapping using a genetic algorithm: a comparison of fitness functions. *Parallel Computing,* 19(11): 1217-1313, Nov. 1993. †

[260] *Proceedings of the First IEEE Conference on Evolutionary Computation,* Orlando, FL, 27.-29. June 1994. IEEE, New York, NY.

[261] *Proceedings of the Second IEEE Conference on Evolutionary Computation,* Perth (Australia), Nov. 1995. IEEE, New York, NY.

[262] *Proceedings of the First IEE/IEEE International Conference on Genetic Algorithms in Engineering Systems: Innovations and Applications,* Sheffield (U.K.), 12.-14. Sept. 1995. IEEE. †

[263] H. Iima and N. Sannomiya. Robustness of crossover operation of genetic algorithm in a production ordering problem. *Trans. Inst. Electr. Eng. Jpn. C (Japan),* 115-C(10):1208-1214, 1995. *

[264] I.M. Ikram. An Occam library for genetic programming on transputer networks. In *Proceedings of the International Conference on Parallel and Distributed Processing Techniques and Applications (PDPTA '96),* Sunnyvale, CA, 9.-11. Aug. 1996. †

[265] C. Jacob. Genetic L-system programming: breeding and evolving artificial flowers with Mathematica. In *Proceedings of the First International Mathematica Symposium,* pages 215-222, Southampton (England), 16.-20. July 1995. Comput. Mech. Publications, Southampton, (U.K.). *

[266] C.Z. Janikow and Z. Michalewicz. An experimental comparison of binary and floating point representation in genetic algorithms. In Belew and Booker [59], pages 31-36.

[267] K.-T. Jean and Y.-Y. Chen. Variable-based genetic algorithm. In *Proceedings of the 1994 IEEE International Conference on Systems, Man, and Cybernetics,* vol. 2, pages 1597-1601, San Antonio, TX, 2.-5. Oct. 1994. IEEE, New York. *

[268] W.M. Jenkins. The estimation of partial string fitnesses in the genetic algorithm. In B.H.V. Topping, editor, *Proceedings of the Developments in Neural Networks and Evolutionary Computing for Civil and Structural Engineering,* pages 137-141, Cambridge, England, 28.-30. Aug. 1995. Civil Comp. Press, Edingburgh. †

[269] L.-M. Jin and S.-P. Chan. Analogue placement by formulation of macrocomponents and genetic partitioning. *International Journal of Electronics,* 73(1):157-173, July 1992.

[270] T. Jones. Crossover, macromutation, and population-based search. In Eshelman [150], pages 73-80. †

[271] T. Jones. *Evolutionary Algorithms, Fitness Landscapes and Search.* Ph.D. thesis, University of New Mexico, 1995. †

[272] K.A.D. Jong and W.M. Spears. An analysis of the interacting roles of population size and crossover in genetic algorithms. In Schwefel and Männer [519], pages 38-47.

[273] K.A.D. Jong and W.M. Spears. A formal analysis of the role of multi-point crossover in genetic algorithms. *Annals of Mathematics and Artificial Intelligence,* 5(1):1-26, Apr. 1992. †

[274] G.F. Joyce. A massively parallel analog system for evolutionary optimization: The wetware approach. In Sebald and Fogel [528]. †

[275] B.A. Julstrom. Very greedy crossover in a genetic algorithm for the traveling salesman problem. In K.M. George, J.H. Carroll, E. Deaton, D. Oppenheim, and J. Hightower, editors, *Proceedings of the 10th ACM Symposium on Applied Computing,* pages 324-328, 1995. ACM Press, New York.

[276] B.A. Julstrom. A simple estimate of population size in genetic algorithms for the traveling salesman problem. In Alander [21], pages 3-14. (ftp: ftp.uwasa.fi: cs/2NWGA/Julstroml.ps.Z)

[277] B.A. Julstrom. Strings of weights as chromosomes in genetic algorithms for combinatorial problems. In Alander [22], pages 33 48. (ftp: ftp.uwasa.fi: cs/3NWGA/Julstrom.ps.Z)

[278] N. Kadaba. XROUTE: *A knowledge-based routing system using neural networks and genetic algorithms.* Ph.D. thesis, North Dakota State University of Agriculture and Applied Sciences, Fargo, 1990.

[279] C.E. Kaiser, G.B. Lamont, L.D. Merkle, G.H. Gates, Jr., and R. Pachter. Exogenous parameter selection in a real-valued genetic algorithm. In *Proceedings of 1997 International Conference on Evolutionary Computation,* pages 569-574, Indianapolis, IN, 13.-16. Apr. 1997. IEEE, New York, NY. *

[280] L. Kallel and M. Schoenauer. A priori comparison of binary crossover operators: No universal statistical measure, but a set of hints. In [141].

[281] G. Kampis. Coevolution in the computer: The necessity and use of distributed code systems. In *Self-organization and life, from simple rules to global complexity, Proceedings of the Second European*

Conference on Artificial Life, pages 537-546, Brussels (Belgium), 24.-26. May 1993. MIT Press, Cambridge, MA.

[282] C. Kappler. Are evolutionary algorithms improved by large mutations. In Voigt et al. [616], pages 346-355.

[283] C. Kappler, T. Bäck, J. Heistermann, A.V. der Velde, and M. Zamparelli. Refueling of a nuclear power plant: Comparison of a naive and a specialized mutation operator. In Voigt et al. [616], pages 829-838.

[284] S. Kawaji and K. Ogasawara. Nonlinear control of dynamic system using genetic algorithms-structurization of search space by switching the fitness function. *Trans. Inst. Electr. Eng. Jpn. D (Japan),* 116D(4):435-440, 1996. (In Japanese.) *

[285] J. Kazimierczak. An approach to evolvable hardware representing the knowledge base in an automatic programming system. In Koza [311]. †

[286] M.J. Keith and M.C. Martin. Advances in genetic programming. In K.E. Kinnear, Jr., editor, *Advances in Genetic Programming,* chapter 13. Genetic programming in C++: Implementation issues, pages 285-310. MIT Press, Cambridge, MA, 1994. †

[287] S. Khuri. Informatic crossover in genetic algorithms. In *1990 IEEE International Symposium on Information Theory,* page 62, San Diego, CA, 14.-19. Jan. 1990. IEEE. *

[288] M.D. Kidwell and D.J. Cook. Genetic algorithm for dynamic task scheduling. In *Proceedings of the 1994 IEEE 13th Annual International Phoenix Conference on Computers and Communications,* pages 61-67, Phoenix, AZ, 12.-15. Apr. 1994. IEEE, New York.

[289] J. Kim and J. Nang. Implementation of parallel genetic algorithm on AP1000 and its performance evaluation. *J. KISS(AJ, Comput. Syst. Theory (South Korea),* 23(2):127-141, 1996. (In Korean.) *

[290] Y.K. Kim, C.J. Hyun, and Y. Kim. Sequencing in mixed model assembly lines: a genetic algorithm approach. *Computers & Operations Research,* 23(12):1131-1145, 1996.

[291] J. Kingdon, J.L. Ribeiro Filho, and P.C. Treleaven. The GAME programming environment architecture. In Stender [568], pages 85-94. †

[292] K.E. Kinnear, Jr. Fitness landscapes and difficulty in genetic programming. In ICEC'94 [260], pages 142-147.

[293] W. Kinnebrock. *Optimierung reit genetischen und selektiven Algorithmen.* Oldenburg Verlag, München (Germany), 1994. †

[294] S. Kitamura and M. Hiroyasu. Genetic algorithm with stochastic automata-controlled, relevant gene-specific mutation probabilities. In ICEC'95 [261], pages 352-355. †

[295] H. Kitano, editor. *Genetic algorithm.* Sangyo Tosho K.K., Tokyo (Japan), 1993, (in Japanesc.) †

[296] C.C. Klimasauskas. An Excel macro for genetic optimization of a portfolio. *Advanced Technology for Developers,* 1(8):11-17, Dec. 1992.

[297] C.C. Klimasauskas. Genetic algorithm optimizes 100-city route in 21 minutes on a PC! *Advanced Technology for Developers,* 2:9 17, Feb. 1993.

[298] L.R. Knight. HYPERGEN – a distributed genetic algorithm on a hypercube. Master's thesis, University of Tulsa, Tulsa, OK, 1993. †

[299] L.R. Knight and R.L. Wainwright. HYPERGEN - a distributed genetic algorithm on hypercube. In *Proceedings, Scalable High Performance Computing Conference SHPCC-92,* pages 232-235, Williamsburg, VA, 26.-29. Apr. 1992. IEEE Computer Society Press, Los Alamitos, CA. *

[300] E.-J. Ko and O.N. Garcia. Adaptive control of crossover rate in genetic programming. In *Proceedings of the Artificial Neural Networks in Engineering (ANNIE'95),* vol. 5, pages 331-337, St. Louis, MO, 12.-15. Nov. 1995. ASME Press, New York, NY. *

[301] M.-S. Ko, T.-W. Kang, and C.-S. Hwang. Adaptive crossover operator based on locality and convergence. In *Proceedings of the 1996 IEEE International Conference on Intelligence and Systems,* pages 18-22, Rockville, MD, 4.-5. Nov. 1996. IEEE Computer Society Press, Los Alamitos, CA. *

[302] K. Kolarov. Landscape ruggedness in evolutionary algorithms. In *Proceedings of 1997 IEEE International Conference on Evolutionary Computation,* pages 19-24, Indianapolis, IN,.13.-16. Apr. 1997. IEEE, New York, NY. *

[303] M. Kolonko. A generalized crossover operation for genetic algorithms. *Complex Systems,* 9(3):177-191, 1995. *

[304] A.H. Konstam, S.J. Hartley, and W.L. Carr. Optimization m a distributed-processing environment using genetic algorithms with multivariate crossover. In *20th Annual Computer Science Symposium, 1992.ACM Computer Science Conference, Proceedings: Communications,* pages 109-116, Kansas City, MO, 3.-5. Mar. 1992. Assoc. Comp. Machinery.

[305] The Korea Science Engineering Foundation, The Australian Academy of Science, The Australian Academy of Technological Sciences and Engineering. *Proceedings of the 1st Korea - Australia Joint Workshop on Evolutionary Computation,* Taejon (Korea), 26-29. Sept. 1995. KAIST, Korea.

[306] E. Koskimäki and J. Göös. Fuzzy fitness function for electric machine design by genetic algorithm. In Alander [21], pages 237-244. (ftp: ftp.uwasa.fi: cs/2NWGA/Koskimaki.ps.Z)

[307] S.V. Kowalski and D. Moldovan. Parallel induction on hypercube. In *Proceedings of the Sixth IASTED/ISMM International Conference. Parallel and Distributed Computing and Systems,* pages 218-221,

Washington, D.C., 3.-5. Oct. 1994. IASTED/ISMM-ACTA Press, Anaheim, CA (U.S.A.). *

[308] J.R. Koza. *Genetic Programming II, Automatic Discovery of Reusable Programs.* MIT Press, Cambridge, MA, 1994. †

[309] J.R. Koza, editor. *Genetic Algorithms and Genetic Programming at Stanford 1997,* Stanford, CA, Winter 1997. Stanford University Bookstore. †

[310] J.R. Koza and D. Andre. Parallel genetic programming on a network of transputer. Report STAN-CS-TR-95-1542, Stanford University, Computer Science Department, 1995. †

[311] J.R. Koza, K. Deb, M. Dorico, D.B. Fogel, M. Garson, H. Iba, and R.L. Riolo, editors. *Genetic Programming 1997: Proceedings of the Second Annual Conference,* Stanford, CA, 13.-16. July 1997. Morgan Kaufmann, San Francisco, CA. †

[312] J.R. Koza, D. Goldberg, D. Fogel, and R.L. Riolo, editors. *Proceedings of the GP-96 Conference,* Stanford, CA, 28.-31. July 1996. MIT Press, Cambridge, MA. †

[313] S. Koziel. Non-uniform and non-stationary mutation in numerical optimization using genetic algorithms. *Kwart. Elektron. Telekomun. (Poland),* 42(3):273-285, 1996. *

[314] V. Kreinovich, C. Quintana, and O. Fuentes. Genetic algorithms: what fitness scaling is optimal *Cybernetics and Systems,* 24(1):9 26, Jan.-Feb. 1993.

[315] K. Krishna Kumar, S. Narayanaswamy, and S. Garg. Solving large parameter optimization problems using a genetic algorithm with stochastic coding. In Winter et al. [639], pages 287-303.

[316] B. Kroger, P. Schwenderling, and O. Vornberger. Genetic packing of rectangles on transputers. In P. Welch, D. Stiles, T.L. Kunii, and A. Bakkers, editors, *Transputing '91. Proceedings of the World Transputer User Group (WOTUG),* pages 593-608, Sunnyvale, CA, 22.-26. Apr. 1991. lOS Press, Amsterdam. †

[317] B. Kröger, P. Schwenderling, and O. Vornberger. Parallel genetic packing on transputers. Technical Report Reihe I Informatik Heft 29, Universität Osnabrück, Fachbereich Matematik/Informatik, 1992.

[318] J. Krone. Ein Evolutionsalgorithmus zur parallelen Bildsegmentierung. In R. Grebe and M. Baumann, editors, *TAT '92, Abstract Volume of the 4th German Transputer Users Group Meeting,* pages 144-145, Aachen, 22.-23. Sept. 1992. Medical School of the Technical University (RWTH), Institute for Physiology. †

[319] V. Kvasnicka and J. Pospíchal. Simple implementation of genetic programming by column tables. In Osmera [430], pages 71-76.

[320] L.L. Lai, J.T. Ma, F. Ndeh-Che, K.P. Wong, and S.Y.W. Wong. Discussion [of [320]]. *IEEE Trans. on Power Systems,* 11(1):136, Feb. 1996.

[321] A. Lane. Programming with genes. *AI Expert,* 8(12):16-19, Dec. 1993.

[322] A. Lane. Walkin' our way through GA. *AI Expert*, 10(2):11-16, Feb. 1995.

[323] C.G. Langton, editor. *Artificial Life, An Overview*. The MIT Press, Cambridge, MA, 1995. †

[324] P. Larrañaga, C.M.H. Kuijpers, R.H. Murga, and Y. Yurramendi. Learning Bayesian network structures by searching for best ordering with genetic algorithm. *IEEE Trans. on Systems, Man, and Cybernetics*, 26(4):487 493, July 1996.

[325] C. Lattaud. Evolution of the chromosomic architecture of genetic agents. In [141].

[326] T.L. Lau and E.P.K. Tsang. Applying a mutation-based genetic algorithm to processor configuration problems. In *Proceedings of the Eighth IEEE International Conference on Tools with Artificial Intelligence*, vol., pages 17-24, Toulouse, France, 16.-19. Nov. 1996. IEEE Computer Society Press, Los Alamitos, CA. *

[327] F. Leclerc and J.-Y. Potvin. A fitness scaling method based on a span measure. In ICEC'95 [261], pages 561-565. †

[328] L. Lemarchand, A. Plantec, B. Pottier, and S. Zanati. An object-oriented environment for specification and concurrent execution of genetic algorithms. *SIGPLAN OOPS Messenger*, 4(2):163-165, Apr. 1993. (Addendum to the proceedings of OOPSLA'92.) †

[329] J. Levenick. Metabits: Generic endogenous crossover control. In Eshelman [150]. †

[330] D.M. Levine. *A parallel genetic algorithm for the set partitioning problem*. Ph.D. thesis, Illinois Institute of Technology, 1994. (Also as [331], ftp: info.mcs.anl.gov:/pub/tech_reports/ANL9423.ps.Z) PUUTTUU (DAI Vol. 55 No. 5.)

[331] D. M. Levine. A parallel genetic algorithm for the set partitioning problem. Report ANL-94/23, Argonne National Laboratory, 1994. (Also as [330], ftp: info.mcs.anl.gov:/pub/tech_reports/ANL9423.ps .Z). †

[332] G. Levinson. Crossovers generate non-random recombinants under Darwinian selection. In *Proceedings of the Fourth International Workshop on the Synthesis and Simulation of Living Systems*, pages 90-101, Cambridge, MA, U.S.A., 6.-8. July 1994. MIT Press, Cambridge, MA. *

[333] Z. Liang-Jie, M. Zhi-Hong, and L. Yan-Da. Mathematical analysis of crossover operator in genetic algorithms and its improved strategy. In ICEC'95 [261], pages 412-417. †

[334] C.E. Liepins and M.D. Vose. Characterizing crossover in genetic algorithms. *Annals of Mathematics and Artificial Intelligence*, 5(1):27 34, 1992. †

[335] F.-T. Lin, C.-Y. Kao, and C.-C. Hsu. Applying the genetic approach to simulated annealing in solving some NP-hard problems. *IEEE Trans. on Systems, Man, and Cybernetics*, 23(6):1752-1767, Dec. 1993.

[336] G. Lin and X. Yao. Analysing crossover operators by search step size. In *Proceedings of 1997 IEEE International Conference on Evolutionary Computation,* pages 107-110, Indianapolis, IN, 13.-16. Apr. 1997. IEEE, New York, NY. *

[337] G. Lin, X. Yao, I. Macleod, L. Kang, and Y. Chen. Parallel genetic algorithms on PVM. In *Proceedings of the International Conference on Parallel Algorithms (ICPA'95),* Wuhan (China), 1995. Cordon and Breach. †

[338] D.S. Linden. Using a real chromosome in a genetic algorithm for wire antenna optimization. In *Proceedings of the 1997 IEEE Antennas and Propagation Society International Symposium Digest,* vol. 3, pages 1704-1707, Que, Montreal, (Canada), 13.-18. July 1997. †

[339] J. Lis. The synthesis of the ranked neural networks applying genetic algorithm with the dynamic probability of mutation. In J. Mira and F. Sandoval, editors, *Proceedings of the International Workshop on Artificial Neural Networks,* pages 498-504, Malaga-Torremolinos, 7.-9. June 1995. Springer-Verlag, Berlin (Germany). †

[340] X. Liu, A. Sakamoto, and T. Shimamoto. Genetic channel router. *IEICE Trans. on Fundamentals of Electronics Communications and Computer Sciences,* E77-A(3):492-501, Mar. 1994.

[341] A.M. Logar, E.M. Corwin, and T.M. English. Implementation of massively parallel genetic algorithm on the MasPar MP-I. In H. Berghel, G. Hedrick, E. Deaton, D. Roach, and R. Wainwright, editors, *SAC'92 Proceedings of the 1992 A CM/SIGAPP Symposium,* vol. II, pages 1015-1020, Kansas City, KS, 1.-3. Mar. 1992. ACM Press, New York. †

[342] S.M. Lucas. Evolving neural network learning behaviours with set-based chromosomes. In *Proceedings of the 4th European Symposium on Artificial Neural Networks,* vol., pages 291-296, Bruges, Belgium, 24.-26. Apr. 1996. D Facto, Brussels. *

[343] C.B. Lucasius and G. Kateman. GATES: genetic algorithm toolbox for evolutionary search, Software library in ANSI C. Technical Report, Catholic University Nijmegen, Laboratory for Analytical Chemistry, 1991. †

[344] C.B. Lucasius and C. Kateman. GATES towards evolutionary largescale optimization: A software-oriented approach to genetic algorithms. I. General perspectives. *Computers & Chemistry,* 18(2):127-136, June 1994. *

[345] C.B. Lucasius and G. Kateman. GATES towards evolutionary largescale optimization: A software-oriented approach to genetic algorithms. II. toolbox description. *Computers & Chemistry,* 18(2):137-156, June 1994. *

[346] S. Luke and L. Spector. A comparison of crossover and mutation in genetic programming. In Koza et al. [311]. †

[347] H.H. Lund. Genetic algorithms with dynamic fitness measures. In Alander [20], pages 69-84. (ftp: ftp.uwasa.fi:cs/1NWGA/Lund1.ps.Z)

[348] D. Macfarlane and I. East. An investigation of several parallel genetic algorithms. In S.J. Turner, editor, *Proceedings of the 12th Occam User Group., Technical Meeting*, pages 60-67, Exeter (U.K.), 2.-4. Apr. 1990. IOS Press, Amsterdam. †

[349] D. Maclay and R.E. Dorey. Application of genetic search techniques to drive-train modeling. In *Proceedings of the 1992 IEEE International Symposium on Intelligent Control*, pages 542-547, Glasgow (Scotland), 11.-13. Aug. 1992. IEEE.

[350] D. Maclay and R.E. Dorey. Applying genetic search techniques to drive-train modeling. *IEEE Control Systems Magazine*, 13(3):50-55, 1993. Also as [349].

[351] M.L. Maher and J. Poon. Co-evolution of the fitness function and design solution for design exploration. In ICEC'95 [261], pages 240-244. †

[352] S.W. Mahfoud. Crossover interactions among niches. In ICEC'94 [260], pages 188-193.

[353] H.S. Maini. *Incorporation of knowledge in genetic recombination.* Ph.D. thesis, Syracuse University, 1994. PUUTTUU (DAI Vol. 56 No. 3)

[354] H.S. Maini, K. Mehrotra, C.K. Mohan, and S. Ranka. Knowledge-based nonuniform crossover. In ICEC'94 [260], pages 22-27.

[355] H.S. Maini, K. Mehrotra, C.K. Mohan, and S. Ranka. Knowledge-based nonuniform crossover. *Complex Systems*, 8(4):257-293, Aug. 1994. PUUTTUU (EEA 37286/95 CCA 36470/95)

[356] J. man Park, J. gue Park, C. hyun Lee, and M. sung Han. Robust and efficient genetic crossover operator: homologous recombination. In *IJCNN '93-NAGOYA Proceedings of 1993 International Joint Conference on Neural Networks*, vol. 3, pages 2975-2978, Nagoya (Japan), 25.-29. Oct. 1993. IEEE.

[357] B. Manderick, M. de Weger, and P. Spiessens. The genetic algorithm and the structure of the fitness landscape. In Belew and Booker [59], pages 143-150.

[358] B. Manderick and P. Spiessens. Computational intelligence imitating life. How to select genetic operators for combinatorial optimization problems by analyzing their fitness landscape, In Jacek M. Zurada, Robert J. Marks, II and Charles J. Robinson editors, *Computational Intelligence Imitating Life*, pages 170-181. IEEE Press, New York, 1994.

[359] S.R. Mangano. Algorithms for directed graphs a unique approach using genetic algorithms. *Dr. Dobb's Journal*, 19(4):92, 94-97, 106-107, 147, Apr. 1994.

[360] R. Männer and B. Manderick, editors. *Parallel Problem Solving from Nature, 2*, Brussels, 28.-30. Sept. 1992. Elsevier Science Publishers, Amsterdam.

[361] J. Mäntykoski. Parallel implementation of genetic algorithms. Master's thesis, Helsinki University of Technology, Department of

Electrical Engineering, Laboratory of Signal Processing and Computer Technology, 1994.

[362] A. Mason. A non-linearity measure of a problem's crossover suitability. In ICEC'95 [261], pages 68-73. †

[363] A.J. Mason. Non-binary codings and partition coefficients for genetic algorithms. Management Studies Research Paper 3/91, University of Cambridge, Engineering Department, 1991. †

[364] T. Masters. *Practical neural network recipes in C++*, chapter 8, Genetic Optimization. Academic Press, Inc., San Diego, CA, 1993. †

[365] T. Masuda, A. Ito, and K. Sato. An acquiring method of fuzzy reasoning rules by genetic algorithm with variable gene length. *Transaction of the Institute of Electrical Engineers of Japan C*, 115-C(11):1265-1272, 1995. *

[366] K.E. Mathias. *Delta coding strategies for genetic algorithms*. Ph.D. thesis, Colorado State University, Fort Collins, 1991. †

[367] K.E. Mathias and D.L. Whitley. Genetic operators, the fitness landscape and the traveling salesman problem. In Männer and Manderick [360], pages 219-228. †

[368] K.E. Mathias and D.L. Whitley. Initial performance comparisons for the delta coding algorithm. In ICEC'94 [260], pages 433-438.

[369] K.E. Mathias and D.L. Whitley. Transforming the search space with gray coding. In ICEC'94 [260], pages 513-518.

[370] D.C. Mattfeld. *Evolutionary Search and the Job Shop*. Physica-Verlag, c/o Springer-Verlag, Berlin, 1996. †

[371] F. Maturana, A. Naumann, and D.H. Norrie. Object-oriented jobshop scheduling using genetic algorithms. *Comput. Ind. (Netherlands)*, 32(3):281-294, 1997. *

[372] B. Mayoh. Artificial life and pollution control: Explorations of a genetic algorithm system on the highly parallel Connection Machine. In vol. 1181 of *Lecture Notes in Computer Science*, pages 68-79, 1996. Springer Verlag, Berlin.

[373] G.D. McClurkin, R.A. Geary, and T.S. Durrani. An investigation into the parallelism of genetic algorithms. In D.J. Pritchard and C.J. Scott, editors, *Proceedings of the Second International Conference on Applications of Transputers*, pages 581-587, Southhampton (U.K.), 11.-13. July 1990. IOS Press, Amsterdam. †

[374] J.R. McDonnell, R.G. Reynolds, and D.B. Fogel, editors. *Evolutionary Programming IV: Proceedings of the Fourth Annual Conference on Evolutionary Programming (EP95)*, San Diego, CA, 1.-3. Mar. 1995. MIT Press. †

[375] C. Medsker and I.Y. Song. ProloGA: a Prolog implementation of a genetic algorithm. In *IEEE International Conference on Developing and Managing Intelligent System Projects*, pages 77-84, Washington, D.C., 29.-31. Mar. 1993. IEEE Computer Society Press, Los Alamitos, CA. *

[376] G.M. Megson and I.M. Bland. Generic systolic array for genetic algorithms. *IEE Proc., Comput. Digit. Tech. (U.K.),* 144(2):107 119, 1997. *

[377] R. Mendes and J. Neves. Genetic algorithms, classifiers and parallelism - an object-oriented approach. In J. Liebowitz, editor, *Moving Toward Expert Systems Globally in the 21st Century, Proceedings of the 2nd World Congress on Expert Systems,* pages 1199-1206, Lisbon (Portugal), 10.-14. Jan. 1994. Cognizant Communications Corp., Elmsford. †

[378] R.E. Mercer and J.R. Sampson. Adaptive search using a reproductive meta-plan. *Kybernetes,* 7:215-228, 1978. †

[379] K.C. Messa. Classification of crossover operators in genetic algorithms. Technical Report CS/CIAKS-89-OO3/TU, Tulane University, Department of Computer Science, 1989.

[380] J.C. Meza, R.S. Judson, T.R. Faulkner, and A.M. Treasurywala. A comparison of a direct search method and a genetic algorithm for conformational searching. *Journal of Computational Chemistry,* 17(9):1142-1151, 15. July 1996.

[381] Z. Michalewicz. *Genetic Algorithms + Data Structures = Evolution Programs.* Artificial Intelligence. Springer-Verlag, New York, 1992.

[382] Z. Michalewicz. *Genetic Algorithms + Data Structures = Evolution Programs.* Springer-Verlag, New York, 2 edition, 1994. (The second edition of [381].) †

[383] Z. Michalewicz. *Genetic Algorithms + Data Structures = Evolution Programs.* Springer Verlag, Berlin, 3rd edition, 1996.

[384] Z. Michalewicz, C.Z. Janikow, and J.R. Krawczyk. A modified genetic algorithm for optimal control problems. *Computers & Mathematics with Applications,* 23(12):83-94, 1992.

[385] E. Michielssen, S. Ranjithan, and R. Mittra. Optimal multilayer filter design using real coded genetic algorithms. *IEEE Proceedings J.: Optoelectronics,* 139(6.):413-420, Dec. 1992.

[386] S. Mikami, M. Wada, and T.C. Fogarty. Learning to achieve cooperation by temporal-spatial fitness sharing. In ICEC'95 [261], pages 803-807. †

[387] M. Milik. An object oriented environment for artificial evolution of protein sequences: The example of rational design of transmembrane sequences. In McDonnell et al. [374]. †

[388] F. Mill, S. Warrington, and R.A. Smith. Component shape encodings for genetic algorithms. In Parmee and Denham [439]. †

[389] M. Mitchell. *An Introduction to Genetic Algorithms.* MIT Press, Cambridge, MA, 1996. †

[390] M. Mitchell, S. Forrest, and J.H. Holland. The royal road for genetic algorithms: Fitness landscapes and GA performance. In Varela and Bourgine [607], pages 245-254. (ftp: ftp.santafe.edu: /pub/Users/mm/sfi-91-10-046.ps.Z).

[391] J. Mitlöhner. Classifier systems and economic modeling. *APL, Quote Quad*, 26(4):77-86, June 1996. (Proceedings of the APL96 Conference)

[392] L. Molgedey. Mean field analysis of tournament selection on a random manifold. In Voigt et al. [616], pages 174-177.

[393] J.H. Moore. Artificial intelligence programming with LabVIEW: genetic algorithms for instrumentation control and optimization. *Computer Methods and Programs in Biomedicine*, 47(1):73 79, June 1995. *

[394] H. Mühlenbein. Parallel genetic algorithms, population genetics and combinatorial optimization. In Schaffer [506], pages 416-421.

[395] H. Mühlenbein. How genetic algorithms really work I. mutation and hill-climbing. In Männer and Manderick [360], pages 15-26.

[396] H. Mühlenbein and J. Kindermann. The dynamics of evolution and learning towards genetic neural networks. In R. Pfeifer, Z. Schreter, F. Fogelmann-Soulie, and L. Stees, editors, *Connectionism in Perspective*, pages 173-198. North-Holland, 1989.

[397] H. Mühlenbein and D. Schlierkamp-Voosen. Optimal interaction of mutation and crossover in the breeder genetic algorithm. In Forrest [170], page 648.

[398] M. Murakawa, S. Yoshizawa, I. Kajitani, T. Furuya, M. Iwata, and T. Higuchi. Hardware evolution at function level. In Voigt et al. [616], pages 62-71.

[399] K. Na, S.-I. Chae, and S. Ann. Modified delta coding algorithm for real parameter optimisation. *Electronics Letters*, 31(14):1169-1171, 1995.

[400] T. Nagao. Homogeneous encoding for genetic algorithm based numerical optimization. *Trans. Inst. Electron. Inf. Commun. Eng. D-II (Japan)*, J80D-II(1):56-62, 1997. In Japanese. *

[401] R. Nakano, Y. Davidor, and T. Yamada. Optimal population size under constant computation cost. In Davidor et al. [122], pages 130-138. *

[402] K. Nara, A. Shiose, M. Kitagawa, and T. Ishihara. Implementation of gcnctic algorithm for distribution systems loss minimum reconfiguration. *IEEE Trans. on Power Systems*, 7(3):1044-1051, Aug. 1992.

[403] M.C. Negoita, A.H. Dediu, and D. Mihaila. Design elements of EHW using GA with local improvement of chromosomes. In *Proceedings of the 5th International Conference Fuzzy Days*, page 604, Dortmund (Germany), 28.-30. Apr. 1997. Springer-Verlag, Berlin (Germany). *

[404] K.M. Nelson. Function optimization and parallel evolutionary programming on the MasPar MP-I. In Sebald and Fogel [528]. †

[405] V. Nemec and J. Schwarz. Parallel genetic algorithms implemented on transputers. In Osmera [129], pages 85 90.

[406] D.J. Nettleton and R. Garigliano. Large ratios of mutation to crossover: The example of the traveling salesman problem. In F.A.

Sadjadi, editor, *Adaptive and Learning Systems II,* vol. SPIE-1962, pages 110-119, Orlando, FL, 12.-13. Apr. 1993. The International Society for Optical Engineering.

[407] A. Neubauer. Real-coded genetic algorithms for bilinear signal estimation. In D. Schipanski, editor, *Tagungsband des .40. Internationalen Wissenschaftlichen Kolloquiums,* vol. 1, pages 347-352. Ilmenau (Germany), 1995. †

[408] A. Neubauer. Adaptive non-uniform mutation for genetic algorithms. In *Proceedings of the International Conference on Computational Intelligence,* Lecture Notes in Computer Science, Dordmund, 28.30. Apr. 1997. Springer-Verlag, Berlin (to appear). †

[409] A. Neubauer. A theoretical analysis of the non-uniform mutation operators for the modified genetic algorithm. In *Proceedings of 1997 IEEE International Conference on Evolutionary Computation,* pages 93-96, Indianapolis, IN, 13.-16. Apr. 1997. IEEE, New York, NY. *

[410] K.V. Nguyen. Improving the crossover operator in genetic algorithms and applications in optimal conference room booking. In Koza [309]. †

[411] V. Nissen. *Evolutionare Algorithmen, Darstellung, Beispiele, betriebswirtschaftliche Anwendungmöglichkeiten.* DUV Deutscher Universitats Verlag, Wiesbaden, 1994. †

[412] A.E. Nix. Comparing finite and infinite population models of a genetic algorithm using the minimum deceptive problem. Goverment report AD-A238679/5, Tennessee University, 1991. †

[413] T. Nomura. An analysis on linear crossover for real number chromosomes in an infinite population size. In *Proceedings of 1997 IEEE International Conference on Evolutionary Computation,* pages 111- 114, Indianapolis, IN, 13.-16. Apr. 1997. IEEE, New York, NY. *

[414] P. Nordin and W. Banzhaf. Genetic reasoning evolving proofs with genetic search. Technical Report SYS-2/96, University of Dortmund, Fachbereich Informatik, 1996.

[415] P. Nordin, F.D. Francone, and W. Banzhaf. Explicitly defined introns and destructive crossover in genetic programming. In *Proceedings of the 12th International Conference on Machine Learning, GP Workshop,* number 95.2, pages 6-22, Tahoe City, 1995. University of Rochester. (Also as [414].)

[416] P. Nordin, F.D. Francone, and W. Banzhaf. Explicitly defined introns and destructive crossover in genetic programming. Internal Report SYS-3/95, University of Dortmund, Fachbereich Informatik, 1995. (Also as [414].)

[417] P. Nordin, F.D. Francone, and W. Banzhaf. Explicitly defined introns and destructive crossover in genetic programming. In P.J. Angeline and K. Kinnear, editors, *Advances in Genetic Programming II,* pages 111-134. MIT Press, Cambridge, CA, 1996.

[418] M.O. Odetayo. Optimal population size for genetic algorithms: an investigation. In *Proceedings of the IEEE Colloquium on Genetic Algorithms for Control and Systems Engineering* [1], pages 2/1-2/4. *

[419] M.O. Odetayo. Relationship between replacement strategy and population size. In Osmera [429], pages 91-96.

[420] K. Ohkura and K. Ueda. A genetic algorithm with neutral mutations for solving nonstationary function optimization problems. In *Proceedings of the 1994 Second Australian and New Zealand Conference on Intelligent Information Systems,* pages 248-252, Brisbane, QLD (Australia), 2. Dec. 1994. IEEE, New York. *

[421] K. Ohkura and K. Ueda. Solving deceptive problems using a genetic algorithm with neutral mutations. In *Proceedings of the Artificial Neural Networks in Engineering (ANNIE95),* vol. 5, pages 345-350, St. Louis, MO, 12.-15. Nov. 1995. ASME Press, New York, NY (U.S.A.). *

[422] S. Oliker, M. Furst, and O. Maimon. Design architectures and training of neural networks with a distributed genetic algorithm. In *1993 IEEE International Conference on Neural Networks,* vol. I, pages 199-202, San Fancisco, CA, 28. Mar.- 1. Apr. 1993. IEEE.

[423] I.M. Oliver, D.J. Smith, and J.R.C. Holland. A study of permutation crossover operators on the traveling salesman problem. In Grefenstette [211], pages 224-230.

[424] F. Oppacher and D. Deugo. Automatic change of representation in genetic algorithm. In Pearson et el. [442], pages 218-222.

[425] F. Oppacher and D. Deugo. The evolution of hierarchical representations. In *Advances in Artificial Life. Proceedings of the Third European Conference on Artificial Life,* vol. 929 of *Lecture Notes in Artificial Intelligence,* pages 302-313, Granada (Spain), 4.- 6..June 1995. Springer-Verlag, Berlin. *

[426] U.-M. O'Reilly and F. Oppacher. Hybridized crossover-based search techniques in program discovery. In ICEC'95 [261], pages 573 578. †

[427] A. Osei, M.P. Schamschula, H.J. Gaulfield, and J. Shamir. Use of quantum indeterminacy in optical parallel genetic algorithms. In *Optical Implementation of Information Processing,* vol. SPIE-2565, pages 192-197, San Diego, CA, 10.-11. July 1995. The International Society for Optical Engineering, Bellingham, WA. *

[428] P. Osmera, editor. *Proceedings of the MENDEL'95,* Brno (Czech Republic), 26.-28. Sept. 1995. Technical University of Brno.

[429] P. Osmera, editor. *Proceedings of the MENDEL '96,* Brno (Czech Republic), June 1996. Technical University of Brno.

[430] P. Osmera, editor. *Proceedings of the 3rd International Mendel Conference on Genetic Algorithms, Optimization problems, Fuzzy Logic, Neural Networks, Rough Sets (MENDEL '97),* Brno (Czech Republic), 25-27 June 1997. Technical University of Brno.

[431] P. Osmera, V. Kvasnicka, and J. Pospíchal. Genetic algorithms with diploid chromosomes. In Osmera [430], pages 111-116.

[432] M. Oussaidène, B. Chopard, and M. Tomassini. Programmarion évolutionniste parallèle. In Dekeyser, Lebert, and Manneback, editors, *Proceedings of the RenPar'7, Actes des 7es Recontres Francopones du Parallélisme*, 30. May- 2. June 1995. PIP-FPMs Mons, Belgium.

[433] K.F. Pal. Genetic algorithms for the traveling salesman problem based on a heuristic crossover. *Biological Cybernetics,* 69(5-6):539 549, 1993. *

[434] C.C. Palmer and A. Kershenbaum. Representing trees in genetic algorithms. In ICEC'94 [260], pages 379-384.

[435] K. Park and B. Carter. On the effectiveness of genetic search in combinatorial optimization. In K.M. George, J.H. Carroll, E. Deaton, B D. Oppenheim, and J. Hightower, editors, *Proceedings of the 10th ACM Symposium on Applied Computing,* pages 329-336, 1995. ACM Press.

[436] L.-J. Park, D.-K. Nam, C.H. Park, and S.-H. Oh. An empirical comparison of simulated annealing and genetic algorithms on NK fitness landscapes. In *Proceedings of 1997 IEEE International Conference on Evolutionary Computation,* pages 147-151, Indianapolis, IN, 13.-16. Apr. 1997. IEEE, New York, NY. *

[437] L.-J. Park and C.H. Park. Application of genetic algorithm to job shop scheduling problems with active schedule constructive crossover. In *Proceedings of the IEEE International Conference on Systems, Man and Cybernetics,* vol. 1, pages 530-535, Vancouver, BC (Canada), 22.-25. Oct. 1995. IEEE, New York, NY. *

[438] L.-J. Park and C.H. Park. Preventing premature convergence in genetic algorithms with adaptive population size. *J. Korea Inst. Telemat. Electron. (South Korea),* 32B(12):136-142, 1995. †

[439] I. Parmee and M.J. Denham, editors. *Adaptive Computing in Engineering Design and Control '96 (ACEDC'96), 2nd International Conference of the Integration of Genetic Algorithms and Neural Network Computing and Related Adaptive Techniques with Current Engineering Practice,* Plymouth (U.K.), 26.-28. Mar. 1996, (To appear.) †

[440] M.A. Pawlowsky. Modified uniform crossover and desegredation in genetic algorithms. Master's thesis, Concordia University, Montreal, Quebec, Canada, 1992. †

[441] M.A. Pawlowsky. Practical handbook of genetic algorithms. In Chambers [88], chapter 4. Crossover operators, pages 101-114.

[442] D.W. Pearson, N.C. Steele, and R.F. Albrecht, editors. *Artificial Neural Nets and Genetic Algorithms,* Ales (France), 19.-21. Apr. 1995. Springer-Verlag, New York.

[443] C.C.B. Pettey, M.R. Leuze, and J.J. Grefenstette. Genetic algorithms on a hypercube multiprocessor. In *Hypercube Multiprocessors 1987*, pages 333-341, 1987. †

[444] D.T. Pham and D. Karaboga. Genetic algorithms with variable mutation rates: application to fuzzy logic controller design. *Proc. Inst. Mech. Eng. I, J. Syst. Control Eng. (U.K.)*, 211(I2):157-167, 1997. *

[445] A.G. Pipe, T.C. Fogarty, and A. Winfield. Balancing exploration with exploitation – solving mazes with real numbered search spaces. In ICEC'94 [260], pages 485-489.

[446] R. Poli and W.B. Langdon. A new schema theory for genetic programming with one-point crossover and point mutation. In Koza et al. [311]. †

[447] P.W. Poon and J.N. Carter. Genetic algorithm crossover operators for ordering applications. *Computers & Operations Research*, 22(1):135-148, 1995.

[448] K. Price and R. Storn. Differential evolution. *Dr. Dobb's Journal*, 22(4):18-20, 22, 24, 78, Apr. 1997.

[449] R.J. Pryor and D.D. Cline. Use of a genetic algorithm to solve two-fluid flow problems on an NCUBE multiprocessor computer. Report SAND-92-2847C, Sandia National Laboratories, Albuquerque, NM, 1992. (Also as [450].) †

[450] R.J. Pryor and D.D. Cline. Use of a genetic algorithm to solve two-fluid flow problems on an NCUBE multiprocessor computer. In *Proceedings of the International Topical Meeting on Mathematical Methods and Supercomputing in Nuclear Applications*, Karlsruhe (Germany), 19.-23. Apr. 1992. (Also as [449].) †

[451] N. Pucello, M. Rosati, M. Celino, G. D'Agostino, F. Pisacane, and V. Rosato. Search of molecular ground state via genetic algorithm; implementation on a hybrid SIMD-MIMD platform. In *Proceedings of the Third European PVM Conference*, pages 339-342, Munich (Germany), 7.-9. Oct. 1996. Springer-Verlag, Berlin (Germany). *

[452] X. Qi and F. Palmieri. The diversification role of crossover in the genetic algorithms. In Forrest [170], pages 132 137.

[453] X. Qi and F. Palmieri. Theoretical analysis of evolutionary algorithms with an infinite population size in continuous space, part I: Basic properties. *IEEE Trans. on Neural Networks*, 5(1):102-119, Jan. 1994. †

[454] X. Qi and F. Palmieri. Theoretical analysis of evolutionary algorithms with an infinite population size in continuous space, part II: Analysis of the diversification role of the crossover. *IEEE Trans. on Neural Networks*, 5(1):120-129, Jan. 1994. †

[455] M. Qingchun. An approach on genetic algorithm with symmetric codes. *Acta Electronica Sinica (China)*, 24(10):27-31, 1996. (In Chinese.) *

[456] F.M. Rabinowitz. Algorithm 744: a stochastic algorithm for global optimization with constraints. *ACM Trans. Math. Softw.,* 21(2):194-213, 1995. *

[457] N.J. Radcliffe and P.D. Surry. Formal memetic algorithms. In T.C. Fogarty, editor, *Evolutionary Computing. Selected Papers of the AISB Workshop,* pages 1-16, Leeds (U.K.), 11.-14. Apr. 1994. Springer-Verlag, Berlin. *

[458] N.J. Radcliffe and P.D. Surry. RPL2: A language and parallel framework for evolutionary computing. In Davidor et al. [122], pages 628-635. *

[459] T. Ragg. Parallelization of an evolutionary neural network optimizer based on PVM. In *Proceedings of the Third European PVM Conference,* pages 351-354, Munich (Germany), 7.-9. Oct. 1996. Springer-Verlag, Berlin (Germany). *

[460] R. Rankin, R. Wilkerson, G. Harris, and J. Spring. A hybrid genetic algorithm for an NP-complete problem with an expensive evaluation function. In K.M.G. Deaton and G.H.H. Berghel, editors, *Proceedings of the 1993 ACM/SICAPP Symposium on Applied Computing,* pages 251-256, Rolla, MO, 14.-16 Feb. 1993. ACM, New York. PUUTTUU (EEA 48825/94 CCA 45425/94 ACM/93).

[461] B.B.P. Rao, L.M. Patnaik, and R.C. Hansdah. A genetic algorithm for channel routing using inter-cluster mutation. In ICEC'94 [260], pages 97-103.

[462] G.J.E. Rawlins, editor. *Foundations of Genetic Algorithms,* Indiana University, 15.-18. July 1991. Morgan Kaufman: San Mateo, CA.

[463] I. Rechenberg. *Evolutionsstrategie: Optimierung technisher Systeme nach Prinzipien der biologischen Evolution.* Frommann-Holzboog Verlag, Stuttgart, 1973, (reprint in [464].) †

[464] I. Rechenberg. *Evolutionsstrategie '94.* Frommann-Holzboog-Verlag, Stuttgart (Germany), 1994. (In German; includes also [463].)

[465] V.G. Red'ko, M.I. Dyabin, V.M. Elagin, N.G. Karpinskii, A.I. Polovyanyuk, V.A. Serechenko, and O.V. Urgant. On microelectronic implementation of an evolutionary optimizer. *Mikroelektronika (Russia),* 24(3):207-210, 1995. PUUTTUU (EEA65081/95)

[466] V.G. Red'ko, M.I. Dyabin, V.M. Elagin, N.G. Karpinskii, A.I. Polovyanyuk, V.A. Serechenko, and O.V. Urgant. On microelectronic implementation of an evolutionary optimizer. *Microelectron. (U.S.A.),* 24(3):182-185, 1995. Translation of: [465]PUUTTUU (EEA65081/95).

[467] G.M. Reese. Parameter estimation by genetic algorithms. In *Proceedings of the 1993 MATLAB Conference,* Boston, MA, 18.-20. Oct. 1993. (Also as [468].) †

[468] G.M. Reese. Parameter estimation by genetic algorithms. Report SAND-93-1298C, Sandia National Laboratories, Albuquerque, NM, 1993. (Also as [467].) †

[469] C.R. Reeves. Using genetic algorithms with small populations. In Forrest [170], pages 92-99.

[470] A. Reinefeld and V. Schnecke. Portability versus efficiency. Parallel applications on PVM and Parix. In *Proceedings of the Workshop on Parallel Programming and Computation (ZEUS95) and the 4th Nordic Transputer Conference (NTUG95)*, pages 35-49, Linköping, Sweden, 1995. IOS Press 1995, Amsterdam, Netherlands. *

[471] J.L. Ribeiro Filho, C. Alippi, and P.C. Treleaven. Genetic algorithm programming environments. In Stender [568], pages 65-84. †

[472] J.L. Ribeiro Filho, P.C. Treleaven, and C. Alippi. Genetic-algorithm programming environments. *Computer,* 27(6):28 45, June 1994.

[473] R.L. Riolo. CFS-C: A package of domain-independent subroutines for implementing classifier systems in arbitrary, user-defined environments. Technical Report, University of Michigan, Department of Computer Science and Engineering, Logic of Computers Group, 1988. †

[474] R.L. Riolo. Survival of the fittest bits. *Scientific American,* 267(1):89-91, July 1992.

[475] P. Robbins. The use of a variable length chromosome for permutation manipulation in genetic algorithms. In Pearson et al. [442], pages 144-147.

[476] S.G. Roberts and M. Turega. Evolving neural network structures: An evaluation of encoding techniques. In Pearson et al. [442], pages 96-99.

[477] G. G. Robertson. Population size in classifier systems. In *Proceedings of the Fifth International Conference on Machine Learning,* pages 142-152, 1988. †

[478] G.G. Robertson. Population size in classifier systems. *Machine Learning,* 5:142 152, 1990. †

[479] S.G. Romaniuk. Applying crossover operators to automatic neural network construction. In ICEC'94 [260], pages 750-752a.

[480] S. Ronald. Robust encodings in genetic algorithm: a survey of encoding issues. In *Proceedings of 1997 IEEE International Conference on Evolutionary Computation,* pages 43-48, Indianapolis, IN, 13.-16. Apr. 1997. IEEE, New York, NY. *

[481] R.S. Rosenberg. Simulation of genetic populations with biochemical properties: II. selection of crossover probabilities. *Mathematical Biosciences,* 8:1-37, 1970. †

[482] B.J. Rosmaita. EXODUS: An extension of the genetic algorithm to problems dealing with permutations. Master's thesis, Vanderbilt University, Nasville, 1985. †

[483] B.J. Rosmaita. EXODUS user's manual (version 1.8). Technical Report No. CS-85-06, Vanderbilt University, Nashville, Department of Computer Science, 1985. †

[484] P. Ross, D.W. Corne, and H. L. Fang. Improving evolutionary timetabling with delta evaluation and directed mutation. In Davidor et al. [122], pages 556-565. *

[485] J. Roupec and J. Krejsa. Dominance and recessivity in genetic algorithms. In Osmera [429], pages 197-199.

[486] J. Rowe and I.R. East. Direct replacement: a genetic algorithm without mutation which avoids deception. In Yao [649], pages 41-48. *

[487] G. Rudolph. On correlated mutations in evolution strategies. In Manner and Manderick [360], pages 105-114. †

[488] G. Rudolph. Asymptotical convergence rates of simple evolutionary algorithms under factorizing mutation distributions. In [141].

[489] C. Ryan. The degree of oneness. In *Proceedings of the First Online Workshop on Soft Computing (WSC1)*, pages 43-48, WWW (World Wide Web), 19.-30. Aug. 1996. Nagoya University.

[490] C. Ryan. Diploidy without dominance. In Alander [22], pages 63-70. (ftp: ftp.uwasa.fi: cs/3NWGA/Ryan.ps.Z)

[491] C. Ryan. Shades – a polygenic inheritance scheme. In Osmera [430], pages 140-147.

[492] C. Ryan and P. Walsh. Paragen II: evolving parallel transformation rules. In *Proceedings of the 5th Fuzzy Days,* page 573, Dortmund (Germany), 28.-30. Apr. 1997. Springer-Verlag, Berlin (Germany). *

[493] M. Ryynänen. The optimal population size of genetic algorithm in magnetic field refinement. In Alander [21], pages 281-282. (ftp: ftp.uwasa.fi: cs/2NWGA/Ryynanen.ps.Z)

[494] J. Sakamoto and J. Oda. Topological optimum design of truss structures using genetic algorithm with biased crossover. In *Technical Papers of the 36th AIAA/ASME/ASCE/AHS/ASC Structures, Strucural Dynamics and Materials Conference,* vol. Pt. 5, pages 3536-3542, New Orleans, LA, 10.-13. Apr. 1995. American Institute of Aeronautics and Astronautics, Washington, D.C. †

[495] H. Sakanashi and Y. Kakazu. Co-evolving genetic algorithm with filtered evaluation function. In *1994 IEEE Symposium on Emerging Technologies and Factory Automation,* pages 454-457, Tokyo (Japan), 6.-10. Nov. 1994. IEEE New York. *

[496] H. Sakanekshi, K. Suzuki, and Y. Kakazu. Controlling dynamics of GA through filtered evaluation function. In Davidor et al. [122]. †

[497] M. Sakawa and M. Tanaka. *Genetic Algorithms.* Asakura Book Publishers Co. Ltd, 1995, (in Japanese.) †

[498] M. Salami and G. Cain. Implementation of genetic algorithms on reprogrammable architectures. In X. Yao, editor, *Proceedings of the Eight Australian Joint Conference on Artificial Intelligence,* page 581, 1995. World Scientific Publishers, Co., Singapore. †

[499] R. Salomon. The influence of different coding schemes on the computational complexity of genetic algorithms in function optimization. In Voigt et al. [616], pages 227-235.

[500] N. Sangalli, Q. Semeraro, and T. Tolio. Performance of genetic algorithms in the solution of permutation flowshop problems. In Pearson et al. [442], pages 495-498.

[501] I. Santibáñez-Koref, H.-M. Voigt, and J. Born. Parallele Evolutionsalgorithmen - Implementierung und Andwendungen auf Transputernetzen. In *Abstraktband des 3. bundesweiten Transputer - Anwendcr - Treffens,* pages 216-218, Klinikum der RWTH Aachen, 1991. †

[502] I. Santibáñez-Koref, H.-M. Voigt, and J. Born. Parallele diploide Evolutionsalgorithmen. In *TAT '92, Abstract Volume of the 4th German Transputer Users Group Meeting,* pages 98-100, Aachen, 22.23. Sept. 1992. Medical School of the Technical University (RWTH), Institute for Physiology. †

[503] N. Saravanan and D.B. Fogel. A bibliography of evolutionary computation & applications. Technical Report FAU-ME-93-100, Florida Atlantic University, Department of Mechanical Engineering, 1993. (ftp: magenta.me.fau.edu: /pub/ep-list/bib/EC-ref.ps.Z)

[504] N. Saravanan, D.B. Fogel, and K.M. Nelson. A comparison of methods for self-adaptation in evolutionary algorithms. *BioSystems,* 36:157-166, 1995.

[505] A. Satyadas and K. KrishnaKumar. Genetic algorithm modules in MATLAB: Design and implementation using software engineering practices. In Winter et al. [639], pages 321-344.

[506] J.D. Schaffer, editor. *Proceedings of the Third International Conference on Genetic Algorithms,* George Mason University, 4.-7. June 1989. Morgan Kaufmann Publishers, Inc.

[507] J.D. Schaffer and L.J. Eshelman. On crossover as an evolutionarily viable strategy. In Belew and Booker [59], pages 61-68.

[508] J.D. Schaffer and A. Morishima. An adaptive crossover distribution mechanism for genetic algorithms. In Grefenstette [211], pages 36-40.

[509] T. Schnier and J.S. Gero. Learning representations for evolutionary computation. In *Proceedings of the Eighth Australian Joint Conference on Artificial Intelligence,* pages 387-394, Canberra, ACT, Australia, 13.-17. Nov. 1995. World Scientific, Singapore. *

[510] A. Schober, M. Thuerk, and M. Eigen. Optimization by hierarchical mutant production. *Biological Cybernetics,* 69(5-6):493 501, 1993.

[511] M. Schoenauer and M. Sebag. Controlling crossover through inductive learning. In Davidor et al. [122]. †

[512] E. Schöneburg, F. Heinzmann, and S. Feddersen. *Genetischer Algoritmen und Evolutionsstrategien.* Addison-Wesley, Verlag, 1994. †

[513] N.N. Schraudolph and R.K. Belew. Dynamic parameter encoding for genetic algorithms. Technical Report CS90-175, University of San Diego, La Jolla, Computer Science and Engineering Department, 1990. †

[514] N.N. Schraudolph and R.K. Belew. Dynamic parameter encoding for genetic algorithms. *Machine Learning,* 9(1):9-21, June 1992.

[515] N.N. Schraudolph and J.J. Grefenstette. A user's guide to GAucsd 1.4. Technical Report CS92-249, University of California, San Diego, 1992. †

[516] H.-P. Schwefel. Subroutines evol, grup, korr, listings and user's guide. Interner Bericht KFA-STE-IB-2/80, Kernforschungsanlage Jülich, Programmgruppe Systemforschung und Technologische Entwicklung, 1980. †

[517] H.-P. Schwefel. Unterprogramme evol, grup, korr, Programme und Benutzeranleitungen. Interner Bericht KFA-STE-IB-3/80, Kernforschungsanlage Julich, Programmgruppe Systemforschung und Technologische Entwicklung, 1980. †

[518] H.-P. Schwefel. *Evolution and Optimum Seeking.* John Wiley & Sons, Inc., New York, 1995.

[519] H.-P. Schwefel and R. Männer, editors. *Parallel Problem Solving from Nature,* vol. 496 of *Lecture Notes in Computer Science,* Dortmund (Germany), 1.-3. Oct. 1991. Springer-Verlag, Berlin. (Proceedings of the 1st Workshop on Parallel Problem Solving from Nature (PPSN1).)

[520] M. Schwehm. Implementation of genetic algorithms on various interconnection networks. In E.O.M. Valero, M. Jane, J.L. Larriba, and B. Suárez, editors, *Parallel Computing and Transputer Applications,* vol. I, pages 195-203. CIMNE, Barcelona & IOS Press, Amsterdam, Barcelona (Spain), 21.-25. Sept. 1992.

[521] M. Schwehm. Massiv parallele genetische Algorithmen. In *Parallel-Algorithmen, -Rechnerstrukturen und -Systemsoftware (PARS-Mitteilungen Nr. 12),* pages 181-191, June 1993. Gesellschaft für Informatik E.V. & Informationstechnische Gesellschaft im VDE.

[522] M. Schwehm. A massively parallel genetic algorithm on the MasPar MP-1. In Albrecht et al. [24], pages 502-507.

[523] M. Schwehm. Massively parallel genetic algorithms. In L. Dekker, W. Smit, and J.C. Zuidervaart, editors, *Massively Parallel Processing Applications and Development,* pages 505-512, 1994. Elsevier Science Publ., Amsterdam. †

[524] M. Schwehm, K.D. Reinartz, T. Walter, S.-S. Gold, C. Schäftner, T. Opaterny, A. Ost, and N. Engst. Massiv parallele genetische Algorithmen, Beiträge zum Tag der Informatik Erlangen 1993. Interner Bericht IMMD VII - 8/93, Friedrich-Alexander-Universität Erlangen-Nürnberg, Institut für Matematische Maschinen und Datenverarbeitung, 1993, (in German.)

[525] S.D. Scott, A. Samal, and S. Seth. HGA: a hardware-based genetic algorithm. In *Proceedings of the 1995 ACM Third International Symposium on Field-Programmable Gate Arrays (FPGA '95),* pages 53-59, Monterey, CA, 12.-14. Feb. 1995. ACM, New York. *

[526] S.D. Scott, A. Samal, and S. Seth. HGA: A hardware-based genetic algorithm. In [2].

[527] M. Sebag and M. Schoenauer. Mutation by imitation in Boolean evolution strategies. In Voigt et al. [616], pages 356-365.

[528] A.V. Sebald and L.J. Fogel, editors. *Proceedings of the Fourth Annual Conference on Evolutionary Programming (EP94)*, San Diego, CA, 24.-26. Feb. 1994. World Scientific, Singapore. †

[529] M.K. Sen and P.L. Stoffa. Rapid sampling of model space using genetic algorithms: Examples from seismic waveform inversion. *Geophysical Journal International*, 108(1):281+, Jan. 1992.

[530] B. Sendhoff and M. Kreutz. Analysis of possible genome-dependence of mutation rates in genetic algorithms. Internal Report 95-07, Ruhr-Universität Bochum, Institut fur Neuroinformatik, 1995. (available via URL:http://www.neuroinformatik.ruhr-uni-bochum.de/ini/VS/PUBLIST/1995/html/irini95.html) †

[531] B. Sendhoff and M. Kreutz. Analysis of possible genome-dependence of mutation rates in genetic algorithms. In Fogarty [163], pages 216-226.

[532] K. Shahookar, W. Khamisani, P. Mazumder, and S.M. Reddy. Genetic beam search for gate matrix layout. *IEEE Proceedings, Computers and Digital Techniques*, 141(2):123-128, Mar. 1994.

[533] Y. Shang and G.-J. Li. New crossover operators in genetic algorithms. In *Proceedings of the 1991 IEEE International Conference on Tools with Artificial Intelligence TAI'91*, pages 150-153, San Jose, CA, 10.-13. Nov. 1991. IEEE Computer Society Press, Los Alamitos, CA. *

[534] B.A. Shapiro and J.C. Wu. An annealing mutation operator in the genetic algorithms for RNA folding. *Computer Applications in the Biosciences (CABIOS)*, 12(3):171-180, 1996.

[535] J. Sheung, A. Fan, and A. Tang. Time tabling using genetic algorithm and simulated annealing. In *Proceedings of the 1993 IEEE Region 10 Conference on Computer, Communication, Control and Power Engineering (TENCON'93)*, vol. 1, pages 448-451, Beijing (China), 19.-21. Oct. 1993. IEEE.

[536] G. Shi, H. Iima, and N. Sannomiya. A new encoding scheme for solving job shop problems by genetic algorithm. In *Proceedings of the 35th IEEE Conference on Decision and Control*, vol. 4, pages 4395-4400, Kobe, Japan, 11.-13. Dec. 1996. IEEE, New York, NY. *

[537] W. Shi and J.H. Chen. Intelligent permuting and its optimization. In *Proceedings of the Conference on Intelligent Manufacturing*, vol. SPIE-2620, pages 514-519, Bellingham, WA (tuskin, ETSI OSOITE), 10 June 1995. Society of Photo-Optical Instrumentation Engineers, Bellingham, WA. *

[538] T. Shibata, T. Abe, K. Tanie, and M. Nose. Motion planning by genetic algorithm for a redundant manipulator using a model of criteria of skilled operators. *Information Sciences*, 102(1-4):171-186, 1997.

[539] H. Shimodaira. A new genetic algorithm using large mutation rates and population-elitist selection (GALME). In *Proceedings of the Eighth IEEE International Conference on Tools with Artificial Intelligence,* pages 25-32, Toulouse, France, 16.-19. Nov. 1996. IEEE Computer Society Press, Los Alamitos, CA. *

[540] A.R. Simpson and S.D. Priest. The application of genetic algorithms to optimization problems in geotechnics. *Computers and Geotechnics,* 15(1):1-19, 1993.

[541] K. Sims. Artificial evolution for computer graphics. *Computer Graphics,* 25(4):319-328, July 1991.

[542] A. Singleton. Genetic programming with C++. *BYTE,* 19(2):171 176, Feb. 1994.

[543] A. Singleton, H. Mills, and A. Patton. DOODLE GARDEN ~ *A screen saver using genetic programming,* 1993.

[544] N. Sitkoff, M. Wazlowski, A. Smith, and H. Silverman. Implementing a genetic algorithm on a parallel custom computing machine. In *Proceedings of the IEEE Symposium on FPGAs for Custom Computing Machines,* pages 180-187, Napa Valley, CA, 19.-21. Apr. 1995. IEEE Computer Society Press, Los Alamitos, CA. *

[545] A.O. Skomorokhov. Genetic algorithms: APL2 implementation and a real life application. *APL Quote Quad,* 26(4):97-106, June 1996. (Proceedings of the APL96 Conference, Lancaster (U.K.) 29. July 1. Aug. *

[546] J. Smith and K. Sugihara. GA toolkit on the Web. In *Proceedings of the First Online Workshop on Soft Computing (WSC1),* pages 93-98, WWW (World Wide Web), 19.-30. Aug. 1996. Nagoya University.

[547] J.E. Smith and T.C. Fogarty. Adaptive parameterised evolutionary systems: Self adaptive recombination and mutation in a genetic algorithm. In Voigt et al. [616], pages 441-450.

[548] R.E. Smith. Adaptively resizing populations: An algorithm and analysis. TCGA Report 93001, University of Alabama, Tuscaloosa, 1993. †

[549] R.E. Smith. Adaptively resizing populations: An algorithm and analysis. In Forrest [170], page 653.

[550] R.E. Smith, D.E. Goldberg, and J.A. Erickson. SGA-C a C language implementation of simple genetic algorithm. TCGA Report 91002, University of Alabama, 1991.

[551] R.E. Smith and E. Smuda. Adaptively resizing populations: Algorithm, analysis, and first results. Report NASA-CR-194277, University of Alabama, Department of Engineering Science and Mechanics, 1993. †

[552] Y.H. Song, G.S. Wang, A.T. Johns, and P.Y. Wang. Improved genetic algorithms with fuzzy logic controlled crossover and mutation. In *Proceedings of the 1996 UKA CC International*

Conference on Control, pages 140-144, Exeter (U.K.), 2.-5. Sept. 1996. IEE, Stevenage (U.K.).

[553] T. Soule, J.A. Foster, and J. Dickinson. Code growth in genetic programming. In Koza et al. [312]. †

[554] W.M. Spears. Adapting crossover in a genetic algorithm. Report AIC-92-025, Naval Research Laboratory AI Center, Washington, 1992. †

[555] W.M. Spears. Crossover or mutation In D.L. Whitley [634], pages 221-238.

[556] W.M. Spears. Adapting crossover in evolutionary algorithms. In McDonnell et al. [374]. †

[557] W.M. Spears and V. Anand. A study of crossover operators in genetic programming. In Z.W. Ras and M. Zemankova, editors, Methodologies for Intelligent Systems, 6th International Symposium, ISMIS '91, pages 409-418, Charlotte, NC., U.S.A., 16.- 19. Oct. 1991. Springer-Verlag.

[558] W.M. Spears and K.A.D. Jong. An analysis of multi-point crossover. In Rawlins [462], pages 301-315, (also AIC Report No. AIC-90-014).

[559] W.M. Spears and K.A.D. Jong. On the virtues of parametrized uniform crossover. In Belew and Booker [59], pages 230-236.

[560] W.M. Spears and K.A.D. Jong. On the virtues of parametrized uniform crossover. NASA Contract Report AD-A293985, Naval Research Laboratory, 1995. †

[561] J. Sprave and H.-P. Schwefel. Evolutionäre Algorithmen auf Transputerfarmen zur Lösung schwieriger Optimierungsprobleme. In R. Grebe and M. Baumann, editors, TAT '92, Abstract Volume of the 4th German Transputer Users Group Meeting, pages 106-109, Aachen, 22.-23. Sept. 1992. Medical School of the Technical University (RWTH), Institute for Physiology.

[562] M. Srinivas and L.M. Patnaik. Adaptive probabilities of crossover and mutation in genetic algorithms. IEEE Trans. on Systems, Man, and Cybernetics, 24(4):656-667, Apr. 1994.

[563] M. Srinivas and L.M. Patnaik. Genetic search: analysis using fitness moments. IEEE Trans. on Knowledge and Data Engineering, 8(1):120-133, Feb. 1996.

[564] L. State. Information theory analysis of the convergence and learning properties of a certain class of genetic algorithms in continuous space and infinite population assumption. In Proceedings of the 1995 IEEE international Conference on Systems, Man and Cybernetics, pages 229-234, Vancouver, BC (Canada), 22.-25. Oct. 1995. IEEE, Piscataway, NJ. *

[565] W.H. Steeb, F. Solms, and T.K. Shi. Genetic algorithms and object oriented programming. Int. J. Mod. Phys. C, Phys. Comput. (Singapore), 6(6):853-869, 1995. *

[566] F.M. Stefanini and A. Camussi. APLOGEN: an object-oriented genetic algorithm performing Monte Carlo optimization. *Comput. Appl. Biosci.*, 11(2):74-91,121-123, June 1993. †

[567] P.A. Stefanski, H.H. Nash, and K.A.D. Jong. An object-oriented toolkit for evolutionary algorithms. In *Proceedings of the Seventh International Conference on Tools with Artificial Intelligence*, pages 156-163, Herndon, VA, 5.-8. Nov. 1995. IEEE Computer Society Press, Los Alamitos, CA. *

[568] J. Stender, editor. *Parallel Genetic Algorithms.* IOS Press, Amsterdam, 1993. †

[569] J. Stender, E. Hillebrand, and J. Kingdon. *Genetic algorithms in optirnization, simulation and modelling.* IOS Press, 1994. †

[570] P.L. Stoffa and M.K. Sen. Nonlinear multiparameter optimization using genetic algorithms - inversion of plane wave seismograms. *Geophysics,* 56(11):1794-1810, Nov. 1991.

[571] C.C.W. Sullivan and A.G. Pipe. Efficient evolution strategies for exploration in mobile robotics. In Fogarty [163], pages 245 259.

[572] P.D. Surry. RPL2 programmer's guide. Technical Report EPCCBG-PAP-RPL2-PG, Edinburgh Parallel Computing Centre, 1993. †

[573] P.D. Surry. RPL2 user guide. Technical Report EPCC-BG-PAP-RPL2-UG, Edinburgh Parallel Computing Centre, 1993. †

[574] P.D. Surry and N.J. Radcliffe. Rp12: A language and parallel framework for evolutionary computing. Technical Report EPCCTR94-10, University of Edinburgh, Parallel Computing Centre, 1994. †

[575] P.D. Surry and N.J. Radcliffe. Formal algorithms + formal representations = search strategies. In Voigt et al. [616], pages 366-375.

[576] G. Syswerda. Uniform crossover in genetic algorithms. In Schaffer [506], pages 2 9.

[577] G. Syswerda. Simulated crossover in genetic algorithms. In D.L. Whitley [634], pages 239-256.

[578] D.S. Szarkowicz. A multi-stage adaptive-coding genetic algorithm for design applications. In D. Page, editor, *Proceedings of the 1991 Summer Computer Simulation Conference,* pages 138-144, Baltimore, MD, 22.-24. July 1991. SCS, San Diego, CA.

[579] D.S. Szarkowicz. A genetic algorithm for mixed-parameter design applications. In *Proceedings of the 1992 Sixth Annual Midwest Computer Conference,* pages 45-53, Hammond, IN, 27. Mar. 1992. Purdue University Calumet, Hammond, IN.

[580] T. Tagami and J. Tanomaru. Enhanced performance of genetic algorithms in non-stationary environments. In *Proceedings of the ISCA International Conference,* pages 51-54, San Francisco, CA, 12.14. June 1995. International Society of Computers and Their Applications (ISCA), Raleigh, NC. *

[581] M. Takeuchi and A. Sakurai. A genetic algorithm with self-formation mechanism of genotype-to-phenotype mapping. *Transaction of the Institute of Electronics, Information and Communication Engineers D-I (Japan)*, J76D-I(6):229-236, June 1993, (in Japanese.) *

[582] E.-G. Talbi. Etude expérimentale d'algorithmes de placement de processus. *Lettre du Transputer et des Calculateurs Distribués*, 15:7-26, Sept. 1992. (in French; ftp: imag.fr:/pub/SYMPA/talbi.LT92.e.ps.Z).

[583] E.-G. Talbi and P. Bessiere. A parallel genetic algorithm applied to the mapping problem. *SIAM News*, 24(4):12-27, July 1991. †

[584] H. Tamaki, H. Kita, N. Shimizu, K. Maekawa, and Y. Nishikawa. A comparison study of genetic codings for the traveling salesman problem. In ICEC'94 [260], page 16.

[585] R. Tanese. Parallel genetic algorithms for a hypercube. In Grefenstette [211], pages 177-183.

[586] J. Tang and D. Wang. An interactive approach based on a genetic algorithm for a type of quadratic programming problems with fuzzy objective and resources. *Computers & Operations Research*, 24(5):413-422, May 1997.

[587] D.M. Tate and A.E. Smith. Expected allele coverage and the role of mutation in genetic algorithms. In Forrest [170], pages 31-37.

[588] S.R. Thangiah. GIDEON: *A genetic algorithm system for vehicle routing with time windows*. Ph.D. thesis, North Dakota State University of Agriculture and Applied Sciences, Fargo, 1991.

[589] S.R. Thangiah, K.E. Nygard, and P.L. Juell. GIDEON: A genetic algorithm system for vehicle routing with time windows. In *Proceedings of the Seventh IEEE Conference on Artificial Intelligence Applications*, vol. 1, pages 322-328, Miami Beach, FL, 24.-28. Feb. 1991. IEEE Computer Society Press, Los Alamitos.

[590] G.M. Thomas, R. Gerth, T. Velasco, and L.C. Rabelo. Using real coded genetic algorithms for Weibull parameter estimation. *Comput. Ind. Eng. (U.K.)*, 29:377-381, 1995. *

[591] S.G. Thompson, M.A. Bramer, and A. Kalus. MPGAIA – A Massively Parallel Genetic Algorithm for Image Analysis. In Fogarty [163], pages 277-290.

[592] M. Tomassini. Massively parallel evolutionary algorithms. In *Proceedings of the 2nd Connection Machine User Meeting*, Paris (France), 1993. †

[593] K. Tomita and N. Tosaka. Optimum design of truss structure by genetic algorithm with variable mutation ratio. *Nippon Kikai Cakkai Ronbunshu A Hen*, 61(585):1096-1101, 1995. *

[594] M. Tommiska and J. Vuori. Implementation of genetic algorithms with programmable logic devices. In Alander [21], pages 71-78. (ftp: ftp.uwasa.fi: cs/2NWGA/Vuori.ps.Z)

[595] K. Tout, J. Ribeiro-Filho, D. Mignot, and N. Idlebi. A cross platform parallel genetic algorithm programming environment. In *Proceedings of the 1994 World Transputer Congress, Transputer Applications and*

Systems '94, pages 79-90, Como (Italy), 5.-7. Sept. 1994. IOS Press, Amsterdam. *

[596] B.C.H. Turton and T. Arslan. A parallel genetic VLSI architecture for combinatorial real-time applications – disc scheduling. In IEE/IEEE Sheffield '95 [262], pages 493-498. †

[597] B.C.H. Turton and T. Arslan. A parallel genetic VLSI architecture for combinatorial real-time applications - disc scheduling. In [2].

[598] B.C.H. Turton, T. Arslan, and D.H. Horrocks. A hardware architecture for a parallel genetic algorithm for image registration. In [2].

[599] G. Ucoluk. A method for chromosome handling of r-permutations of n-element set in genetic algorithms. In *Proceedings of 1997 IEEE International Conference on Evolutionary Computation*, pages 55-58, Indianapolis, IN, 13.-16. Apr. 1997. IEEE, New York, NY. *

[600] R. Unger and J. Moult. A genetic algorithm for 3D protein folding simulations. In Forrest [170], pages 581-588.

[601] R. Unger and J. Moult. Genetic algorithms for protein folding simulations. *Journal of Molecular Biology*, 231(1):75-81, May 1993.

[602] U. Utrecht and K. Trint. Mutation operators for structure evolution of neural networks. In Davidor et al. [122], pages 492-501. †

[603] R.J.M. Vaessens, E.H.L. Aarts, and J.H. van Lint. Genetic algorithms in coding theory - a table for $A_3(n, d)$. *Discrete Applied Mathematics*, 45(1):71-87, Aug. 1993.

[604] F.H. van Batenburg. An APL-programmed genetic algorithm for the prediction of RNA secondary structure. *Journal of Theoretical Biology*, 174(3):269-280, 7. June 1995. †

[605] A.H.C. van Karopen and L.M.C. Buydens. The effectiveness of recombination in the genetic algorithm methodology. A comparison to simulated annealing. In Alander [21], pages 115-130. (ftp: ftp.uwasa.fi: cs/2NWGA/Kampen.ps.Z)

[606] H.F. VanLandingham and S. Sampan. Evolutionary algorithms for design. In *Proceedings of the IEEE SOUTHEASTCON 97*, pages 191-195, Blacksburg, VA (U.S.A.), 12.-14. Apr. 1997. IEEE, New York, NY. PUUTTUU (EEA101465/97)

[607] F.J. Varela and P. Bourgine, editors. *Toward a Practice of Autonomous System: Proceedings of the First European Conference on Artificial Life*, Paris, 11.-13. Dec. 1991. MIT Press, Cambridge, MA.

[608] F. Vavak, K. Jukes, and T.C. Fogarty. Learning the local search range for genetic optimisation in nonstationary environments. In *Proceedings of 1997 IEEE International Conference on Evolutionary Computation*, pages 355-360, Indianapolis, IN, 13.-16. Apr. 1997. IEEE, New York, NY. *

[609] A.R. Venkatachalam. An analysis of an embedded crossover scheme on GA-hard problem. *Computers & Operations Research,* 22(1):149-157, 1995.

[610] R. Venkateswaran, Z. Obradovic, and C.S. Raghavendra. Cooperative genetic algorithm for optimization problems in distributed computer systems. In WEC2 [629], pages 49-52.

[611] G. Venturini. AGIL: Solving the exploration versus exploitation dilemma in a simple classifier system applied to simulated robotics. In D. Sleeman and P. Edwards, editors, *Machine Learning, Proceedings of the Ninth International Workshop (ML92),* pages 458-463. Morgan Kaufmann Publishers, July 1992.

[612] S. Voget. Theoretical analysis of genetic algorithms with infinite population size. Hildesheimer Informatikberichte 31/95, Universitat Hildesheim, Institut für Matematik, 1995.

[613] H.-M. Voigt and T. Anheyer. Modal mutations in evolutionary algorithms. In ICEC'94 [260], pages 88-92.

[614] H.-M. Voigt, J. Born, and I. Santibáñez-Koref. A multivalued evolutionary algorithm. Technical Report TR-93-022, International Computer Science Institute (ICSI), Berkeley, CA, 1993.

[615] H.-M. Voigt, J. Born, and J. Treptow. The evolution machine - manual, v. 2.1. Informatik Informationen Reporte iir 7(1991), Akademie der Wissenschaft der DDR, Institute for Informatics and Computing Techniques, 1991. (ftp: ftp-bionik.fb10.tu-berlin.de: /pub/software/Evolution-Machine/em-man.ps.Z)

[616] H.-M. Voigt, W. Ebeling, I. Rechenberg, and H.-P. Schwefel, editors. *Parallel Problem Solving from Nature - PPSN IV,* vol. 1141 of *Lecture Notes in Computer Science,* Berlin (Germany), 22.-26. Sept. 1996. Springer-Verlag, Berlin.

[617] M.D. Vose. A closer look at mutation in genetic algorithms. In S.S. Chen, editor, *Neural and Stochastic Methods in Image and Signal Processing II,* vol. SPIE-2032, pages 48-54, San Diego, CA, 12.13. July 1993. The International Society for Optical Engineering. †

[618] M.D. Vose. A closer look at mutation in genetic algorithms. *Annals of Mathematics and Artificial Intelligence,* 10(4):423-434, 1994. *

[619] M.D. Vose and G. Wright. Simple genetic algorithms with linear fitness. *Evolutionary Computation,* 2(4):347-368, 1994. †

[620] K.-N. Wada, H. Doi, C.-I. Tanaka, and Y. Wada. A neo-Darwinian algorithm: Asymmetrical mutations due to semiconservative DNA-type replication promote evolution. *Proceedings of the National Academy of Sciences of the United States of America,* 90(24):11934-11938, Dec. 1993.

[621] T. Wagner, C. Kueblbeck, and C. Schittko. Genetic selection and generation of textural features with PVM. In *Proceedings of the Third European PVM Conference,* pages 305-310, Munich (Germany), 7.-9. Oct. 1996. Springer-Verlag, Berlin (Germany). PUUTTUU (EEA25668/97)

[622] R.L. Wainwright. A family of genetic algorithm packages on a workstation for solving combinatorial optimization problems. *SIGICE Bulletin,* 19(3):30-36, Feb. 1994. *

[623] J. Wakunda and A. Zell. EVA – a tool for optimization with evolutionary algorithms. In *Proceedings of the 23rd Euromicro Conference New Frontiers of Information Technology,* pages 644-652, Budapest (Hungary), 1.-4. Sept. 1997. IEEE Computer Society Press, Los Alamitos, CA. †

[624] B.C. Wallet, D.J. Marchette, and J.L. Solka. A matrix representation for genetic algorithms. In *Automatic Object Recognition VI, B* vol. SPIE-2756, pages 206-214, Orlando, FL, 9.-10. Apr. 1996. The International Society for Optical Engineering, Bellingham, WA. †

[625] X. Wang, X. Shi, and Z. Lu. The self-encoding genetic algorithm. In *Proceedings of the International Conference on Neural Information Processing,* vol. 2, pages 832-837, Hong Kong, 24.-27. Sept. 1996. Springer-Verlag, Berlin (Germany). PUUTTUU (EEA21114/97)

[626] A.H. Watson and I.C. Parmee. Steady state genetic programming with constrained complexity crossover. In Koza et al. [311]. †

[627] M. Watson. *C++ Power Paradigms.* McGraw-Hill, Inc., New York, 1995. †

[628] T. Watson. Genetic algorithms and the representation of hierarchies. In *Proceedings of the 2nd Singapore International Conference on Intelligent Systems (SPICIS'94),* pages B165-B170, Singapore, 14.-17. Nov. 1994. Japan-Singapore AI Centre, Singapore. *

[629] *Proceedings of the Second Online Workshop on Evolutionary Computation (WEC2),* Nagoya (Japan), 4.-22. Mar. 1996.

[630] E. Weinberger. A more rigorous derivation of some properties of uncorrelated fitness landscapes. *Journal of Theoretical Biology,* 134:125-129, 1988. †

[631] E. Weinberger. Correlated and uncorrelated fitness landscapes and how to tell the difference. *Biological Cybernetics,* 63:325-336, 1990.

[632] T.C. Wesselkamper and J. Danowitz. Some new results for multiple valued genetic algorithms. In *Proceedings of the 1995 25th International Symposium on Multiple-Valued Logic,* pages 264-269, Bloomington, IN, 23.-25. May 1995. IEEE, Los Alamitos, CA. *

[633] T. White and F. Oppacher. Adaptive crossover using automata. In Davidor et al. [122], pages 229-238. *

[634] D.L. Whitley, editor. *Foundations of Genetic Algorithms 2 (FOGA-92),* Vail, CO, 24.-29. July 1992 1993. Morgan Kaufmann: San Mateo, CA.

[635] D.L. Whitley, K.E. Mathias, and P. Fitzhorn. Delta coding: An iterative search strategy for genetic algorithms. In Belew and Booker [59], pages 77-84.

[636] P.A.I. Wijkman. The purpose of sex. In *Proceedings of the 1994 Second Australian and New Zealand Conference on Intelligent*

Information Systems, pages 273-277, Brisbane, QLD, 29. Nov.- 2. Dec. 1994. IEEE, New York. *

[637] S. Wilson. How to grow a starship pilot [genetic algorithms for space probes]. *AI Expert,* 8(12):20-26, Dec. 1993.

[638] M. Wineberg and F. Oppacher. A representation scheme to perform program induction in a canonical genetic algorithm. In Davidor et al. [122]. †

[639] G. Winter, J. Periaux, M. Galgán, and P. Cuesta, editors. *Genetic Algorithms in Engineering and Computer Science (EUROGEN95),* Las Palmas (Spain), Dec. 1995. John Wiley & Sons, New York.

[640] K.P. Wong and Y.W. Wong. Floating-point number coding method for genetic algorithms. In *ANZIIS-93 Proceedings of the Inaugural Australian and New Zealand Conference on Intelligent Information Systems,* Perth (Australia), 1.-3. Dec. 1993, (to appear). †

[641] Y.L.L. Xiao and D.E. Williams. GAME: Genetic algorithm for minimization of energy, an interactive FORTRAN program for three-dimensional intermolecular interactions. *Computers & Chemistry,* 18(2):199-201, June 1994. †

[642] M. Yagiura and T. Ibaraki. On genetic crossover operators for sequencing problems. *Trans. of the Institute of Electrical Engineers of Japan C,* 114-C(6):713-720, June 1994, (in Japanese). *

[643] M. Yagiura and T. Ibaraki. Use of dynamic programming in genetic algorithms for permutation problems. *Eur. J. Oper. Res.,* 92(2):387 401, 1996.

[644] T. Yamada and R. Nakano. Scheduling by genetic local search with multi-step crossover. In Voigt et al. [616], pages 960-969.

[645] M. Yamamura, H. Satoh, and S. Kobayashi. An analysis of crossover's effect in genetic algorithms. In ICEC'94 [260], pages 613-618.

[646] W. Yan and Z. Zhu. A real-valued genetic algorithm for optimization problem with continuous variables. *Nanjing University of Aeronautics & Astronautics, Trans.,* 14(1):1-5, 1997. †

[647] M. Yanagiya. A simple mutation-dependent genetic algorithm. In Forrest [170], page 659.

[648] H.-T. Yang, P.-C. Yang, and C.-L. Huang. A parallel genetic algorithm approach to solving the unit commitment problem: Implementation on the transputer networks. *IEEE Trans. on Power Systems,* 12(2):661-668, May 1997. (Proceedings of the IEEE/PES Summer Meeting, July 28 - August 1, 1996 Denver, CO.)

[649] X. Yao, editor. *Progress in Evolutionary Computation. Proceedings of the AI'93 and AI'94 Workshops on Evolutionary Computation,* vol. 956 of *Lecture Notes in Artificial Intelligence,* Melbourne and Armidale (Australia), 16. Nov. 1993 and 21.-22. Nov. 1994 1995. Springer Verlag, Berlin. †

[650] M. Yasunaga. Genetic algorithms implemented by wafer scale integration - wafer scale integration by LDA (leaving defects alone) approach. *Transaction of the Institute of Electronics, Information and*

Communication Engineers D-I (Japan), J77D-I(2):141 148, Feb. 1994, (in Japanese). *

[651] J. Ye, M. Tanaka, and T. Tanino. Genetic algorithm with evolutionary chain-based mutation and its applications. *Memoirs of the Faculty of Engineering, Okayama University*, 30(1) :111-120, Dec. 1995.

[652] S. Yeralan and C.-S. Lin. Genetic search with dynamic operating disciplines. *Computers & Operations Research*, 21(8):941-954, Oct. 1994.

[653] T. Yoshikawa, T. Furuhashi, and Y. Uchikawa. The effects of combination of DNA coding method with pseudo-bacterial GA. In *Proceedings of 1997 IEEE International Conference on Evolutionary Computation*, pages 285-290, Indianapolis, IN, 13.-16. Apr. 1997. IEEE, New York, NY. PUUTTUU (EEA56313/97)

[654] Y. Yukiko and A. Nobue. A diploid genetic algorithm for preserving population - pseudo-meiosis GA. In Davidor et al. [122], pages 36-45.

[655] Q. Yulu and N.N. Win. An adaptive mutation rate approach in genetic algorithm. In *Proceedings of the 3rd Pacific Rim International Conference on Artificial Intelligence (PRICAI-94)*, vol. 1, pages 409-414, Beijing (China), 15.-18. Aug. 1994. International Academic Publishers, Beijing. *

[656] W.-M. Yun and Y.-G. Xi. Optimum motion planning in joint space for robots using genetic algorithms. *Robotics and Autonomous Systems*, 18(4):373-393, Oct. 1996.

[657] J. Zeanah. Naturally selective Axelis Evolver 2.1. *AI Expert*, 9(9):22-23, Sept. 1994.

[658] W. Zhai, P. Kelly, and W.-B. Gong. Genetic algorithms with noisy fitness. *Math. Comput. Model. (U.K.)*, 23(11-12):131-142, 1996. *

[659] B.-T. Zhang and H. Mühlenbein. Evolving optimal neural networks using genetic algorithms with Occam's razor. *Complex Systems*, 7(3):199-220, June 1993. *

[660] L.-J. Zhang, Z.-H. Mao, and Y.-D. Li. Mathematical analysis of mutation operator in genetic algorithms and its improved strategy. In *Proceedings of International Conference on Neural Information Processing (ICONIP95)*, vol. 1, pages 267-270, Beijing (China). 30-Oct 2-Nov. 1995. Publishing House of Electron. Ind., Beijing, China. *

[661] Y. Zhou. *Genetic algorithm with qualitative knowledge enchancement for layout design under continuous space formulation*. Ph.D. thesis, University of Illinois at Chicago, 1993. PUUTTUU (DAI 54/12)

Notations

†(ref) = the bibliography item does not belong to Vaasa collection of genetic papers.

* = only abstract seen.

Printed and bound by CPI Group (UK) Ltd, Croydon, CR0 4YY

24/10/2024

01778278-0020